WINTER'S GUIDE TO WARSHIPS 1890

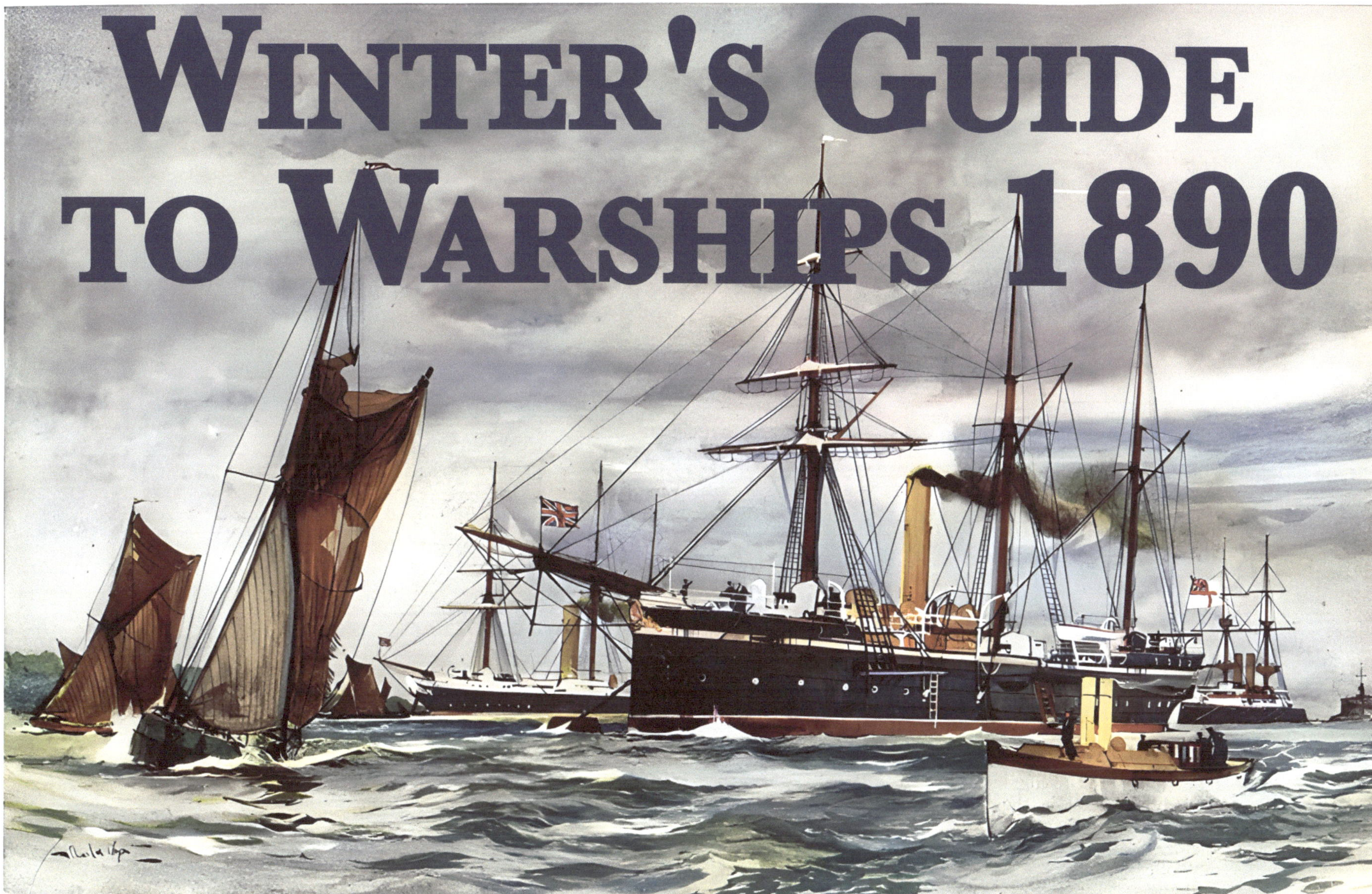

DIMITRY MALKOV

*N*IMBLE BOOKS LLC
NimbleBooks.com

Nimble Books LLC
2846 S. Knightsbridge Circle
Ann Arbor, Michigan, USA
NimbleBooks.com

ISBN: 978-1-60888-080-5

Publisher's Notes

An editorial note: the interior art and text were created by Dimitry Malkov using a CAD/CAM software system connected to Dimitry's personal database of 65,000 warships from the year 1858 to present. Dimitry is a native Russian speaker and the data is stored using numeric and nautical conventions familiar to him. We made a effort to convert figures and abbreviations to more familiar Anglospheric conventions, but some incongruities remain. For example, book titles and ship names are found in double quotes, rather than italics as is normal in US publishing. Similarly, because the book must present a great deal of information in tight spaces, the reader may encounter some moments of surprising terseness or abbreviation.

A note on chronology:

- Dates are always presented in YYYY MM DD format, ISO 8601.
- The year in the format 1889 01 01 always means the year 1890 (month and day are unknown).
- The year in the format 1889 07 01 always means 1889-July (the day is unknown).
- The year in the format 1889 09 09 always means 1889 autumn (1889 03 03 - 1889 spring, etc.)

A legal note: this book is written by a Russian author at a time of high tension between Russia and the countries that comprise most of the naval-history-buying book audience of the world. To ease any concerns that readers may have, I verified with the US Commerce Department that this publication is permissible under the sanctions regime.

A personal note: like Dimitry Malkov, I spent my formative years obsessed with Fred T. Jane's "Fighting Ships." So it is with special personal pleasure that I introduce this authoritative reference work devoted to the warships of the world at the turn of the year 1890.

—Fred Zimmerman
Publisher, Nimble Books LLC

Dedication

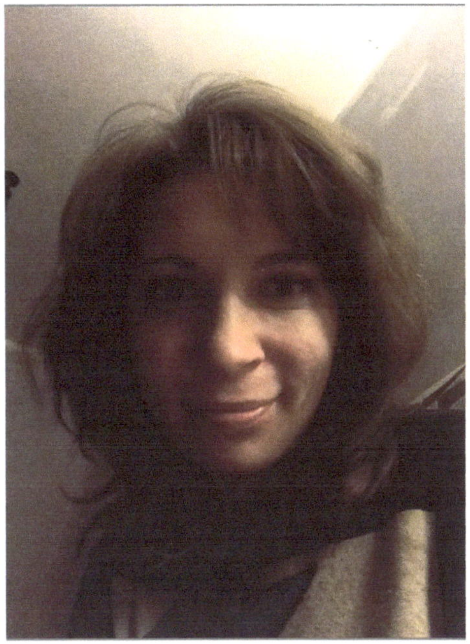

Dedicated to my lovely wife Svetlana Yudaeva.
For her patience, belief, and support ...
With all my tenderness, Dimitry

Foreword

When the author of this book first contacted me I saw considerable merit in his research and the data he showed me. Consequently I forwarded him to a publisher for consideration as I felt the book certainly deserved to be put into print. Since then the work has proceeded and I recently saw completed chapters. The finished work did not disappoint me and were actually even better than the material I had seen during its early development. Dimitry Malkov has provided pages of incredibly detailed information that will surely become the number one "go to" reference source for warships of the pre-dreadnought era. From the largest navies to the smallest the pages are full of wonderful material for the reader that is often very difficult to find without access to difficult to find sources, but would require naval enthusiasts to engage in lengthy, time consuming research. Every ship has diagrams showing its appearance and in many cases the actual paint scheme is also illustrated if known. Not only are the statistics of ships of all sizes shown, the reader is treated to information on the arcs of fire for the guns each ship carried and armour spread, penetration and range. The diagrams are clear and easy to understand. As well, Malkov has included maps and detail of the naval shipyards where vessels were built, rather in the form of the older versions of Jane's Fighting Ships, yet somewhat clearer. Be they British colonial or very minor national navies, the ships are all covered. In Volume 2, the fledgling US Navy is fully shown to a depth matched by few.

As an author used to research, I am truly amazed at the amount of effort the contents of this book would have required the writer to engage in. I confess that there are many ships shown that I have previously not seen much detail off, and in the case of small vessels some I have never even heard of at all. Only a very serious author would even attempt to produce the material in this book and I therefore commend it to anyone with a love of the pre-Dreadnought era when naval technology was developing very fast, and required warship designers to find solutions to the problems posed by new ideas, new materials, and advancing gun technology. Some ships found their guns to be already obsolete during the time it took to design and launch them because the science of naval construction was developing at such a rapid rate. Many new ships were experimental with the result that they were rarely built in large classes such as during World War I and World War II, when designs became more entrenched. Yet, it is all covered within this book, leaving the reader with a stunning access to the era. I highly commend the work of Dimitry Malkov to the serious naval researcher, model builder or naval wargamer.

Mal Wright

Imperial Russian Navy torpedo boat destroyer "Strashni" (Mal Wright).

Contents

PREFACE

This edition became possible with the grace of two persons: Malcolm Wright and Fred Zimmerman. Being already among the greatest naval authors, Malcolm noticed my materials and informed his editor Fred Zimmerman about it, who invested in the whole project. Now I am proud to present the result.

This reference book was inspirited by the author's love of ships (yes, I really think that naval architecture is one of the most beautiful of all human creations), curiosity and the commitment to integrity that, all together, multiplied by 30 years of work. To write this book I had to create simultaneously two things:

- The first: a data base of world's naval ships and boats from 1858 01 01 to now. The database gives the possibility to make the time cut on any date from 1858 01 01 by all the 65,935 warships contained in it. This includes not just the list of ships, but also ship renamings, technical data and armament changes, that are always appropriate to his concrete chosen date.
- The second: I created a gallery of ship images (already partially exposed in Internet where I suppose many of you already saw it) and I tried to make every image maximally reliable, compared with the shipbuilder's plans. Also I consulted plans drawn by such great artists as Allan Raven and A.D.Baker III and some national naval artists less famous but not less reliable, like Peter Kovasc, who much surprised me with his very detailed and reliable plans of Austria-Hungarian warships. In some cases, when there was not any graphic information for the ship, I used photos for the ship reconstruction. For that I invented a proprietary algorithm for details triangulation; I have checked the results against builder plans and am completely satisfied by the result achieved. I tried not only to achieve the mentioned naval artists level, but to overcome it. I believe I achieved that goal; even if my drawings are of the same level accuracy, they are more detailed and they are colorized. Victorian ship colors are something! Volume 1 contains more than 440 ship plans; Volume 2 will contain not fewer.

There are two serial books that inspired my thirty years work: Jane's "Fighting Ships" and "Conway's All the Worlds Fighting Ships." I wanted to create something that will combine together their best features yet avoid the shortcomings of each. I tried also to incorporate in all other available and reliable (to my regard) sources to make my information integral and complex. Existing sources were already numerous like the famous and well known series of the "Die deutschen Kriegsschiffe 1815-1945" by Groner Erich, "Warships of the Imperial Japanese Navy, 1869-1945" by Hansgeorg Jentschura & Dieter Jung & Peter Mickel, "Schwarze Gesellen. Band 1: Torpedoboote bis 1914" by Harald Fock etc. During this thirty-year period many new and very reliable editions on national navies appeared, such as "The Ottoman Steam Navy 1828-1923" by Bernd Langensiepen & Ahmet Guleryuz, "French Warships in the Age of Steam 1859-1914" by Stephen S. Roberts, "British Warships in the Age of Steam 1817-1863" by Rief Winfield, "British Cruisers of the Victorian Era" by Norman Friedman, "British Warship Recognition, The Perkins Identification Albums" by Richard Perkins, and a series of reference books on the Russian and Soviet Navies by Sergey S. Berejnoy (some of which I illustrated), altogether containing dozens of thousands of pages. Also I found a huge amount of eclectic material (builders ship plans and ship fates) on the Internet. It was a huge effort to collect and compose all this information in a single reference book. The result is "Winter's Guide to Warships 1890."

The "brand" or "class" name, if you will, for this series of books is "Winter's" because "winter" in Russian means Зима—this is my nickname, which is well known in online naval circles. I told Fred about this and he volunteered that his father was also nicknamed "Zim"! Winter" is also appropriate because the entire manuscript is keyed to the specific winter day of 1 January 1890.

Why is 1890 chosen? I thoroughly discussed which period of time for the book to choose with my editor Fred Zimmerman. He initially proposed 1870 or 1920. I see his absolutely professional logic—these years are separated from today by approximately 150 and 100 years respectively. But I managed to convince him to start at 1890. The choice of 1870 would have been not good from the visual point of view —at that date almost all ships were alike—sailing, masted, with the artillery located on the sides of the ship; or monitors with a single turret amidships, nothing special except for HMS "Devastation", "Thunderer" and "Dreadnought" and Russian Empire Ship "Petr Velikiy", seaworthy 10,000-ton pioneers whose axial fire main turrets foreshadowed future standards. Despite these outliers, the fleets of that year were mostly composed of ships of the previous Age of Sail, well covered by such publications as "French Warships in the Age of Sail 1786-1861" by Rif Winfield and Stephen S. Roberts and "British Warships in the Age of Sail 1817-1863" by Winfield. Of course ships from the Age of Sail were still present in 1890, but not to the extent of 1870—just a few ships were kept in the main fleet lists, mostly as fleet auxiliaries. As far as 1920 is concerned, that was the second year of a dramatic naval contraction—ships had already vanished by the hundreds from naval lists, world navies were already deprived some of their most good looking and quite modern ships and ship classes, and the navies of the German, Austria-Hungarian and Russian Empires had ceased to exist. A comparison with 1890 is really amazing. From that year forward was a time of extensive growth and major naval architectural decisions from numerous national projects emerged many amazing ships.

I have made the concrete fleet lists in this book to reflect the situation EXACTLY on 1890 01 01. Every information is specifically for that day. Even the day after or a previous one will produce some changes. That's why you will not find any information here that is beyond date 1890 01 01 (save a couple of maps and some uncredited photos). Everything photos, ship plans, city maps where shipbuilding plants were located are all prior to 1890 01 01. Even if I had a very informative ship photo taken later 1890 01 01, I omitted it, even if it meant there were no photos at all.

The 1890 edition is divided into two volumes. The information in both is "equal" (so to say). I took all national fleets by their importance and matched with the British fleet of the Volume 1 French and Russian (2nd and 3rd fleets of the world by displacement) of the Volume 2 and so on. So if you like Volume 1 you will be interested in Volume 2 as well.

Volume 1 contains fourteen fleets that are presented in the book in the following order (by the Fleet importance):

- Britain 1,060,888 tons (including 3 colonial navies of Australia 12,774 tons, India 6,625, New Zealand 52 tons), Italy 241,056 tons, Turkey 113,845 tons, Netherlands 102,621 tons, Japan 45,724 tons, Denmark 40,660 tons, Sweden 29,608 tons, Norway 16,939 tons, Chile 16,536 tons, Greece 10,907 tons, Siam 2,230 tons, Mexico 1,850 tons, Uruguay 1,330 tons, San Domingo 389 tons. (The displacement figures are to a certain extent approximate. The displacement is count ONLY for the ships constructed after 1857 12 31. To my knowledge the accuracy is 99%).
- Britain, Australia, India, New Zealand: 116 pages • 147 photos • 199 ship plans • 20 maps; Chile: 9 pages • 9 photos • 10 ship plans; Denmark: 19 pages • 30 photos • 36 ship plans • 1 map; Greece: 8 pages • 8 photos • 14 ship plans; Italy: 40 pages • 49 photos • 53 ship plans • 8 maps; Japan: 18 pages • 14 photos • 24 ship plans • 2 maps; Netherlands: 23 pages • 27 photos • 36 ship plans • 2 maps; Norway: 10 pages • 7 photos • 10 ship plans • 1 map (recent); Sweden: 15 pages • 17 photos • 23 ship plans • 2 maps; Turkey: 23 pages • 29 photos • 29 ship plans • 1 map; Uruguay, Siam, San Domingo, Mexico: 4 pages (1 page each) • 1 ship plan. In total 287 pages • 338 photos • 440 ship plans. Also there are 36 maps taken prior and circa 1890 of shipbuilding cities.

For Volume 2 the illustration materials will be of the same "density" level. Volume 2 will contain nineteen fleets in the following order:

- France 707,241 tons, Russia 274,382 tons, Germany 222,631 tons, US 151,484, Spain 147,309, Austria-Hungary 109,173, China 80,018 tons, Brazil 38,367 tons, Portugal 17,584 tons, Argentine 11,586 tons, Egypt 11,430 tons, Peru 7,572 tons, Haiti 4,886 tons, Romania 2,086 tons, Iraq 1,200 tons, Belgium 684 tons, Sarawak 593 tons, Ecuador 69 tons, Bulgaria 32 tons.
- France: 83 pages • 115 photos • 139 ship plans • 13 maps; Russia: 57 pages • 70 photos • 99 ship plans • 6 maps; Germany: 35 pages • 47 photos • 55 ship plans • 6 maps; US: 29 pages • 38 photos • 40 ship plans • 4 maps; Spain: 34 pages • 45 photos • 51 ship plans • 6 maps; Austria-Hungary • 26 pages • 36 photos • 46 ship plans • 2 maps;

China: 20 pages • 28 photos • 30 ship plans • 2 maps; Brazil: 13 pages • 17 photos • 20 ship plans • 1 map; Portugal: 13 pages • 17 photos • 15 ship plans • 1 map; Argentine: 9 pages • 8 photos • 14 ship plans ; Egypt : 1 page; Peru: 1 page; Haiti: 1 page; Romania : 4 pages • 4 photos • 4 ship plans; Belgium: 1 page; Sarawak: 1 page; Ecuador: 1 page; Bulgaria: 1 page • 1 ship plan.

Every national fleet section is divided into the following subsections:

- National shipbuilding if any (with presentation of all "effective" ventures building the fleet);
- Ship recognition (with the entire fleet recognition ship silhouettes in uniform scale);
- Ship alphabet indexes;
- Naval guns;
- Pages dedicated to the concrete ship types (with the ship images all in the uniform scale).

In the ship section you will find something completely new: the disposition of the artillery and sectors of fire for every drawn ship. In spite of the very schematic presentation—the ship herself is shown just by 2 lines of her waterline (to draw the exact deck plans at the level of my side ones would have taken me another 30 years)—all guns are in their proper places and the guns and sectors of fire are uniformly scaled. All that gives a good visualization of the ship's fire power, and, even more important, where it could be effectively directed.

The ship section is arranged in the following order: capital ships, cruising ships, torpedo ships, coastal ships. Every section begins with the ships under construction and the most modern ships and is closed by obsolete effective ships. Auxiliaries are indicated in the ship recognition section and are not shown in the main ships section. A statistical summary is found at page xii. The reference pages contain many abbreviations and graphics. A list of abbreviations and a key to the graphics are found beginning at page xiii.

I wish you many hours of enjoyable reading, looking and comparing details.

Dimitry G. Malkov
Moscow, Russia

HMS "Hercules" painted by Henry J. Morgan in 1869.

Towards the Conventional Design as of 1890 01 01

My editor Fred Zimmerman asked me to add a naval trend analysis for the years preceding 1890, similar to the annual essays found in Jane's, and I do it gladly.

Before I proceed I must make a couple of digressions and explanations:

- All the following figures are generated by my data base "All the World's Fighting Ships since 1858 01 01". The data base does not cover the classes smaller than corvettes prior to 1858 01 01. That's why in my analyses I omit ALL ships present in my data base and laid down prior to 1858 01 01 despite the fact there are already 986 units. So my analyses begin, so to say, "scratch +2 years".
- Although the database makes it possible to make analysis at daily scales, I prefer not to overload readers with figure, so I will examine only 4 time cuts, namely dates 1860 01 01, 1870 01 01, 1880 01 01 and of course 1890 01 01.
- Figures in brackets (>7, +21, =28) mean (>quantity of ships in the previous time cut, +laid down/-deleted or perished/-changed the class (mostly to the auxiliaries) ships balance, ="current" time quantity of ships).

In 1858 a new revolutionary ship class emerged: battery ironclads, first built by France. And that started a huge naval armament race in the world, comparable only with the dreadnought fever of 1905. In most cases the new vessels were just armoured large sailing frigates of the Age of Sail with one deck battery of heavy guns. There was also a gunnery revolution—smooth bore guns were replaced by rifled bore and breech loading guns. Both those facts defined the world naval ideology for at least 20 years and put a total end to the Age of Sail, despite the fact that iron and steel cruising ships did not fully replace the wooden ships for another twelve years, with the last wooden vessels ordered even at the end of the Eighties.

It is surprising, but Britain, while possessing the largest wooden sail and steam fleet and the largest naval design school in the world, did not immediately respond to this revolution in naval affairs. During the period 1858 to 1860, Britain started to build twelve wooden screw battleships. Their destiny was more than poor. Most of them were cancelled, while some reverted to ironclads with a large loss of time and resources. Of twelve large frigates and corvettes also laid down from 1858 01 01 to 1860 01 01 few were commissioned. And taking in consideration the cancellation of ships laid down prior to 1858 01 01 these poor decisions inflicted on Britain the "loss" of a whole fleet of approximately forty capital vessels—almost a Trafalgar- or Scapa Flow-scale catastrophe.

By 1860 01 01, after 2 years of naval construction there had been laid down and launched 199 effective ships of 14 countries. There were 9 ship classes: battery ironclads (>0, +11, =11), screw sailing battleships (>110, +12), screw sailing cruisers (>141, +36), gunboats (>341, +128), floating batteries (>5, +4), paddle sailing ships (>146, +2), auxiliaries (>24, +1), landing vessels (>7, +4), sailing ships (>164, +1). Except ironclads all ships were masted and wooden. The global construction rate was 100 ships per year.

The battery ironclads had three huge shortcomings that defined their decline in some twenty years.

- The battery could not deploy guns with more than 11 inches calibre.
- Battery gun deployment put severe restriction on the arcs of fire.
- The heavy guns had no axial fire.

None of those shortcomings could be solved within the broadside concept of gunnery deployment. The appearance of the turret battleship during the Sixties was an absolutely logical naval concept development. That decade saw multiple attempts to find the most reliable turret disposition with approaches that were sometimes very far from what became, by 1890 01 01, the conventional design entrenched everywhere, with two bow and rear turrets or barbettes.

In the second half of the Seventies a revolutionary new weapon emerged—the torpedo. The possibility of remote underwater attack proved such an efficient threat that it led to two new ship classes—torpedo boats and submarines.

On 1870 01 01 to the existing 9 classes 11 new classes were added: battleships (>0, +7, =7), coastal battleships (>0, +12, =12), monitors (>0, +96, =96), armoured cruisers (>0, +1, =1), unprotected cruisers (>0, +8, =8), torpedo boats (>0, +1, =1), submarines (>0, +2, =2), river monitors (>0, +10, =10), minelayers (>0, +6, =6), armed auxiliaries (>0, +9, =9). Existing class figure changes were: battery ironclads (>11, +126, =137), screw sailing battleships (>12, -6, =6), screw sailing cruisers (>36, +87, =123), gunboats (>128, +255, =353), floating batteries (>4, +5, =9), paddle sailing ships (>2, +14, =16), auxiliaries (>1, +17, =18), landing vessels (>4, +7, =11), sailing ships (>1, -0, =1). The construction rate for the previous 10 years was 66 ships per year. There were twenty-four countries involved.

On 1880 01 01 the existing 19 ship classes were: battleships (>7, +21, =28), coastal battleships (>12, +7, =19), battery ironclads (>137, +10, =147), monitors (>96, -3, =93), armoured cruisers (>1, +10, =11), unprotected cruisers (>8, +40, =48), screw sailing battleships (>6, -2, =4), screw sailing cruisers (>123, +32, =155), torpedo boats (>1, + 328, =329), submarines (>2, +1, =3), gunboats (>353, +261, =614), floating batteries (>9, -2, =7), landing vessels (>10, +5, =15), paddle sailing ships (>16, -2, =14), river monitors (>10, +10, =20), minelayers (>6, -3, =3), armed auxiliaries (>9, +8, =17), auxiliaries (>17, +22, =39), sailing ships (>1, -0, =1). 2 new classes were added: protected cruisers (>0, +13, =13) and torpedo gunboats (>0, +3, =3), the class that was to withstand and to eliminate the sufficient growth of the torpedo boat threat. The construction rate for the previous 10 years was 73 ships per year. There were thirty-one countries involved.

From 1880 01 01 to 1890 01 01, no new ship classes were added to the existing 21, which shows that the world naval fleet classes concept was for the time being defined. Classes were: battleships (>28, +46, =74), battleships coastal (>19, +15, =34), battery ironclads (>147, -53, =94), monitors (>93, -5, =88), armoured cruisers (>11, +24, =35), protected cruisers (>13, +135, =148), unprotected cruisers (>48, +41, =89), screw sailing battleships (>4, -0, =4), screw sailing cruisers (>155, -47, =108), torpedo gunboats (>3, + 74, =77), torpedo boats (>329, +700, =1,029), submarines (>3, +6, =9), gunboats (>614, +18, =632), floating batteries (>7, -6, =1), landing vessels (>15, +4, =19), paddle sailing ships (>14, -1, =13), river monitors (>20, -4, =16), minelayers (>3, +2, =5), armed auxiliaries (>17, +1, =18), auxiliaries (>39, +57, =96), sailing ships (>1, +2, =3). The construction rate for the previous 10 years reached 101 ships per year. There were thirty-seven countries involved.

Genesis of the Naval Design Concept

The previous Age of Sail had a profound impact at the World's capital ship design. They were masted sail ships with the broadside gun deployment.

The Age of Steel with the new modern technologies objectively required a change in naval construction ideology. Two national approaches to armored capital ship design emerged: the British and the French. The French naval design concept will be treated in Volume 2. As for Britain, changes were started with the battleship HMS "Monarch" (below) laid down in 1866. The turret concept was adopted from the US Navy's "Monitor", which operated only in littoral waters. The British requirement to have a seaworthy oceangoing ship generated a high freeboard masted vessel with turrets in place of the central battery and thus without axial fire.

The following four images are not shown at consistent scales.

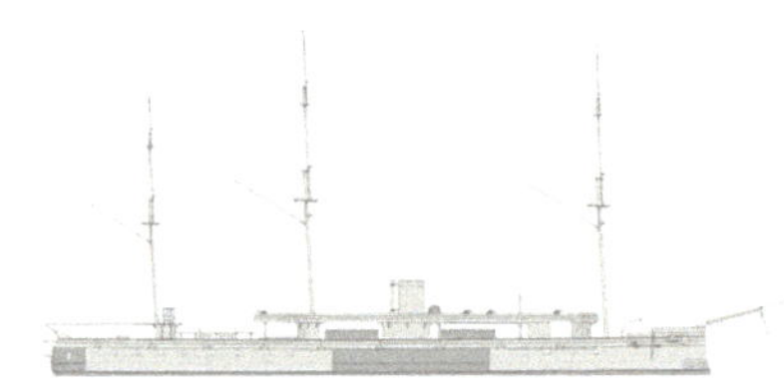

The need to have axial main guns fire led to research on better turret disposition. In 1869 the problem was fixed with the HMS "Devastation" design (below). What is quite interesting is that 1869 Russia also produced a very similar design, the battleship "Petr Velikiy." The need to have axial main guns fire required the omission of sails. This became a rule with British capital ships from the “Admirals” class on, despite some attempts to somehow compose these two incompatible things in the HMS "Inflexible".

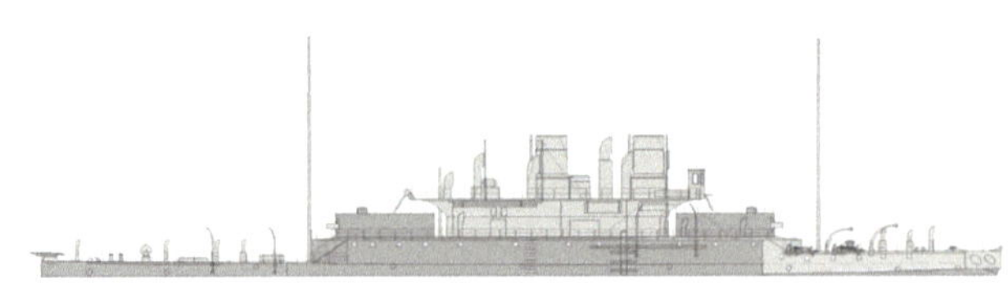

As for the cruising ships the Age of Sail still defined their design for some twenty years. The most significant cruising ships concepts were derived from the Elswick ships built for export. The first, a small Chilean unprotected cruiser, "Arturo Prat" (below), was laid down in 1879. The ship had two heavy guns forward and aft with the bow and stern clear for the main guns fire. Significantly, she also omitted sails.

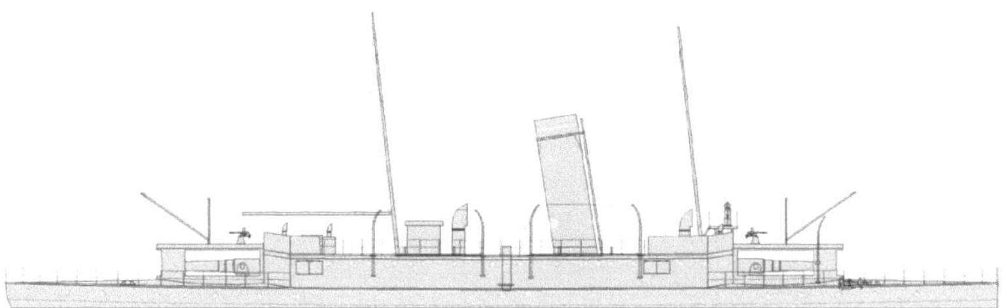

During the Eighties, cruising ships of the Brirtish navy started to take the form considered effective in “Arturo Prat”. The only difference between "Arturo Prat" (1879) and HMS "Blake" (1888) (below) was the size for the much more powerfull machinery and secondary armament deployment.

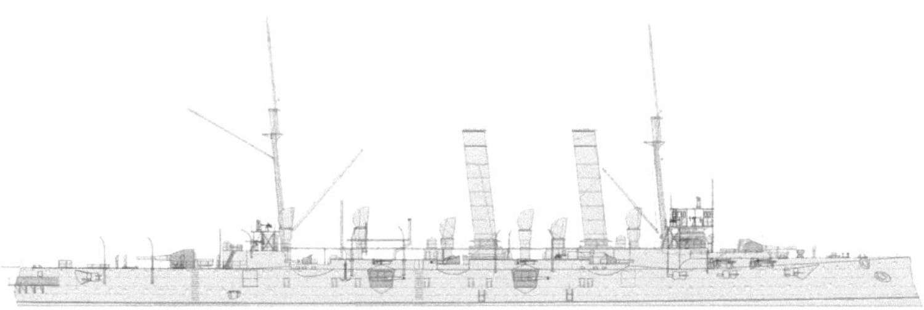

As for the minor warships (sloops and gunboats), the Age of Sail design concept of masted ship with the broadside guns deployment remained absolutely intact throughout the period 1858-1890.

All other countries but for Russia and Germany (and those, in just a couple of designs) followed the designs proposed by Britain and France, especially since they were the two major naval warship exporters in the world.

About the Author

Dimitry Malkov was born in Moscow the capital of the USSR in 1966. His higher education was in the foreign affairs domain and had nothing in common with naval history, nor the naval service, nor with naval construction. But, at the age of six years old, Dimitry had drawn his first ship. That defined his next fifty years of consistent "unprofessional" interest and collecting efforts, which resulted in creating the gallery of approximately 3,000 highly detailed ship plans and the data base “All the World's Fighting Ships 1858-2022.” The current book “Winter's Guide to Warships 1890” in two volumes is the synthesis of these labors and provides the most accurate access to the world’s navies at the peak of the Gilded Age. As a secondary goal, the work also underlines the Russian/Soviet navy’s important and integral role in the world's naval history.

Dimitry’s professional domain is private currency board investing. He lives now in Moscow, the capital of the Russian Federation.

In 1996-1998 Dimitry issued two series of the naval books “Russian navy in 2D” (five books) namely “Russian Black Sea Fleet 1871-1917” in two volumes, “German projects in the Russian fleet 1861-1917”, “French projects in the Russian fleet 1861-1917”, “Russian projects in the Russian fleet 1861-1917” and “Russian navy in 3D” (three books) namely: “Armoured cruiser “Ryurik”, “Protected cruiser “Varyag”, “Battleship “Tsesarevich”. In 2010-2011 he issued two books: “Pantera” class submarines”, and “Baltiyskiy Iron Works” 1861-1917”. He participated in the illustration of naval books: in 2001 “Russian battleships 1861-1917” by S. Berejnoy, in 2010 “Gunboat ‘Gilyak’” by A. Tsarkov. In 2009, 2010, 2012 he became a coauthor (2D and 3D ship graphics) of the book “Russian-Japanese naval war 1904-1905” in three volumes. In 2014 the “Russian-Japanese naval war 1904-1905” book series was reissued in a single volume.

WORLD'S SHIP CLASSES LAID DOWN BY YEAR TOTAL FIGURES

Class \ Year	1858	1859	1860	1861	1862	1863	1864	1865	1866	1867	1868	1869	1870	1871	1872	1873	1874	1875	1876	1877	1878	1879	1880	1881	1882	1883	1884	1885	1886	1887	1888	1889
Battleships							1		1	1		4	1	2		3	2	1	4	3	2	5	2	3	8	7	4	6	3		5	9
Battleships coastal					3	2		8		1				2	1			2	1			1				1	3				2	3
Battery ironclads	3	20	7	42	33	27	21	16	6	7	13	4	4		2	7	3	6	2			1				1						
Monitors				1	30	31	5	5	2	11	8	4	7	1	1	6	6	2	2													1
Armoured cruisers													2			1	2		1	3			1	3			1	7	3		2	5
Protected cruisers																			8		1	3	4	4	2	9	11	14	14	20	25	33
Unprotected cruisers				1		1		1	1	3			1	1	1	6	3	6	6	4	3	9	7	10	6	9	6	3	1	2	2	1
Screw sailing battleships	4	2	1																													
Screw sailing cruisers	17	16	22	20	4	19	6	5	13	10	8	9	4	7	9	9	5	4	5	7	2	2	3	1	2	1	1				1	
Torpedo gunboats																1		1			1					7	1	7	8	9	20	19
Torpedo boats						1								6		5	4	13	5	76	136	44	48	35	48	56	85	114	160	75	57	72
Submarines					1	1																	3		1				2	2	1	
Gunboats	23	89	33	62	25	27	20	15	30	85	10	13	25	26	27	53	43	35	32	28	33	30	24	18	31	39	23	30	30	28	16	6
Landing vessels		4	6	1	1				1						1	1			2	1	2		1	1	2			1	1			
Paddle sailing ships	4	1	2	12	28	10	2	3	1			1	1			1																
River monitors					3			3		1	2		2	1	6	2	1															
Minelayers							1									2												1			1	1
Armed auxiliaries		1	3			1											1		1	3		1	2			1	1	2	1		1	1
Auxiliaries					1	1	2	2				2		1	2	1				1						4	2	2	1	1		

WORLD'S SHIP CLASSES DELETIONS BY YEAR TOTAL FIGURES

Class \ Year	1858	1859	1860	1861	1862	1863	1864	1865	1866	1867	1868	1869	1870	1871	1872	1873	1874	1875	1876	1877	1878	1879	1880	1881	1882	1883	1884	1885	1886	1887	1888	1889
Battleships										1		1																1	2			
Battleships coastal																														1		
Battery ironclads					5	1	14	17	4	4		2		12	3	2		7			1	2	1	1	5	3	4	11	9	2	2	3
Monitors					1	1	1	2									18	6		1					1	2	3					
Protected cruisers																														1		
Unprotected cruisers								1					1													2		1				
Screw sailing battleships			1	4		1		1		4	1	1	4	1	2	1	3	2		2				1	2	2		1	1	1	2	
Screw sailing cruisers			1	2	2	12	9	4	3	7	5	14	5	8	9		8	11	5	6	3	3	4	7	5	11	12	10	8	14	6	5
Torpedo boats													1								1	1	1	5		1		9	3	3		3
Submarines														1																		
Gunboats		4	1	6	18	16	49	33	36	34	19	43	26	37	17	11	15	15	13	13	26	8	17	12	16	17	17	60	17	19	23	30
Landing vessels								1			1				1		2										1		1			
Paddle sailing ships			3	5	5	5	16	21	19	26	19	11	6	2	2	2	4	5	2	2	4	5	4	3	2	8	3	3	1	3	2	2
River monitors																				1							2		1			1
Minelayers																							1									
Armed auxiliaries						1	2	1					1				1	4			2							2		1		
Auxiliaries										1																					2	
Sailing ships		1	2	1	6	5	10	10	6	7	6	5	18	3		3	4	10	2		3	6	1			5	3	1	2	6	2	2

Abbreviations

A
accomS, accommodation ship
AH, Austria-Hungary
app, approximately
Ar, Argentine
Au, Australia

B
BB, battleship pre-dreadnought
Be, Belgium
BO, casemate/battery ironclad
Br, Britain
Bu, Bulgaria
BW, wooden screw battleship
Bz, Brazil
BZ, floating battery
BZA, armoured battery

C
CA, armoured cruiser
CB, belted cruiser
CDB, coast defence battleship
coastGV, coast guard vessel
commandS, command staff ship
Ci, Chile
Ch, China
CP, protected cruiser
CS, unprotected cruiser
CT, conning tower
CW, wooden screw cruiser
CWP, wooden paddle cruiser
CWS, wooden sailing cruiser
CZ, screw brig/sloop
CZP, paddle sloop

D
DC, torpedo gunboat
debarq, floating moorage
deptS, depot ship
Dm, Denmark
drilS, drill ship

E
E, 1 engine
EE, 2 engines etc.
Eg, Egypt
Eq, Equador

F
Fr, France

G
GB, gunboat
GBA, armoured gunboat
GBP, paddle gunboat
Gm, Germany
Gr, Greece

H
Ha, Haiti
harbV, harbour vessel
hospS, floating hospital
HPn, normal horse power
HPf, forced horse power
hulk, hulk
hydgV, hydrographic vessel

I
In, India
Ir, Iran
It, Italy

J
Jp, Japan

L
lighH, floating lighthouse

M
minTr, mine transport
ML, minelayer
MM, monitor
MR, river monitor
Mx, Mexico

N
NAWI, North America and West Indies
Ne, Netherlands
No, Norway
NZ, New Zealand

O
olr, oilier

P
Pe, Peru
PS, patrol vessel
Po, Portugal

R
Ro, Romania
Rs, Russia / Soviet Union
rect, rectangular

S
Sa, Sarawak
SalTS, sailing training ship
SD, San Domingo
Si, Siam / Thailand
Sp, Spain
SS, submarine
Sw, Sweden
SZ, midget submarine

T
targtS, target ship
TB1, torpedo boat 1 class
TB2, torpedo boat 2 class
tendr, tender
TM, military transport
traiS, training ship
triaV, trials vessel
Tu, Turkey
tug, tug

U
Ur, Urguay
US, United States

W
WLBelt, armoured belt by waterline
WLDeck, armoured deck by waterline
WiP, work in progress

Y
YA, yacht

Z
ZCD, despatch vessel
ZL, liner

Shipbuilder's page keys

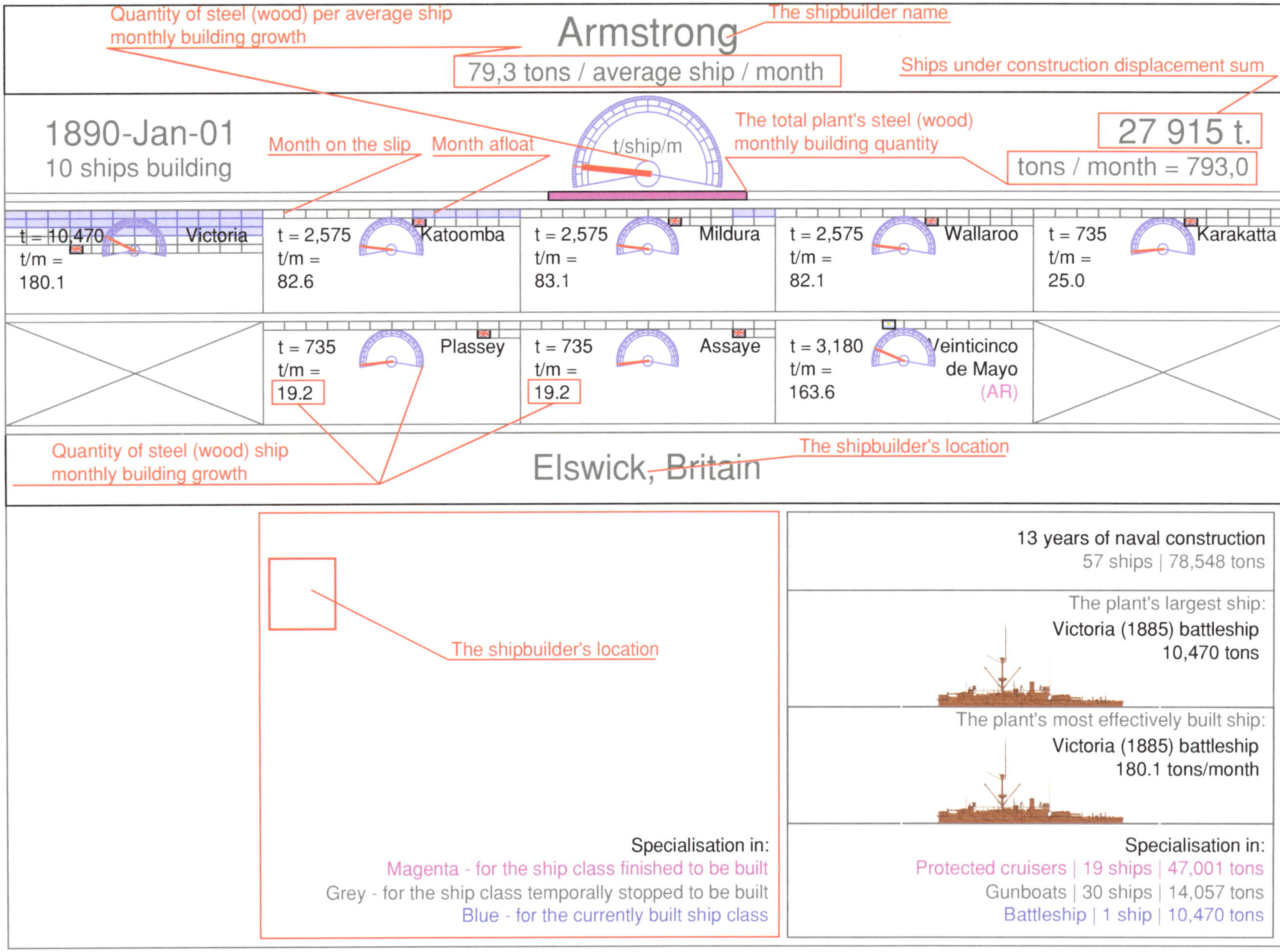

Ship page keys

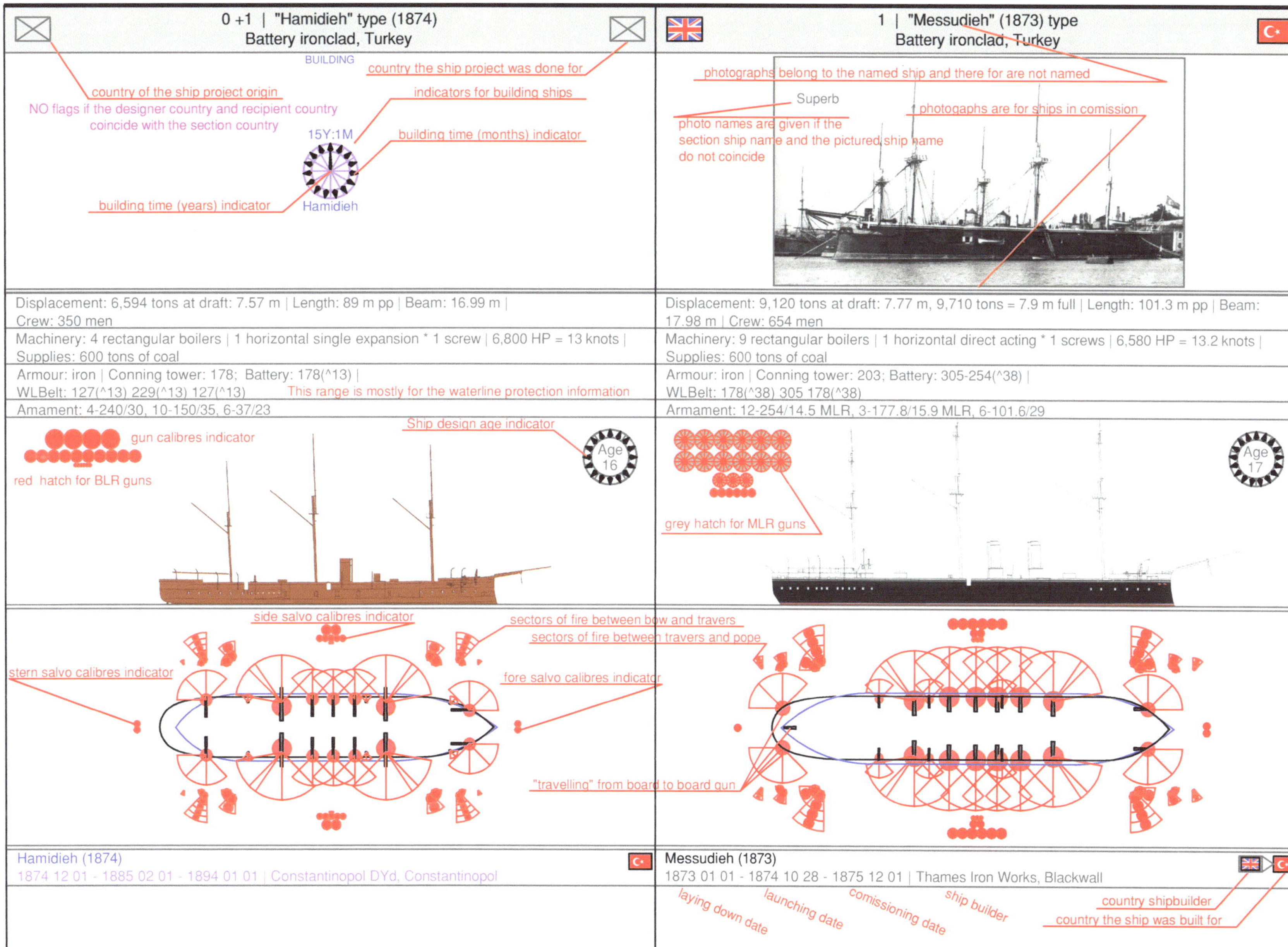

BRITISH FLEET

and Australian, Indian, New Zealand fleets

National flag

Fleet age 7

Ensign

Jack

Admiral

Months derivation +9

Warships built from 1858 01 01 to 1890 01 01

569 warships = 1,060,595 tons = 1,865 tons / average ship

Fleet age - 1883 06 30

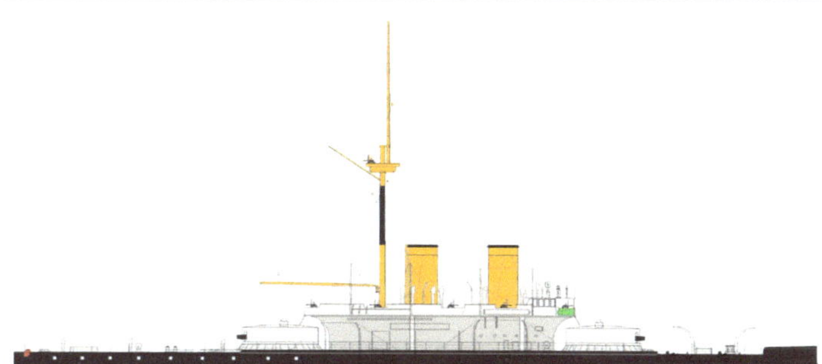

1889 fleet largest ship: Dreadnought (1870) battleship

British naval construction

Armstrong, Elswick
Chathem DYd
Devonport DYd
Earle, Hull
Laird, Birkenhead
Leslie, Hebburn
London & Glasgow, Govan
Napier, Govan
Naval Construction & Armament, Barrow
Palmer, Jarrow
Pembroke DYd
Portsmouth DYd
Samuda, Poplar
Scott's SB, Greenock
Sheereness DYd
Stephenson, Newcastle
Thames IW, Blackwall
Thompson, Clydebank
Thornycroft, Woolston
White, Cowes
Yarrow, Poplar

Armstrong

79.3 tons / average ship / month

1890-Jan-01
10 ships building

t/ship/m

27,915 t.
tons / month = 793.0

Ship	t	t/m	Engine
Victoria	10,470	180.1	Humphrys
Katoomba	2,575	82.6	Leslie
Mildura	2,575	83.1	Leslie
Wallaroo	2,575	82.1	Leslie
Karakatta	735	25.0	Belliss
Boomerang	735	24.6	Belliss
Plassey	735	19.2	Maudslay
Assaye	735	19.2	Maudslay
Veinticinco de Mayo (Ar)	3,180	163.6	Humphrys
Spartan	3,600	113.7	Maudslay

Elswick, Britain

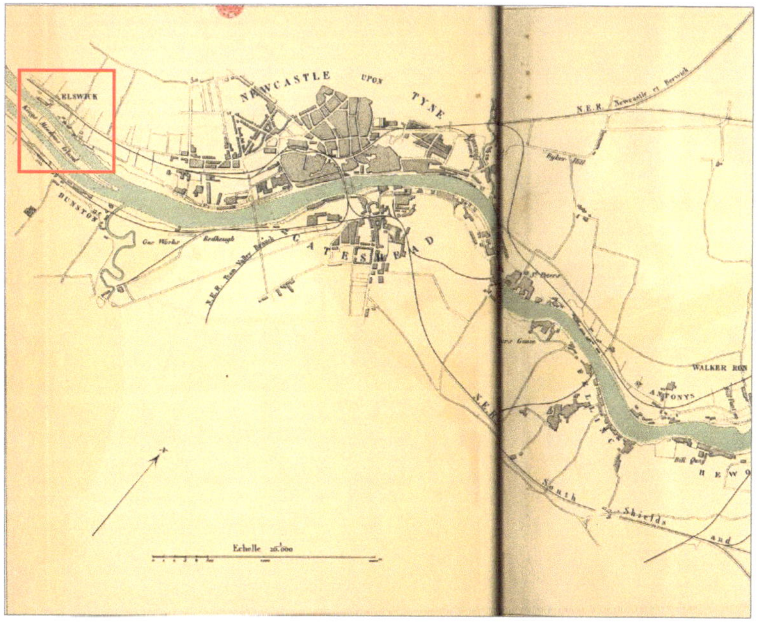

13 years of naval construction
57 ships | 78,548 tons

The plant's largest ship:
Victoria (1885) battleship
10,470 tons

The plant's most effectively built ship:
Victoria (1885) battleship
180.1 tons/month

Specialisation in:
Protected cruisers | 19 ships | 47,001 tons
Gunboats | 30 ships | 14,057 tons
Battleship | 1 ship | 10,470 tons

Chatham dockyard

110.6 tons / average ship / month

1890-Jan-01

9 ships building

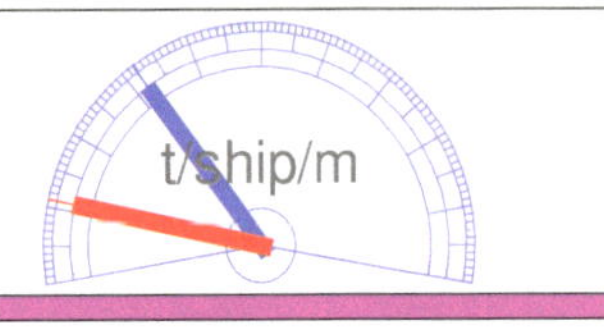

40,390 t.

tons / month = 995.5

t = 735 t/m = 22.8 **Seagull**

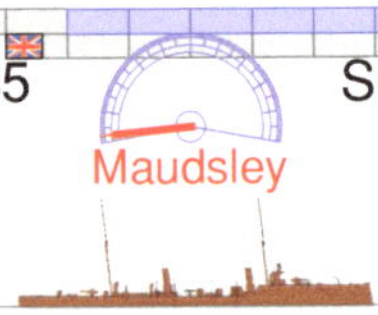

t = 735 t/m = 19.3 **Salamander**

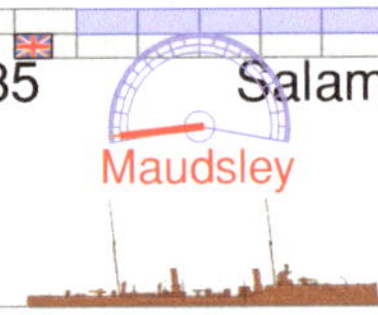

t = 9,150 t/m = 212.9 **Blake**

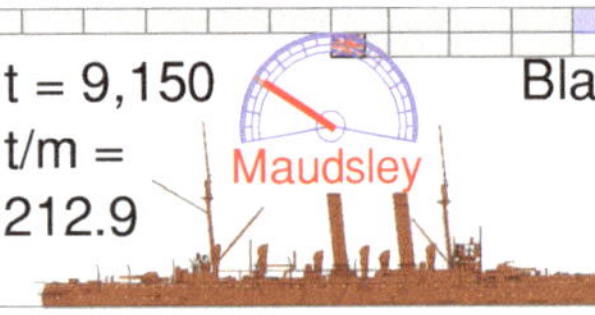

t = 735 t/m = 20.5 **Skipjack**

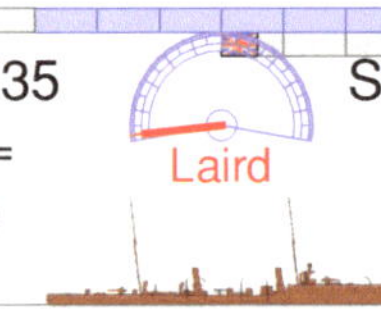

t = 735 t/m = 36.0 **Sheldrake**

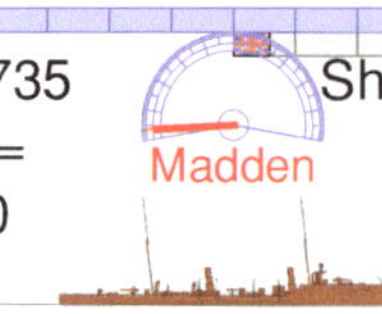

t = 3,400 t/m = 109.6 **Andromache**

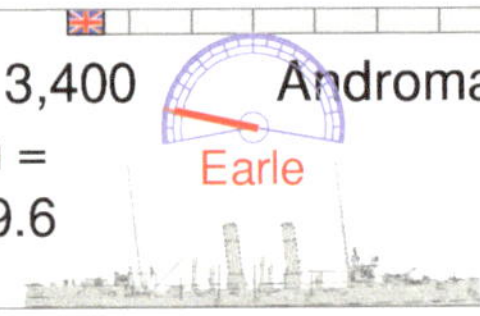

t = 3,400 t/m = 99.7 **Apollo**

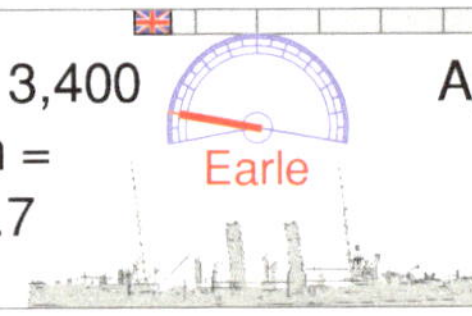

t = 7,350 t/m = 156.9 **Hawke**

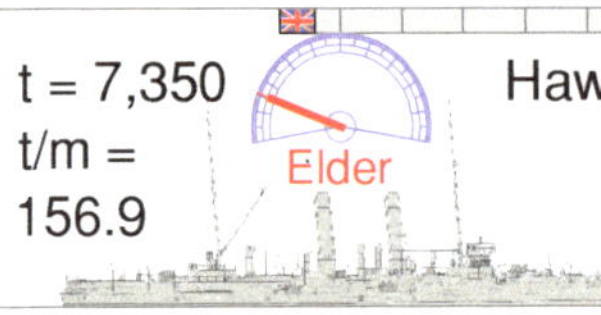

t = 14,150 t/m = 317.8 **Hood**

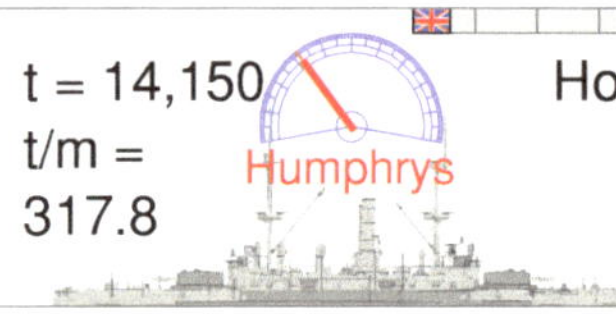

Chatham, Britain

32+ years of naval construction

63 ships | 247,943 tons

The plant's largest ship:

Hood (1889) battleship

14,150 tons

The plant's most effectively built ship:

Hood (1889) battleship

317.8 tons/month

Specialisation in:

Battery ironclads | 8 ships | 67,577 tons

Protected cruisers | 10 ships | 42,150 tons

Battleships | 4 ships | 41,282 tons

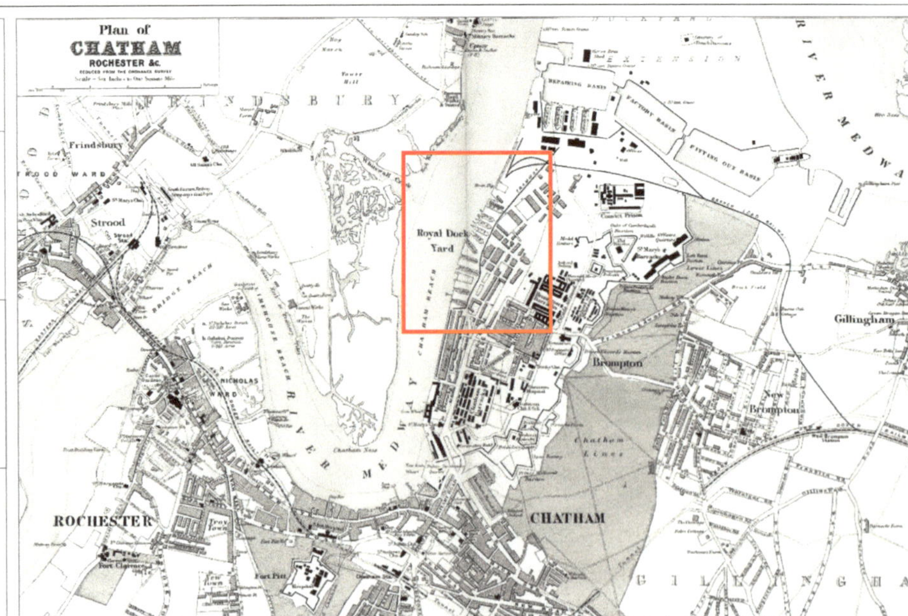

Devonport dockyard

62.7 tons / average ship / month

1890-Jan-01
6 ships building

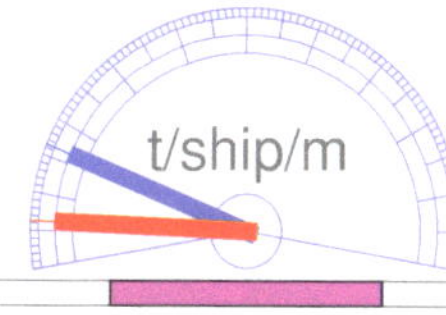

14,780 t.
tons / month = 376.5

Ship	t	t/m	Builder
Spanker	735	24.4	Maudslay
Speedwell	735	27.9	Maudslay
Phoebe	2,575	59.6	builder
Philomel	2,575	85.7	Earle
Edgar	7,350	163.9	Elder
Antelope	810	14.9	Yarrow

Devonport, Britain

32+ years of naval construction
54 ships | 92,029 tons

The plant's largest ship:
Edgar (1889) protected cruiser
7,350 tons

The plant's most effectively built ship:
Edgar (1889) protected cruiser
163.9 tons/month

Specialisation in:
Gunboats | 31 ships | 31,006 tons
Screw wooden cruisers | 9 ships | 22,136 tons
Protected cruisers | 5 ships | 16,040 tons

Earle

136.1 tons / average ship / month

1890-Jan-01
1 ship building

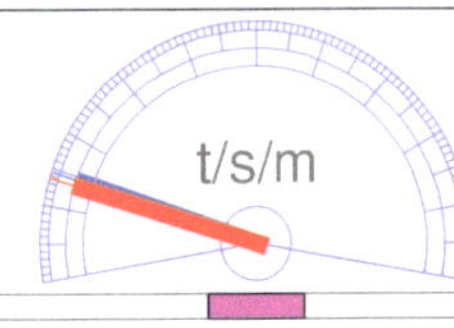

7,350 t.
tons / month = 136.1

t = 7,350
t/m =
136.1

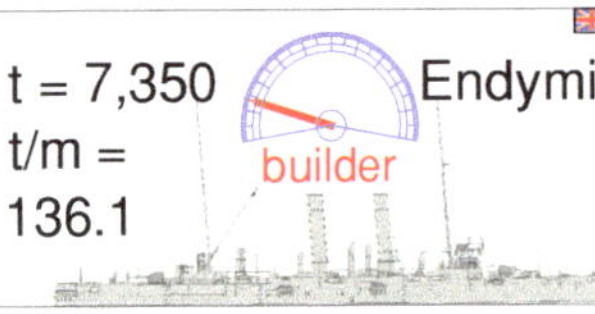

Hull, Britain

17 years of naval construction
10 ships | 27,495 tons

The plant's largest ship:
Endymion (1889) protected cruiser
7,350 tons

The plant's most effectively built ship:
Almirante Cochrane (1873) battery ironclad (Ci)
144.8 tons/month

Specialisation in:
Battery ironclads | 3 ships | 8,940 tons
Protected cruiser | 1 ship | 7,350 tons
Armoured cruiser | 1 ship | 5,600 tons

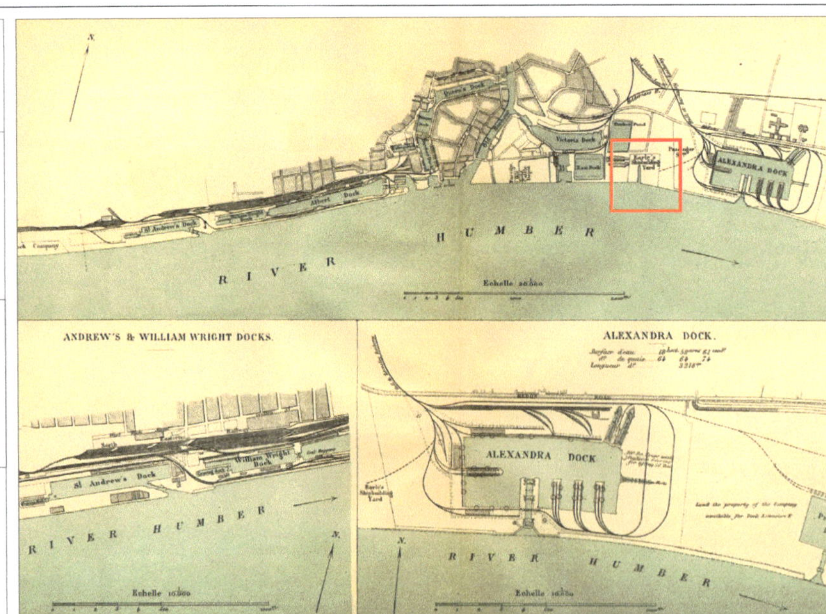

Laird

53.1 tons / average ship / month

1890-Jan-01
6 ships building

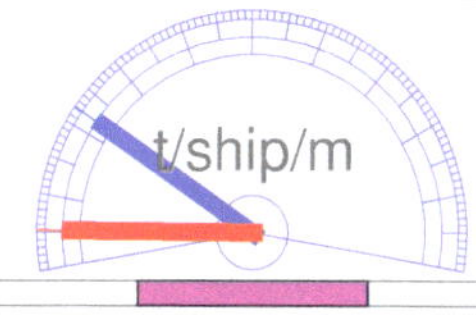

7,126 t.
tons / month = 318.7

t = 713 t/m = 30.6	Almirante Condell (CI)	
t = 713 t/m = 30.6	Almirante Lynch (CI)	
t = 520 t/m = 22.3	Espora (AR)	
t = 2,330 t/m = 115.4	Rosales (AR)	
t = 2,330 t/m = 97.3	Independencia (AR)	
t = 2,330 t/m = 115.4	Nueve de Julio (AR)	

Birkenhead, Britain

32+ years of naval construction
50 ships | 67,826 tons

The plant's largest ship:
Agincourt (1861) battery ironclad
10,600 tons

The plant's most effectively built ship:
Captain (1867) battleship
222.0 tons/month

Specialisation in:
Battery ironclads | 2 ships | 16,509 tons
Coast defence battleships | 5 ships | 13,482 tons
Gunboats | 26 ships | 12,516 tons

Leslie

56.5 tons / average ship / month

1890-Jan-01
1 ship building

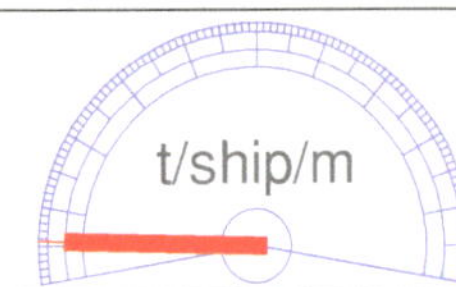

1,830 t.
tons / month = 56.5

t = 1,830
t/m =
56.5

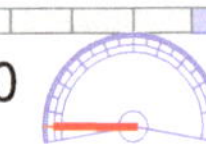

Bellona

Hebburn, Britain

2 years of naval construction
1 ship | 1,830 tons

The plant's largest ship:
Bellona (1888) protected cruiser
1,830 tons

The plant's most effectively built ship:
Bellona (1888) protected cruiser
56.5 tons/month

Specialisation in:
Protected cruiser | 1 ship | 1,830 tons

London and Glasgow

101.2 tons / average ship / month

1890-Jan-01
2 ships building

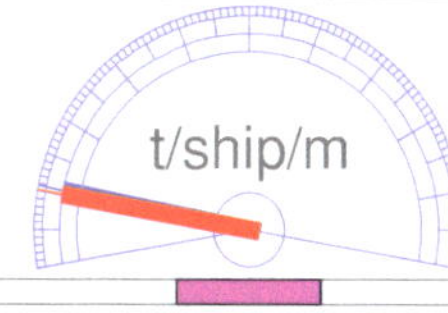

7,200 t.
t / month = 202.3

t = 3,600 t/m = 107.0 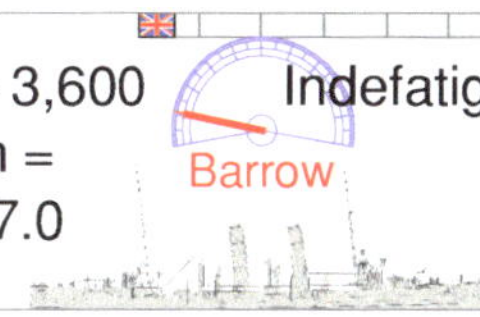 Barrow — Indefatigable	t = 3,600 t/m = 95.3 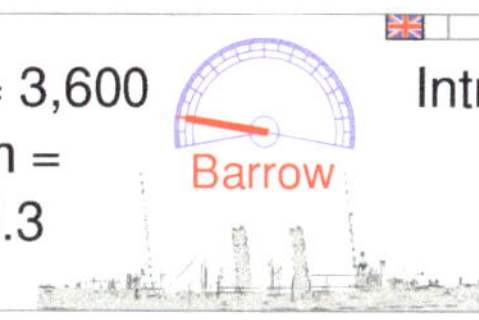 Barrow — Intrepid

Govan, Britain

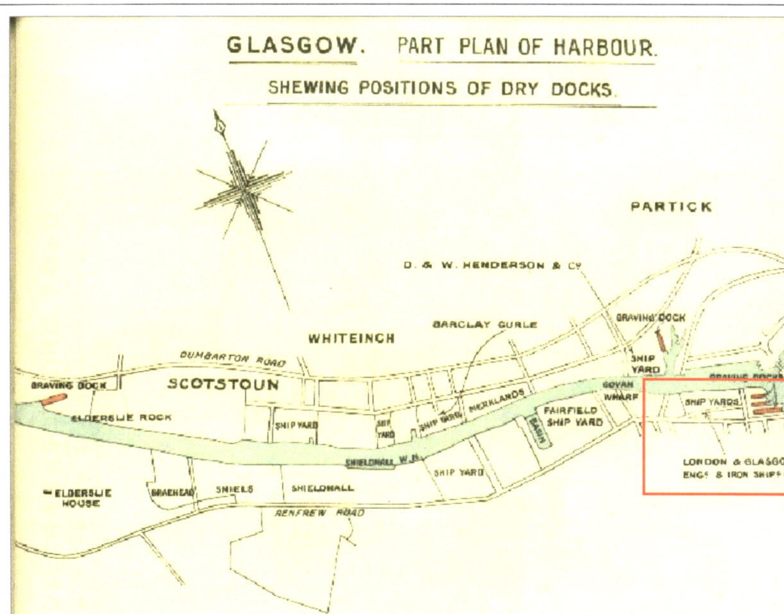

1 year of naval construction
2 ships | 7,200 tons

The plant's largest ship:
Indefatigable (1889) protected cruiser
3,600 tons

The plant's most effectively built ship:
Indefatigable (1889) protected cruiser
107.0 tons/month

Specialisation in:
Protected cruisers | 2 ships | 7,200 tons

Napier

124.9 tons / average ship / month

1890-Jan-01
1 ship building

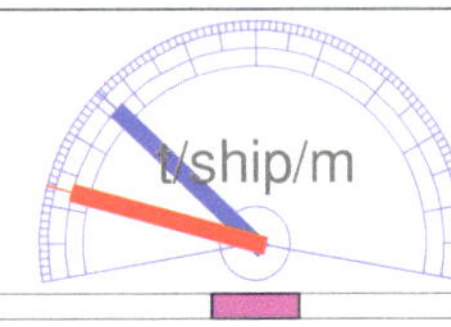

7,350 t.
tons / month = 124.9

t = 7,350
t/m =
124.9

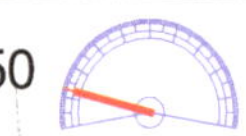

Gibraltar

Govan, Britain

32+ years of naval construction
26 ships | 98,218 tons

The plant's largest ship:
Black Prince (1859) battery ironclad
9,250 tons

The plant's most effectively built ship:
Black Prince (1859) battery ironclad
264.7 tons/month

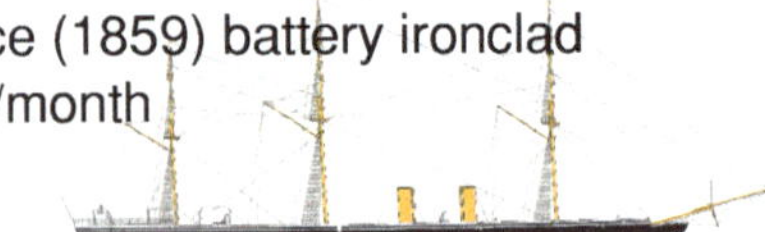

Specialisation in:
Battery ironclads | 7 ships | 46,978 tons
Protected cruisers | 4 ships | 20,250 tons
Armoured cruisers | 2 ships | 13,230 tons

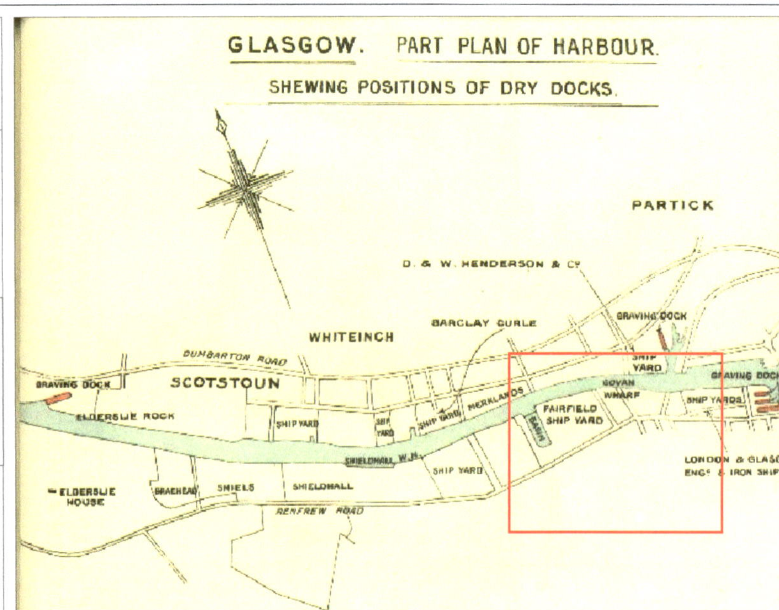

Naval Construction & Armament

110.4 tons / average ship / month

1890-Jan-01
3 ships building

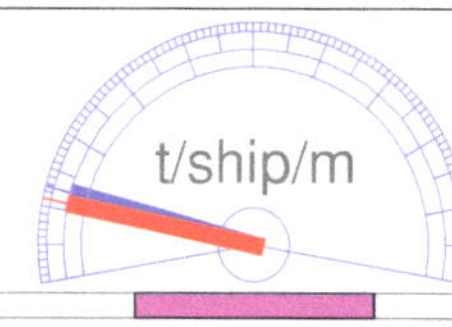

10,200 t.
tons / month = 331.1

t = 3,400 t/m = 78.7 Barrow Latona	t = 3,400 t/m = 126.0 Barrow Melampus	t = 3,400 t/m = 126.5 Barrow Naiad

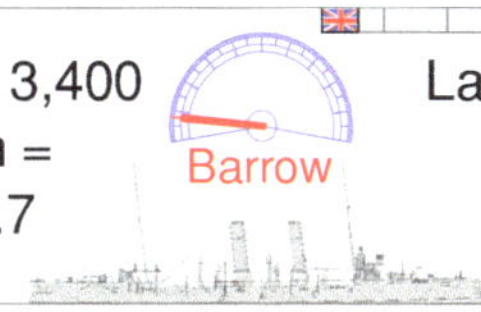
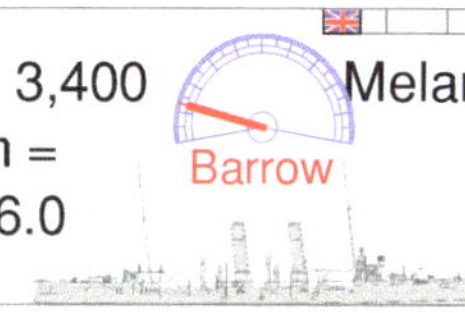
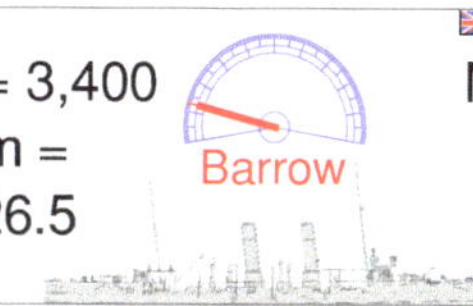

Barrow, Britain

6 years of naval construction
4 ships | 11,780 tons

The plant's largest ship:
Naiad (1889) protected cruiser
3,400 tons

The plant's most effectively built ship:
Naiad (1889) protected cruiser
126.5 tons/month

Specialisation in:
Protected cruisers | 4 ships | 11,780 tons

Palmer

70.7 tons / average ship / month

1890-Jan-01
3 ships building

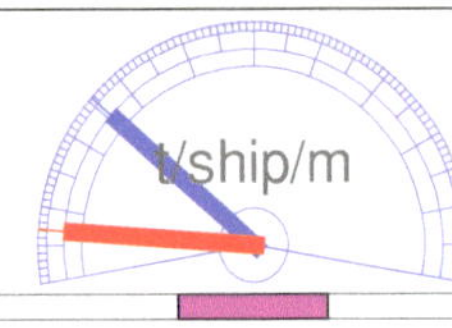

7,725 t.
tons / month = 212.1

t = 525 t/m = 21.9	Planet
t = 3,600 t/m = 90.1	Pique
t = 3,600 t/m = 100.0	Rainbow

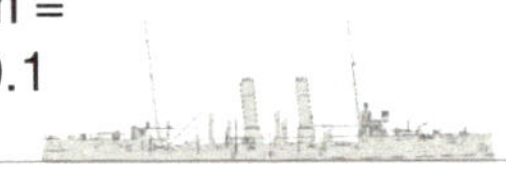
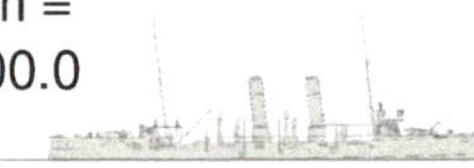

Jarrow, Britain

32+ years of naval construction
24 ships | 53,481 tons

The plant largest ship:
Swiftsure (1868) battery ironclad
6,910 tons

The plant most effectively built ship:
Defence (1859) battery ironclad
256.3 tons/month

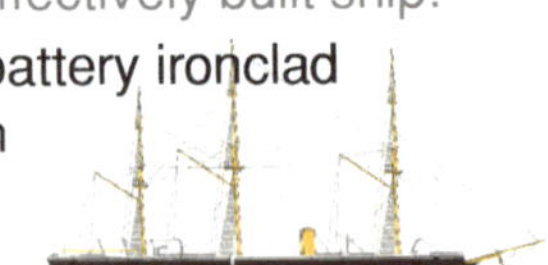

Specialisation in:
Battery ironclads | 3 ships | 19,700 tons
Armoured cruisers | 2 ships | 11,200 tons
Protected cruisers | 2 ships | 7,200 tons

Pembroke dockyard

131.8 tons / average ship / month

1890-Jan-01
8 ships building

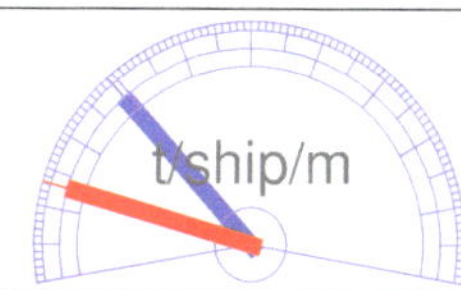

48,235 t.
tons / month = 1,054.0

Nile t = 12,590 t/m = 200.1 Maudslay	Redpoll t = 805 t/m = 33.6 Earle	Widgeon t = 805 t/m = 32.2 Rennie	Blanche t = 1,580 t/m = 79.0 Earle	Blonde t = 1,580 t/m = 79.0 Earle
	Pearl t = 2,575 t/m = 61.4 Earle	Empress of India t = 14,150 t/m = 290.8 Humphrys	Repulse t = 14,150 t/m = 278.3 Humphrys	

Pembroke, Britain

32+ years of naval construction
70 ships | 231,632 tons

The plant's largest ship:
Empress of India (1889) battleship
14,150 tons

The plant's most effectively built ship:
Empress of India (1889) battleship
290.8 tons/month

Specialisation in:
Battleships | 11 ships | 118,766 tons
Battery ironclads | 6 ships | 32,800 tons
Gunboats | 38 ships | 25,890 tons

Portsmouth dockyard

213.73 tons / average ship / month

1890-Jan-01
7 ships building

t/ship/m

47,275 t.
tons / month = 1,496.1

t = 12,590 t/m = 255.5 Trafalgar Humphrys

t = 2,950 t/m = 110.5 Melpomene Palmer

t = 1,580 t/m = 79.0 Barrosa Palmer

t = 6,600 t/m = 358.2 Vulcan Humphrys

t = 1,830 t/m = 128.0 Barham Leslie

t = 2,575 t/m = 107.7 Pallas Leslie

t = 14,150 t/m = 457.2 Royal Sovereign Humphrys

Portsmouth, Britain

32+ years of naval construction
47 ships | 137,362 tons

The plant's largest ship:
Royal Sovereign (1889) battleship
14,150 tons

The plant's most effectively built ship:
Royal Sovereign (1889) battleship
457.2 tons/month
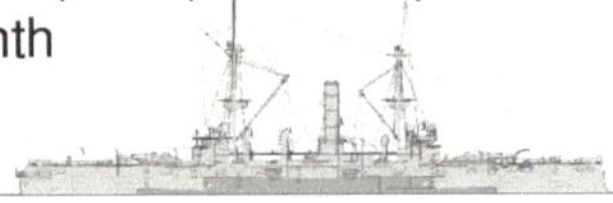

Specialisation in:
Battleships | 5 ships | 57,060 tons
Protected cruisers | 8 ships | 23,065 tons
Unprotected cruisers | 3 ships | 13,920 tons

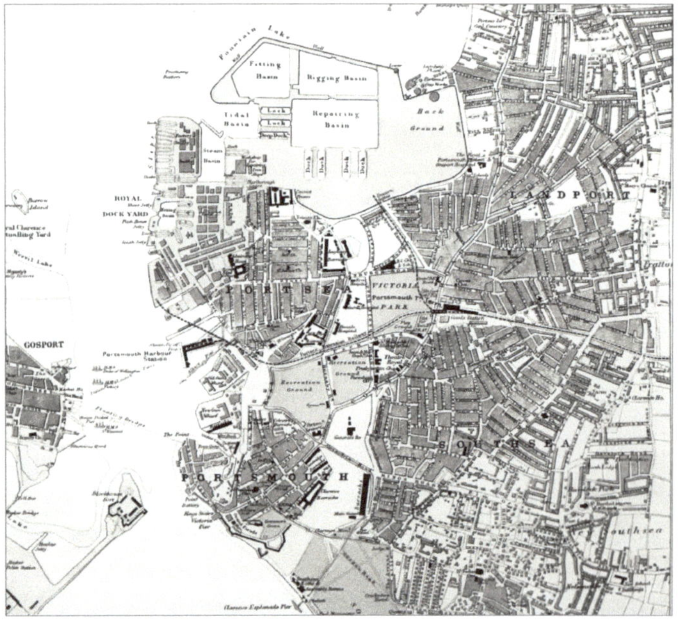

Samuda

85.0 tons / average ship / month

1890-Jan-01
2 ships building

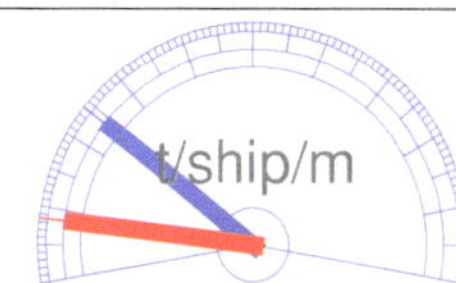

6,800 t.
tons / month = 170.0

Scylla
t = 3,400
t/m =
83.0

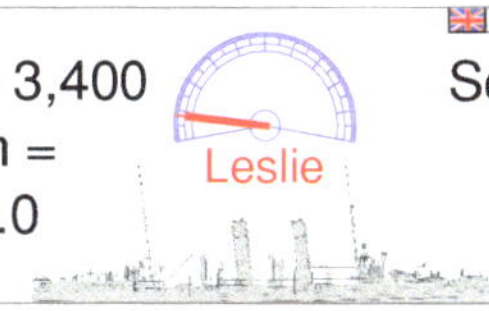

Sappho
t = 3,400
t/m =
87.1

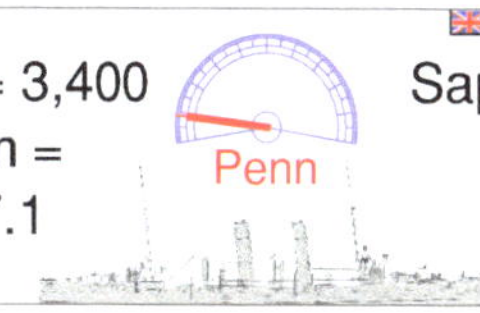

Poplar, Britain

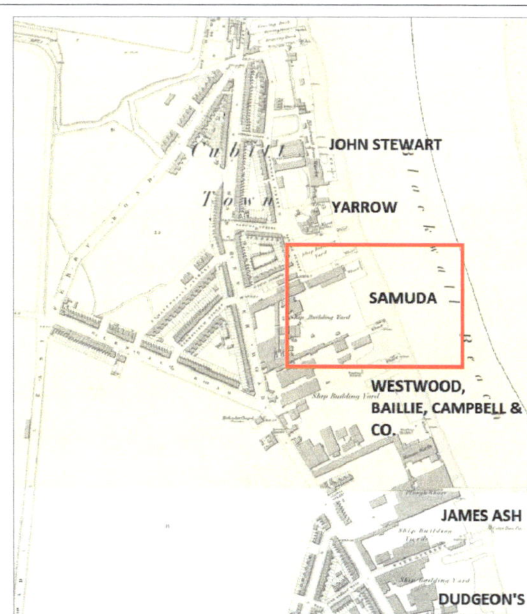

28 years of naval construction
22 ships | 75,109 tons

The plant's largest ship:
Kaiser (1872) battery ironclad (Gm)
7,645 tons

The plant's most effectively built ship:
Riachuelo (1882) battleship (Bz)
241.0 tons/month

Specialisation in:
Battery ironclads | 9 ships | 44,576 tons
Battleships | 2 ships | 10,531 tons
Protected cruisers | 2 ships | 6,800 tons

Scott's shipbuilding

30.5 tons / average ship / month

1890-Jan-01
2 ships building

t/ship/m

1,610 t.
tons / month = 61.0

t = 805
t/m =
28.9

Sparrow

t = 805
t/m =
32.2

Thrush

Greenock, Britain

2 years of naval construction
2 ships | 1,810 tons

The plant's largest ship:
Thrush (1888) gunboat
805 tons

The plant's most effectively built ship:
Thrush (1888) gunboat
32.2 tons/month

Specialisation in:
Gunboats | 2 ships | 1,610 tons

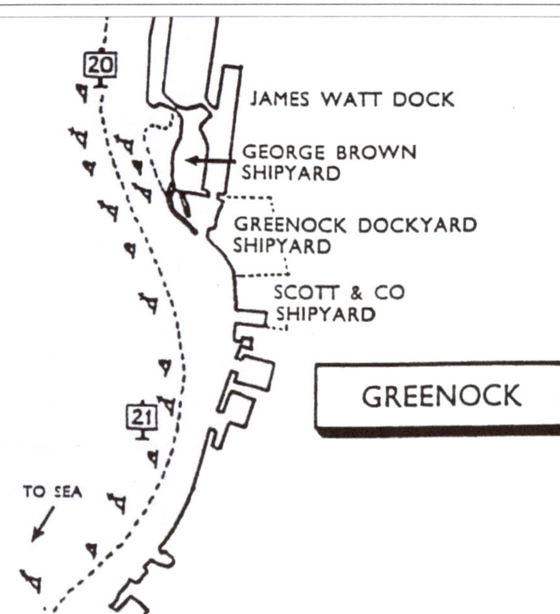

Sheerness dockyard

36.6 tons / average ship / month

1890-Jan-01
3 ships building

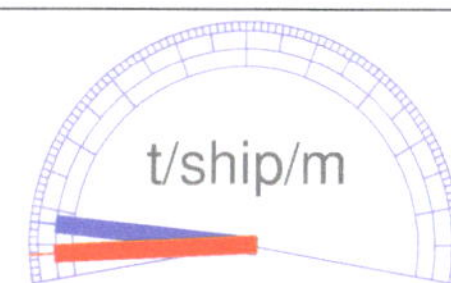

3,050 t.
tons / month = 110.0

t = 1,580 t/m = 65.9 Barracuda Palmer	t = 735 t/m = 21.1 Gleaner builder	t = 735 t/m = 23.2 Gossamer builder

Sheerness, Britain

32+ years of naval construction
23 ships | 28,867 tons

The plant's largest ship:
Orestes (1858) screw corvette
2,431 tons

The plant's most effectively built ship:
Encounter (1871) screw corvette
80.9 tons/month

Specialisation in:
Gunboats | 14 ships | 15,310 tons
Screw sailing cruisers | 7 ships | 13,669 tons
Topedo-gunboats | 3 ships | 2,020 tons

Stephenson

62.4 tons / average ship / month

1890-Jan-01
1 ship building

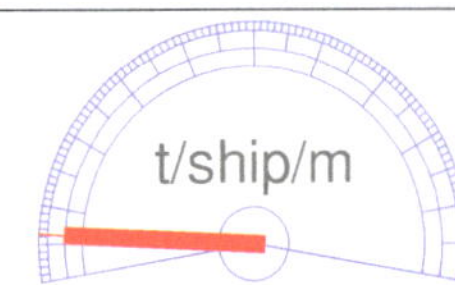

3,400 t.
tons / month = 62.4

t = 3,400
t/m =
62.4

Sybille

Newcastle, Britain

1 years of naval construction
1 ship | 3,400 tons

The plant's largest ship:
Sybille (1889) protected cruiser
3,400 tons

The plant's most effectively built ship:
Sybille (1889) protected cruiser
62.4 tons/month

Specialisation in:
Protected cruiser | 1 ship | 3,400 tons

Thames iron works

104.0 tons / average ship / month

1890-Jan-01
4 ships building

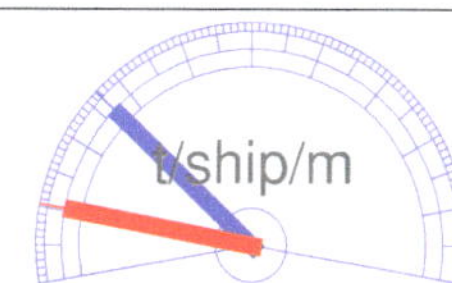

27,258 t.
tons / month = 416.1

t = 10,470 Sans Pareil t/m = 140.7 Humphrys	t = 9,150 Blenheim t/m = 135.3 Humphrys	t = 288 Limpopo (Po) t/m = 12.0	t = 7,350 Grafton t/m = 128.0 Humphrys

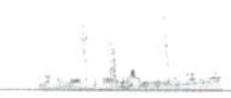

Blackwall, Britain

32+ years of naval construction
47 ships | 132,736 tons

The plant's largest ship:
Minotaur (1861) battery ironclad
10,690 tons

The plant's most effectively built ship:
Mahmudieh (1864) battery ironclad (Tu)
267.0 tons/month

Specialisation in:
Battery ironclads | 11 ships | 64,780 tons
Battleships | 2 ships | 21,070 tons
Protected cruisers | 2 ships | 16,500 tons

Thompson

85.8 tons / average ship / month

1890-Jan-01
8 ships building

t/ship/m

21,290 t.
tons / month = 686.2

Tartar t = 1,770 t/m = 23.4 	Mohawk t = 1,770 t/m = 25.5 	Chiyoda (Jp) t = 2,400 t/m = 96.3 	Tauranga t = 2,575 t/m = 99.8 	Ringarooma t = 2,575 t/m = 99.5 
	Terpsichore t = 3,400 t/m = 109.4 	Thetis t = 3,400 t/m = 117.2 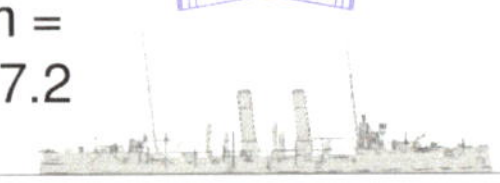	Tribune t = 3,400 t/m = 115.1 	

Clydebank, Britain

32+ years of naval construction
20 ships | 41,304 tons

The plant's largest ship:
Reina Regente (1886) protected cruiser (Sp)
4,725 tons

The plant's most effectively built ship:
Reina Regente (1886) protected cruiser (Sp)
257.3 tons/month

Specialisation in:
Protected cruisers | 13 ships | 32,275 tons
Battery ironclad | 1 ship | 4,670 tons
Cruisers armoured | 1 ship | 2,400 tons

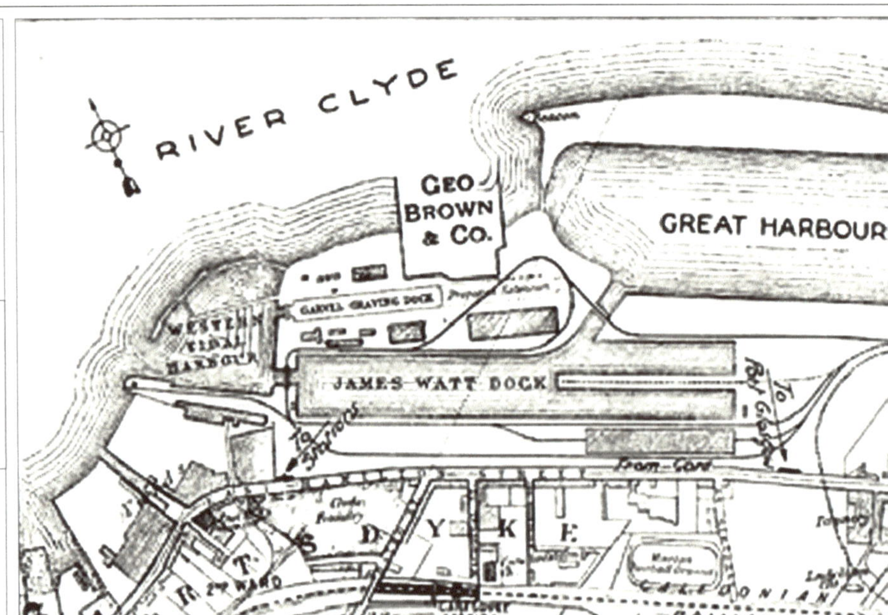

Thornycroft

4.9 tons / average ship / month

1890-Jan-01
8 ships building

t/ship/m

829 t.
tons / month = 39.3

t = 110
t/m = 6.4
Commodore Py (Ar)

t = 106
t/m = 6.2
Iguatemi (Bz)

t = 110
t/m = 6.4
Muratore (Ar)

t = 97
t/m = 4.1
7 (Ci)

t = 97
t/m = 4.1
8 (Ci)

t = 97
t/m = 4.1
9 (Ci)

t = 106
t/m = 3.6
Araguary (Bz)

t = 106
t/m = 4.4
Bento Gonsalves (Bz)

Woolston, Britain

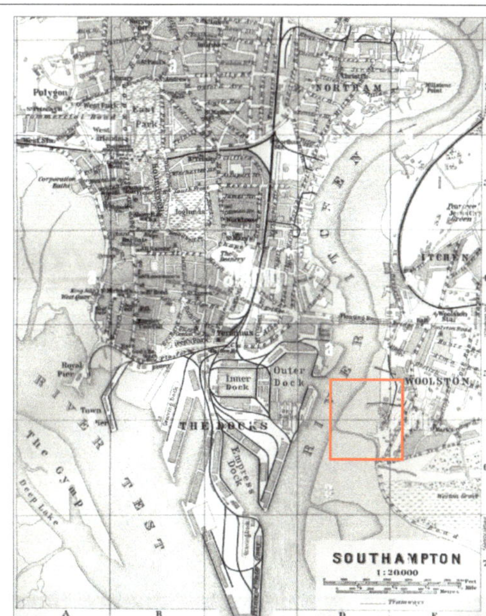

32+ years of naval construction
206 ships | 6,787 tons

The plant's largest ship:
Muratore (1889) torpedo boat (Ar)
110 tons

The plant's most effectively built ship:
Suhum (1882) torpedo boat (Rs)
11.3 tons/month

Specialisation in:
Torpedo boats | 206 ships | 6,787 tons

White

4.0 tons / average ship / month

1890-Jan-01

3 ships building

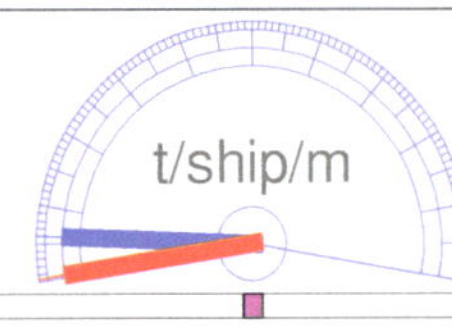

285 t.

tons / month = 11.8

t = 95 t/m = 0.4 Mahratta	t = 95 t/m = 0.4 Rajput	t = 95 t/m = 0.4 Sikh

Cowes, Britain

32+ years of naval construction

37 ships | 5,807 tons

The plant's largest ship:

Chiangtzu (1862) paddle frigate (CH)

1,289 tons

NO PLAN AVAILABLE

The plant's most effectively built ship:

Chiangtzu (1862) paddle frigate (CH)

55.4 tons/month

NO PLAN AVAILABLE

Specialisation in:

Gunboats | 6 ships | 3,162 tons

Paddle sailing aviso | 1 ship | 1,289 tons

Torpedo boats | 29 ships | 955 tons

Yarrow

2.1 tons / average ship / month

1890-Jan-01
25 ships building

t/ship/m

792 t.
tons / month = 52.0

t = 17 t/m = 0.7 39	t = 17 t/m = 0.7 40	t = 17 t/m = 0.7 41	t = 17 t/m = 0.7 42	t = 17 t/m = 0.7 43
t = 17 t/m = 0.7 44	t = 17 t/m = 0.7 45	t = 17 t/m = 0.7 46	t = 17 t/m = 0.7 47	t = 17 t/m = 0.7 48
t = 82 t/m = 3.4 Countess of Hopetown	t = 75 t/m = 6.3 Bathurst (Ar)	t = 75 t/m = 6.3 Buchardo (Ar)	t = 75 t/m = 6.3 Jorge (Ar)	t = 75 t/m = 6.3 King (Ar)
t = 75 t/m = 6.3 Pinedo (Ar)	t = 75 t/m = 6.3 Thorne (Ar)	t = 12 t/m = 0.5 3 (Ar)	t = 12 t/m = 0.5 4 (Ar)	t = 12 t/m = 0.5 5 (Ar)
t = 12 t/m = 0.5 6 (Ar)	t = 12 t/m = 0.5 7 (Ar)	t = 12 t/m = 0.5 8 (Ar)	t = 12 t/m = 0.5 9 (Ar)	t = 12 t/m = 0.5 10 (Ar)

Polpar, Britain

17 years of naval construction
145 ships | 6,860 tons

The plant's largest ship:
Kotaka (1886) torpedo boat (Jp) | 203 tons

The plant's most effectively built ship:
80 (1886) torpedo boat | 8.8 tons/month

Specialisation in:
Torpedo boats | 145 ships | 6,860 tons

British fleet effective list

16 battleships

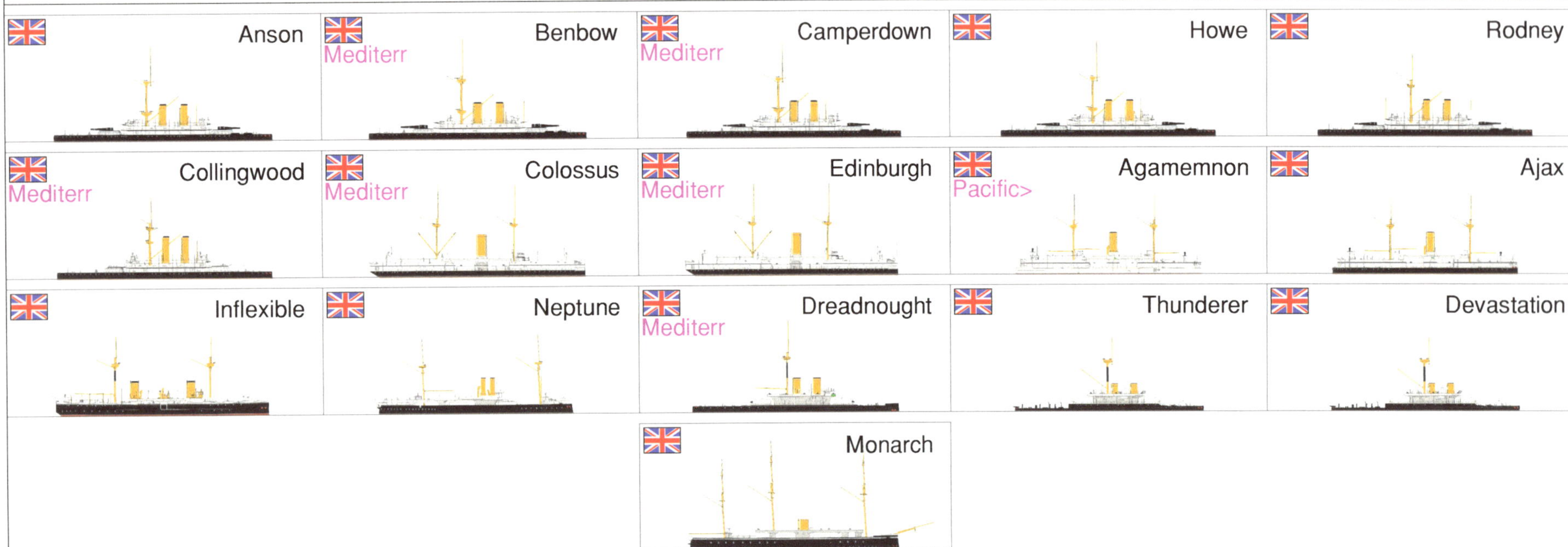

21 boardside ironclads

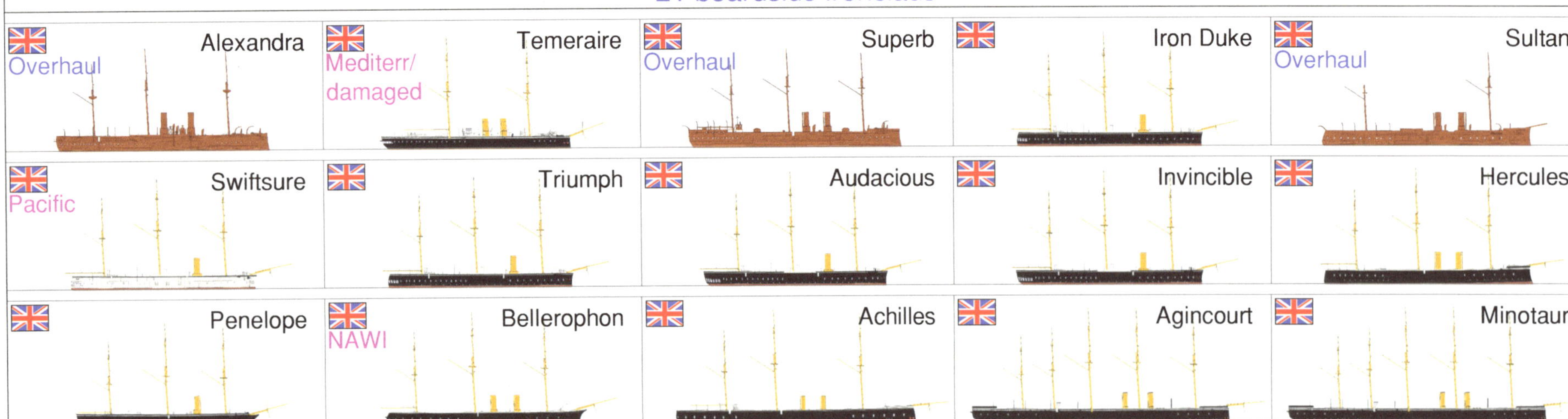

>>> boardside ironclads (continued)

Northumberland

Hector

Valiant
damaged

Black Prince

Warrior

Resistance
damaged

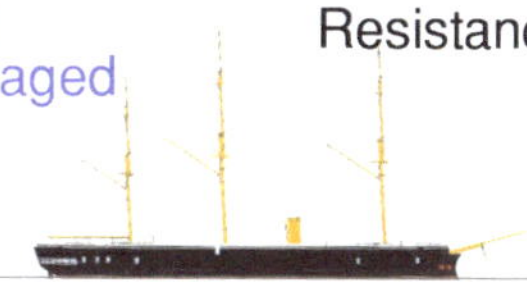

14 coast defence armoured ships

Hero

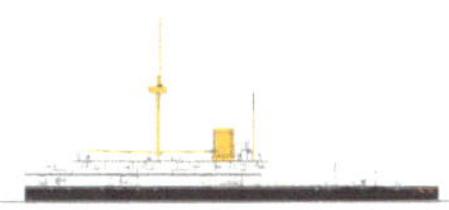

Conqueror

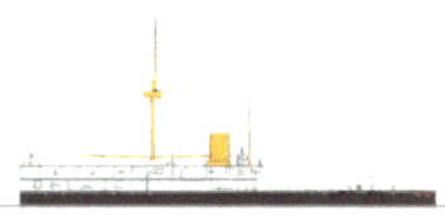

Rupert

Cyclops

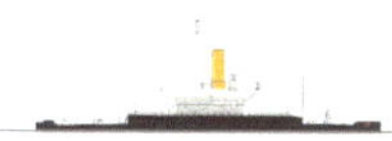

Hydra

Gorgon

Hecate

Glatton

Hotspur

Orion
Pacific

Belleisle

Wivern
Pacific

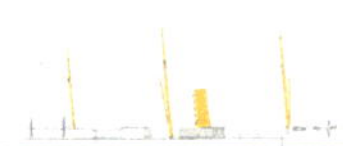

Scorpion
Bermudas

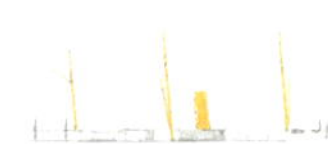

Prince Albert

12 armoured cruisers

Aurora

Australia

Immortalite

Galatea

Narcissus

Orlando
Australia

Undaunted

Imperieuse
Pacific

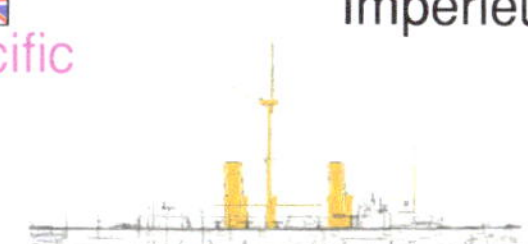

Warspite

Nelson

>>> armoured cruisers (continued)

Northampton 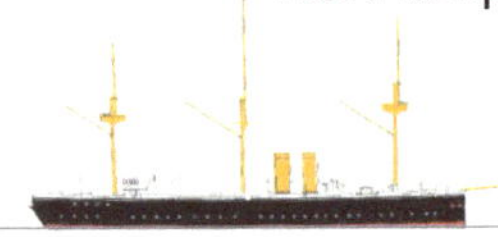	Shannon 

31 protected cruisers

Marathon 	Magicienne 	Medea 	Medusa 	Archer Good Hope 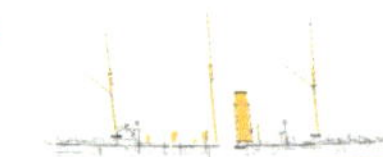
Racoon 	Porpoise Pacific 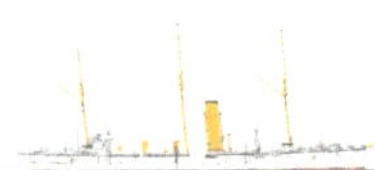	Serpent 	Cossack India 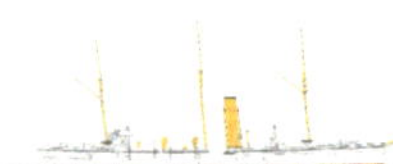	Brisk India 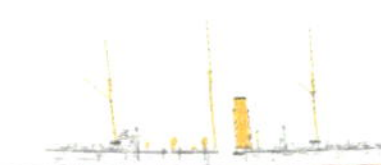
Fearless 	Scout 	Severn Pacific 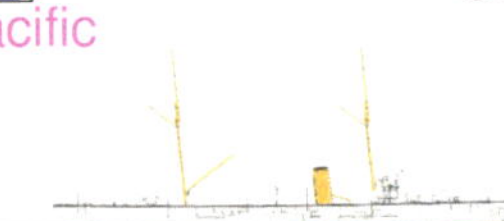	Thames 	Forth 
Mersey 	Amphion Pacific> 	Calliope Australia> 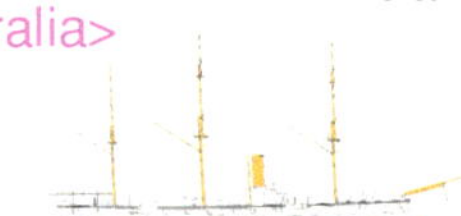	Calypso 	Arethusa 
Leander Pacific 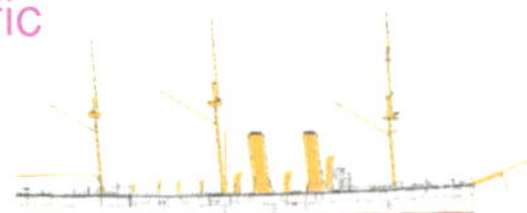	Phaeton Mediterr 	Canada NAWI 	Cordelia Pacific 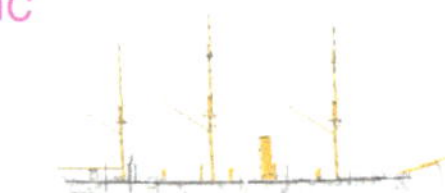	Constance 
Carysfort Mediterr 	Champion Pacific 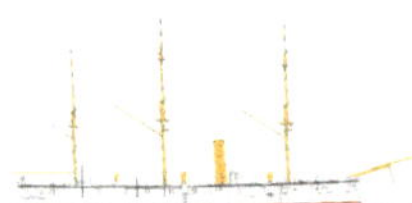	Cleopatra Bermudas 	Comus NAWI 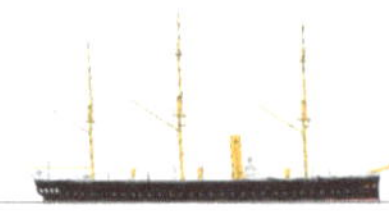	Conquest India 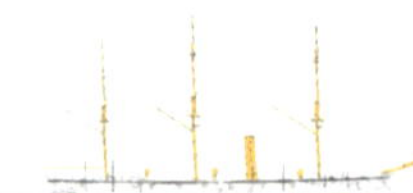

>>> protected cruisers (continued)

Curacoa — >Australia

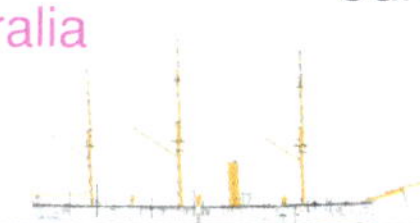

23 unprotected & wooden cruisers

Alacrity — Pacific

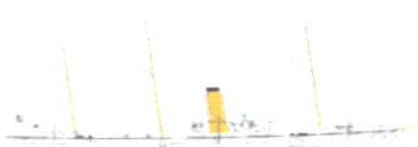

Surpise

Mercury — >Pacific

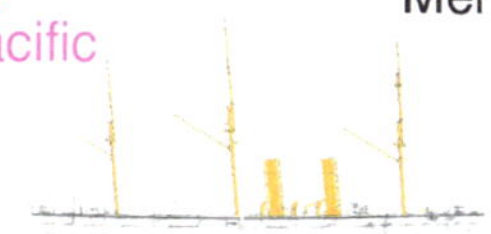

Iris

Garnet — India>

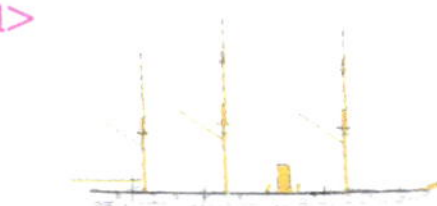

Emerald — Bermudas

Ruby

Tourmaline — Bermudas

Turquoise — India

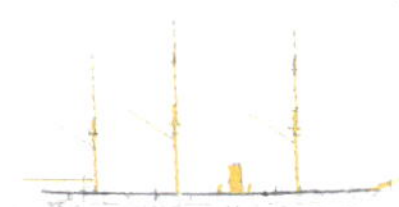

Bacchante

Boadicea — India

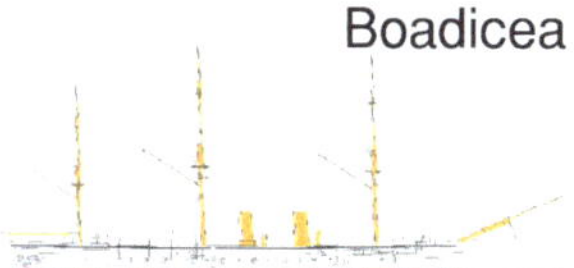

Euryalus

Opal — Australia>

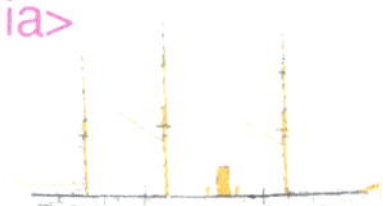

Sapphire

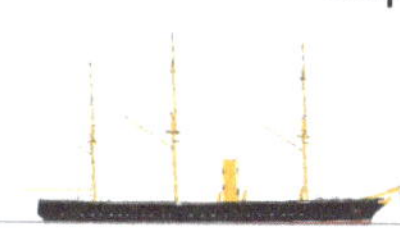

Rover

Shah

Raleigh — Good Hope

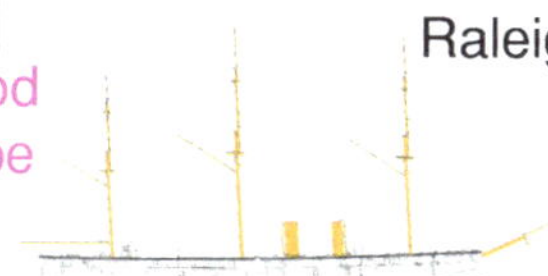

Active

Volage

Inconstant

Danae

Eclipse

Thalia

32 sloops

Caroline — Pacific

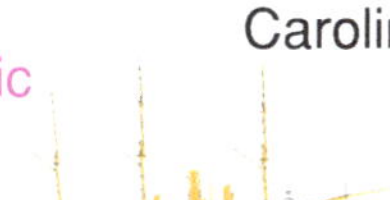

Heroine

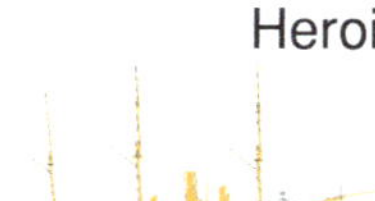

Hyacinth — Pacific

Pylades — Bermudas

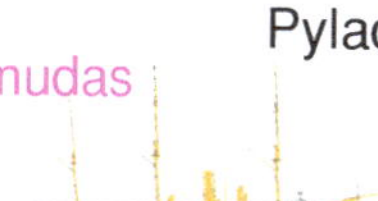

Rapid — Australia

Royalist — Australia

>>> sloops (continued)

Basilisk	Beagle Bermudas	Buzzard NAWI	Daphne Pacific	Nymphe Pacific	Swallow Good Hope
Acorn Pacific	Icarus Pacific	Mariner India	Melita Mediterr	Racer	Reindeer India
Dolphin Mediterr	Wanderer Pacific	Dragon	Espiegle Pacific	Gannet Mediterr	Kingfisher India
Miranda	Mutine Pacific	Pegasus	Cormorant Mediterr	Osprey Bermudas	Pelican NAWI
Penguin Pacific	Wild Swan	Egeria			

102 gunboats

Goldfinch	Lapwing	Magpie Good Hope	Redbreast	Ringdove	Partridge NAWI	Peacock Good Hope	Pheasant Good Hope	Pigeon India	Pigmy Pacific
Plover Pacific	Bramble Bermudas	Lizard Australia	Rattler Pacific	Curlew	Landrall Mediterr	Albacore Mediterr	Misletoe	Watchful	Algerine India
Rambler	Ranger	Linnet Pacific	Swift Pacific	Falcon	Flamingo Bermudas	Griffon	Ready NAWI	Rifleman	Avon
Elk	Midge Pacific	Viper Bermudas	Vixen Bermudas	Water-witch	Nimble	Drudge	Handy	Bouncer	Insolent

>>> gunboats (continued)

Banterer	Bullfrog	Cock- chafer	Espoir	Grappler	Raven Australia	Redwing	Starling	Stork Mediterr	Wrangler
Gadfly Good Hope	Griper Good Hope	Pincher	Tickler Good Hope	Firebrand Pacific	Firefly	Firm	Forester	Forward NAWI	Foxhound
Sheldrake	Dee Mediterr	Don Mediterr	Esk Pacific	Medina	Medway	Sabrina	Slaney	Spey	Tay
Tees	Trent	Tweed Pacific	Goshawk Mediterr	Merlin	Swinger Australia	Ant	Arrow	Badger	Blazer
odhound	Bonetta	Bulldog	Bustard	Comet	Cuckoo	Fidget	Hyaena	Kite	Mastiff
Pickle	Pike	Scourge	Snake	Snap	Weazel	Plucky	Staunch	Britomart	Cherub

Minstrel	Tyrian

6 torpedo gunboats

Sharpshooter	Grasshopper	Sandfly	Spider	Rattlesnake

Polyphemus

152 torpedo boats

Vesuvius	79	80	25	34	35	36	37	38	41

>>> torpedo boats (continued)

42	43	44	45	46	47	48	49	50	51
52	53	54	55	56	57	58	59	60	30
31	32	33	61	62	63	64	65	66	67
68	66	69	65	70	71	72	73	74	75
76	77	78	23	24	26	27	28	29	39
40	81	82	83	84	85	86	87	21	22
20 NO PLAN AVAILABLE	13 NO PLAN AVAILABLE	14	15 NO PLAN AVAILABLE	2	3	4	5	6	7
8	9	10	11	12	17	18	19 NO PLAN AVAILABLE	1	49
50	99	100	1	2	3	4	5	6	7
8	9	10	11	12	98	74 NO PLAN AVAILABLE	75 NO PLAN AVAILABLE	96 NO PLAN AVAILABLE	97 NO PLAN AVAILABLE
64	65	66	67	68	69	70	71	72	73
76	77	78	79	80	81	82	83	84	85
86	87	88	89	90	91	92	93	94	95
51	52	53	54	55	56	57	58	59	60
				61	62				

Fleet auxiliaries

Defence — Target ship

Newcastle — Port reserve

Defiance — Demonstration ship

Grampian — Demonstration ship

Revenge — Demonstration ship

Marlborough — Demonstration ship

12 ex-screw battleships		
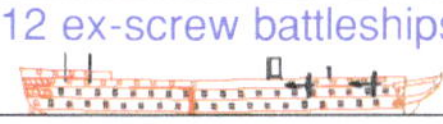		
2 floating batteries		
Asia	Indus	
2 accomodation ships		
Pembroke	Victor Emmanuel	
1 depot ship		
Irresistible		
1 hulk		
Hannibal		
6 training stations		
Britannia	Marlborough	Exmouth
Impregnable	Cambridge	Vernon
24 ex-sail battleships		
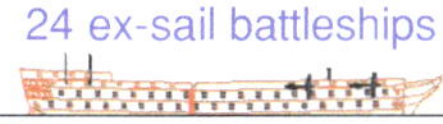		
1 hospital ship		
Meeanee		
3 accomodation ships		
Hibernia	Pembroke	Royal Adelaide
1 depot ship		
Carnatic		
2 hulks		
Vengeance	Wellesley	

1 target ship		
Nettle		
15 training stations		
Trafalgar	Formidable	Hindostan
Lion	Foudroyant	St. Vincent
Mars	Ganges	Warspite
Excellent	Implacable	Wellesley
Impregnable	Royal William	Wellington
1 debacader		
Cornwallis		
3 ex-screw frigates		

1 accomodation ship		
Forte		
2 depot ships		
Liffey	Ariadne	
27 ex-sail frigates		

2 hulks		
Monmouth	Thisbe	
2 hospital ships		
Pique	Hamadryad	
3 accomodation ships		
Nankin	Circe	Belvidera

5 depot ships		
Flora	Lavinia	Mercury
Melampus	Leonidas	
3 hulks		
Africaine	Cambrian	Minerva
1 harbor vessel		
Leda		
11 training stations		
Arethusa	Indefatigable	Active
Actaeon	Unicorn	Southampton
Daedalus	Clyde	Trincomalee
Mount Edgecumbe		Briton
5 ex-screw corvettes		

1 training ship		
Cruizer		
1 depot ship		
Peterel		
1 hulk		
Dido		
2 training stations		
Clio	Challenger	

8 ex-sail corvettes		

7 depot ships		
Enterprise	Dido	Columbine
Alarm	Racehorse	Favourite
Talbot		
1 Hulk		
Helena		
13 ex-sail brigs and gunboats		

1 patrol ship		
Rambler		
4 depot ships		
Dapper	Fancy	Skylark
Clinker		
1 target ship		
Resistance		
5 hulks		
Pet	Opossum	Snapper
Procris	Swan	
2 harbor vessels		
Perseus	Volcano	
1 paddle frigate		Valorous

Detained ships

Diogenes

Sea Serpent

British colonial fleets effective list

INDIAN FLEET

2 coast defence armoured ships

Abyssinia

Magdala

4 torpedo boats

Baluchi

Karen

Pathan

Gurkha

All 4 Indian torpedo boats though listed in the Indain list did not left Britain

AUSTRALIAN FLEET

1 coast defence armoured ship / 2 training stations

Cerberus

Wolverine — Training station
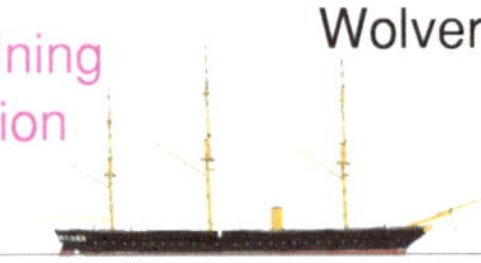

Nelson — Training station

7 gunboats / 8 gunboats (ex-barges)

Protector

Albert
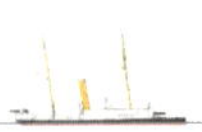

Gayundah

Paluma

Lady Loch — NO PLAN AVAILABLE

Otter — NO PLAN AVAILABLE

Victoria
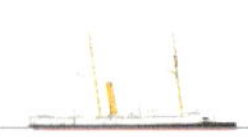

Gannet

Batman

Fawkner

Bonito

Bream

Dolphin

Pumba

Stingaree

8 torpedo boats (1 Tasmanian)

Childers

Gordon

Mosquito

Lonsdale

Nepean

Acheron

Avernus

191

NEW ZEALANDIAN SHIPS

4 torpedo boats

1

2

3

4

British section ship names glossary (1-H)

1 (TB2)
10 (TB1)
10 (TB2)
10 (TB2)
100 (TB2)

11 (TB1)
11 (TB2)
12 (TB1)
12 (TB2)
13 (TB1)
14 (TB1)
15 (TB1)
17 (TB1)
18 (TB1)
19 (TB1)

2 (TB1)
2 (TB2)
20 (TB1)
21 (TB1)
22 (TB1)
23 (TB1)
24 (TB1)
25 (TB1)
26 (TB1)
27 (TB1)
28 (TB1)
29 (TB1)

3 (TB1)
3 (TB2)
3 (TB2)
30 (TB1)
31 (TB1)
32 (TB1)
33 (TB1)
34 (TB1)
35 (TB1)
36 (TB1)
37 (TB1)
38 (TB1)
39 (TB1)
39 (TB2)

4 (TB1)
4 (TB2)
4 (TB2)
40 (TB1)
40 (TB2)
41 (TB1)
41 (TB2)
42 (TB1)
42 (TB2)
43 (TB1)
43 (TB2)
44 (TB1)
44 (TB2)
45 (TB1)
45 (TB2)
46 (TB1)
46 (TB2)
47 (TB1)
47 (TB2)
48 (TB1)
48 (TB2)
49 (TB1)
49 (TB2)

5 (TB1)
5 (TB2)
5 (TB2)
50 (TB1)
50 (TB2)
51 (TB1)
51 (TB2)
52 (TB1)
52 (TB2)
53 (TB1)
53 (TB2)
54 (TB1)
54 (TB2)
55 (TB1)
55 (TB2)
56 (TB1)
56 (TB2)
57 (TB1)
57 (TB2)
58 (TB1)
58 (TB2)
59 (TB1)
59 (TB2)

6 (TB1)
6 (TB2)
6 (TB2)
60 (TB1)
60 (TB2)
61 (TB1)
61 (TB2)
62 (TB1)
62 (TB2)
63 (TB1)
64 (TB1)
64 (TB2)
65 (TB1)
65 (TB2)
66 (TB1)
66 (TB2)
67 (TB1)
67 (TB2)
68 (TB1)
68 (TB2)
69 (TB1)
69 (TB2)

7 (TB1)
7 (TB1)
7 (TB2)
7 (TB2)
70 (TB1)
70 (TB2)
71 (TB1)
71 (TB2)
72 (TB1)
72 (TB2)
73 (TB1)
73 (TB2)
74 (TB1)
74 (TB2)
75 (TB1)
75 (TB2)
76 (TB1)
76 (TB2)
77 (TB1)
77 (TB2)
78 (TB1)
78 (TB2)
79 (TB1)
79 (TB2)

8 (TB1)
8 (TB1)
8 (TB2)
8 (TB2)
80 (TB1)
80 (TB2)
81 (TB1)
81 (TB2)
82 (TB1)
82 (TB2)
83 (TB1)
83 (TB2)
84 (TB1)
84 (TB2)
85 (TB1)
85 (TB2)
86 (TB1)
86 (TB2)
87 (TB1)
87 (TB2)
88 (TB2)
89 (TB2)

9 (TB1)
9 (TB1)
9 (TB2)
9 (TB2)
90 (TB2)
91 (TB2)
92 (TB2)
93 (TB2)
94 (TB2)
95 (TB2)
96 (TB2)
97 (TB2)
98 (TB2)
99 (TB2)

A
Achilles (BO)
Actaeon (drilS)
Active (CS)
Active (drilS)
Africaine (hulk)
Agamemnon (BB)
Agincourt (BO)
Ajax (BB)
Alacrity (CS)
Alarm (deptS)
Albacore (GB)
Alexandra (BO)
Algerine (GB)
Algiers (BW)
Almirante Condell (DC)
Almirante Lynch (DC)
Amphion (CP)
Andromache (CP)
Anson (BB)
Ant (GB)
Antelope (DC)
Apollo (CP)
Araguary (TB1)
Archer (CP)
Arethusa (CP)
Arethusa (drilS)
Ariadne (drilS)
Arrow (GB)
Asia (BZ)
Assaye (DC)
Audacious (BO)
Aurora (CA)
Australia (CA)
Avon (GB)

B
Bacchante (CS)
Badger (GB)
Banterer (GB)
Barham (CP)
Barracuda (CP)
Barrosa (CP)
Basilisk (CZ)
Bathurst (TB1)
Beagle (CZ)
Belleisle (BO)
Bellerophon (BO)
Bellona (CP)
Belvidera (accomS)
Benbow (BB)
Black Prince (CA)
Blake (CP)
Blanche (CP)
Blazer (GB)
Blenheim (CP)
Blonde (CP)
Bloodhound (GB)
Boadicea (CS)
Bonetta (GB)
Bouncer (GB)
Bramble (GB)
Brisk (CP)
Britannia (drilS)
Britomart (GB)
Briton (drilS)
Buchardo (TB1)
Bulldog (GB)
Bullfrog (GB)
Bustard (GB)
Buzzard (CZ)

C
Calliope (CP)
Calypso (CP)
Cambrian (hulk)
Cambridge (drilS)
Camperdown (BB)
Canada (CP)
Carnatic (deptS)
Caroline (CW)
Carysfort (CP)
Challenger (drilS)
Champion (CP)
Cherub (GB)
Chiyoda (CA)
Circe (accomS)
Cleopatra (CP)
Clinker (deptS)
Clio (drilS)
Clyde (drilS)
Cockchafer (GB)
Collingwood (BB)
Colossus (BB)
Columbine (deptS)
Comet (GB)
Commodore Py (TB1)
Comus (CP)
Conqueror (CDB)
Conquest (CP)
Constance (CP)
Cordelia (CP)
Cormorant (CZ)
Cornwallis (debarq)
Cossack (CP)
Countess of Hopetown (TB1)
Cruizer (trailS)
Cuckoo (GB)
Curacoa (CP)
Curlew (GB)
Cyclops (MM)

D
Daedalus (drilS)
Danae (CW)
Daphne (CZ)
Dapper (tendr)
Dee (GB)
Defiance (BW)
Devastation (BB)
Dido (deptS)
Dido (hulk)
Diogenes (CS)
Dolphin (CZ)
Don (GB)
Dragon (CZ)
Dreadnought (BB)
Drudge (GB)

E
Eclipse (CW)
Edgar (CP)
Edinburgh (BB)
Egeria (CZ)
Elk (GB)
Emerald (CW)
Endymion (CP)
Enterprise (deptS)
Esk (GB)
Espiegle (CZ)
Espoir (GB)
Espora (DC)
Euryalus (CS)
Excellent (drilS)
Exmouth (drilS)

F
Falcon (GB)
Fancy (tendr)
Favourite (deptS)
Fearless (CP)
Fidget (GB)
Firebrand (GB)
Firefly (GB)
Firm (GB)
Flamingo (GB)
Flora (deptS)
Forester (GB)
Formidable (drilS)
Forte (accomS)
Forth (CP)
Forward (GB)
Foudroyant (drilS)
Foxhound (GB)
Frolic (drilS)

G
Gadfly (GB)
Galatea (CA)
Ganges (drilS)
Gannet (CZ)
Garnet (CW)
Gibraltar (CP)
Glatton (MM)
Gleaner (DC)
Goldfinch (GB)
Gorgon (MM)
Goshawk (GB)
Gossamer (DC)
Grafton (CP)
Grampian (BW)
Grappler (GB)
Grasshopper (DC)
Griffon (GB)
Griper (GB)

H
Hamadryad (hospS)
Handy (GB)
Hannibal (hulk)
Harpy (targtS)
Hawke (CP)
Hecate (MM)
Hecla (tendr)
Hector (BO)
Helena (hulk)

British section ship names glossary (H-W)

Hercules (BO)
Hero (CDB)
Heroine (CW)
Hibernia (accomS)
Hindostan (drilS)
Hood (BB)
Hotspur (MM)
Howe (BB)
Hyacinth (CW)
Hyaena (GB)
Hydra (MM)

I
Icarus (CZ)
Iguatemi (TB1)
Immortalite (CA)
Imperieuse (CA)
Implacable (drilS)
Impregnable (drilS)
Impregnable (drilS)
Inconstant (CS)
Indefatigable (CP)
Indefatigable (drilS)
Independencia (CDB)
Indus (BZ)
Inflexible (BB)
Insolent (GB)
Intrepid (CP)
Invincible (BO)
Iris (CS)
Iron Duke (BO)
Irresistible (deptS)

J
Jorge (TB1)

K
King (TB1)
Kingfisher (CZ)
Kite (GB)

L
Landrall (GB)
Lapwing (GB)
Latona (CP)
Lavinia (deptS)
Leander (CP)
Leda (harbV)
Leonidas (deptS)
Liffey (deptS)
Lightning (TB1)
Limpopo (GB)
Linnet (GB)
Lion (drilS)
Lizard (GB)
Locust (tug)

M
Magicienne (CP)
Magpie (GB)
Mahratta (TB1)
Marathon (CP)
Marcilio Dias (TB1)
Mariner (CZ)
Marlborough (BW)
Marlborough (drilS)
Mars (drilS)
Mastiff (GB)
Medea (CP)
Medina (GB)
Medusa (CP)
Medway (GB)
Meeanee (hospS)
Melampus (CP)
Melampus (deptS)
Melita (CZ)
Melpomene (CP)
Mercury (CS)
Mercury (deptS)
Merlin (GB)
Mersey (CP)
Midge (GB)
Minerva (hulk)
Minotaur (BO)
Minstrel (GB)
Miranda (CZ)
Misletoe (GB)
Mohawk (CP)
Monarch (BB)
Monmouth (tendr)
Mount Edgecumbe (drilS)
Muratore (TB1)
Mutine (CZ)

N
Naiad (CP)
Nankin (accomS)
Narcissus (CA)
Nelson (CA)
Neptune (BB)
Nettle (targtS)
Newcastle (CW)
Nicochea (CP)
Nile (BB)
Nimble (GB)
Northampton (CA)
Northumberland (BO)
Nueve de Julio (CDB)
Nymphe (CZ)

O
Opal (CW)
Opossum (hulk)
Orion (BO)
Orlando (CA)
Osprey (CZ)
Otter (tug)

P
Pallas (CP)
Pandora (CP)
Partridge (GB)
Peacock (GB)
Pearl (CP)
Pegasus (CZ)
Pelican (CZ)
Pelorus (CP)
Pembroke (accomS)
Pembroke (accomS)
Penelope (BO)
Penguin (CZ)
Perseus (harbV)
Persian (CP)
Pet (hulk)
Peterel (deptS)
Phaeton (CP)
Pheasant (GB)
Philomel (CP)
Phoebe (CP)
Phoenix (CP)
Pickle (GB)
Pigeon (GB)
Pigmy (GB)
Pike (GB)
Pincher (GB)
Pinedo (TB1)
Pique (CP)
Pique (hospS)
Planet (DC)
Plassey (DC)
Plover (GB)
Plucky (GB)
Polyphemus (DC)
Porpoise (CP)
Prince Albert (CDB)
Procris (hulk)
Psyche (CP)
Pylades (CW)

R
Racehorse (deptS)
Racer (CZ)
Racoon (CP)
Rainbow (CP)
Rajput (TB1)
Raleigh (CS)
Rambler (PS)
Ranger (GB)
Rapid (CW)
Rattler (GB)
Rattlesnake (DC)
Raven (GB)
Ready (GB)
Redbreast (GB)
Redpoll (GB)
Redwing (GB)
Reindeer (CZ)
Renown (BB)
Repulse (BB)
Resistance (targtS)
Revenge (BW)
Rifleman (GB)
Ringdove (GB)
Rodney (BB)
Rosales (DC)
Rover (CS)
Royal Adelaide (accomS)
Royal Sovereign (BB)
Royal William (drilS)
Royalist (CW)
Ruby (CW)
Rupert (MM)

S
Sabrina (GB)
Salamander (DC)
Sandfly (DC)
Sans Pareil (BB)
Sapphire (CW)
Sappho (CP)
Satellite (CW)
Scorpion (MM)
Scourge (GB)
Scout (CP)
Scylla (CP)
Sea Serpent (DC)
Seagull (DC)
Serpent (CP)
Severn (CP)
Shah (CS)
Shannon (CDB)
Sharpshooter (DC)
Sheldrake (DC)
Sheldrake (GB)
Sikh (TB1)
Skipjack (DC)
Skylark (tendr)
Slaney (GB)
Snake (GB)
Snap (GB)
Snapper (hulk)
Southampton (drilS)
Spanker (DC)
Sparrow (GB)
Spartan (CP)
Speedwell (DC)
Spey (GB)
Spider (DC)
St. Vincent (drilS)
Starling (GB)
Staunch (GB)
Stork (GB)
Sultan (BO)
Superb (BO)
Surprise (CS)
Swallow (CZ)
Swan (hulk)
Swift (GB)
Swiftsure (BO)
Swinger (GB)
Sybille (CP)

T
Talbot (deptS)
Tartar (CP)
Tay (GB)
Tees (GB)
Temeraire (BO)
Terpsichore (CP)
Thalia (CW)
Thames (CP)
Thetis (CP)
Thisbe (tendr)
Thorne (TB1)
Thrush (GB)
Thunderer (BB)
Tickler (GB)
Tourmaline (CW)
Trafalgar (BB)
Trafalgar (drilS)
Trent (GB)
Tribune (CP)
Trincomalee (drilS)
Triumph (BO)
Turquoise (CW)
Tweed (GB)
Tyrian (GB)

U
Undaunted (CA)
Unicorn (drilS)

V
Valorous (CWP)
Vengeance (hulk)
Vernon (drilS)
Vesuvius (DC)
Victor Emmanuel (accomS)
Victoria (BB)
Victoria (BW)
Viper (GB)
Vixen (GB)
Volage (CS)
Volcano (harbV)
Vulcan (CP)

W
Wanderer (CZ)
Warren Hastings (tug)
Warrior (CA)
Warspite (CA)
Warspite (drilS)
Watchful (GB)
Waterwitch (GB)
Weazel (GB)
Wellesley (drilS)
Wellesley (hulk)
Wellington (drilS)
Whiting (DC)
Widgeon (GB)
Wild Swan (CZ)
Wivern (MM)
Wizard (DC)
Wrangler (GB)

Indian section ship names glossary

Abyssinia (MM), Baluchi (TB1), Gurkha (TB1), Karen (TB1), Magdala (MM), Pathan (TB1)

Australian section ship names glossary

1
191 (TB2)

A
Acheron (TB2)
Albert (GB)
Avernus (TB2)

B
Batman (GB)
Bonito (GB)
Bream (GB)

C
Cerberus (MM),
Childers (TB1)

D
Dolphin (GB)

F
Fawkner (GB)

G
Gannet (GB),
Gayundah (GB)
Gordon (TB1)

L
Lady Loch (GB)
Lonsdale (TB2)

M
Mosquito (TB1)

N
Nelson (CWS)
Nepean (TB2)

O
Otter (GB)

P
Paluma (GB)
Protector (GBA)
Pumba (GB)

S
Stingaree (GB)

V
Victoria (GB)

W
Wolverine (drilS)

New Zealandian section ship names glossary

1 (TB2), 2 (TB2), 3 (TB2), 4 (TB2)

Y1889 British fleet stricken and lost ships

Actaeon (Sailing sloop), stricken, BU
Albatross (Screw sloop), stricken, BU
Arab (Gunboat ocean-going), stricken, BU
Ariel (Gunboat), stricken, BU
Chichester (Sailing frigate), stricken, BU
Condor (Gunboat ocean-going), stricken, BU
Contest (Gunboat), stricken, BU
Coquette (Gunboat), stricken, BU
Cracker (Gunboat ocean-going), stricken, BU
Cumberland (Drill ship), +++++ in accident
Cygnet (Gunboat), stricken, BU
Daring (Screw sloop), stricken, BU
Diamond (Screw corvette), stricken, BU
Express (Gunboat), stricken, BU
Fantome (Screw sloop), stricken, BU
Hornet (Gunboat ocean-going), stricken, BU
Lily (Gunboat ocean-going), +++++ in accident
Lord Warden (Battery battleship 1), stricken, BU
Mallard (Gunboat), stricken, BU
Myrmidon (Patrol vessel), stricken, BU
Repulse (Battery battleship 2), stricken, BU
Speedy (Gunboat), stricken, BU
Sylvia (Patrol vessel), stricken, BU
Zephyr (Gunboat), stricken, BU

British fleet naval guns

Calibre (mm)	Mark	Calibre (in)/pdr	Gun type	First Introduced	Weight (kg)	Gun lenght (in/cal)	Bore lenght (cal)	Charge weight (lbs)	Muzzle energy (ft/t)	Projectile weight (lbs)	Muzzle velocity (ft/s)	Wrought iron perforation (in)
317.5	I	12,5"	MLR	1877	38 610	222.8 / 17.82	15.84	165.0	11 820	818.0	1 442	17.4
317.5	II	12,5"	MLR	1877	38 610	222.8 / 17.82	15.84	210.0	14 070	818.0	1 575	19.3
304.8	I (35 t)	12"	MLR	1879	35 562	195.0 / 16.25	13.54	110.0	9 566	714.0	1 390	15.9
304.8	II (25 t)	12"	MLR	1869	25 401	182.5 / 15.21	12.09	85.0	7 190	614.0	1 292	13.4
279.4	II	11"	MLR	1876	25 401	180.0 / 16.36	13.18	85.0	7 028	548.0	1 360	14.2
254.0	II	10"	MLR	1868	18 289	180.0 / 18.00	14.55	70.0	5 408	410.0	1 379	13.1
228.6	V	9"	MLR	1865	12 193	156.0 / 17.33	13.89	50.0	3 681	256.0	1 440	11.4
203.2	III	8"	MLR	1862	9 144	144.0 / 18.00	14.75	35.0	2 398	179.0	1 390	9.7
177.8	III	7"	MLR	1865	6 604	133.0 / 19.00	15.86	30.0	1 848	114.6	1 525	9.2
160.0	III	64 cwt	MLR	1864	3 250	118.0 / 18.73	15.47	10.0	865	64.5	1 390	-
120.0	II, L	4,7"	MLR	1865	1 778	120.0 / 25.53	22.00	6.8	312	38.8	1 425	-
412.8	I	16,25"	BLR	1887	112 273	524.0 / 32.25	30.00	960.0	54 390	1 800.0	2 087	34.2
342.9	I, II - III, IV	13,5"	BLR	1883	68 075 - 70 107	433.0 / 32.08	30.00	630.0	35 230	1 250.0	2 016	30.1
304.8	III, IV - V, VI	12"	BLR	1886	45 722 - 46 738	328.5 / 27.38	25.25	259.0	18 130	714.0	1 914	22.6
254.0	II, III, IV	10"	BLR	1887	29 465	342.4 / 34.24	32.00	252.0	14 430	500.0	1 914	22.3
233.9	I - II	9,2"	BLR	1883	21 337 - 22 353	255.8 / 27.20	25.56	140.0	8 622	380.0	1 809	17.9
233.9	III - IV	9,2"	BLR	1886	22 353 - 23 369	310.0 / 33.70	31.50	166.0	10 910	380.0	2 035	20.3
203.2	III	8"	BLR	1887	13 209	226.3 / 28.29	25.60	104.0	5 554	210.0	1 953	15.4
203.2	IV - VI	8"	BLR	1885	14 225 - 15 241	254.5 / 31.81	29.60	118.0	6 730	210.0	2 150	17.1
177.8	-	7"	BLR	1864	4 166	120.0 / 17.14	14.21	11.0	909	91.3	1 100	-
152.4	-	6" / 80 pr	BLR	1872	4 064	165.6 / 27.60	26.00	34.0	1 961	78.4	1 880	10.5
152.4	IIIa	6"	BLR	1873	4 521	170.7 / 28.45	25.53	36.0	1 672	100.0	1 672	10.6
152.4	IIIb	6"	BLR	1873	5 080	170.7 / 28.45	25.53	48.8	1 938	100.0	1 960	12.4
152.4	IV	6"	BLR	1873	5 080	173.5 / 28.92	26.00	48.8	2 665	100.0	1 960	12.4
152.4	VI	6"	BLR	1873	5 080	173.5 / 28.92	26.00	45.0	2 424	100.0	1 870	11.9
127.0	II, III - IV, V	5"	BLR	1870	1 931 - 2 032	129.2 / 25.83	25.00	15.5	1 124	50.0	1 770	8.7
120.0	II, L	4,7"	BLR	1884	2 083	194.1 / 41.09	40.00	12.0	486	45.0	2150	8.5
101.6	II, III - IV, V, VI	4"	BLR	1859	1 169 - 1 321	120.0 / 30.00	27.00	12.0	625	25.0	1 900	7.2
95.3	-	20 pr	BLR	-	762	66.1 / 17.63	14.43	2.5	183	21.8	1 100	-
95.3	I	20 pr	BLR	-	610	107.5 / 28.65	27.00	6.0	380	19.5	1 677	-
76.2	I	12 pr	BLR	1865	356	92.4 / 30.78	28.00	4.0	249	12.5	1 530	-
57.0	I & II	6 pr	BLR	1876	406	96.5 / 43.00	40.00	1.2	345	6.0	1 818	3.7
47.0	I & II	3 pr	BLR	1876	254	79.6 / 43.00	40.00	1.8	321	3.3	1 873	3.1
37.0	Hotchkiss Rev.	1 pr	BLR	1873	209	33.5 / 23.00	19.13	0.2	-	1.1	1 450	-
37.0	Hotchkiss	1 pr	BLR	1884	72	33.5 / 23.00	19.15	0.2	-	1.1	1 450	-

BRITISH FLEET WARSHIPS SECTION

Ship classes	Building (exports including)	Active
Battleships	8	16
Boardside ironclads		21
Armoured cruisers	1	11
Coast defence armoured ships	2	13
Protected cruisers	43	31
Unprotected cruisers		13
Torpedo gunboats	18	7
Torpedo boats 1 class	18	86
Torpedo boats 2 class	18	63
Submarines		
Wooden screw cruisers		17
Wooden screw sloops		26
Gunboats	5	35
Gunboats coastal		66

BATTLESHIPS

0 +1 | "Hood" type (1889)
Battleship, Britain

BUILDING

0Y:5M

Hood

Displacement: 14,150 tons at draft: 8.38 m, 15,590 tons full | Length: 124.97 m oa, 115,82 m pp | Beam: 22.86 m | Crew: 690 men

Machinery: 8 cylindrical boilers | 2 * 4,500 HPn / 5,500 HPf steam EE * 2 screws | 9,000 HPn = 15.5 knots / 11,000 HPf = 16.5 knots | Supplies: 900 / +550 tons of coal

Armour: compound | CT: 356/152; Turrets: 423; Barbettes: 423/179; Casemates: 152; UPBelt: 0 =76= 76 102 76 =102= 0 | WLBelt: 0(v64) =356= 356>203(^76) 452>203(^76) 356>203(^76) =406= 0(v64)

Armament: 4-343/30, 10-152.4/40, 12-57/43, 12-47/43, 7-457.2 TT (4 ^ | 3 v)

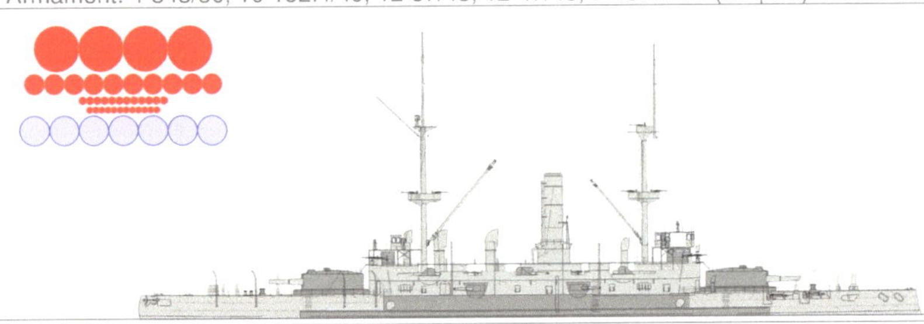

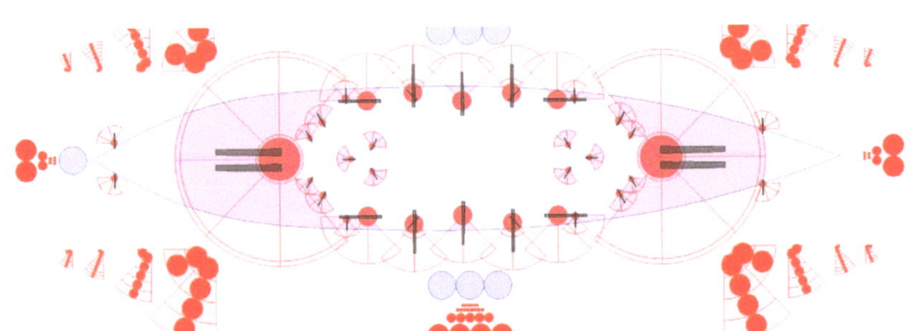

Hood (1889)
2 Humphrys vertical triple expansion EE |
1889 08 12 - ***** - ***** | Chatham DYd, Chatham

0 +3 | "Royal Sovereign" type (1889)
Battleships, Britain

BUILDING

0Y:0M 0Y:4M 0Y:6M

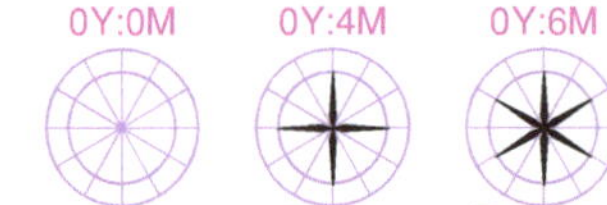

Repulse Royal Sovereign Renown

Displacement: 14,150 tons at draft: 8.38 m | 15,580 tons full | Length: 124.97 m oa, 115.82 m pp | Beam: 22.86 m | Crew: 712 men

Machinery: 8 cylindrical boilers | 2 * 4,500 HPn / 5,500 HPf steam EE * 2 screws | 9,000 HP = 15.5 knots / 11,000 HPf = 16,5 knots | Supplies: 900 / +500 tons of coal

Armour: compound | CT: 356/152; Barbettes: 423/179; Casemates: 152; UPBelt: 0 =76= 76 102 76 =102= 0 | WLBelt: 0(v64) =356= 356>203(^76) 452>203(^76) 356>203(^76) =406= 0(v64)

Armament: 4-343/30, 10-152.4/40, 16-57/43, 12-47/43, 7-457.2 TT (4 ^ | 3 v)

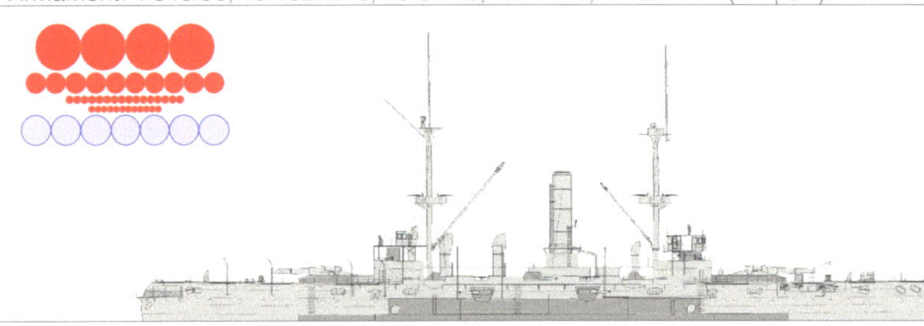

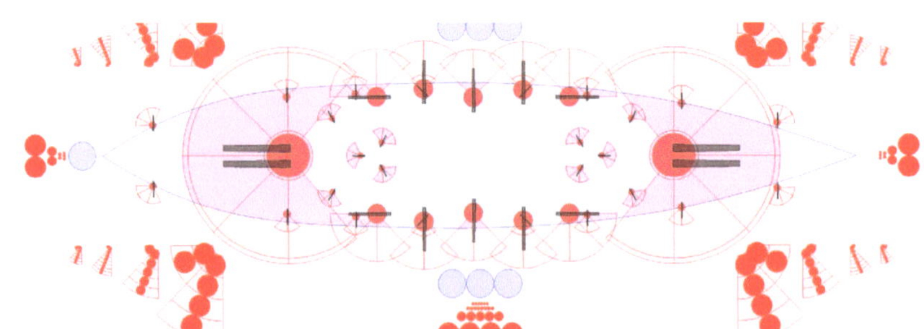

Empress of India (1889)
2 Humphrys vertical triple expansion EE |
1889 07 09 - ***** - ***** | Pembroke DYd, Pembroke

Repulse (1890)
2 Humphrys vertical triple expansion EE |
1890 01 01 - ***** - ***** | Pembroke DYd, Pembroke

0 +3 | "Royal Sovereign" type (1889)
Battleships, Britain (continued)

Royal Sovereign (1889)
2 Humphrys vertical triple expansion EE |
1889 09 30 - ***** ***** | Portsmouth DYd, Portsmouth

0 +2 | "Nile" type (1886)
Battleships, Britain (continued)

Trafalgar

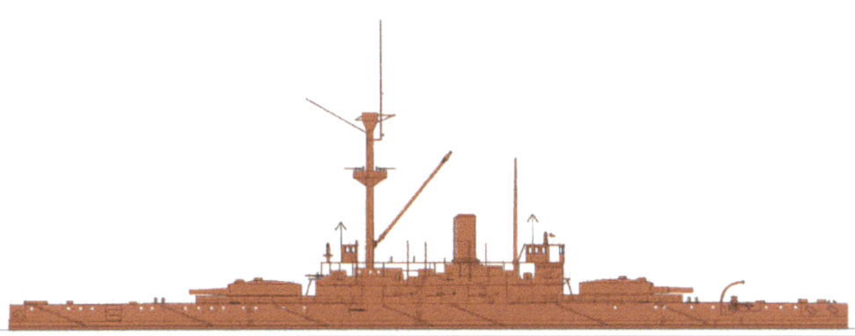

0 +2 | "Nile" type (1886)
Battleships, Britain

BUILDING

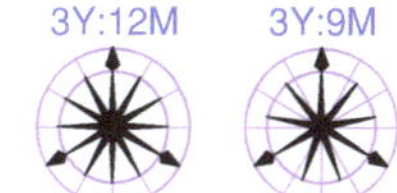

Displacement: 12,590 tons at draft: 8.69 m | Length: 105.15 m pp | Beam: 22.25 m | Crew: 577 men
Machinery: 6 cylndrical boilers | 2 * 3,750 HPn / 6,000 HPf steam EE * 2 screws | 7,500 HPn = 15 knots / 12,000 HPf = 17 knots | Supplies: 900 / +300 tons of coal
Armour: compound | CT: 356; Battery: =127= 102 =127=; Turrets: 457; UPBelt: 457(^76) 406(^76) 457(^76) | WLBelt: 0(^76) =356= 356(^76) 457 508 457 356(^76) =406= 0(^76)
Armament: 4-343/30, 6-120/40, 8-57/43, 9-47/43, 4-356 TT (2 ^ | 2 v)

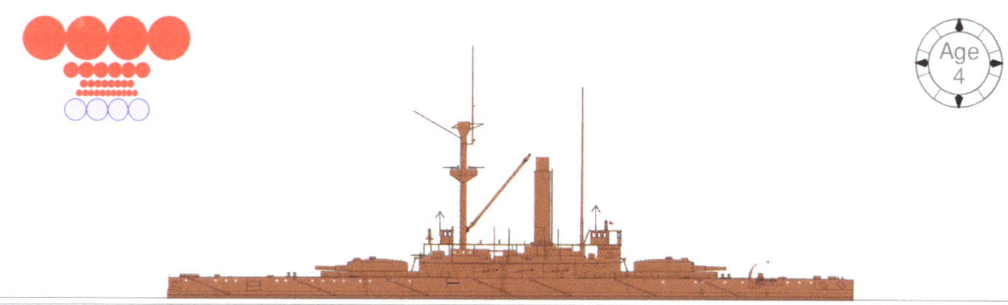

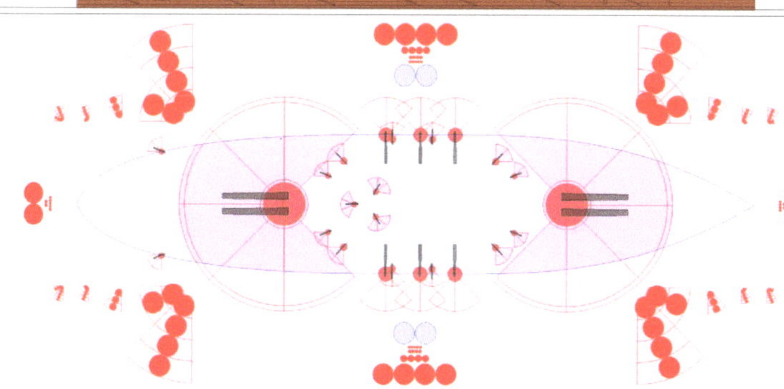

Nile (1886)
2 Maudslay vertical triple expansion EE |
1886 04 08 - 1888 03 27 - ***** | Pembroke DYd, Pembroke

Trafalgar (1886)
2 Humphrys vertical triple expansion EE |
1886 01 18 - 1887 09 20 - ***** | Portsmouth DYd, Portsmouth

0 +2 | "Victoria" type (1885)
Battleships, Britain

BUILDING

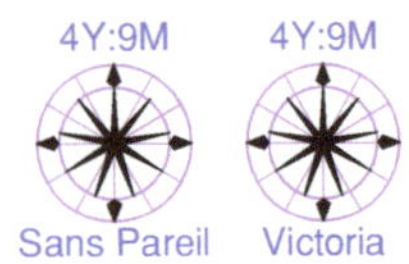

Displacement: 10,470 tons at draft: 8.15 m | Length: 110.64 m oa | Beam: 21.34 m | Crew: 550 men

Machinery: 8 cylindrical boilers | 2 * 3,750 HPn / 7,000 HPf steam EE * 2 screws | 7,500 HPn = 15.2 knots / 14,000 HPf = 17.2 knots | Supplies: 750 / +250 tons of coal

Armour: compound | CT: 356 / 51; UPBelt: 0 =152= 76 =152=; Turret: 432; Barbette: 457 | WLBelt: 00(v76) =406= 457(^76) =406= 00(v76)

Armament: 2-412.75/30, 1-254/32, 12-152.4/26, 12-57/43, 9-47/43, 6-25.4/40, 6-356 TT (4 ^, 2 v)

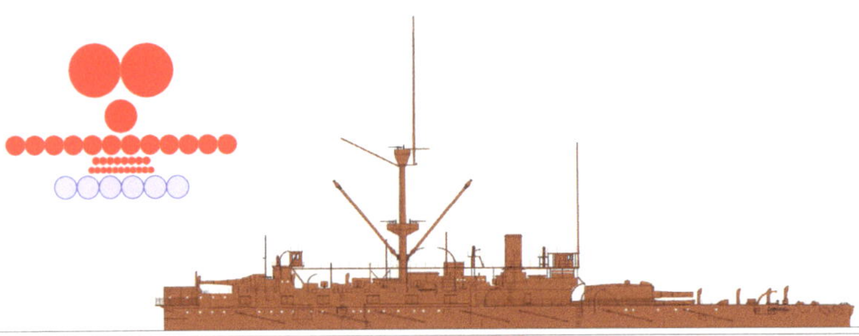

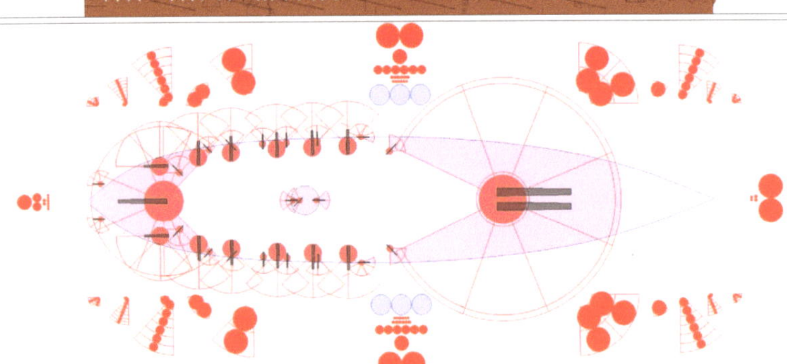

Sans Pareil (1885)
2 Humphrys vertical triple expansion EE
1885 04 21 - 1887 05 09 - ***** | Thames IW, Blackwall

Victoria (1885)
2 Humphrys vertical triple expansion EE
1885 04 23 - 1887 04 09 - ***** | Armstrong, Elswick

1 | "Benbow" type (1882)
Battleship, Britain

Displacement: 10,600 tons at draft: 8.48 m | Length: 100.58 m pp | Beam: 20.88 m | Crew: 523 men

Machinery: 12 cylindrical boilers | 2 * 3,750 HPn / 5,750 HPf steam EE * 2 screws | 7,500 HPn = 15.7 knots / 11,500 HPf = 17.4 knots | Supplies: 900 / +300 tons of coal

Armour: compound | CT: 305(^51); Barbettes: 356 / 305(v76); Hoists: 305 / 254 | WLBelt: 0(v63) =457>178= 457>203(^76) =457>178= 0(v63)

Armament: 2-412.75/30, 10-152.4/26, 12-57/43, 7-47/43, 5-356 TT (^)

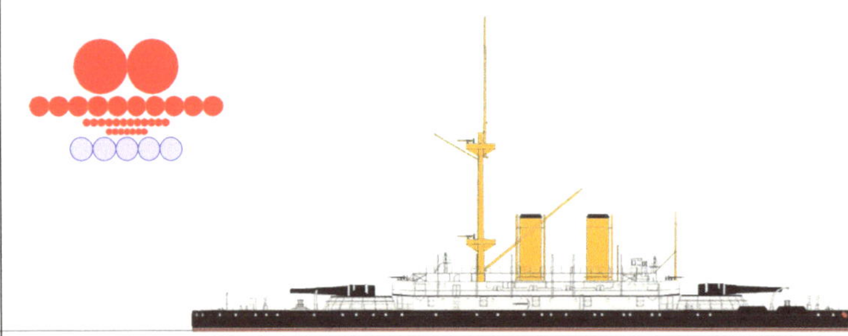

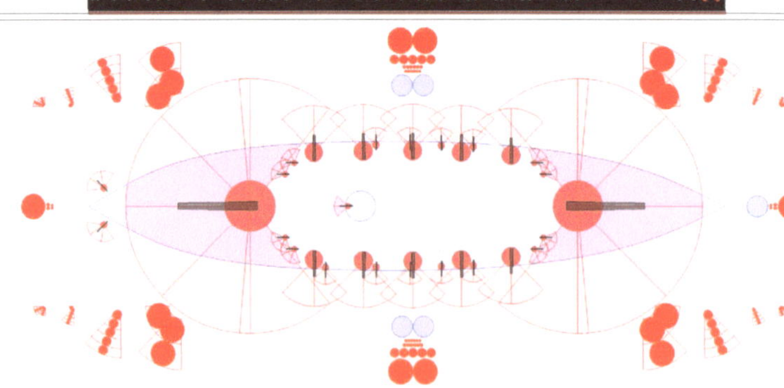

Benbow (1882)
2 Maudslay inverted compound EE | 8,658 HPn = 15.7 knots / 10,853 HPf = 17.7 knots
1882 11 01 - 1885 06 15 - 1888 06 01 | Thames IW, Blackwall

2 | "Camperdown" type (1882)
Battleships, Britain

Displacement: 10,300 tons at draft: 8.48 m | Length: 100.58 m pp | Beam: 20.88 m | Crew: 530 men

Machinery: 12 cylindrical boilers | 2 * 3,750 HPn / 5,750 HPf steam EE * 2 screws | 7,500 HPn = 15.7 knots / 11,500 HPf = 17.4 knots | Supplies: 900 / +300 tons of coal

Armour: compound | CT: 305(^51); Barbettes: 356 / 305(v76); Hoists: 305 / 254 | WLBelt: 0(v63) =457>178= 457>203(^76) =457>178= 0(v63)

Armament: 4-343/30, 6-152.4/26, 12-57/43, 10-47/43, 5-356 TT (^)

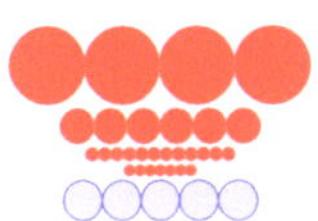

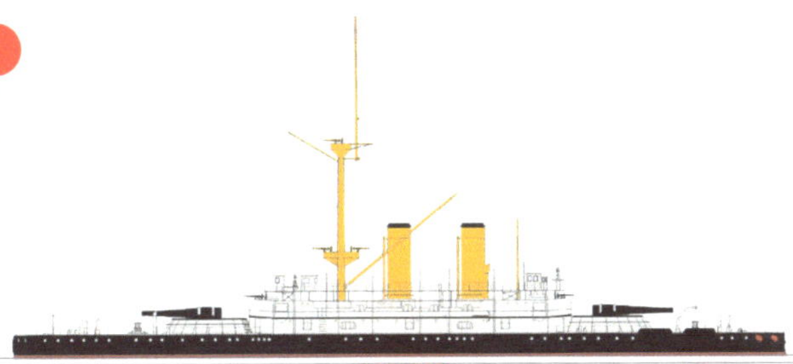

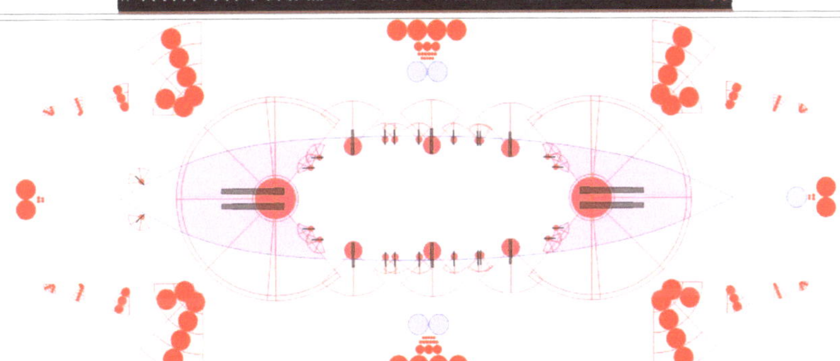

Anson (1883)
2 Humphrys inverted compound EE | 12,568 HPf = 17,4 knots
1883 04 24 - 1886 02 17 - 1889 05 01 | Pembroke DYd, Pembroke

Camperdown (1882)
2 Maudslay inverted compound EE | 8,491 HPn = 16,5 knots / 11,741 HPf = 17,14 knots
1882 12 18 - 1885 11 24 - 1889 07 01 | Portsmouth DYd, Portsmouth

2 | "Camperdown" type (1882)
Battleships, Britain (continued)

Camperdown

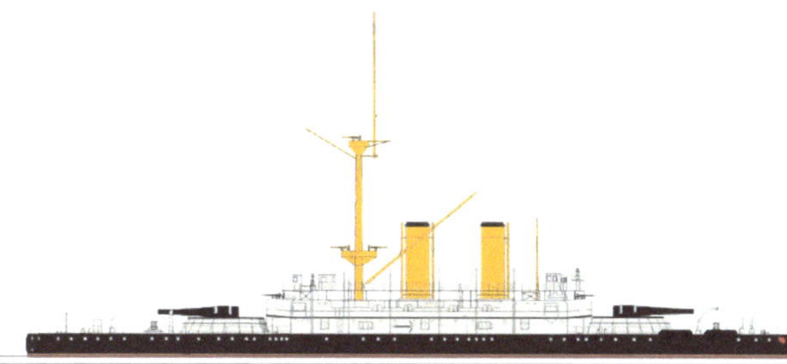

2 | "Rodney" type (1882)
Battleships, Britain (continued)

Howe

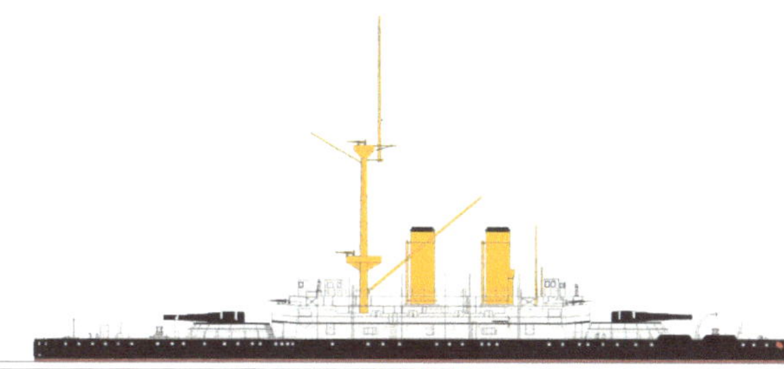

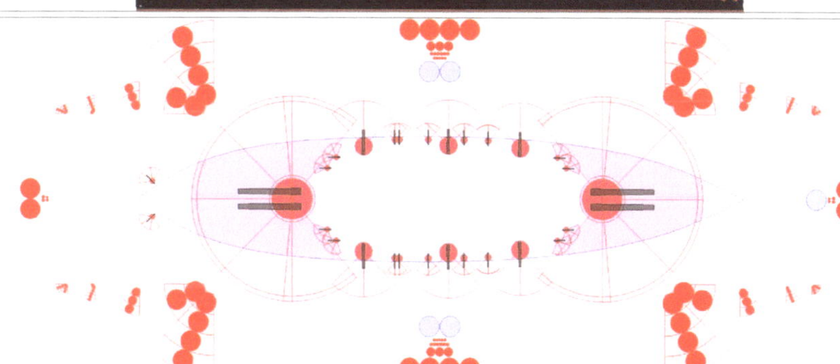

2 | "Rodney" type (1882)
Battleships, Britain

Displacement: 10,300 tons at draft: 8.48 m | Length: 99.06 m pp | Beam: 20.73 m | Crew: 530 men

Machinery: 12 cylindrical boilers | 2 * 3,750 HPn / 5,750 HPf steam EE * 2 screws | 7,500 HPn = 15,7 knots / 11,500 HPf = 17.4 knots | Supplies: 900 / +300 tons of coal

Armour: compound | CT: 305(^51); Barbettes: 292 / 254(v76); Hoists: 305 / 254 | WLBelt: 0(v63) =457>178= 457>203(^76) =457>178= 0(v63)

Armament: 4-343/30, 6-152.4/26, 12-57/43, 10-47/43, 4-356 TT (^) | in "Howe" 5-356 TT (^)

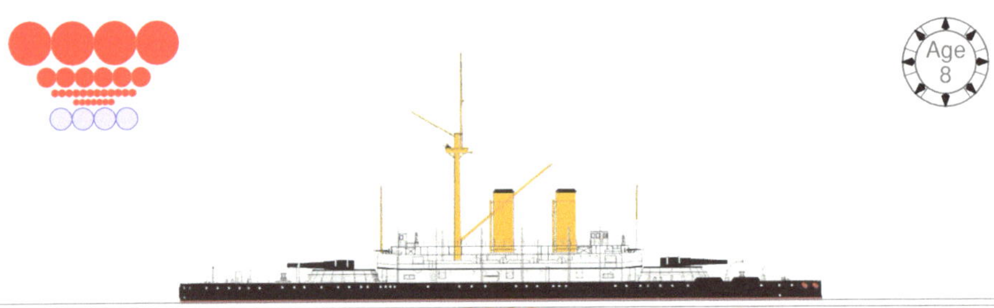

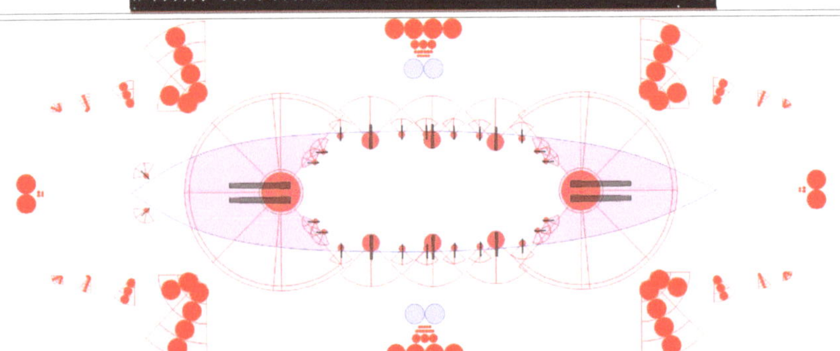

Howe (1882)
2 Humphrys inverted compound EE | 7,733.4 HPn = 15.87 knots / 11,711.6 HPf = 16.94 knots
1882 06 07 - 1885 04 28 - 1889 07 01 | Pembroke DYd, Pembroke

Rodney (1882)
2 Humphrys inverted compound EE | 9,760 HPn = 16 knots / 11,157 HPf = 16.9 knots
1882 02 06 - 1884 10 08 - 1888 06 01 | Chatham DYd, Chatham

1 | "Collingwood" type (1880)
Battleship, Britain

Displacement: 9,500 tons at draft: 8.03 m | Length: 99.06 m pp | Beam: 20.73 m | Crew: 498 men

Machinery: 12 cylindrical boilers | 2 * 3,500 HPn / 4,800 HPf steam EE * 2 screws | 7,000 HPn = 15.5 knots / 9,600 HPf = 16.8 knots | Supplies: 900 / +300 tons of coal

Armour: compound | CT: 305(^51); Barbettes: 292 / 254(v76); Hoists: 305 / 254 | WLBelt: 0(v63) =457>178= 457>203(^76) =457>178= 0(v63)

Armament: 4-304.8/25, 6-152.4/26, 12-57/43, 8-47/43, 4-356 TT (^)

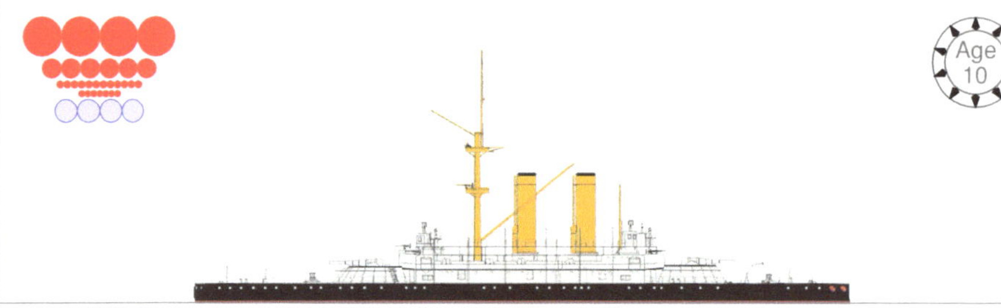

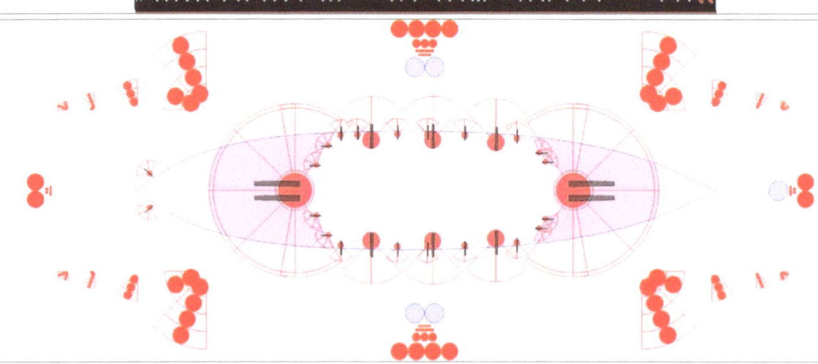

Collingwood (1880)
2 Humphrys inverted compound EE | 8,369 HPn = 16.6 knots / 9,573 HPf = 16.84 knots
1880 07 12 - 1882 11 22 - 1887 07 01 | Pembroke DYd, Pembroke

2 | "Colossus" type (1879)
Battleships, Britain

Displacement: 9,420 tons at draft: 7.85 m | Length: 99.06 m pp | Beam: 20.73 m | Crew: 396 men

Machinery: 9 cylindrical boilers | 2 * 3,500 HP steam EE * 2 screws | 7,000 HP = 16 knots | Supplies: 850 / +250 tons of coal

Armour: compound | CT: 356; Turrets: 406-356; Citadel: =406= 457(^63) =406= |
WLBelt: 0(v63) =406= 406>325 457>356 406>325 =406= 0(v76)

Armament: 4-304.8/25, 5-152.4/26, 4-57/43, 2-356 TT (^)

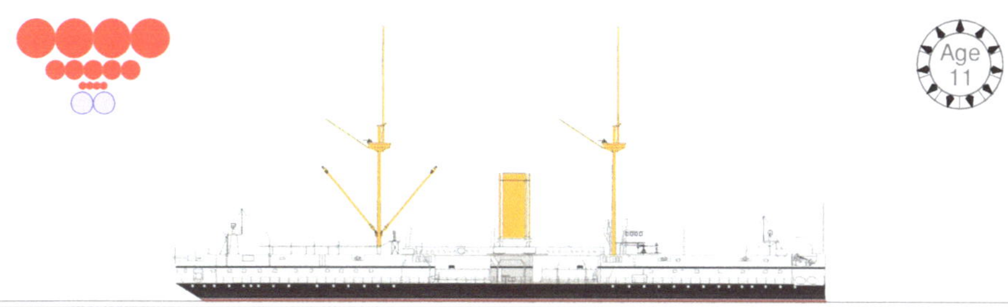

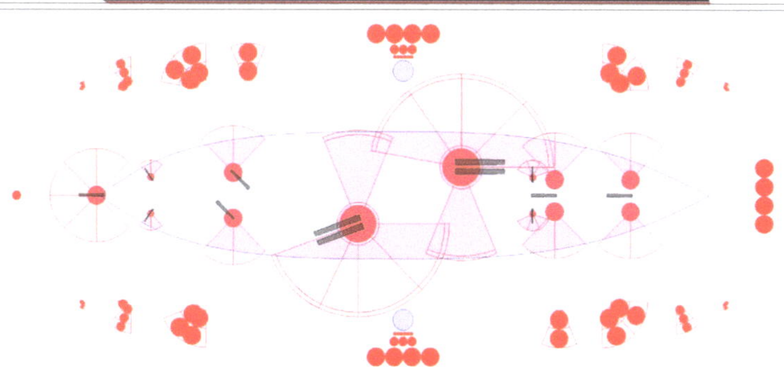

Colossus (1879)
2 Maudslay inverted compound EE | 7,488 HP = 16.5 knots |
1879 06 06 - 1882 03 21 - 1887 10 31 | Portsmouth DYd, Portsmouth

Edinburgh (1879)
2 Humphrys inverted compound EE | 6,808 HP = 16 knots |
1879 03 20 - 1882 03 18 - 1887 07 08 | Pembroke DYd, Pembroke

2 | "Ajax" type (1876)
Battleships, Britain

Displacement: 8,510 tons at draft: 7.16 m | Length: 91.67 m oa | Beam: 20.12 m | Crew: 345 men

Machinery: 10 tubular boilers | 2 * 3,000 HP steam EE * 2 screws | 6,000 HP = 13 knots | Supplies: 700 / +260 tons of coal

Armour: compound | CT: 305; Turrets: 406-356; Citadel: =381= 381(^76) =381= |
WLBelt: 0(v76) =381= 381>203(^76) 457>229(^76) 381>203(^76) =229= 0(v76)

Armament: 4-317.5/15.8MLR, 2-152.4/26, 6-57/43, 2-356 TT (^)

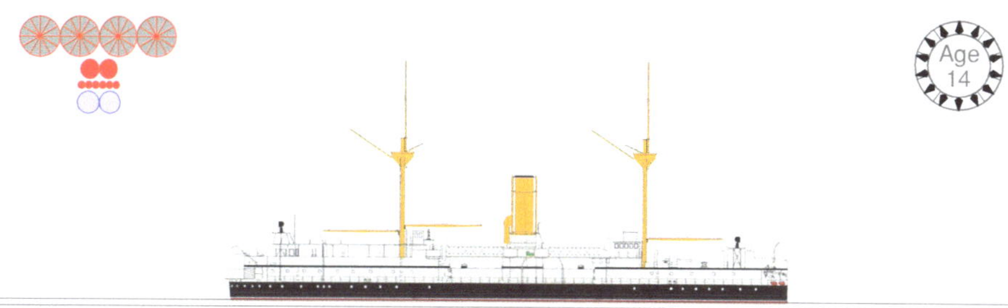

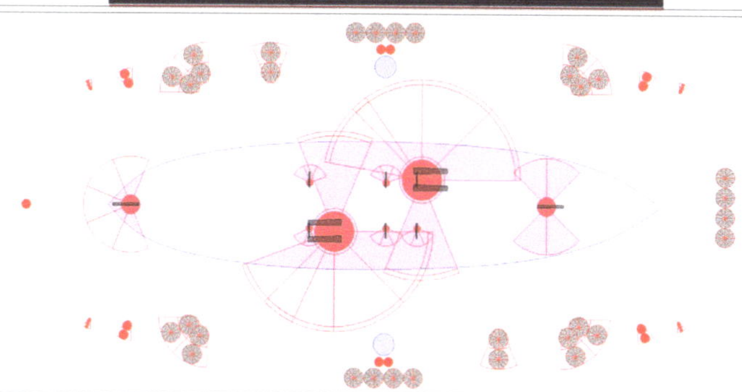

Agamemnon (1876)
2 Penn inverted compound EE |
1876 05 09 - 1879 09 17 - 1883 03 29 | Chatham DYd, Chatham

Ajax (1876)
2 Penn inverted compound EE |
1876 03 21 - 1880 03 10 - 1883 03 30 | Pembroke DYd, Pembroke

1 | "Inflexible" type (1874)
Battleship, Britain

Displacement: 10,300 tons at draft: 7.77 m, 11,880 tons full | Length: 104.85 m oa, 97.54 m pp | Beam: 22.86 m | Crew: 440 men

Machinery: 12 cylindrical boilers | 2 * 4,200 HP steam EE * 2 screws | 8,400 HP = 14 knots | Supplies: 1,200 / +100 tons of coal

Armour: compound | CT: 305; Turrets: 89+178 / 292; UPBelt: 0 =508= 508(^76) =508= 0 | WLBelt: 0(v76) =607>406= 607>406 =607>406= 0(v76)

Armament: 4-406.4/18 MLR, 8-95.3/21 MLR, 4-356 TT (^)

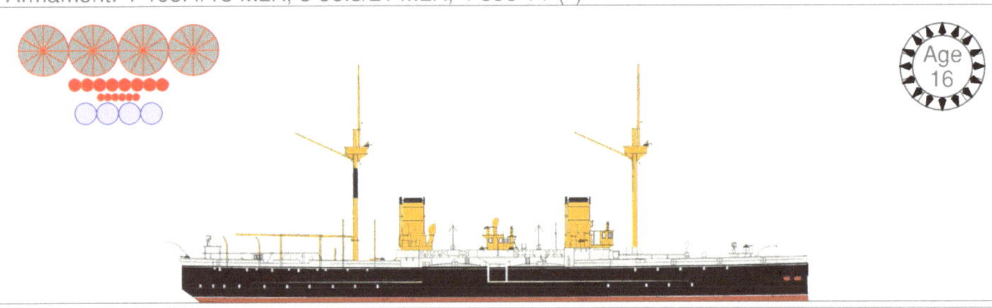

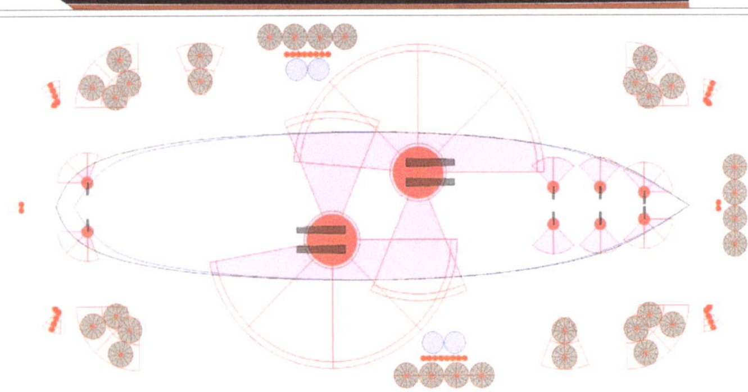

Inflexible (1874)
2 Elder inverted compound expansion EE | 8,407 HP = 14.75 knots |
1874 02 24 - 1876 04 27 - 1881 10 18 | Portsmouth DYd, Portsmouth

1 | "Independencia"/"Neptune" type (1873)
Battleship, Brazil

Displacement: 9,130 tons at draft: 7.62 m | Length: 91.44 m. pp | Beam: 19.2 m | Crew: 541 men

Machinery: 8 rectangular boilers | 1 steam E * 1 screw | 8,000 HP = 14 knots | Supplies: 670 tons of coal

Armour: steel | CT: 203/152; Belt: 0 152 0; Turrets: 330/280; UPBelt: =203= 254(^76) =203= | WLBelt: 229(^51) 254 305 254 229(^51)

Armament: 4-304.8/12 MLR, 2-228.9/13.9 MLR, 6-57/43, 8-47/43, 2-356 TT (^)

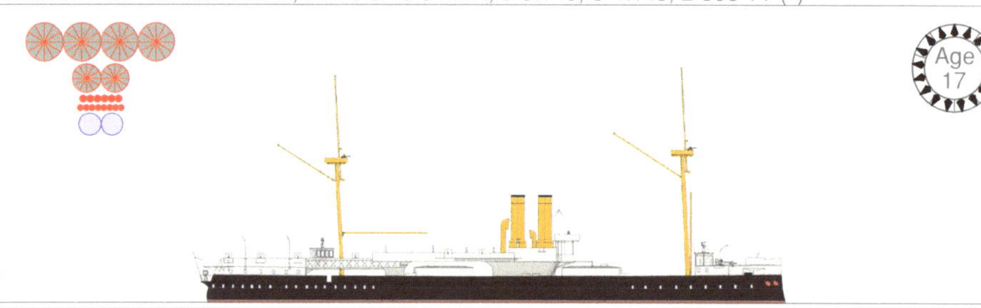

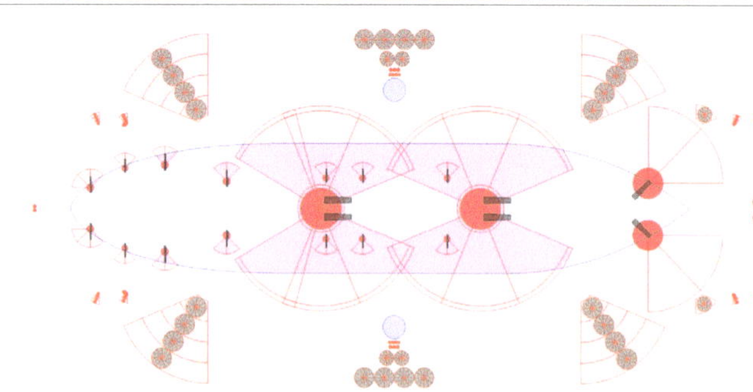

Neptune | ex-Indepedencia (1873)
1 Penn horizontal trunk E | 8,832 HP = 14.65 knots |
1873 01 01 - 1874 09 10 - 1881 09 03 | Dudgeon, Poplar > Samuda, Poplar > Portsmouth DYd

1 | "Dreadnought" type (1870)
Battleship, Britain

Displacement: 10,886 tons at draft: 8.08 m | Length: 104.6 m oa | Beam: 19.46 m | Crew: 369 men
Machinery: 12 cylindrical boilers | 2 * 4,100 HP steam EE * 2 screws | 8,200 HP = 14.5 knots |
Supplies: 1,800 tons of coal
Armour: iron | CT: 229; Turrets: 356; UPBelt: 0 =330= 280(^76) 356(^76) 280(^76) =330= 0 |
WLBelt: 203(^63) 356>203 203(^63)
Armament: 4-317.5/15.8 MLR, 12-25.4/40 Mg

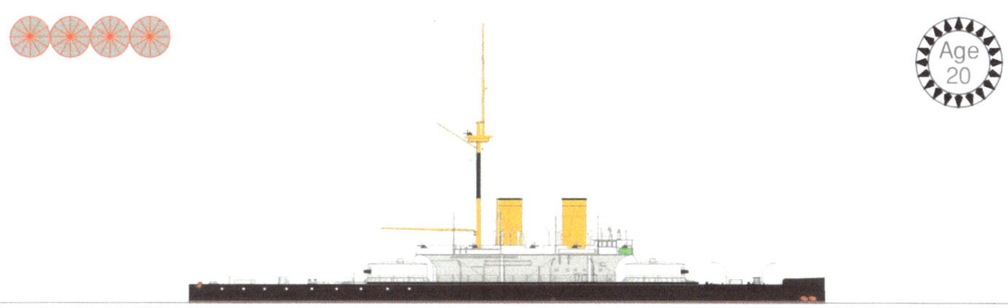

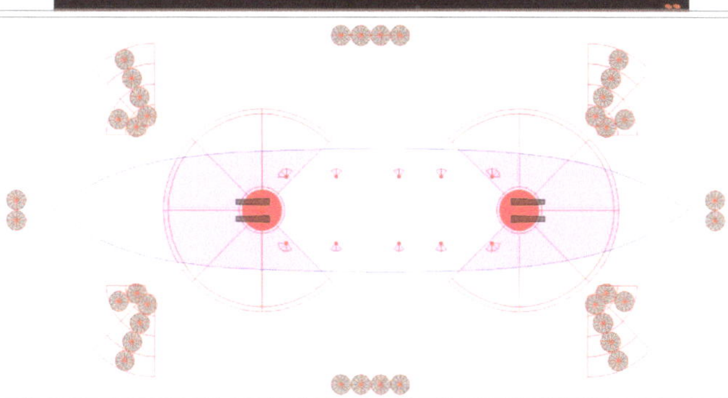

Dreadnought (1870)
2 Elder vertical compound expansion EE | 8,206 = 14.52 knots |
1870 09 10 - 1875 03 08 - 1879 02 15 | Pembroke DYd, Pembroke

2 | "Devastation" type (1869)
Battleships, Britain

Displacement: 9,330 tons at draft: 8.03 m | Length: 93.57 m oa | Beam: 18.97 m | Crew: 358 men
Machinery: 8 rectangular boilers | 2 * 3,250 HP steam EE * 2 screws | 6,500 HP = 13.5 knots |
Supplies: 1,600 tons of coal
Armour: iron | CT: 229; Turrets: 356 / 305; Breastwork: =305= 305(^51) 254(^51) 305(^51) =305= |
WLBelt: 203(^76) 254(^76) 203(^76)
Armament: 4-304.8/12 MLR, 8-25.4/40 Mg, 2-356 TT (^)

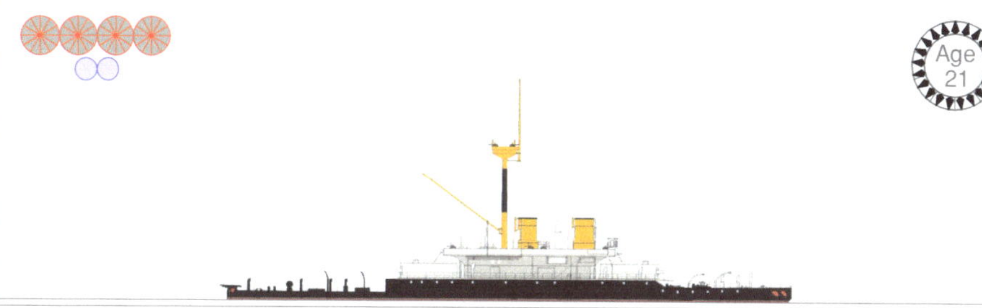

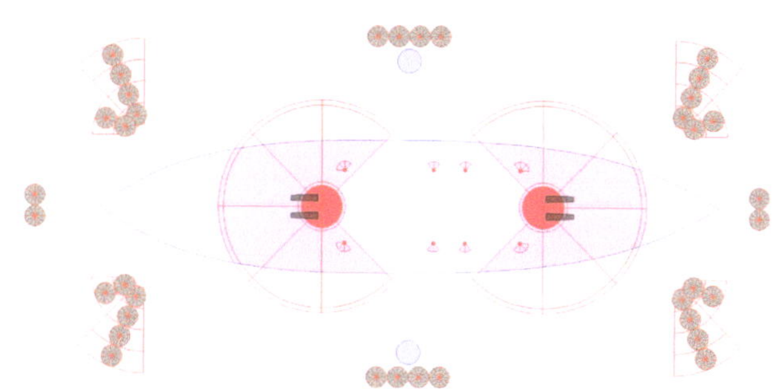

Devastation (1869)
2 Penn trunk EE | 6,650 HP = 13.84 knots |
1869 11 12 - 1871 07 12 - 1873 04 19 | Pembroke DYd, Pembroke

Thunderer (1869)
2 Humphrys horizontal direct acting EE | 6,270 HP = 13.4 knots |
1869 06 26 - 1872 03 25 - 1877 05 26 | Pembroke DYd, Pembroke

1 | "Monarch" type (1866)
Battleship, Britain

Displacement: 8,322 tons at draft: 7.39 m | Length: 100.58 m pp | Beam: 17.53 m | Crew: 575 men

Machinery: 9 cylindrical boilers | 1 steam E * 1 screw | 7,800 HP = 14.94 knots | Supplies: 600 tons of coal

Armour: iron | CT: 203; Turrets: 254 / 203; UPBelt: 0 =102= 178 =114= 0 | WLBelt: 114 152 178 152 114

Armament: 4-304.8/12 MLR, 3-177.8/15.9 MLR

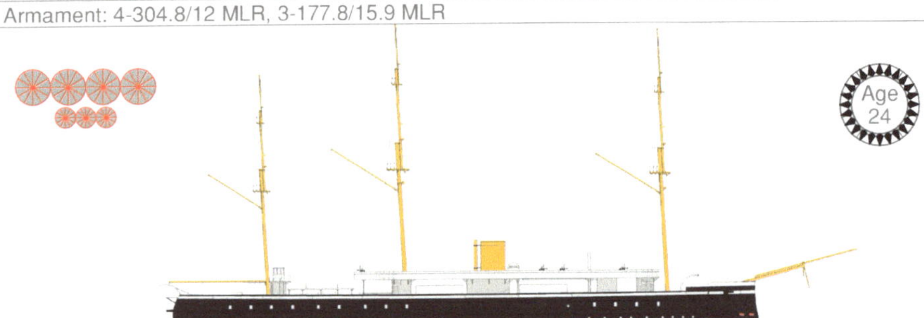

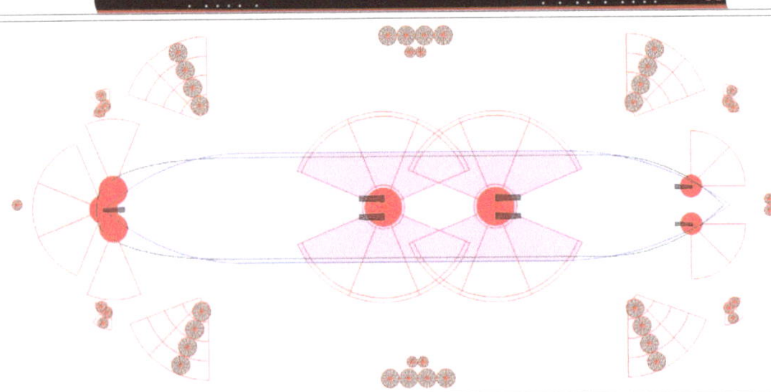

Monarch (1866)
1 Humphrys horizontal return connecting rod E | 7,842 HP = 14.94 knots |
1866 06 01 - 1868 05 25 - 1869 06 12 | Chatham DYd, Chatham

2 | "Peyki Shereef"/"Belleisle" type (1874)
Battery ironclads, Turkey

BATTERY IRONCLADS

Displacement: 4,870 tons at draft: 6.4 m | Length: 74.68 m pp | Beam: 15.85 m | Crew: 250 men

Machinery: 4 cylindrical boilers | 2 * 2,000 HP steam EE * 2 screws | 4,000 HP = 13 knots | Supplies: 510 tons of coal

Armour: iron | CT: 229; Battery: 203-267-203(^51), UPBelt: 203-229-203 | WLBelt: 152(^76) 305 178(^76)

Armament: 4-304.8/14 MLR, 4-95.3/21, 2-356 TT (^)

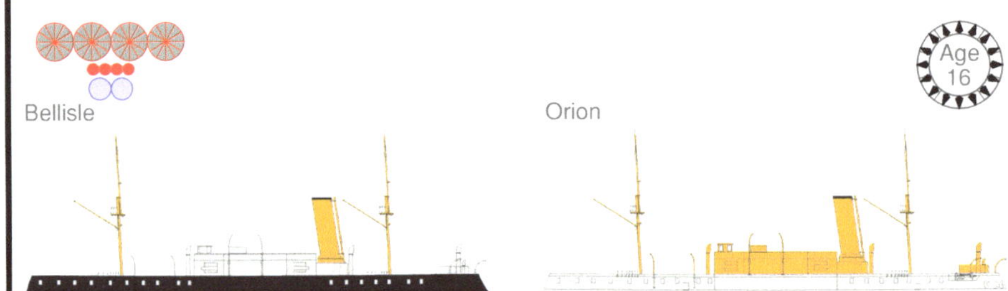

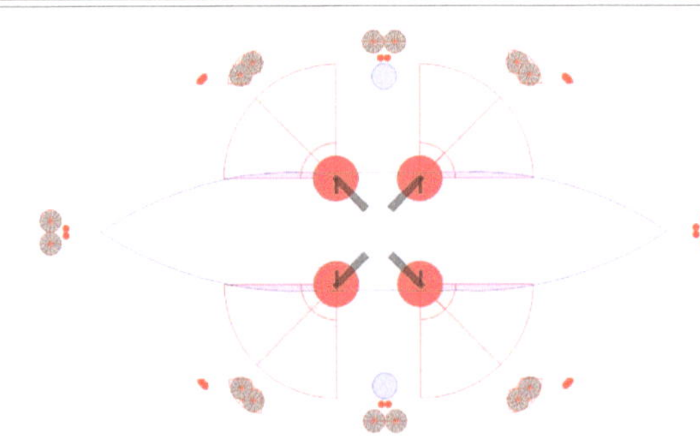

Belleisle | ex-Peyki Shereef (1874)
2 Maudslay horizontal direct acting EE | 4,040 HP = 12.99 knots |
1874 01 01 - 1876 02 12 - 1878 07 19 | Samuda, Poplar

Orion | ex-Boodhi Zaffer (1876)
2 Maudslay horizontal direct acting EE |
1876 01 01 - 1879 01 23 - 1882 07 03 | Samuda, Poplar

1 | "Temeraire" type (1873)
Battery ironclad, Britain

Displacement: 8,540 tons at draft: 8.23 m | Length: 86.87 m pp | Beam: 18.9 m | Crew: 580 men
Machinery: 12 cylindrical boilers | 2 * 3,750 HP steam EE * 2 screws | 7,500 HP = 14 knots | Supplies: 620 tons of coal
Armour: iron | CT: 305; Barbettes: 254-203; Battery: =203-127= 203 =203= | WL Belt: 140(^38) 152(^38) 280>229 152(^38) 127(^38)
Armament: 4-279.4/12, 4-254/14.5, 4-95.3/21, 2-356 TT (^)

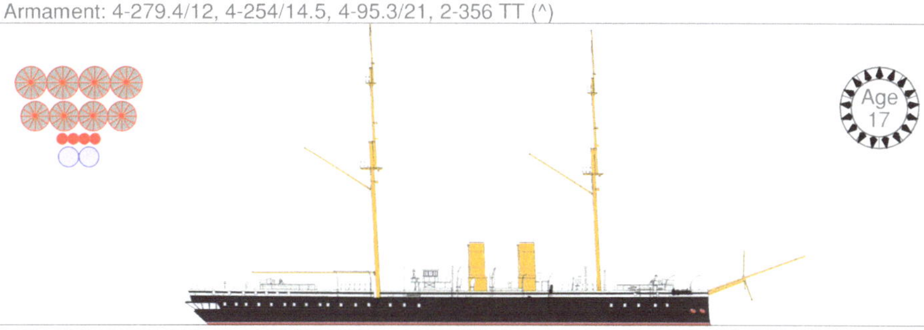

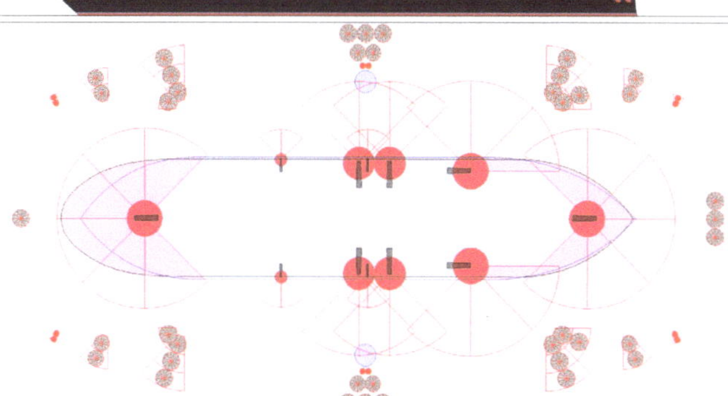

Temeraire (1873)
2 Humphrys vertical inverted compound EE | 7,697 HP = 14.65 knots |
1873 08 18 - 1876 05 09 - 1877 08 31 | Chatham DYd, Chatham

1 | "Alexandra" type (1873)
Battery ironclad, Britain

Displacement: 9,492 tons at draft: 8.0 m | Length: 104.85 m oa | Beam: 19.46 m | Crew: 674 men
Machinery: 12 cylindrical boilers | 2 * 4,200 HP steam EE * 2 screws | 8,400 HP = 15 knots | Supplies: 500 / +180 tons of coal
Armour: steel | CT: 305; UPBattery: =152= 203(^25) =152=; Battery: =203= 305 =203= | WLBelt: 152(^38) 254(^38) 305 254(^38) 152(^38)
Armament: 2-279.4/12 MLR, 10-254/14.5 MLR, 6-95.3/21, 4-356 TT (^)

At overhaul

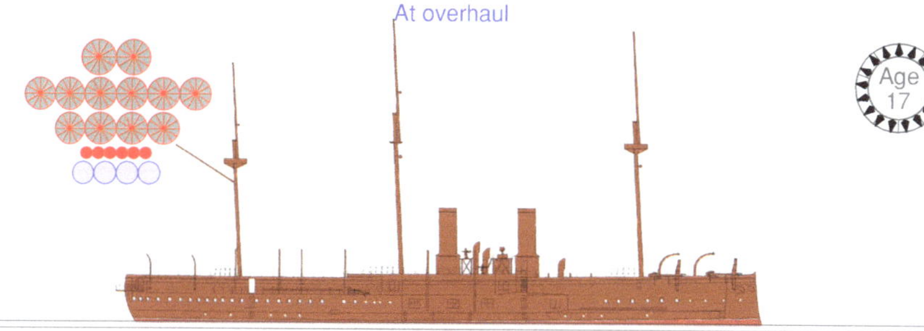

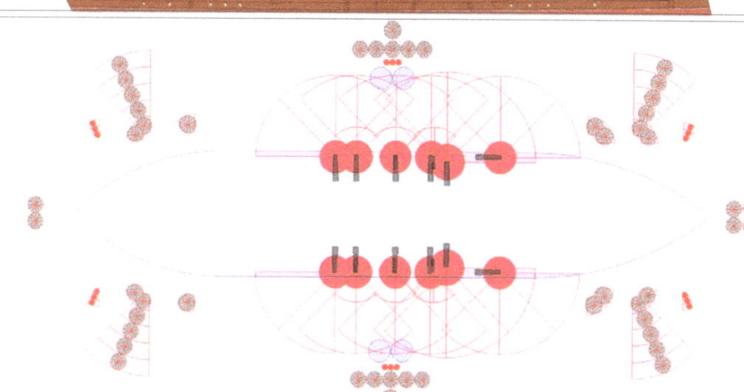

Alexandra (1873)
2 Humphrys vertical inverted compound EE | 8,498 HP = 15.09 knots |
1873 03 05 - 1875 04 07 - 1877 01 31 | Chatham DYd, Chatham

1 | "Messudieh"/"Superb" type (1873)
Battery ironclad, Turkey

Displacement: 9,710 tons full at draft: 7.9 m | Length: 101.3 m pp | Beam: 17.98 m | Crew: 654 men

Machinery: 9 rectangular boilers | 1 steam E * 1 screw | 6,500 HP = 13 knots | Supplies: 600 tons of coal

Armour: iron | CT: 203; Battery: 305-254(^38) | WLBelt: 178(^38) 305 178(^38)

Armament: 16-254/14.5 MLR, 6-95.3/21, 4-356 TT (^)

At overhaul

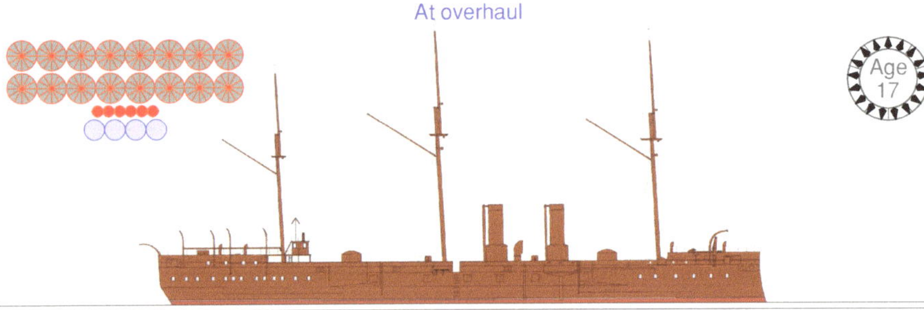

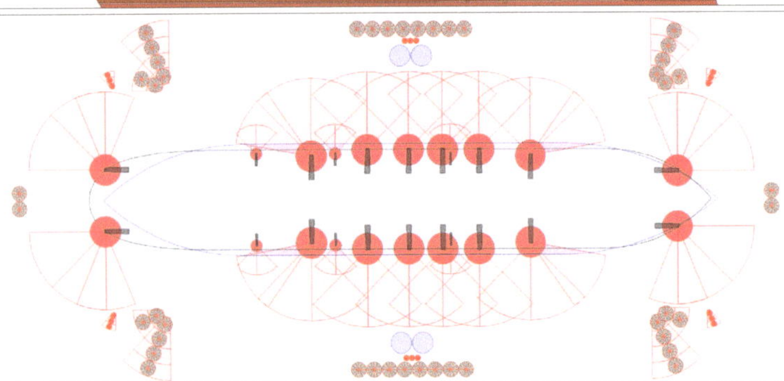

Superb (1873)
1 Maudslay horizontal direct acting E | 6,580 HP = 13.2 |
1873 01 01 - 1875 11 16 - 1880 11 15 | Thames IW, Blackwall

Sister ship can be found in the British section

1 | "Sultan" type (1868)
Battery ironclad, Britain

Displacement: 9,290 tons at draft: 8.05 m | Length: 99.06 m pp | Beam: 17.98 m | Crew: 633 men

Machinery: 8 rectangular boilers | 1 steam E * 1 screw | 7,700 HP = 14 knots | Supplies: 810 tons of coal

Armour: iron | CT: 203; UPBattery: =114= 203 =114=; Battery: =127= 229(^38) =152= | WLBelt: 152(^38) 203(^38) 229 203(^38) 152(^38)

Armament: 8-254/14.5 MLR, 4-228.6/13.9 MLR, 7-95.3/21

At overhaul

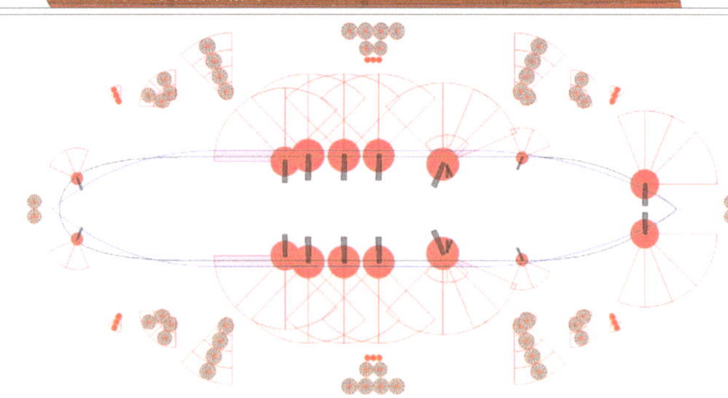

Sultan (1868)
1 Penn horizontal trunk E | 7,720 HP = 14.13 knots |
1868 02 29 - 1870 05 31 - 1871 10 10 | Chatham DYd, Chatham

2 | "Swiftsure" type (1868)
Battery ironclads, Britain

Displacement: 6,910 tons at draft: 7.62 m | Length: 85.34 m pp | Beam: 16.76 m | Crew: 450 men

Machinery: 6 rectangular boilers | 1 steam E * 1 screw | 4,900 HP = 14 knots | Supplies: 600 tons of coal

Armour: iron | CT: 76; UPBattery: 152; Battery: =102= 152 =127= |
WLBelt: 152 203 152

Armament: 10-228.6/13.9 MLR, 4-160/16 MLR, 6-95.3/21

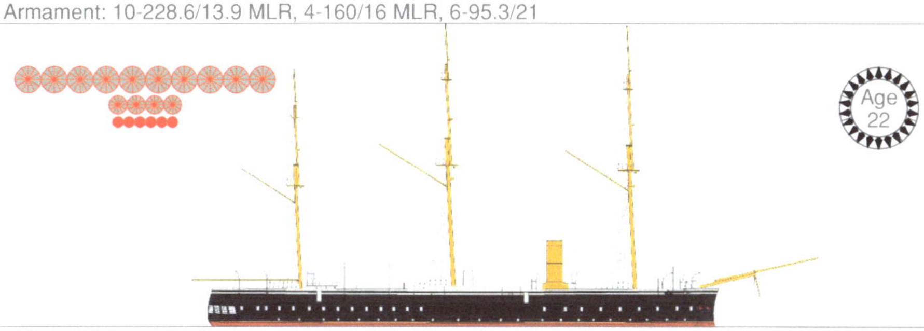

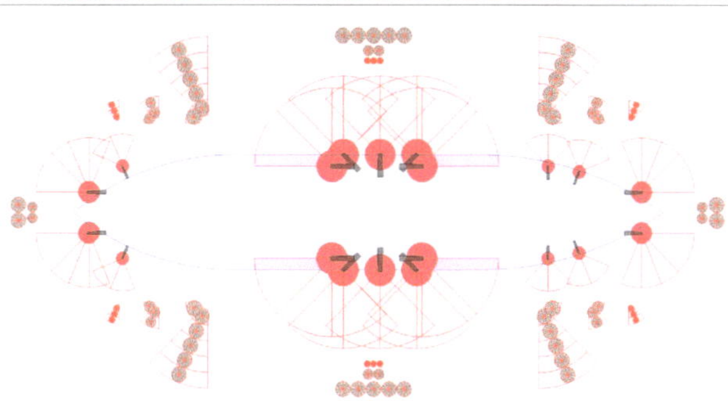

Swiftsure (1868)
1 Maudslay horizontal return connecting rod E | 4,910 HP = 13.75 knots |
1868 08 31 - 1870 06 15 - 1872 06 27 | Palmer, Jarrow

Triumph (1868)
1 Maudslay horizontal return connecting rod E | 4,890 HP = 14.07 knots |
1868 08 31 - 1870 09 27 - 1873 04 08 | Palmer, Jarrow

2 | "Swiftsure" type (1868)
Battery ironclads, Britain (continued)

Triumph

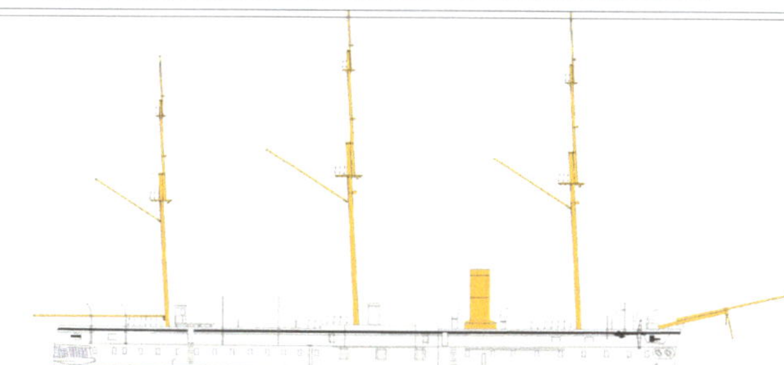

3 | "Audacious" type (1867)
Battery ironclads, Britain (continued)

Iron Duke (1868)
2 Ravenhill horizontal return connecting rod EE | 4,270 HP = 13.64 knots |
1868 08 23 - 1870 03 01 - 1871 01 21 | Pembroke DYd, Pembroke

Iron Duke

3 | "Audacious" type (1867)
Battery ironclads, Britain

Displacement: 5,909 tons at draft: 6.88 m | Length: 85.34 m pp | Beam: 16.46 m | Crew: 450 men

Machinery: 6 rectangular boilers | 2 * 2,400 HP steam EE * 2 screws | 4,800 HP = 13 knots | Supplies: 600 tons of coal

Armour: iron | CT: 76; UPBattery: 152; Battery: =102= 152 =127= | WLBelt: 152 203 152

Armament: 10-228.6/13.9 MLR, 4-160/16 MLR, 6-95.3/13

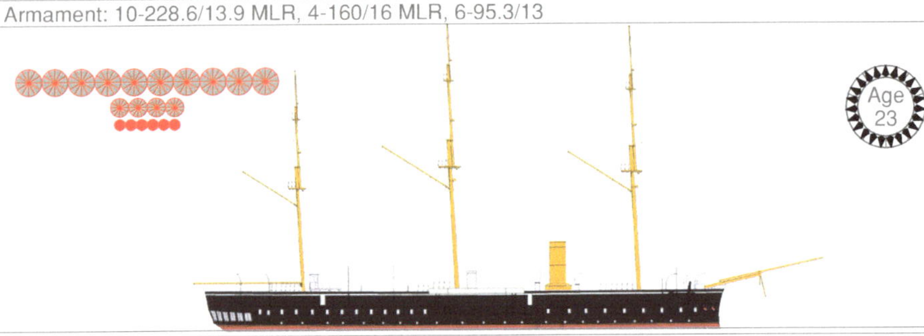

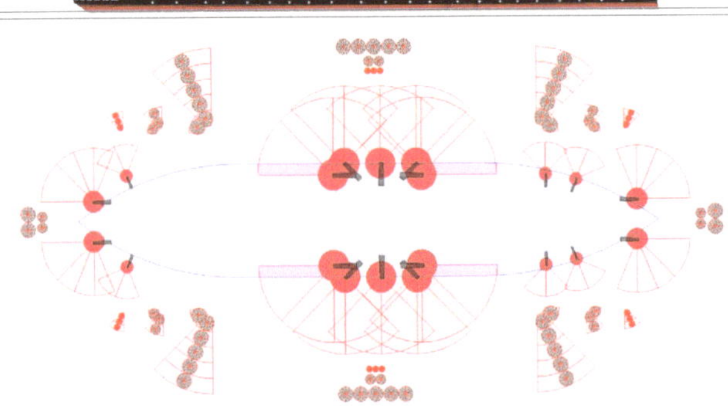

Audacious (1867)
2 Ravenhill horizontal return connecting rod EE | 4,040 HP = 13.02 knots |
1867 06 26 - 1869 02 27 - 1870 09 10 | Napier, Govan

Invincible (1867)
2 Napier horizontal return connecting rod EE | 4,830 HP = 14.09 knots |
1867 06 28 - 1869 05 29 - 1870 10 01 | Napier, Govan

1 | "Hercules" type (1866)
Battery ironclad, Britain

Displacement: 8,667 tons at draft: 7.72 m | Length: 99.06 m pp | Beam: 18 m | Crew: 638 men

Machinery: 9 rectangular boilers | 1 steam E * 1 screw | 6,750 HP = 14.7 knots | Supplies: 610 tons of coal

Armour: iron | Battery: =127= 203(^38) =152= | WLBelt: 152(^38) 203(^38) 229 203(^38) 152(^38)

Armament: 8-254/14.5 MLR, 2-228.6/13.9 MLR, 4-177.8/15.9 MLR

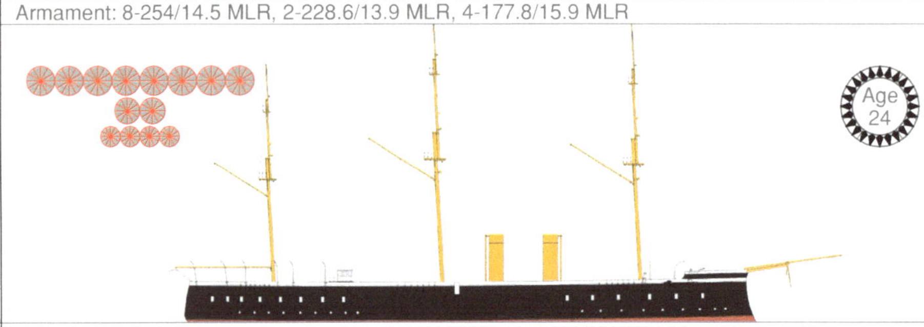

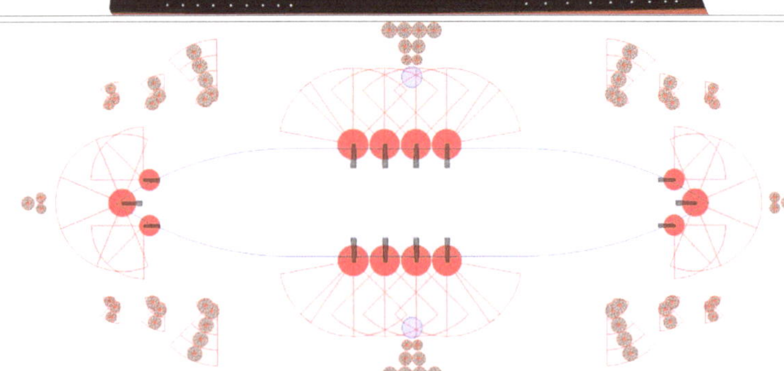

Hercules (1866)
1 Penn horizontal trunk E | 7,178 HP = 14.69 knots |
1866 02 01 - 1868 02 10 - 1868 11 21 | Chatham DYd, Chatham

1 | "Penelope" type (1865)
Battery ironclad, Britain

Displacement: 4,470 tons at draft: 5.1 m | Length: 79.25 m pp | Beam: 15.24 m | Crew: 350 men

Machinery: 4 rectangular boilers | 2 * 2,350 HP steam EE * 2 screws | 4,700 HP = 13 knots | Supplies: 500 tons of coal

Armour: iron | Battery: =114= 152 =114=; UPBelt: 0 152 0 | WLBelt: 127(^29) 152(^29) 127(^29)

Armament: 8-203,2/14,8 MLR, 3-127/15, 6-95,3/21

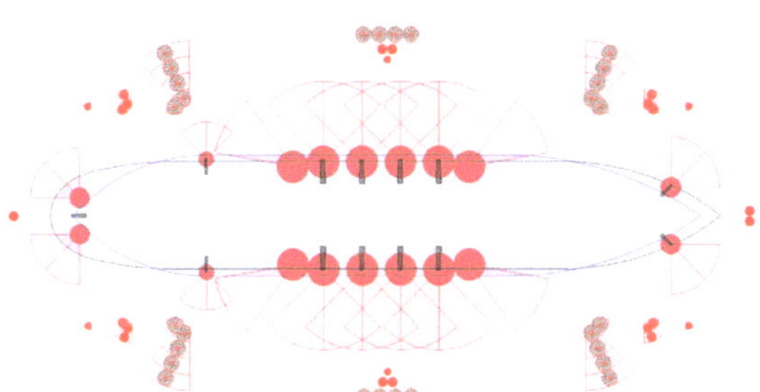

Penelope (1865)

2 Maudslay horizontal ruturn connecting rod EE | 4,763 HP = 12.76 knots |
1865 09 04 - 1867 06 18 - 1868 06 27 | Pembroke DYd, Pembroke

1 | "Bellerophon" type (1863)
Battery ironclad, Britain

Displacement: 7,551 tons at draft: 7.52 m | Length: 91.44 m pp | Beam: 17.09 m | Crew: 650 men

Machinery: 8 rectangular boilers | 1 steam E * 1 screw | 6,500 HP = 14 knots | Supplies: 640 tons of coal

Armour: iron | CT: 203; Battery: =127= 152 =127= | WLBelt: 114 127 =127= 152 =127= 127 76

Rearmed (1885): 10-203.2/25, 4-152.4/26, 4-57/43, 12-25.4/40 Mg, 2-406.4 TT (^)

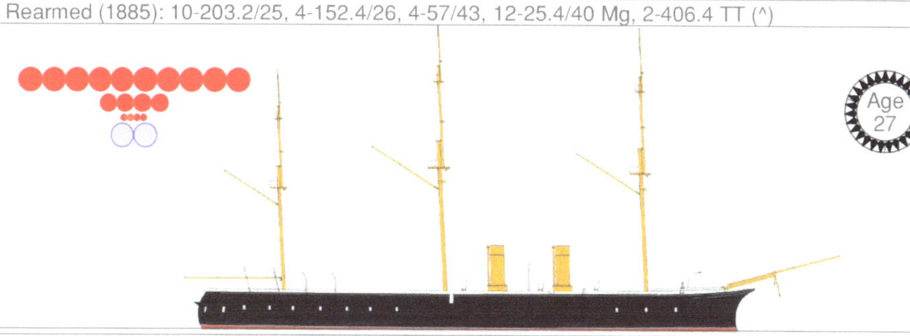

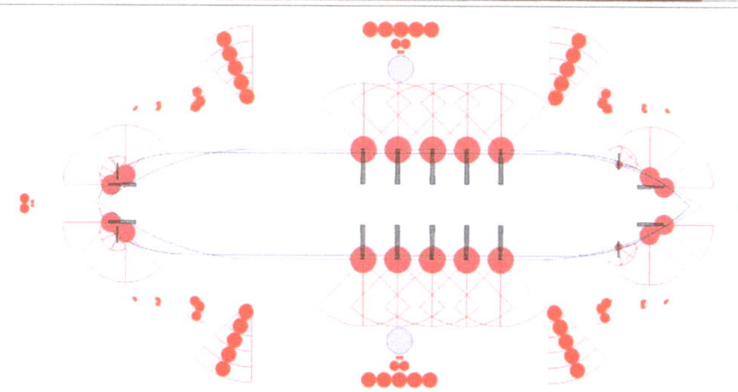

Bellerophon (1863)

1 Penn horizontal trunk E | 6,521 HP = 14.17 knots |
1863 12 23 - 1865 05 26 - 1866 04 11 | Chatham DYd, Chatham

1 | "Northumberland" type (1861)
Battery ironclad, Britain

Displacement: 10,784 tons at draft: 8.46 m | Length: 124 m oa | Beam: 18.4 m | Crew: 800 men

Machinery: 10 rectangular boilers | 1 steam E * 1 screw | 6,700 HP = 14 kts | Supplies: 750 / +650 tons of coal

Armour: iron | CT: 114; Battery: 0 =140= 140 =140= 0 |
WLBelt: 114 140 114

Rearmed: 4-228.6/13.9 MLR, 22-203.2/14.8 MLR, 2-177.8/15.9 MLR, 8-47/43

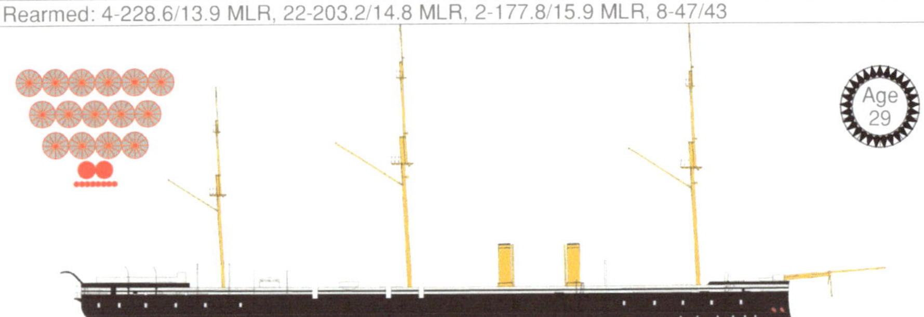

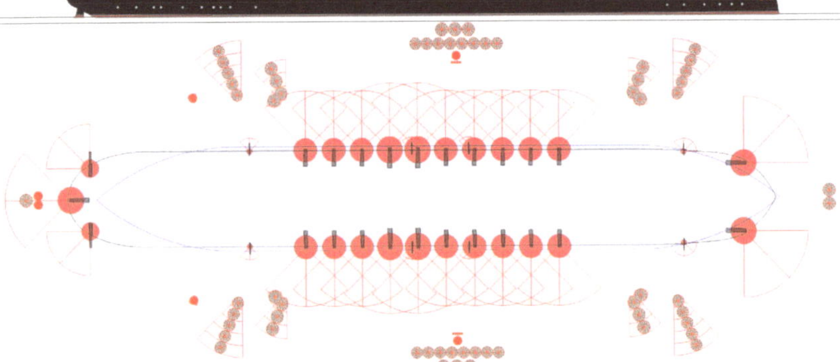

Northumberland (1861)
1 Penn horizontal trunk E | 6,560 HP = 14.13 knots |
1861 10 10 - 1866 04 17 - 1868 10 08 | Millwall Iron Works, Millwall

2 | "Agincourt" type (1861)
Battery ironclads, Britain

Displacement: 10,600 tons at draft: 8.46 m | Length: 124 m oa | Beam: 18.4 m | Crew: 800 men

Machinery: 10 rectangular boilers | 1 steam E * 1 screw | 6,700 HP = 14 knots | Supplies: 750 / +650 tons of coal

Armour: iron | Battery: 114 140 114 =140= 0 |
WLBelt: 114 140 114

Rearmed (1881): 15-228.6/13.9 MLR, 2-152.4/25, 8-47/43

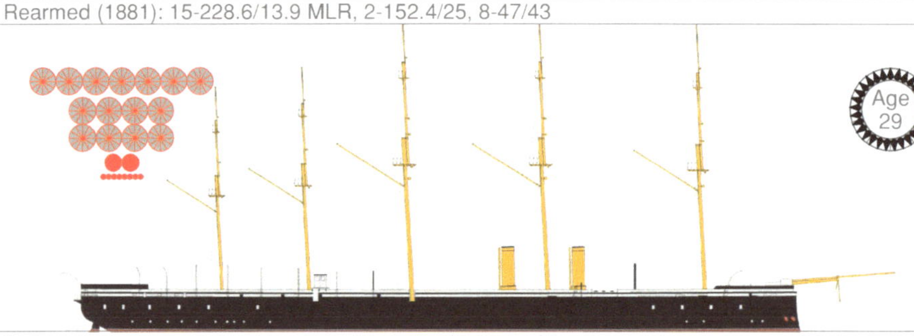

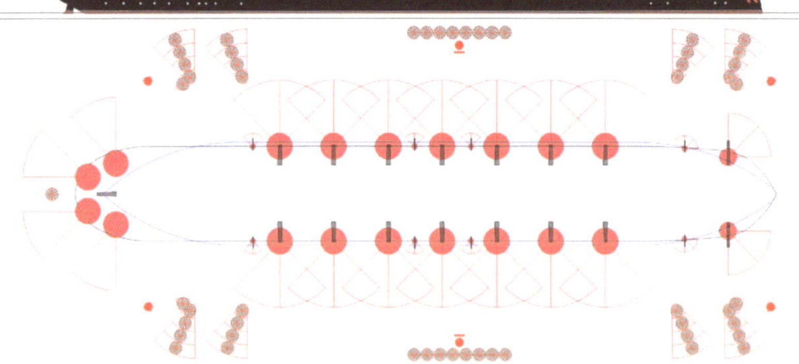

Agincourt (1861)
1 Maudslay return connecting rod E | 6,867 HP = 14,8 knots |
1861 10 30 - 1865 03 27 - 1867 06 01 | Laird, Birkenhead

Minotaur (1861)
1 Penn horizontal trunk E | 6,700 HP = 14,32 knots |
1861 09 12 - 1863 12 12 - 1868 12 19 | Thames IW, Blackwall

1 | "Achilles" type (1861)
Battery ironclad, Britain

Displacement: 9,829 tons at draft: 8.31 m | Length: 115.82 m pp | Beam: 17.77 m | Crew: 709 men
Machinery: 10 rectangular boilers | 1 steam E * 1 screw | 5,700 HP = 14 knots | Supplies: 1,000 tons of coal
Armour: iron | Battery: 0 =114= 114 =114= 0; UPBelt: 0 =114= 114 =114= 0 | WLBelt: 114
Armament: 14-228.6/13.9 MLR, 2-152.4/26, 8-47/43

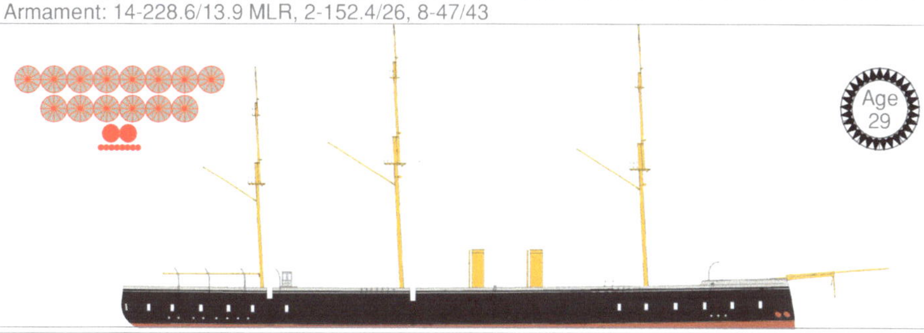

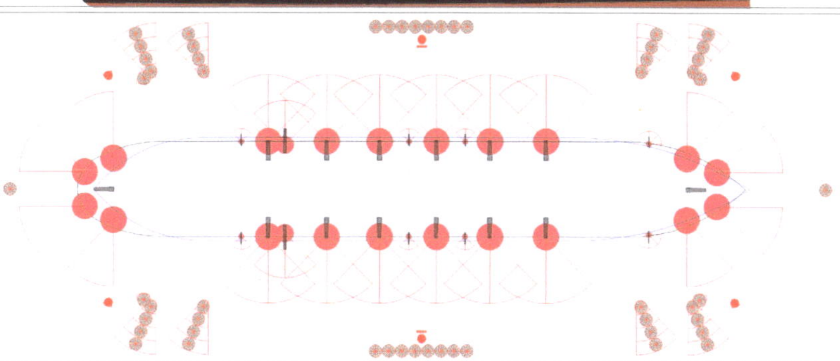

Achilles (1861)
1 Penn horizontal single expansion trunk E | 5,720 HP = 14.32 knots |
1861 08 01 - 1863 12 23 - 1864 11 26 | Chatham DYd, Chatham

2 | "Hector" type (1861)
Battery ironclads, Britain

Displacement: 6,710 tons at draft: 7.62 m | Length: 85.39 m pp | Beam: 17.14 m | Crew: 530 men
Machinery: 6 rectangular boilers | 1 steam E * 1 screw | 3,500 HP = 12.5 knots | Supplies: 450 tons of coal
Armour: iron | Battrey: 63 114 63 |
WLBelt: 0 =114= 114 =114= 0
Armament: Disarmed

"Hector" disarmed 1886.
"Valiant" laid up due to the accident with "Defence" 1884 07 20. Disarmed 1885.

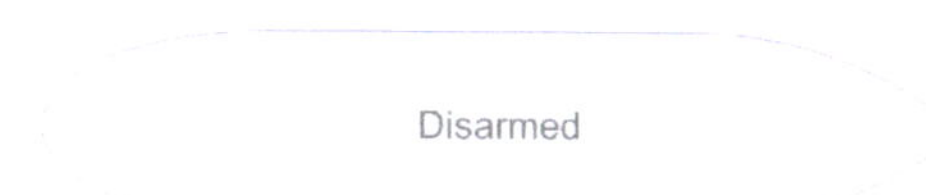

Hector (1861)
1 Maudslay horizontal connecting rod E | 3,260 HP = 12.65 knots |
1861 03 01 - 1862 09 26 - 1864 02 22 | Napier, Govan

Valiant (1861)
1 Maudslay horizontal connecting rod E | 3,560 HP = 12.36 knots |
1861 02 01 - 1863 10 14 - 1868 09 15 | Westwood & Baillie, Millwall > Thames IW, Blackwall

1 | "Defence" type (1859)
Battery ironclad, Britain

Displacement: 6,150 tons at draft: 7.62 m | Length: 92.04 m oa | Beam: 16.51 m | Crew: 460 men

Machinery: 4 rectangular boilers | 1 steam E * 1 screw | 2,500 HP = 12 knots | Supplies: 460 tons of coal

Armour: IR. Battery: 114 |
WLBelt: =114= 114 =114=

Armament: disarmed

Age 31

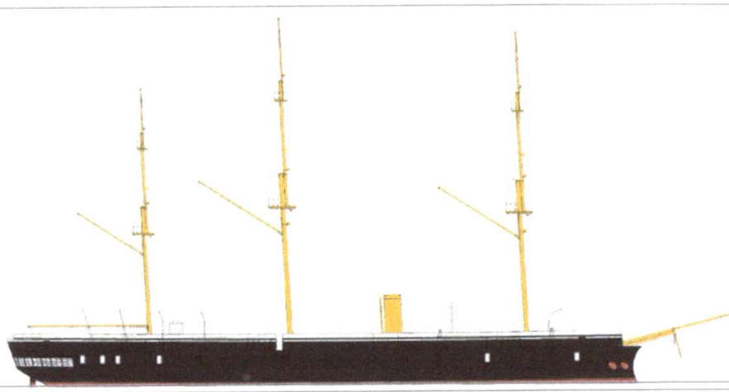

"Defence" disarmed 1879. Laid up due to the accident with "Valiant" 1884 07 20.

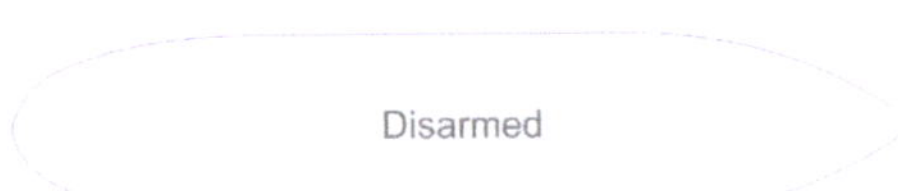

Defence (1859)
1 Penn horizontal single expansion trunk E | 2,540 HP = 11.62 knots |
1859 12 01 - 1861 04 24 - 1861 12 02 | Palmer, Jarrow

ARMOURED CRUISERS

7 | "Australia" type (1885)
Belted cruisers, Britain

Displacement: 5,600 tons at draft: 6.86 m | Length: 91.44 m pp | Beam: 17.07 m | Crew: 484 men

Machinery: 4 2-ended cylindrical boilers | 2 * 2,550 HPn / 4,250 HPf steam EE * 2 screws | 5,500 HPn = 17 knots / 8,500 HPf = 18 knots | Supplies: 900 tons of coal

Armour: compound | CT: 305 |
WLBelt: 0(v76) =406= 254>127(^51) =406= 0(v76)

Armament: 2-234/32, 10-152.4/26, 6-57/43, 10-47/43, 8-25.4/40 Mg, 6-457 TT (4 ^, 2 v)

Age 5

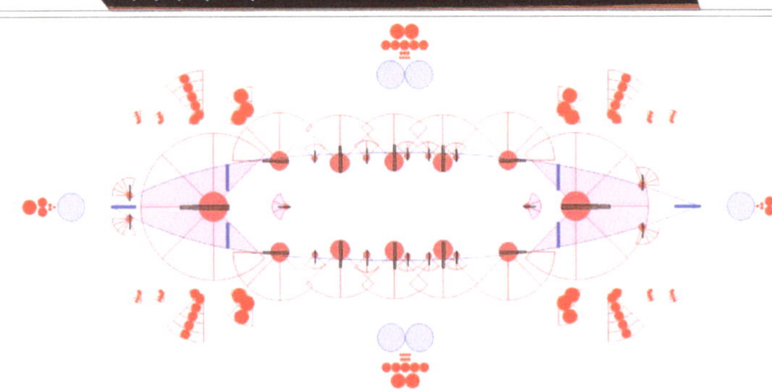

Aurora (1886)
2 Thompson vertical triple expansion EE |
1886 02 01 - 1887 10 28 - 1889 07 01 | Pembroke DYd, Pembroke

Australia (1885)
2 Fairfield vertical triple expansion EE |
1885 04 21 - 1886 11 25 - 1888 10 01 | Fairfield, Glasgow

7 | "Australia" type (1885)
Belted cruisers, Britain (continued)

Galatea (1885)
2 Napier vertical triple expansion EE |
1885 04 21 - 1887 03 10 - 1889 05 01 | Napier, Govan

Immortalite (1886)
2 Thompson vertical triple expansion EE |
1886 01 18 - 1887 07 07 - 1889 07 01 | Chatham DYd, Chatham

Narcissus (1885)
2 Earle vertical triple expansion EE |
1885 04 27 - 1886 12 15 - 1889 07 01 | Earle, Hull

Orlando (1885)
2 Palmer vertical triple expansion EE |
1885 04 23 - 1886 08 03 - 1888 06 01 | Palmer, Jarrow

Undaunted (1885)
2 Palmer vertical triple expansion EE |
1885 04 23 - 1886 11 25 - 1889 07 01 | Palmer, Jarrow

Orlando

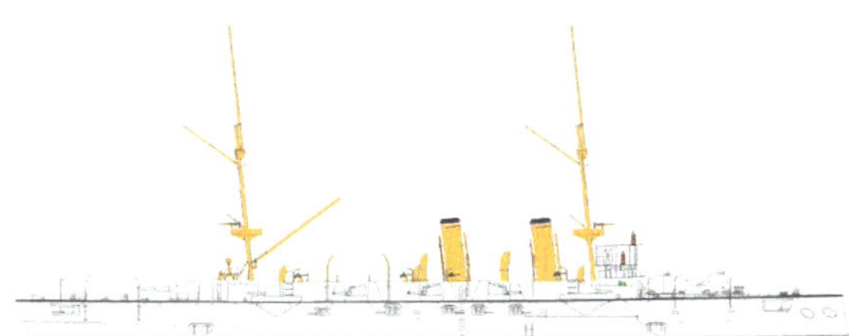

2 | "Imperieuse" type (1881)
Belted cruisers, Britain (continued)

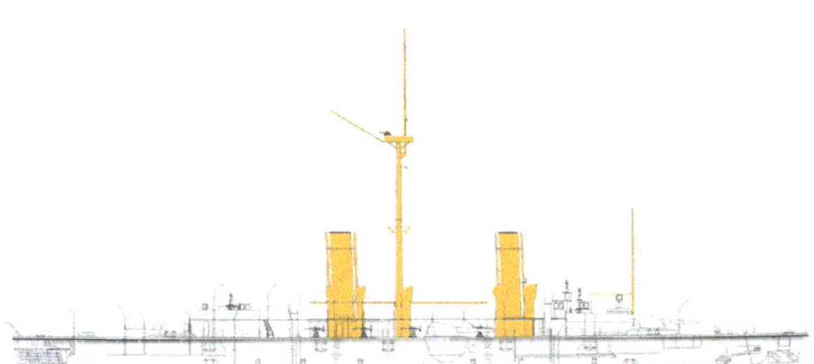

2 | "Imperieuse" type (1881)
Belted cruisers, Britain

Displacement: 8,500 tons at draft: 8.15 m | Length: 96.01 m pp | Beam: 18.9 m | Crew: 555 men
Machinery: 12 cylindrical and oval boilers | 2 * 4,000 HPn / 5,000 HPf steam EE * 2 screws | 8,000 HPn = 16 knots / 10,000 HPf = 16.75 knots | Supplies: 1,130 tons of coal
Armour: compound | CT: 229; Barbettes: 203; Hoists: 76 |
WLBelt: 0(v102) =229= 254(^51) =229= 0(v102)
Armament: 4-234/32, 10-152.4/26, 12-57/43, 6-457 TT (^)

Age 9

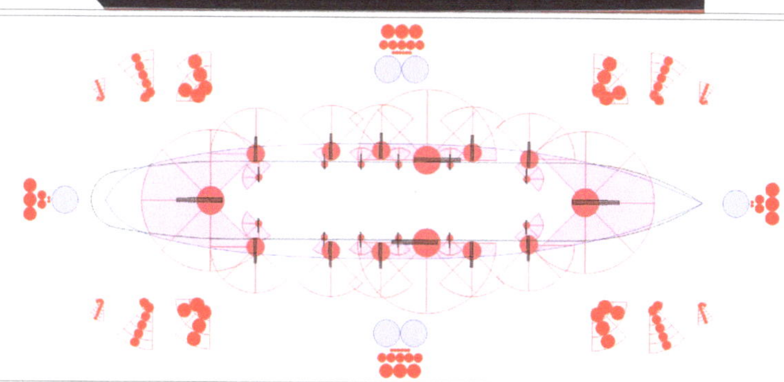

Imperieuse (1881)
2 Maudslay inverted compound EE |
1881 08 10 - 1883 12 18 - 1886 09 01 | Portsmouth DYd, Portsmouth

Warspite (1881)
2 Penn inverted compound EE |
1881 10 25 - 1884 01 29 - 1888 06 01 | Chatham DYd, Chatham

2 | "Northampton" type (1874)
Belted cruisers, Britain

Displacement: 7,473 tons at draft: 7.57 m | Length: 85.34 m pp | Beam: 18.29 m | Crew: 560 men

Machinery: 10 oval boilers | 2 * 3,250 HP steam EE * 2 screws | 6,500 HP = 14 knots | Supplies: 1,150 tons of coal

Armour: iron | CT: 229; Bulkheads: 25 |
WLBelt: 0(v76) =152= 152(^51) 229(^51) 152(^51) =229= 0(v76)

Armament: 4-254/14.5 MLR, 8-228.6/13.9 MLR, 4-121/22, 6-95.3/13, 14-47/23, 2-356 TT (^)

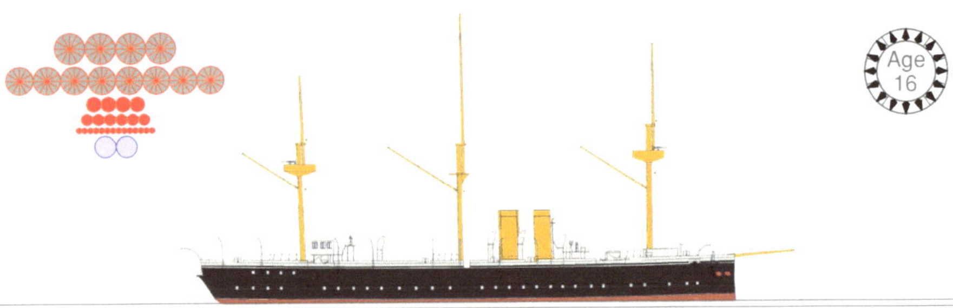

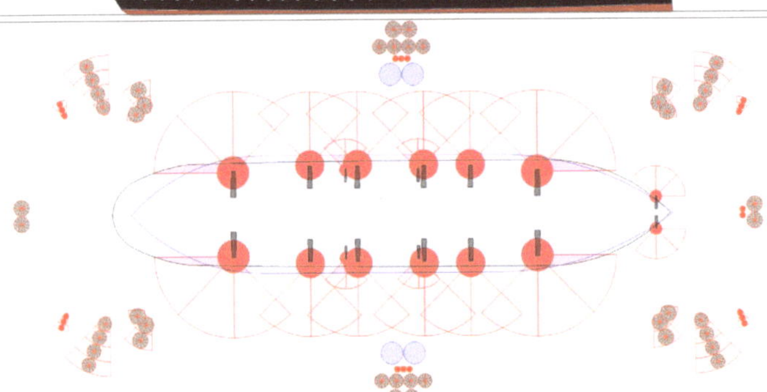

Nelson (1874)
2 Elder inverted compound EE | 6,624 HP = 14 knots |
1874 11 02 - 1876 11 04 - 1881 07 26 | Elder, Glasgow

Northampton (1874)
2 Penn inverted compound EE | 6,073 HP = 13.17 knots |
1874 10 26 - 1876 11 18 - 1878 12 07 | Napier, Govan

1 | "Shannon" type (1873)
Belted cruiser, Britain

Displacement: 5,670 tons at draft: 6.78 m | Length: 79.25 m pp | Beam: 16.46 m | Crew: 452 men

Machinery: 8 cylindrical boilers | 1 steam E * 1 screw | 3,400 HP = 13 knots | Supplies: 560 tons of coal

Armour: iron | CT: 229; Bulkheads: =203= |
WLBelt: 0(v51) 152(^51) 229(^51) 152(^51) =229= 0(v76)

Armament: 2-254/14.5 MLR, 7-228.6/13.9 MLR, 6-95.3/13, 4-356 TT (^)

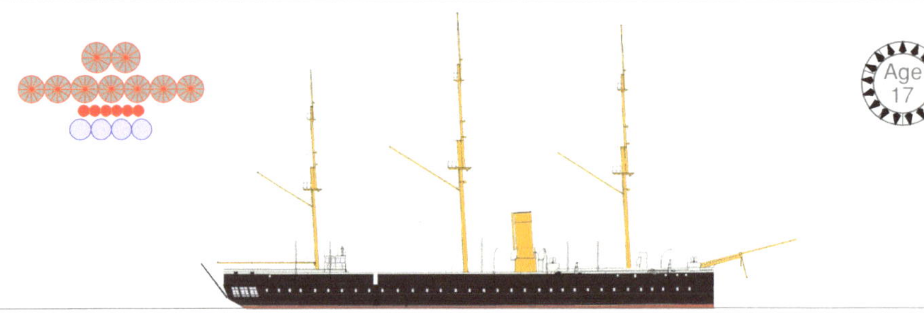

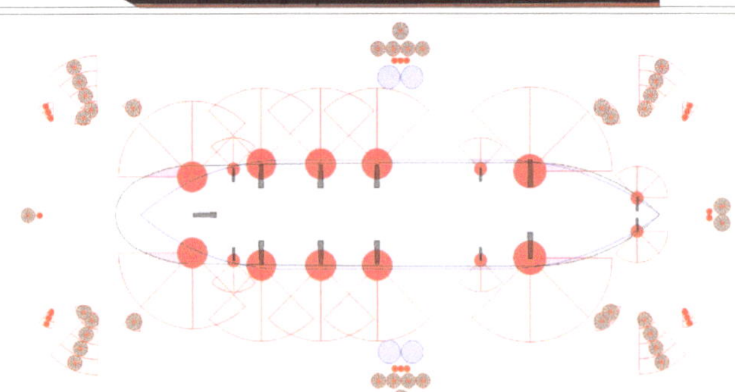

Shannon (1873)
1 Laird horizontal return connecting rod E | 3,370 HP = 12.25 knots |
1873 08 29 - 1875 11 11 - 1877 09 17 | Pembroke DYd, Pembroke

2 | "Warrior" type (1859)
Battery ironclads, Britain

Displacement: 9,137 tons at draft: 7.92 m | Length: 128 m oa | Beam: 17.78 m | Crew: 707

Machinery: 10 rectangular boilers | 1 steam E * 1 screw | 5,500 HP = 14 knots | Supplies: 850 tons of coal

Armour: iron | Battery: 0 =114= 114 =114= 0; UPBelt: 0 =114= 114 =114= 0 | WLBelt: 0 =114= 114 =114= 0

Rearmed (1884): 4-203.2/14.8 MLR, 28-177.8/15.8 MLR, 4-95.3/21

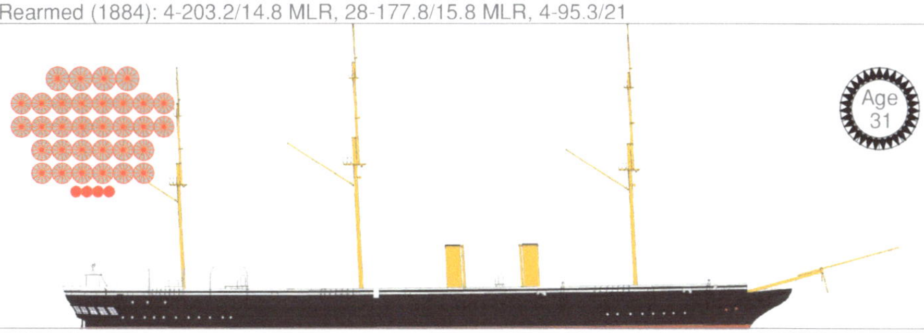

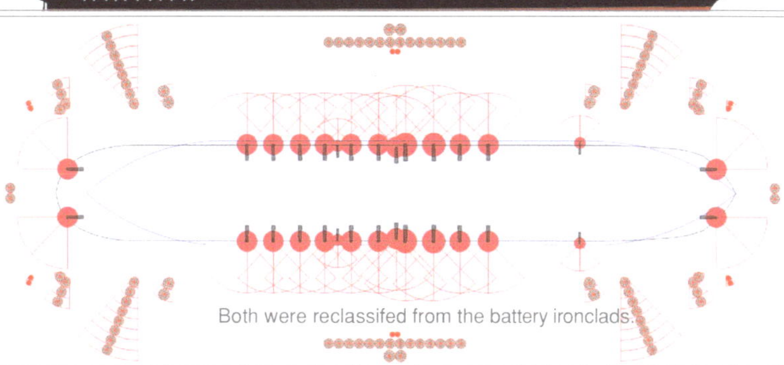

Both were reclassifed from the battery ironclads.

Black Prince (1859)
1 Penn horizontal single expansion trunk E | 5,770 HP = 13.6 knots |
1859 10 12 - 1861 02 27 - 1862 09 12 | Napier, Govan

Warrior (1859)
1 Penn horizontal single expansion trunk E | 5,267 HP = 14,08 knots |
1859 05 25 - 1860 12 29 - 1861 10 24 | Ditchburn & Mare, Blackwall

COAST DEFENCE BATTLESHIPS

2 | "Conqueror" type (1879)
Coast defence battleships, Britain

Displacement: 6,220 tons at draft: 7.16 m | Length: 87.78 m oa | Beam: 17.68 m | Crew: 330 men

Machinery: 8 cylindrical boilers | 2 * 2,250 HP steam EE * 2 screws | 4,500 HP = 14 knots | Supplies: 500 tons of coal

Armour: compound | CT: 305-152; Turret: 356-305; UPBelt: 0 =267= 305 =267= 0 | WLBelt: 0(^64) =279= 203(^38) 305(^38) 203(^38)

Armament: 2-304.8/25, 4-152.4/26, 6-57/43, 7-47/43, 6-356 TT (^)

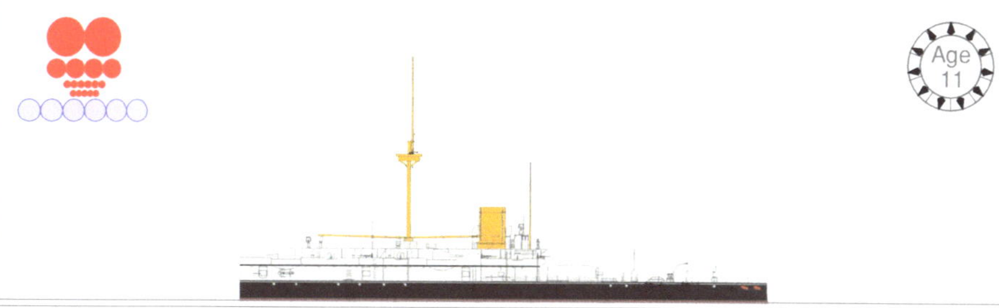

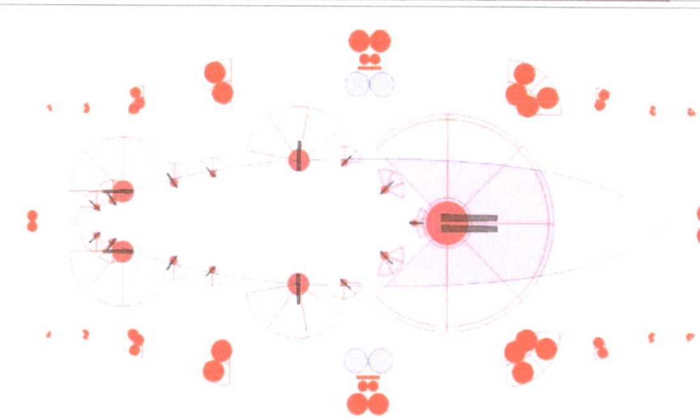

Hero (1884)
2 Rennie inverted compound EE |
1884 04 11 - 1885 10 27 - 1888 05 01 | Chatham DYd, Chatham

2 | "Conqueror" type (1879)
Coast defence battleships, Britain

Displacement: 6,220 tons at draft: 7.16 m | Length: 87.78 m oa | Beam: 17.68 m | Crew: 330 men
Machinery: 8 cylindrical boilers | 2 * 2,250 HP steam EE * 2 screws | 4,500 HP = 14 knots | Supplies: 500 tons of coal
Armour: compound | CT: 305-152; Turret: 356-305; UPBelt: 0 =267= 305 =267= 0 | WLBelt: 0(^64) =279= 203(^38) 305(^38) 203(^38)
Armament: 2-304.8/25, 4-152.4/26, 6-57/43, 7-47/43, 6-356 TT (^)

Age 11

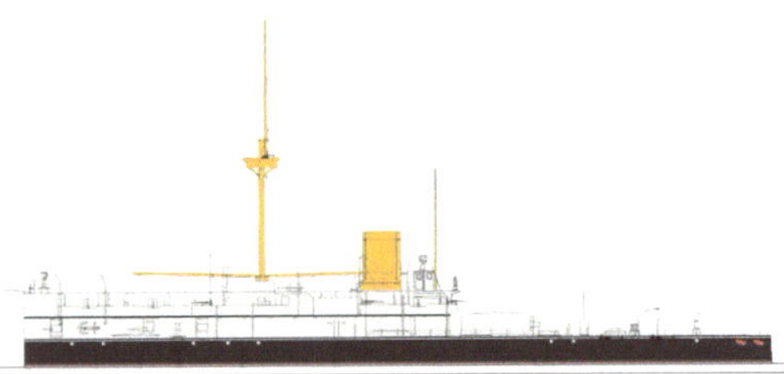

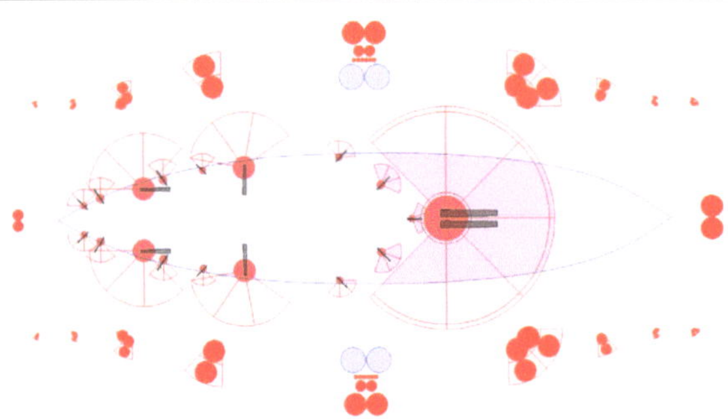

Conqueror (1879)
Humphrys inverted compound EE |
1879 04 28 - 1881 09 08 - 1886 03 01 | Chatham DYd, Chatham

1 | "Rupert" type (1870)
Monitor, Britain

Displacement: 5,440 tons at draft: 6.86 m | Length: 80.48 m pp | Beam: 16.15 m | Crew: 217 men
Machinery: 6 rectangular boilers | 2 * 2,100 HP steam EE * 2 screws | 4,200 HP = 14 knots | Supplies: 550 tons of coal
Armour: iron | CT: 305; Turret: 356-305; Breastwork: 305 | WLBelt: 229(^51) 280(^76) 229(^51)
Armament: 2-254/14.5 MLR, 2-160/15.5 MLR

Age 20

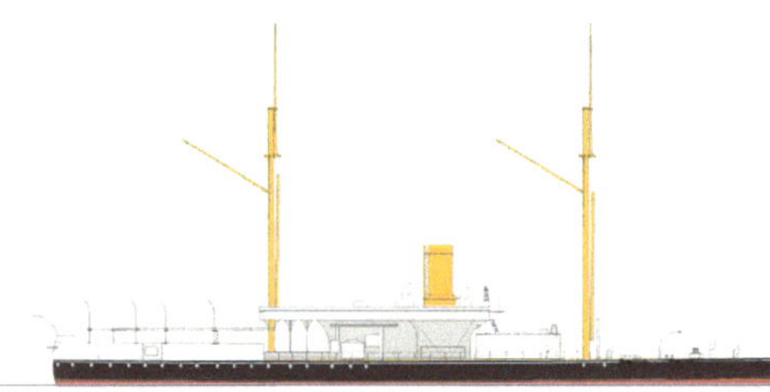

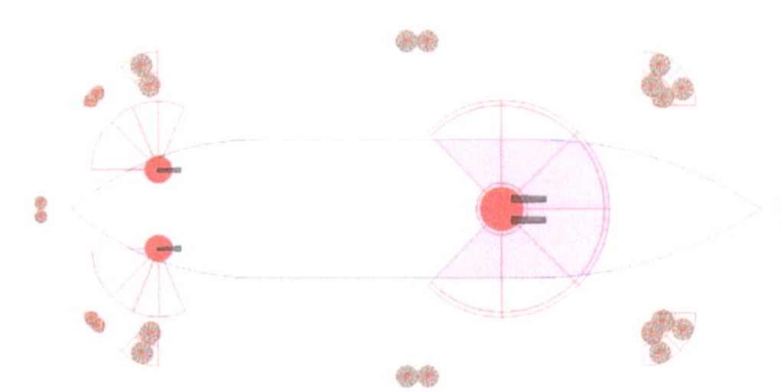

Rupert (1870)
Napier (type uncertain) engines | 4,630 HP = 13.56 knots |
1870 06 06 - 1872 03 12 - 1874 07 01 | Chatham DYd, Chatham

4 | "Cyclops" type (1870)
Monitors, Britain

Displacement: 3,480 tons at draft: 4.95 m | Length: 68.58 m pp | Beam: 13.72 m | Crew: 150 men
Machinery: 4 rectangular boilers | 2 * 800 HP steam EE * 2 screws | 1,600 HP = 11 knots | Supplies: 270 tons of coal
Armour: iron | CT: 229-203; Turrets: 254-229; Breastwork: 229-203 |
WLBelt: 152(^38) 203(^38) 152(^38)
Armament: 4-254/14 MLR, 4-47/43

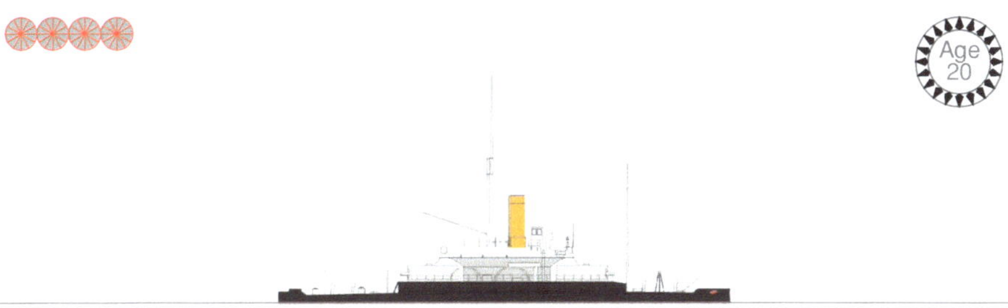

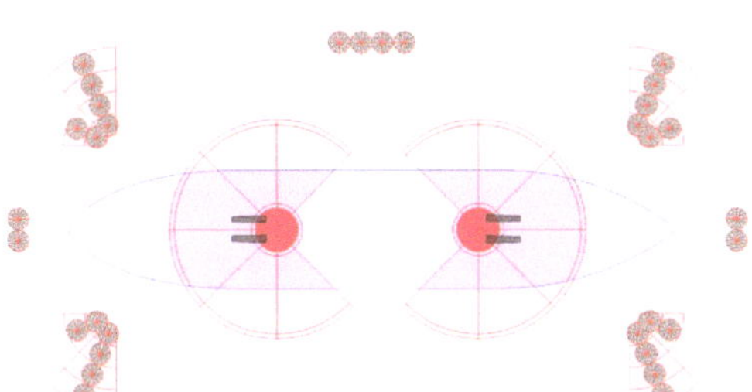

Cyclops (1870)
2 Elder compound EE | 1,660 HP = 11 knots |
1870 09 10 - 1871 07 18 - 1877 05 04 | Thames IW, Blackwall

Gorgon (1870)
2 Ravenhill horizontal direct acting EE | 1,670 HP = 11 knots |
1870 09 05 - 1871 10 14 - 1874 03 19 | Palmer, Jarrow

4 | "Cyclops" type (1870)
Monitors, Britain (continued)

Hydra (1870)
2 Elder compound EE | 1,472 HP = 11.2 knots |
1870 09 05 - 1871 12 28 - 1876 05 31 | Napier, Govan

Hecate (1870)
2 Ravenhill horizontal direct acting EE | 1,755 HP = 10.9 knots |
1870 09 05 - 1871 09 30 - 1877 05 24 | Dudgeon, Poplar

Gorgon

Hecate

Hydra

1 | "Hotspur" type (1868)
Monitor, Britain

Displacement: 4,331 tons at draft: 6.05 m | Length: 71.63 m pp | Beam: 15.24 m | Crew: 209 men
Machinery: 4 rectangular boilers | 2 * 1,750 HP steam EE * 2 screws | 3,500 HP = 12 knots |
Supplies: 300 tons of coal
Armour: iron | CT: 254-152; Turret (compound): 216; UPBelt: =203= 203 =203= |
WLBelt: 203(^38) 279(^38) 203(^38)
Rearmed (1884): 2-304.8/12 MLR, 2-152.4/26, 3-76.2/22, 7-25.4 Mg, 1-356 TT

Age 22

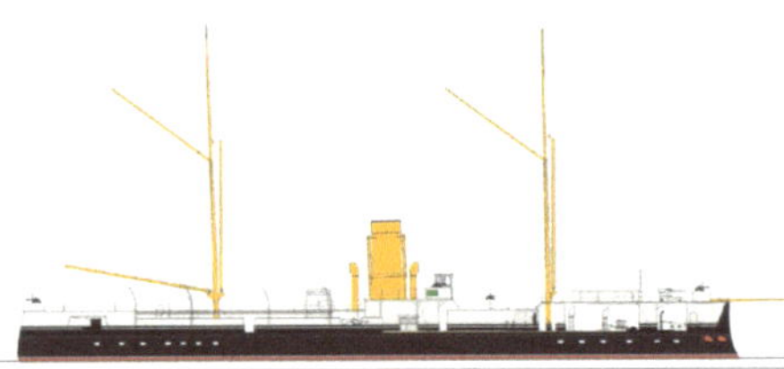

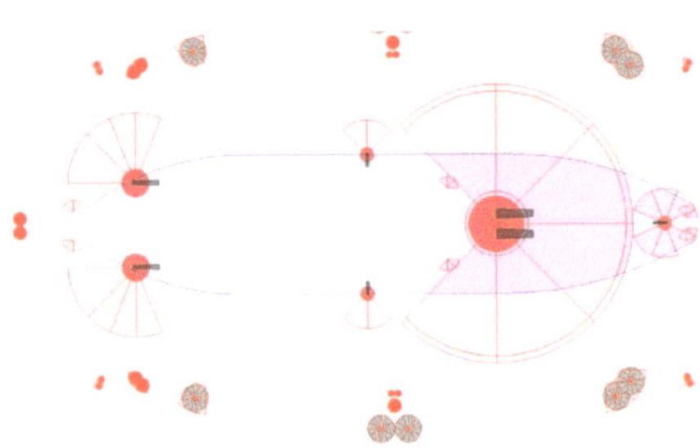

Hotspur (1868)
2 Napier horizontal single expansion EE | 12.65 knots |
1868 10 02 - 1870 03 19 - 1871 11 17 | Napier, Govan

1 | "Glatton" type (1868)
Monitor, Britain

Displacement: 4,912 tons at draft: 5.79 m | Length: 74.68 m pp | Beam: 16.46 m | Crew: 185 men
Machinery: 4 rectangular boilers | 2 * 1,400 HP steam EE * 2 screws | 2,800 HP = 12 knots |
Supplies: 540 tons of coal
Armour: iron | CT: 229-152; Turret: 356-305; Breastwork: 305 |
WLBelt: 254(^76-38) 305(^76-38) 254(^76-38)
Armament: 2-304.8/12 MLR

Age 22

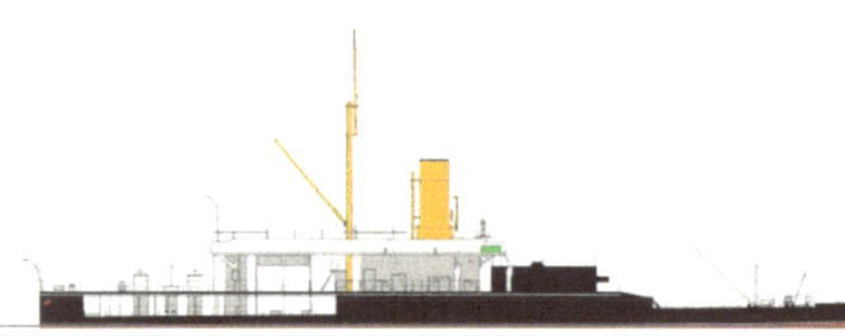

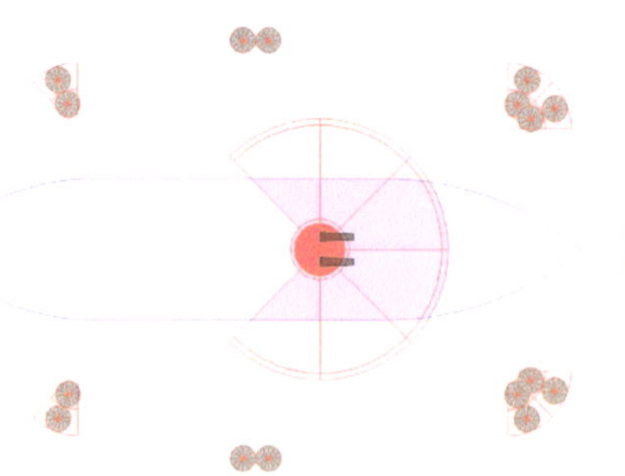

Glatton (1868)
2 Laird horizontal return connecting rod EE | 2,870 HP = 12.11 |
1868 08 10 - 1871 03 08 - 1872 02 24 | Chatham DYd, Chatham

2 | unnamed/"Wivern" type (1862)
Monitors, Confederate States of America

Displacement: 2,751 tons at draft: 4.95 m | Length: 68.43 m pp | Beam: 12.9 m | Crew: 153 men

Machinery: 4 rectangular boilers | 1 steam E * 1 screw | 1,450 HP = 10.5 knots | Supplies: 320 tons of coal

Armour: iron | CT: 76; Turrets: 254-127 |
WLBelt: 51(^25) 114(^25) 76(^25)

Armament: 4-228,6/13,9 MLR

Age 28

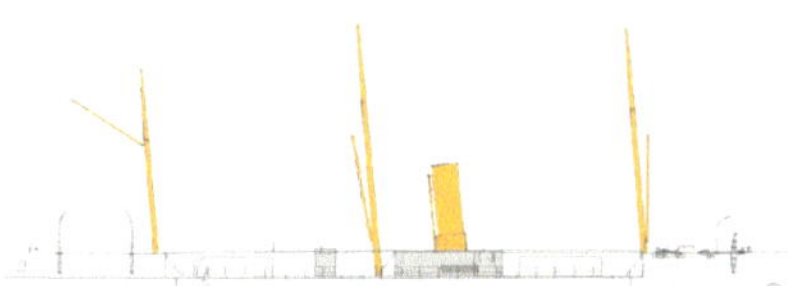

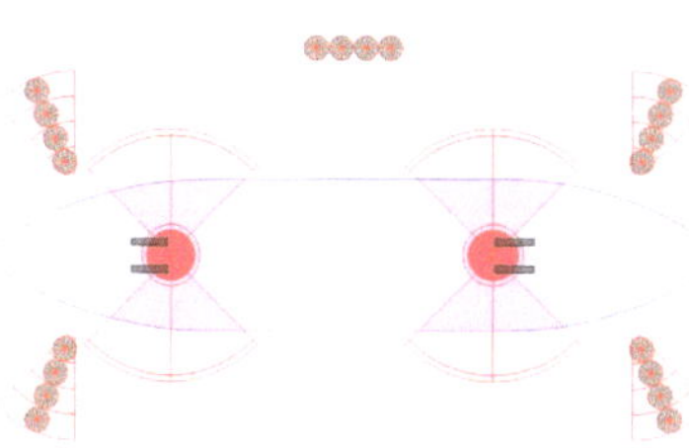

Scorpion (1862)
1 Laird horizontal direct acting E |
1862 04 01 - 1863 07 01 - 1865 10 10 | Laird, Birkenhead

Wivern (1862)
1 Laird horizontal direct acting E |
1862 04 01 - 1863 08 29 - 1865 10 10 | Laird, Birkenhead

1 | "Prince Albert" type (1862)
Coast defence armoured ship, Britain

Displacement: 3,687 tons at draft: 5.99 m | Length: 73.15 m pp | Beam: 14.66 m | Crew: 201 men

Machinery: 4 rectangular boilers | 1 steam E * 1 screw | 2,100 HP = 11 knots | Supplies: 300 tons of coal

Armour: iron | Turrets: 267-140 |
WLBelt: 89(^29) 114(^29) 89(^29)

Armament: 4-228.6/13.9 MLR, 2*25.4/40 Mg

Age 28

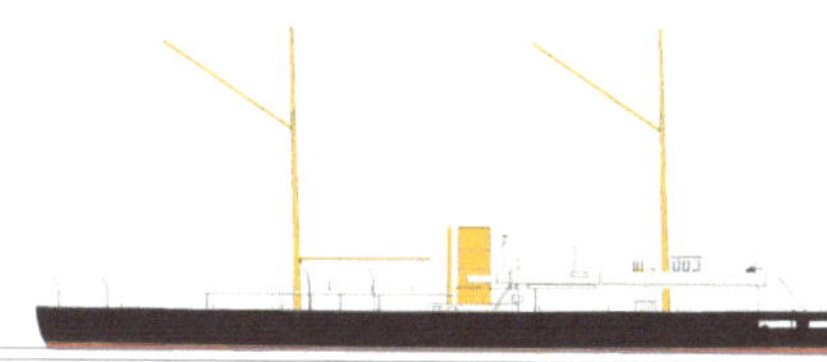

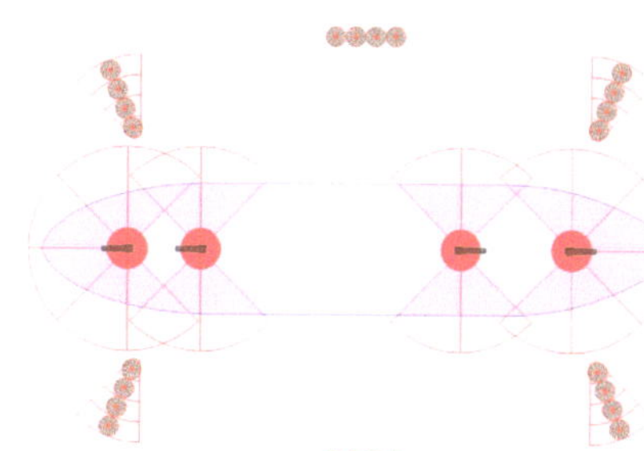

Prince Albert (1862)
1 Humprys direct acting E | 2,128 HP = 11.26 knots |
1862 04 29 - 1864 05 23 - 1866 02 23 | Samuda, Poplar

0 +5 | "Edgar" type (1889)
Protected cruisers 1 class, Britain

PROTECTED CRUISERS

BUILDING

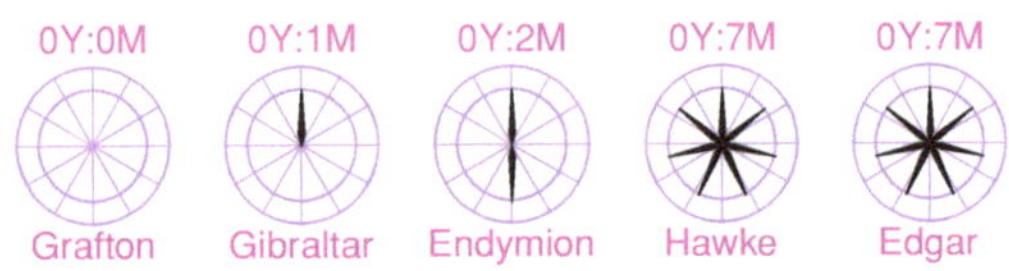

Displacement: 7,700 tons at draft: 7.24 m | Length: 118.11 m oa | Beam: 18.29 m | Crew: 544 men

Machinery: 4 cylindrical boilers | 2 * 5,000 HPn / 6,000 HPf steam EE * 2 screws | designed 10,000 HPn = 18 knots | 12,000 HPf = 20 knots | Supplies: 1,250 tons of coal

Armour: compound | CT: 305(^51); Casemates: 152; Gun shields: 76 / 51; Glacis: 178; Hoists: 178 | WLDeck: 0(/127^76)

Armament: 1-234/30, 12-152.4/40, 12-57/43, 5-47/43, 4-457 TT (^)

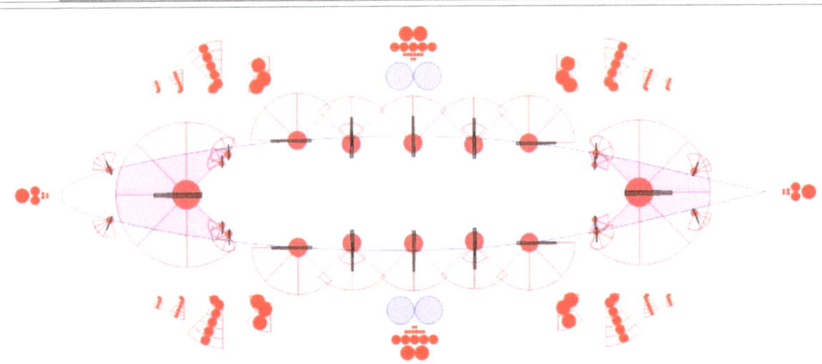

Edgar (1889)
2 Elder vertical triple expansion EE |
1889 06 03 - ***** - ***** | Devonport DYd, Devonport

Endymion (1889)
2 Earle vertical triple expansion EE |
1889 11 21 - ***** - ***** | Earle, Hull

0 +5 | "Edgar" type (1889)
Protected cruisers 1 class, Britain (continued)

Gibraltar (1889)
2 Napier vertical triple expansion EE |
1889 12 02 - ***** - ***** | Napier, Govan

Grafton (1890)
2 Humphrys vertical triple expansion EE |
1890 01 01 - ***** - ***** | Thames IW, Blackwall

Hawke (1889)
2 Elder vertical triple expansion EE |
1889 06 17 - ***** - ***** | Chatham DYd, Chatham

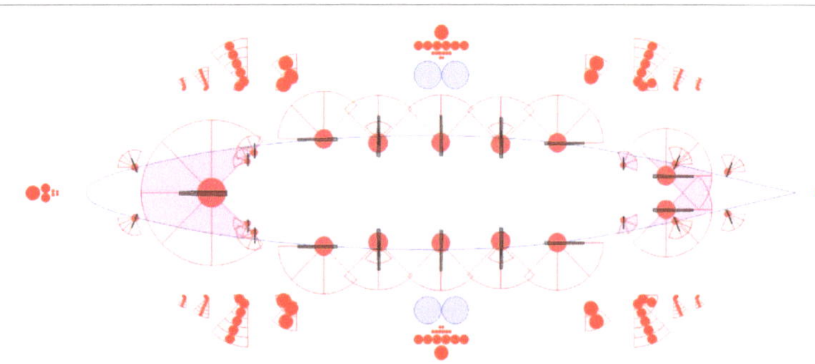

The revised "Edgar" protected cruiser project artillery deployment

0 +2 | "Blake" type (1888)
Protected cruisers 1 class, Britain

BUILDING

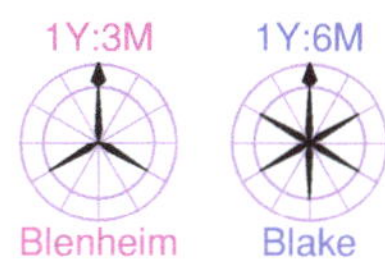

Displacement: 9,150 tons at draft: 7.32 m | Length: 121.84 m oa | Beam: 19.81 m | Crew: 570 men

Machinery: 6 cylindrical boilers | 2 * 6,500 HPn / 10,000 HPf steam EE * 2 screws | 13,000 HPn = 19 knots / 20,000 HPf = 22 knots | Supplies: 1,800 tons of coal

Armour: compound | CT: 305(^51); Casemates: 152; Gun shields: 114; Glacis: 203 | WLDeck: 0(/152^76)

Armament: 2-234/30, 10-152.4/26, 16-47/43, 4-356 TT (^)

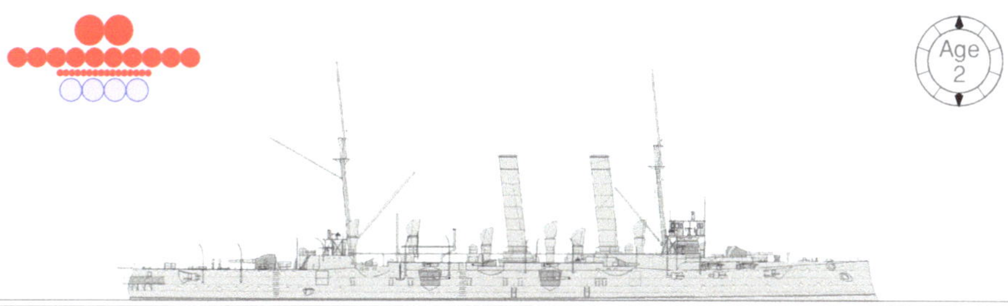

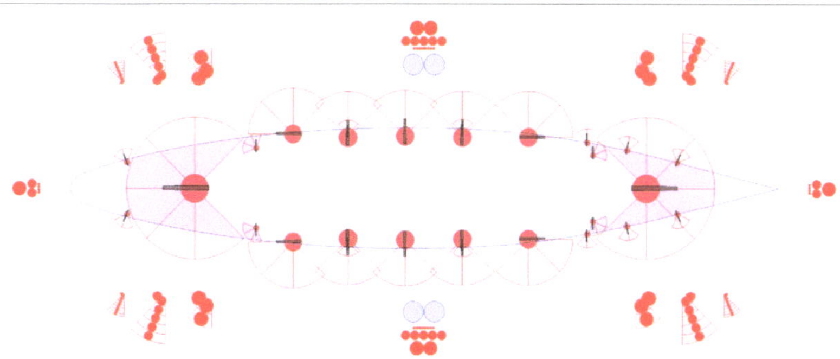

Blake (1888)
2 Maudslay vertical triple expansion EE |
1888 07 01 - 1889 11 23 - ***** | Chatham DYd, Chatham

Blenheim (1888)
2 Humphrys vertical triple expansion EE |
1888 10 01 - ***** - ***** | Thames Iron Works, Blackwall

0 +11 | "Apollo" type (1889)
Protected cruisers 2 class, Britain

BUILDING

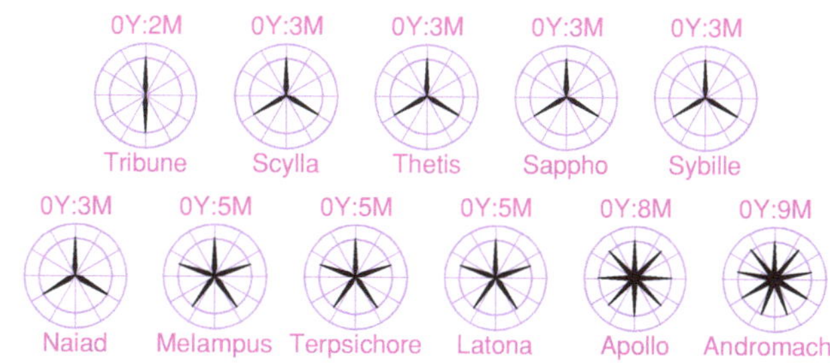

Displacement: 3,400 tons at draft 5.33 m | Length: 95.7 m oa | Beam: 13.11 m | Crew: 273 men

Machinery: 5 cylindrical boilers | 2 * 3,500 HPn / 4,500 HPf steam EE * 2 screws | 7,000 HPn = 18.5 knots / 9,000 HPf = 20 knots | Supplies: 535 tons of coal

Armour: compound | CT: 76(^25); Gun shields: 114; Glacis: 127 | WLDeck: 0(/51^31)

Armament: 2-152.4/26, 6-120/40, 8-57/43, 1-47/43, 4 MG, 4-356 TT (^)

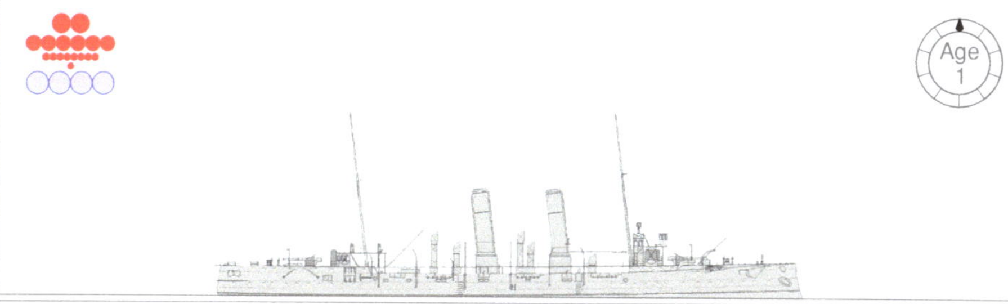

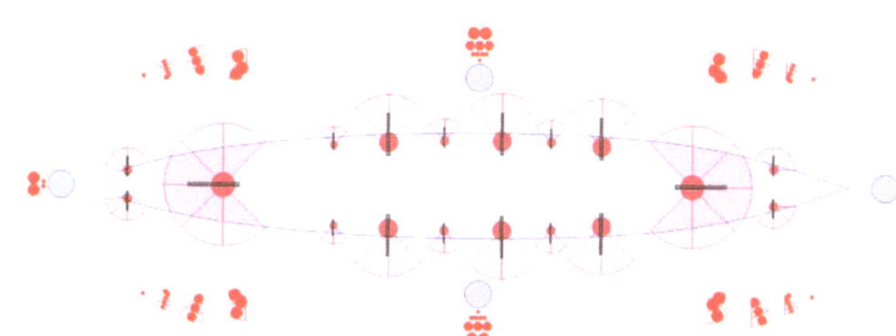

Andromache (1889)
2 Earle vertical triple expansion EE |
1889 04 29 - ***** - ***** | Chatham DYd, Chatham

Apollo (1889)
2 Earle vertical triple expansion EE |
1889 05 27 - ***** - ***** | Chatham DYd, Chatham

0 +11 | "Apollo" type (1889)
Protected cruisers 2 class, Britain (continued)

Latona (1889)
2 Barrow vertical triple expansion EE |
1889 08 22 - ***** - ***** | Naval Construction & Armament, Barrow

Melampus (1889)
2 Barrow vertical triple expansion EE |
1889 08 30 - ***** - ***** | Naval Construction & Armament, Barrow

Naiad (1889)
2 Barrow vertical triple expansion EE |
1889 10 03 - ***** - ***** | Naval Construction & Armament, Barrow

Sappho (1889)
2 Penn vertical triple expansion EE |
1889 10 29 - ***** - ***** | Samuda, Poplar

Scylla (1889)
2 Leslie vertical triple expansion EE |
1889 10 29 - ***** - ***** | Samuda, Poplar

Sybille (1889)
2 Leslie vertical triple expansion EE |
1889 10 11 - ***** - ***** | Stephenson, Newcastle

Terpsichore (1889)
2 builder's vertical triple expansion EE |
1889 08 27 - ***** - ***** | Thompson, Clydebank

Thetis (1889)
2 builder's vertical triple expansion EE |
1889 10 29 - ***** - ***** | Thompson, Clydebank

Tribune (1889)
2 builder's vertical triple expansion EE |
1889 11 12 - ***** - ***** | Thompson, Clydebank

0 +6 | "Apollo" sheathed type (1889)
Protected cruisers 2 class, Britain (continued)

Rainbow (1889)
2 builder's vertical triple expansion EE |
1889 12 30 - ***** - ***** | Palmer, Jarrow

Pique (1889)
2 builder's vertical triple expansion EE |
1889 10 30 - ***** - ***** | Palmer, Jarrow

Sirius (1889)
2 Maudslay vertical triple expansion EE |
1889 08 01 - ***** - ***** | Armstrong, Elswick

Spartan (1889)
2 Maudslay vertical triple expansion EE |
1889 08 01 - ***** - ***** | Armstrong, Elswick

0 +6 | "Apollo" sheathed type (1889)
Protected cruisers 2 class, Britain

BUILDING

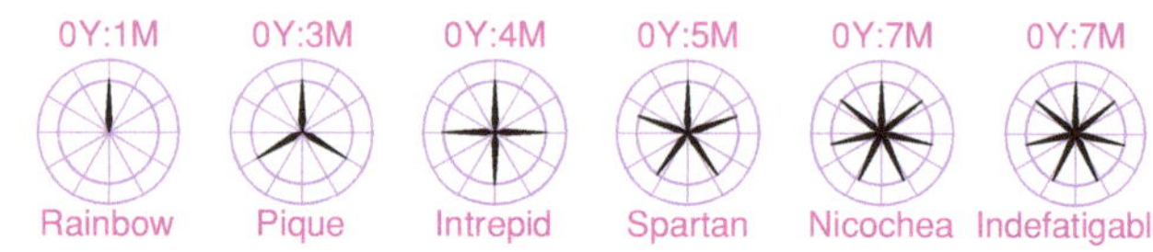

Displacement: 3,400 tons at draft 5.33 m | Length: 95.7 m oa | Beam: 13.31 m | Crew: 273 men
Machinery: 5 cylindrical boilers | 2 * 3,500 HPn / 4,500 HPf steam EE * 2 screws | 7,000 HPn = 18.5 knots / 9,000 HPf = 19.8 knots | Supplies: 535 tons of coal
Armour: compound | CT: 76(^25); Gun shields: 114; Glacis: 127 |
WLDeck: 0(/51^31)
Armament: 2-152.4/26, 6-120/40, 8-57/43, 1-47/43, 4 MG, 4-356 TT (^)

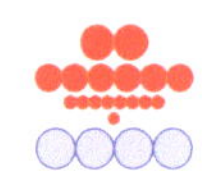

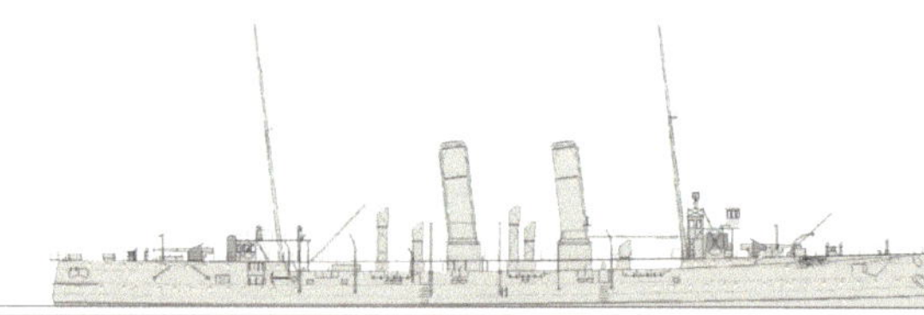

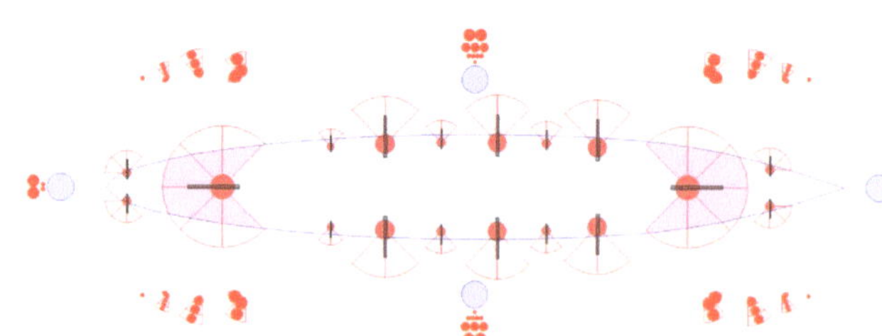

Indefatigable (1889)
2 Barrow vertical triple expansion EE |
1889 06 10 - ***** - ***** | London & Glasgow, Glasgow

Intrepid (1889)
2 Barrow vertical triple expansion EE |
1889 09 06 - ***** - ***** | London & Glasgow, Glasgow

0 +1 | "Vulcan" type (1888)
Protected cruiser 2 class, Britain

BUILDING

Displacement: 6,600 tons at draft: 6.71 m | Length: 113.69 m oa | Beam: 17.68 m | Crew: 432 men

Machinery: 5 locomotive boilers | 2 * 6,000 HP steam EE * 2 screws | 12,000 HP = 20 knots | Supplies: 1,000 tons of coal

Armour: compound | CT: 152(^25); Glacis: 127 | WLDeck: 0(/51^51) 0(/127^51) 0(/51^51)

Armament: 8-120/30, 12-47/43, 4-356 TT (^); 9-second class TBs

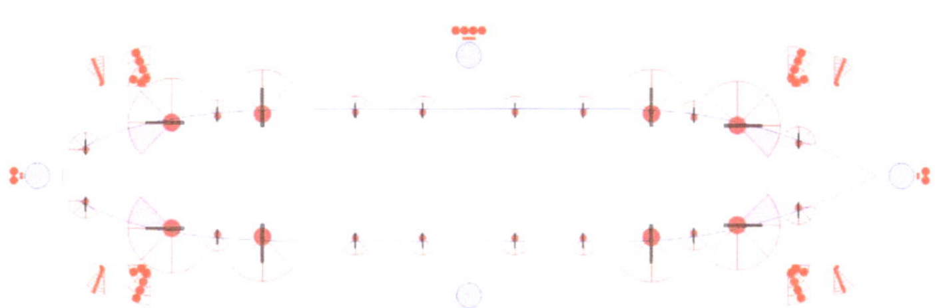

Vulcan (1888)
2 Humphrys vertical triple expansion EE |
1888 06 18 - 1889 06 13 - ***** | Portsmouth DYd, Portsmouth

4 +1 | "Medea" type (1887)
Protected cruisers 2 class, Britain

Displacement: 2,800 tons at draft: 5.03 m | Length: 80.77 m pp | Beam: 12.54 m | Crew: 218 men

Machinery: 6 cylindrical boilers | 2 * 4,500 HP steam EE * 2 screws | 9,000 HP = 20 knots | Supplies: 400 tons of coal

Armour: steel | CT: 76; Shields 114 | WLDeck: 0(/51^25)

Armament: 6-152.4/25, 8-57/43, 1-47/43, 4-25/40, 4-356 TT (^)

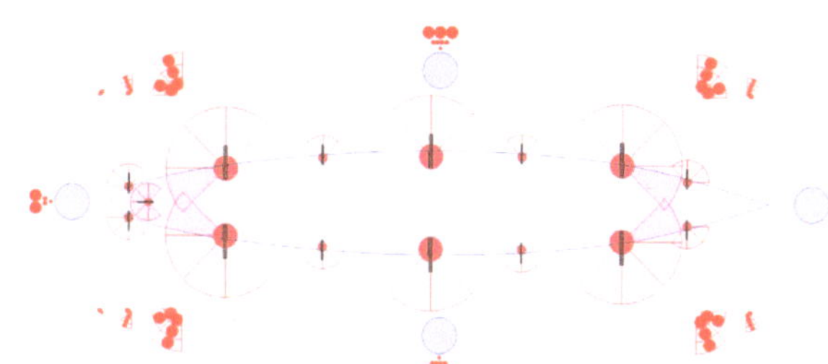

Medea (1887)
2 Humphrys vertical compound EE | 20.0 knots |
1887 04 25 - 1888 06 09 - 1889 01 01 | Chatham DYd, Chatham

Medusa (1887)
2 Humphrys vertical compound EE | 20.0 knots |
1887 08 25 - 1888 08 11 - 1889 01 01 | Chatham DYd, Chatham

4 +1 | "Medea" type (1887)
Protected | sheathed cruisers 2 class, Britain (continued)

BUILDING

Displacement: 2,950 tons at draft: 5.03 m | Length: 80.77 m pp | Beam: 12.8 m | Crew: 218 men
Machinery: 6 cylindrical boilers | 2 * 4,500 HP steam EE * 2 screws | 9,000 HP = 19.75 knots | Supplies: 400 tons of coal
Armour: steel | CT: 76; Shields 114 |
WLDeck: 0(/51^25)
Armament: 6-152.4/25, 8-57/43, 1-47/43, 4-25/40, 4-356 TT (^)

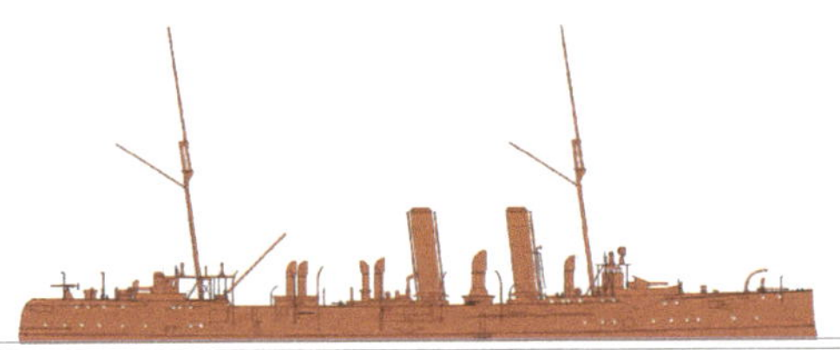

Magicienne (1887)
2 Leslie horizontal double acting compound EE |
1887 08 10 - 1888 05 12 - 1889 01 01 | Fairfield, Glasgow

Marathon (1887)
2 Leslie horizontal double acting compound EE |
1887 08 10 - 1888 08 23 - 1889 01 01 | Fairfield, Glasgow

Melpomene (1887)
2 Palmer horizontal double acting compound EE |
1887 10 10 - ***** - ***** | Portsmouth DYd, Portsmouth

4 | "Mersey" type (1883)
Protected cruisers 2 class, Britain (continued)

Severn (1884)
2 Humphrys horizontal direct acting compound EE |
1884 01 01 - 1885 09 29 - 1888 01 01 | Chatham DYd, Chatham

Thames (1884)
2 Penn horizontal direct acting compound EE |
1884 04 14 - 1885 12 03 - 1888 01 01 /// Pembroke DYd, Pembroke

4 | "Mersey" type (1883)
Protected cruisers 2 class, Britain

Displacement: 4,050 tons at draft: 5.94 m | Length: 96,01 m oa | Beam: 14.02 m | Crew: 325 men
Machinery: 12 cylindrical boilers | 2 * 2,250 HPn / 3,000 HPf steam EE * 2 screws | 4,500 HPn = 17 knots / 6,000 HPf = 18 knots | Supplies: 900 tons of coal
Armour: steel | CT: 229 |
WLDeck: 0/76(^51)
Armament: 2-203.2/30, 10-152.4/26, 3-57/43, 3-47/43, 9-25.4/40, 2-356 TT (v)

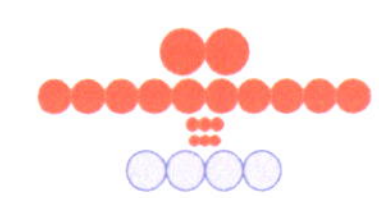

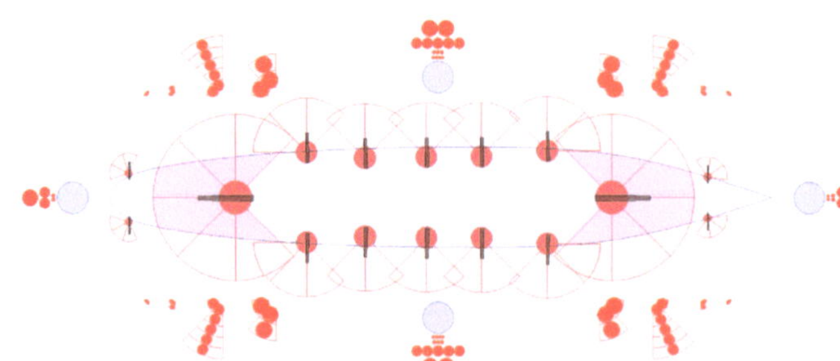

Forth (1884)
2 Leslie horizontal direct acting compound EE |
1884 12 01 - 1886 10 23 - 1889 01 01 | Pembroke DYd, Pembroke

Mersey (1883)
2 Humphrys horizontal direct acting compound EE |
1883 07 09 - 1885 03 31 - 1887 01 01 | Chatham DYd, Chatham

4 | "Mersey" type (1883)
Protected cruisers 2 class, Britain (continued)

Severn

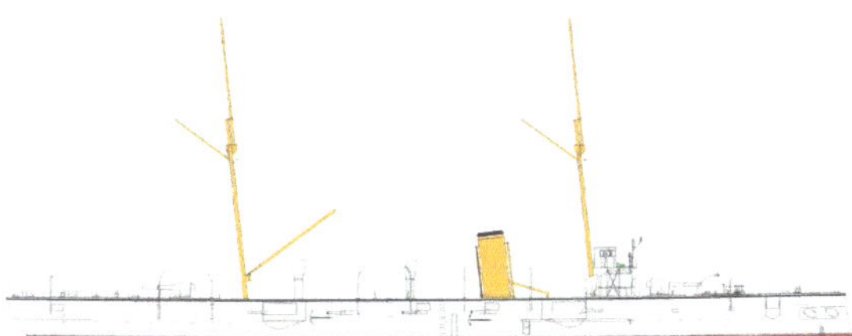

Forth

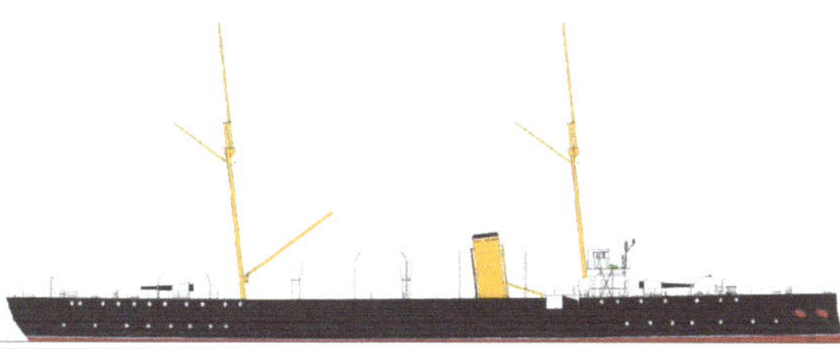

4 | "Leander" type (1880)
Protected cruisers, Britain (continued)

Leander (1880)
2 builder's horizontal direct acting compound EE |
1880 06 14 - 1882 10 28 - 1885 03 29 | Napier, Glasgow

Phaeton (1880)
2 builder's horizontal direct acting compound EE |
1880 06 14 - 1883 02 27 - 1886 04 20 | Napier, Glasgow

4 | "Leander" type (1880)
Protected cruisers, Britain

Displacement: 4,300 tons at draft: 6.25 m | Length: 96.01 m oa | Beam: 14.02 m | Crew: 278 men
Machinery: 12 cylindrical boilers | 2 * 2,750 HP steam EE * 2 screws | 5,500 HP = 16.5 knots |
Supplies: 1,016 tons of coal
Armour: iron |
WLDeck: 0(^38)
Armament: 10-152.4/26, 16-25.4/40 Mg, 4-356 TT (^)

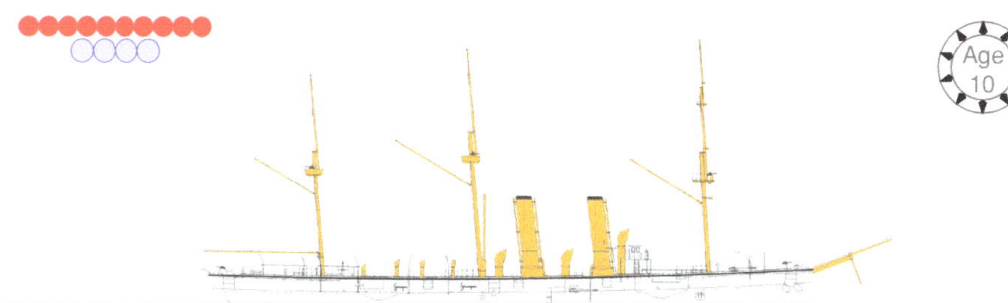

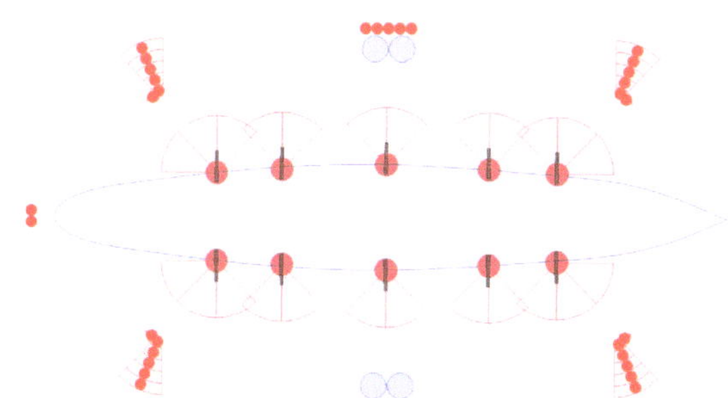

Amphion (1881)
2 Maudslay horizontal direct acting compound EE |
1881 04 25 - 1883 10 13 - 1887 01 01 | Pembroke DYd, Pembroke

Arethusa (1880)
2 builder's horizontal direct acting compound EE |
1880 06 14 - 1882 12 23 - 1886 01 01 | Napier, Glasgow

4 | "Leander" type (1880)
Protected cruisers, Britain (continued)

Phaeton

Amphion

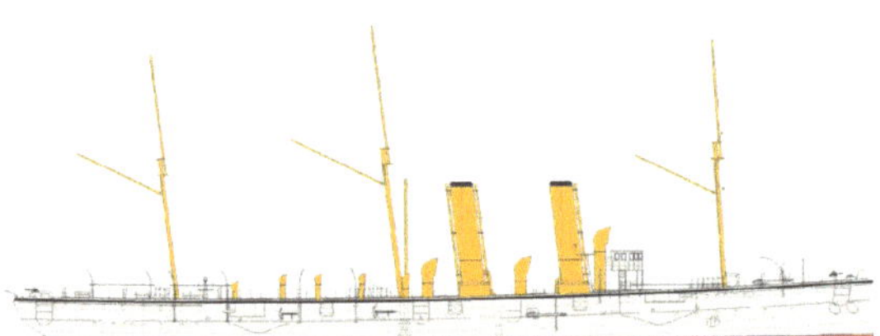

2 | "Calypso" type (1881)
Protected cruisers, Britain (continued)

Calliope

2 | "Calypso" type (1881)
Protected cruisers, Britain

Displacement: 2,770 tons at draft: 6.07 m | Length: 71.62 m pp | Beam: 13.59 m | Crew: 317 men

Machinery: 6 rectangular boilers | 1 steam E * 1 screw | 3,000 HP = 13.75 knots | Supplies: 550 tons of coal

Armour: steel |
WLDeck: 0(^38)

Armament: 4-152.4/26, 12-127/25, 2-93.5/13.6, 10-25.4/50, 2-356 TT (^)

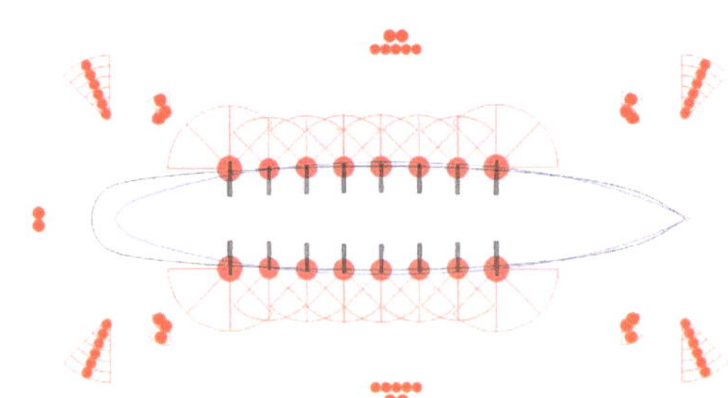

Calliope (1881)
1 Rennie horizontal compound expansion E |
1881 09 01 - 1884 07 24 - 1887 01 25 | Portsmouth DYd, Portsmouth

Calypso (1881)
1 Rennie horizontal compound expansion E |
1881 09 01 - 1883 06 07 - 1885 10 01 | Chatham DYd, Chatham

9 | "Comus" type (1876)
Protected cruisers, Britain

Displacement: 2,380 tons at draft: 5.86 m | Length: 68.58 m pp | Beam: 13.56 m | Crew: 265 men

Machinery: 6 cylindrical boilers | 1 steam E * 1 screw | 2,400 HP = 13 knots |

Supplies: 470 tons of coal

Armour: steel | WLDeck: 0(^38)

Arm.: 4-152.4/26, 8-160/16, 2-93.5/14.2, 8-25.4/40, 2-356 TT

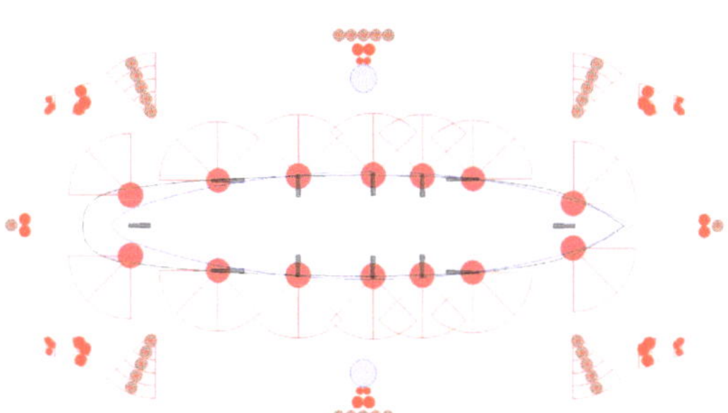

Comus (1876)
1 Elder horizontal compound expansion E | 2,450 HP = 12.93 knots |
1876 08 17 - 1878 04 03 - 1879 10 23 | Elder, Glasgow

9 | "Comus" type (1876)
Protected cruisers, Britain (continued)

Displacement: 2,380 tons at draft: 5.86 m | Length: 68.58 m pp | Beam: 13.56 m | Complement: 265

Machinery: 6 cylindrical boilers | 1 steam E * 1 screw | 2,400 HP = 13 knots |

Supplies: 470 tons of coal

Armour: steel | WLDeck: 0(^38)

Armament: 2-177.8/15.9, 12-160/16

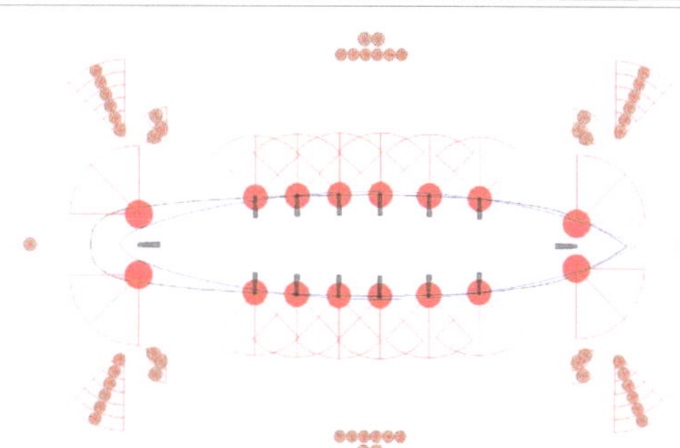

Carysfort (1876)
1 Elder horizontal compound expansion E | 2,403 HP = 12.96 knots |
1876 08 17 - 1878 09 26 - 1880 09 15 | Elder, Glasgow

Cleopatra (1876)
1 Humphrys horizontal compound expansion E | 2,610 HP = 13 kts |
1876 08 17 - 1878 08 01 - 1880 08 24 | Elder, Glasgow

9 | "Comus" type (1876)
Protected cruisers, Britain (continued)

Constance

Cleopatra

Conquest

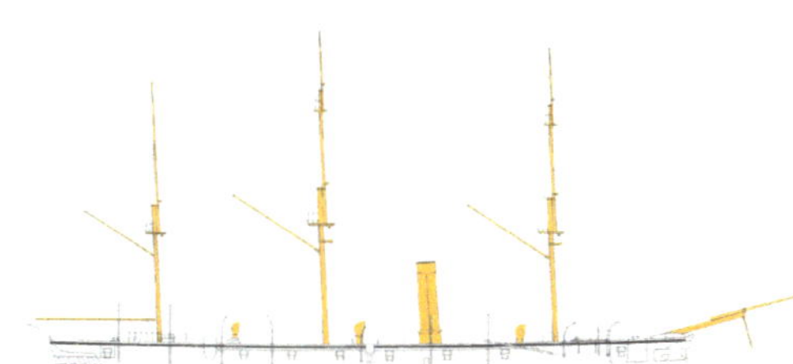

Conquest (1876)
1 Humphrys horizontal compound expansion E | 2,670 HP = 13 kts |
1876 08 17 - 1878 10 28 - 1885 04 18 | Elder, Glasgow

Constance (1878)
1 Penn horizontal compound expansion E | 2,590 HP = 13.7 knots |
1878 09 14 - 1880 06 09 - 1882 10 03 | Chatham DYd

9 | "Comus" type (1876)
Protected cruisers, Britain (continued)

Displacement: 2,380 tons at draft: 5.86 m | Length: 68.58 m pp | Beam: 13.56 m | Crew: 265 men
Machinery: 6 cylindrical boilers | 1 steam E * 1 screw | 2,400 HP = 13 knots |
Supplies: 470 tons of coal
Armour: steel | WLDeck: 0(^38)
Armament: 2-177.8/15.9, 12-160/16

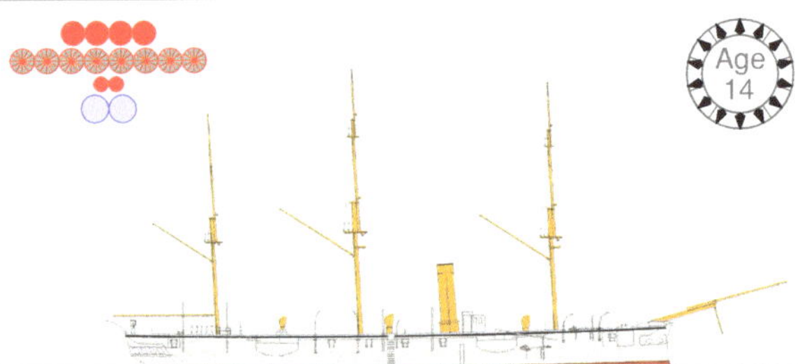

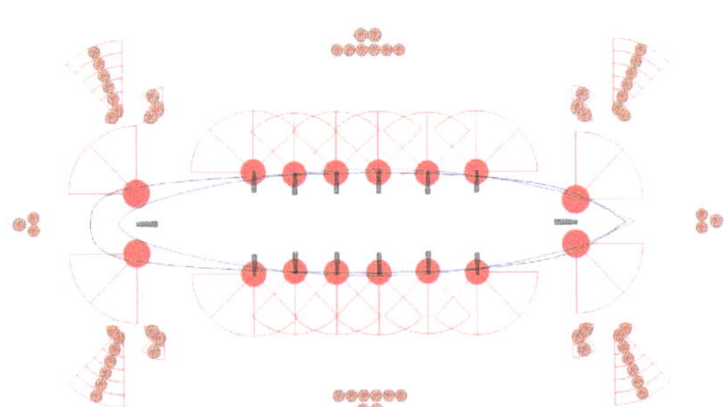

Champion (1876)
1 Elder horizontal compound expansion E | 2,310 HP = 13 knots |
1876 08 17 - 1878 07 01 - 1880 12 07 | Elder, Glasgow

Curacoa (1876)
1 Humphrys horizontal compound expansion E | 2,540 HP = 13 kts |
1876 08 17 - 1878 04 18 - 1880 02 24 | Elder, Glasgow

9 | "Comus" type (1876)
Protected cruisers, Britain (continued)

Champion

Canada

9 | "Comus" type (1876)
Protected cruisers, Britain (continued)

Displacement: 2,380 tons at draft: 5.86 m | Length: 68.58 m pp | Beam: 13.56 m | Crew: 265 men
Machinery: 6 cylindrical boilers | 1 steam E * 1 screw | 2,400 HP = 13 knots |
Supplies: 470 tons of coal
Armour: steel | WLDeck: 0(^38)
Armament: 10-152.4/26, 2-93.5/13.5, 10-25.4/40, 2-356 TT (^)

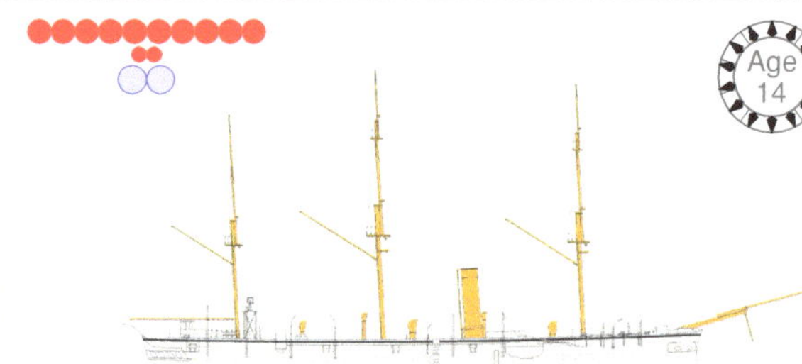

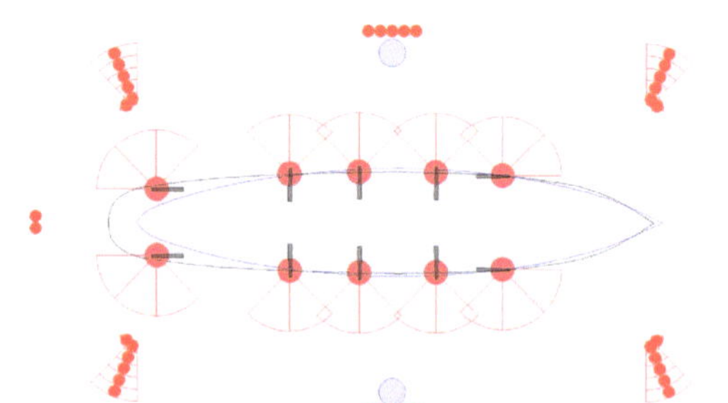

Canada (1879)
1 Rennie horizontal compound expansion E | 2,430 HP = 13 knots |
1879 07 07 - 1881 08 26 - 1883 05 01 | Portsmouth DYd

Cordelia (1879)
1 Rennie horizontal compound expansion E | 2,420 HP = 13 knots |
1879 07 17 - 1881 10 25 - 1887 01 25 | Portsmouth DYd

0 +4 | "Phoenix"/"Pallas" type (1888)
Protected cruisers 3 class, Australia

BUILDING

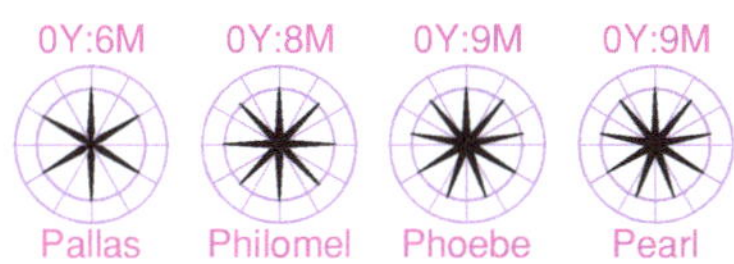

Displacement: 2,575 tons at draft: 4.72 m | Length: 84.73 m | Beam: 12.5 m | Crew: 217 men

Machinery: 4 cylindrical boilers | 2 * 2,250 HPn / 3,750 HPf steam EE * 2 screws | 4,500 HPn = 17.5 knots / 7,500 HPf = 19 knots | Supplies: 300 / +142 tons of coal

Armour: steel | CT: 76(^25); Gun shields: 51 |
WLDeck: 0(/51^25)

Armament: 8-120/40, 8-47/43, 2-25.4/40 Mg, 2-356 TT (^)

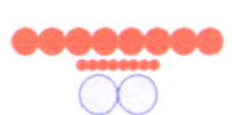

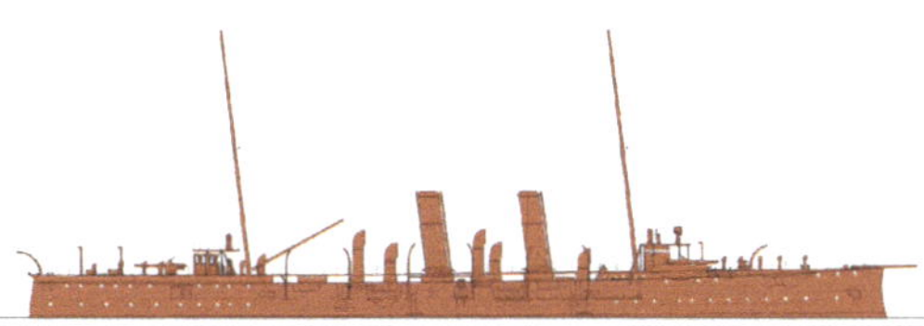

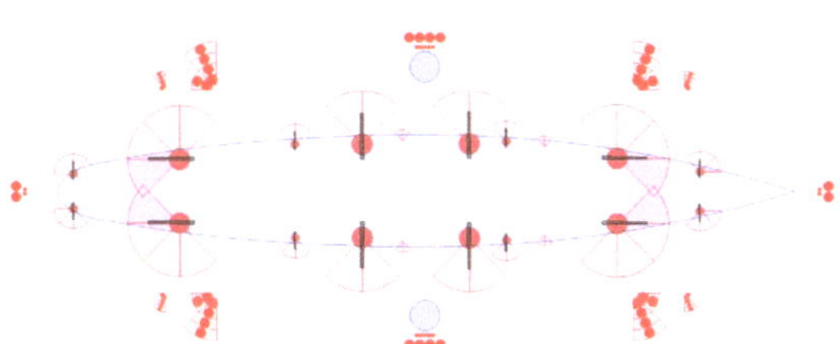

Pallas (1889)
2 Leslie vertical triple expansion EE |
1889 07 01 - ***** - ***** | Portsmouth DYd, Portsmouth

Pearl (1889)
2 Earle vertical triple expansion EE |
1889 04 01 - ***** - ***** | Pembroke DYd, Pembroke

0 +4 | "Phoenix" type (1888)
Protected cruisers 3 class, Australia (continued)

Philomel (1889)
2 Earle vertical triple expansion EE |
1889 05 08 - ***** - ***** | Devonport DYd, Devonport

Phoebe (1889)
2 builder's? vertical triple expansion EE |
1889 04 23 - ****** - ***** | Devonport DYd, Devonport

5 more cruisers are built for the Australian navy and can be found in the Australian section

0 +4 | "Barracuda" type (1888)
Protected cruisers 3 class, Britain (continued)

Blanche (1888)
2 Earle vertical triple expansion EE |
1888 05 01 - 1889 09 06 - ***** | Pembroke DYd, Pembroke

Blonde (1888)
2 Earle vertical triple expansion EE |
1888 05 01 - 1889 10 22 - ***** | Pembroke DYd, Pembroke

0 +4 | "Barracuda" type (1888)
Protected cruisers 3 class, Britain

BUILDING

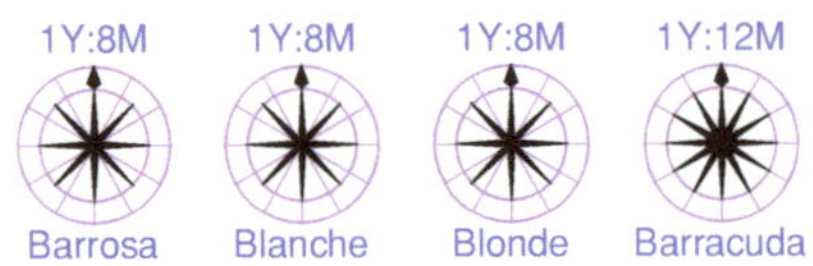

Displacement: 1,580 tons at draft: 4.6 m | Length: 85.34 m pp | Beam: 10.67 m | Crew: 160 men

Machinery: 4 cylindrical boilers | 2 * 1,375 HPn / 1,500 HPf steam EE * 2 screws | 1,750 HPn = 15 knots / 3,000 HPf = 16.5 knots | Supplies: 160 tons of coal

Armour: steel | CT: 51(^25); Gun shields: 51 |
WLDeck: 0(/51^25)

Armament: 6-120/40, 4-47/43, 2-356 TT (^)

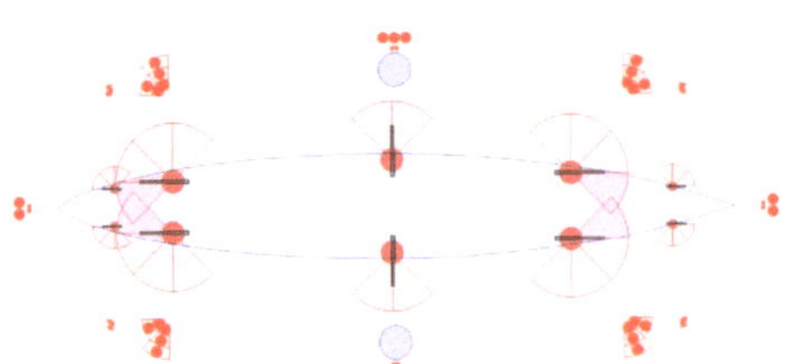

Barracuda (1888)
2 Palmer vertical triple expansion EE |
1888 01 01 - 1889 05 16 - ***** | Sheerness DYd, Sheerness

Barrosa (1888)
2 Palmer vertical triple expansion EE |
1888 05 01 - 1889 04 16 - ***** | Portsmouth DYd, Portsmouth

0 +2 | "Barham" type (1888)
Protected cruisers 3 class, Britain

BUILDING

Displacement: 1,830 tons at draft: 4.04 m | Length: 85.34 m pp | Beam: 10.67 m | Crew: 170 men

Machinery: 6 locomotive boilers | 2 * 1,800 HPn / 3,000 HP steam EE * 2 screws | 3,600 HPn = 16.5 knots / 6,000 HPf = 19.5 knots | Supplies: 140 tons of coal

Armour: steel | CT: 51(^25); Gun shields: 51 |
WLDeck: 0(/51^25)

Armament: 6-120/40, 4-47/43, 2-356 TT (^)

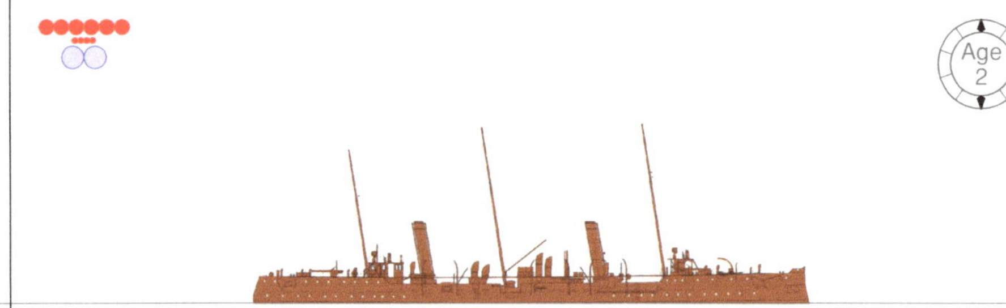

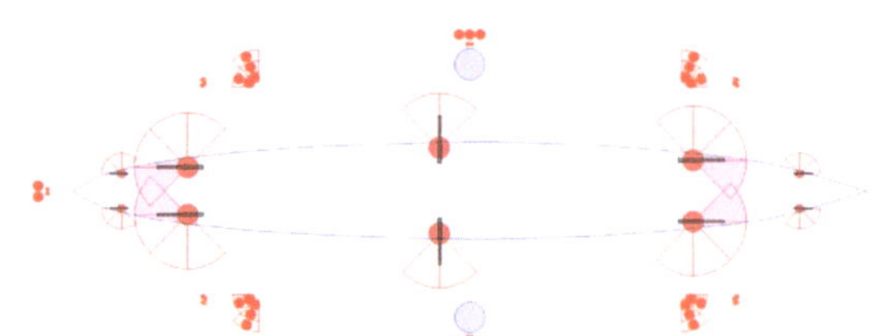

Barham (1888)
2 Leslie vertical triple expansion EE |
1888 10 22 - 1889 09 11 - ***** | Portsmouth DYd, Portsmouth

Bellona (1888)
2 builder's vertical triple expansion EE |
1888 10 16 - ***** - ***** | Leslie, Hebburn

6 +2 | "Archer" type (1885)
Protected cruisers 3 class, Britain

BUILDING

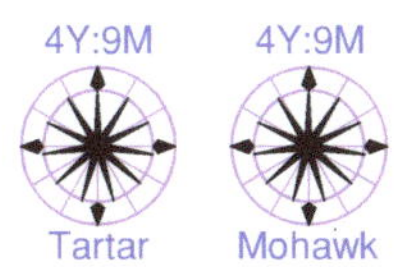

Displacement: 1,770 tons at draft: 4.42 m, 1950 tons full | Length: 73.15 m oa | Beam: 10.97 m | Crew: 176 men

Machinery: 4 boilres | 2 * 1,250 HPn / 1,750 HPf steam EE * 2 screws | 2,500 HPn = 15 knots / 3,500 HPf = 16.5 knots | Supplies: 475 tons of coal

Armour: steel | CT: 76 | WLDeck: 0(^15)

Armament: 6-152.4/26, 8-47/43, 2-25.4/40, 5-356 TT (^)

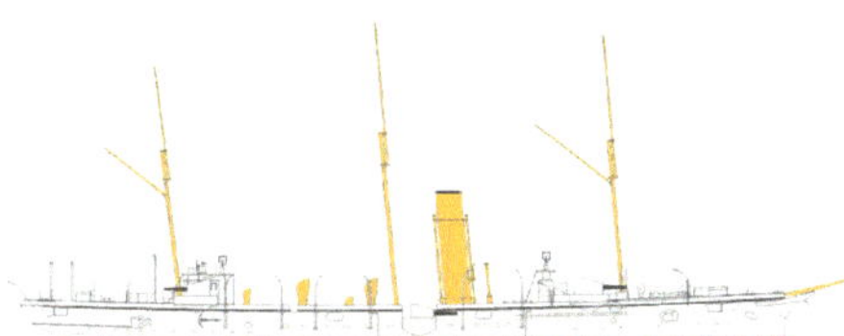

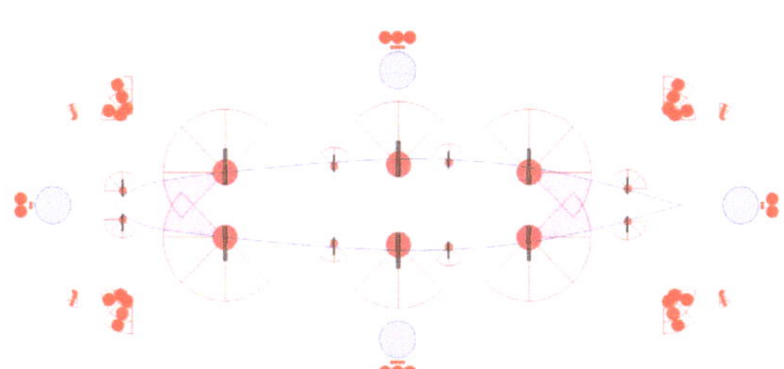

Archer (1885)
2 builder's horizontal direct acting compound EE |
1885 03 02 - 1885 12 23 - 1888 12 11 | Thompson, Clydebank

Brisk (1885)
2 builder's horizontal direct acting compound EE |
1885 03 02 - 1886 04 08 - 1888 03 20 | Thompson, Clydebank

6 +2 | "Archer" type (1885)
Protected cruisers 3 class, Britain (continued)

Cossack (1885)
2 builder's horizontal direct acting compound EE |
1885 03 02 - 1886 06 03 - 1889 01 01 | Thompson, Clydebank

Mohawk (1885)
2 builder's horizontal direct acting compound EE |
1885 03 02 - 1886 02 06 - ***** | Thompson, Clydebank

Porpoise (1885)
2 builder's horizontal direct acting compound EE |
1885 03 02 - 1886 05 07 - 1888 02 12 | Thompson, Clydebank

Racoon (1886)
2 Harland horizontal direct acting compound EE | 4,500 HPf = 17.5 knots |
1886 02 01 - 1887 05 06 - 1888 03 01 | Devonport DYd, Devonport

Serpent (1885)
2 Harland horizontal direct acting compound EE | 4,500 HPf = 17.5 knots
1885 11 09 - 1887 03 10 - 1888 03 01 | Devonport DYd, Devonport

Tartar (1885)
2 builder's horizontal direct acting compound EE |
1885 03 02 - 1886 10 28 - ***** | Thompson, Clydebank

Serpent

UNPROTECTED CRUISERS

2 | "Scout" type (1884)
Protected cruisers 3 class, Britain

Displacement: 1,580 tons at draft: 4.42 m | Length: 68.58 m. wl | Beam: 10.44 m | Crew: 147 men

Machinery: 4 cylindrical boilers | 2 * 1,000 HPn / 1,600 HPf steam EE * 2 screws | 2,000 HPn = 16 knots / 3,200 HPf = 17 knots | Supplies: 450 tonns of coal

Armour: steel | CT: 76 |

WLDeck: 0(^15)

Armament: 4-127/25, 8-47/43, 2-25/40 Mg, 3-356 TT (aw)

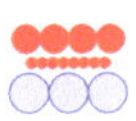

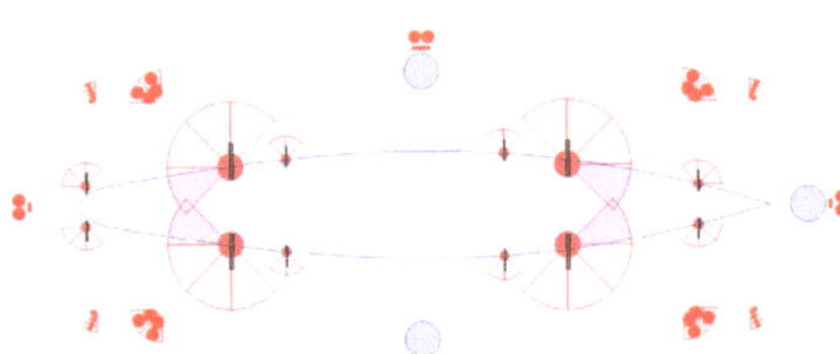

Fearless (1884)
2 Barrow horizontal direct acting compound EE |
1884 09 22 - 1886 03 20 - 1887 07 01 | Naval Construction & Armament, Barrow

Scout (1884)
2 builder's horizontal direct acting compound EE |
1884 01 08 - 1885 07 30 - 1885 08 20 | Thompson, Clydebank

1 | "Raleigh" type (1871)
Unprotected cruiser 1 class, Britain

Displacement: 5,200 tons at draft: 7.48 m | Length: 90.83 m pp | Beam: 14.93 m | Crew: 340 men

Machinery: 10 rectangular boilers | 1 steam E * 1 screw | 5,600 HP = 16.2 knots | Supplies: 890 tons of coal

Armour: nil

Armament: 2-228.6/13.9 MLR, 14-177.8/15.9 MLR, 6-160/15.5 MLR

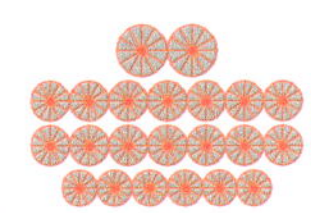

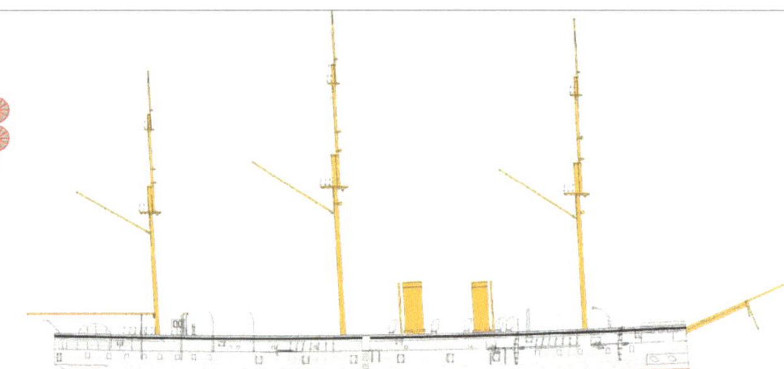

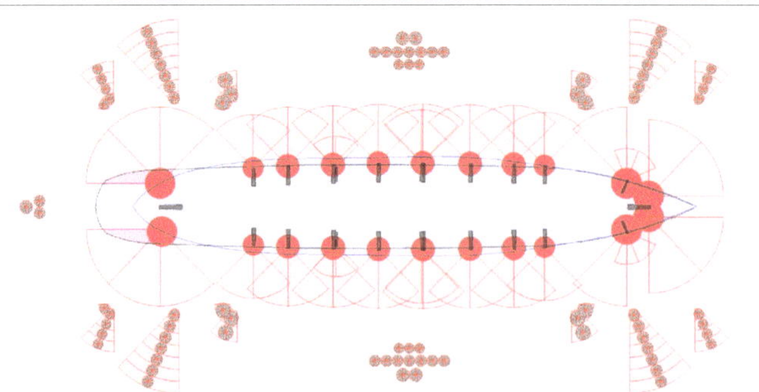

Raleigh (1871)
1 Humphrys horizontal return connecting rod EE |
1871 02 08 - 1873 03 01 - 1874 06 01 | Chatham DYd, Chatham

2 | "Inconstant" type (1866)
Unprotected cruisers 1 class, Britain

Displacement: 6,250 tons at draft: 8.06 m | Length: 101.8 m pp | Beam: 15.84 m | Crew: 340 men

Machinery: 10 rectangular boilers | 1 steam E * 1 screw | 7,400 HP = 16 knots | Supplies: 890 tons of coal

Armour: nil

Armament: 2-228.6/13.9 MLR, 16-177.8/15.9 MLR, 8-160/15.5 MLR, 4-95.3/21, 12-25.4/40 Mg

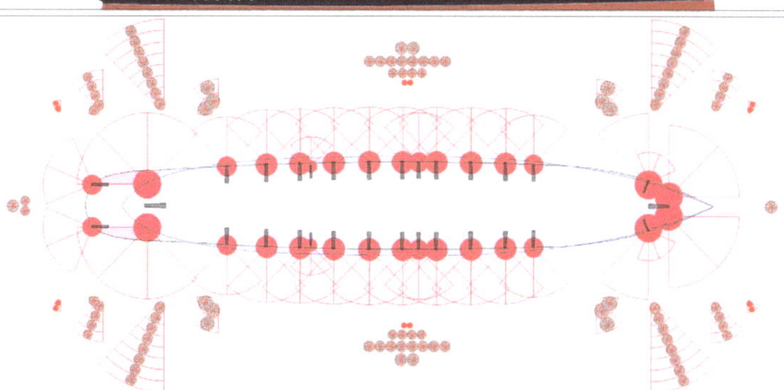

Shah (1870)

1 Penn horizontal single expansion trunk E | 7,480 HP = 16,2 knots |
1870 03 07 - 1873 09 01 - 1875 12 01 | Portsmouth DYd, Portsmouth

2 | "Inconstant" type (1866)
Unprotected cruisers 1 class, Britain (continued)

Displacement: 5,780 tons at draft: 7.77 m | Length: 102.81 m pp | Beam: 15.31 m | Crew: 340 men

Machinery: 10 rectangular boilers | 1 steam E * 1 screw | 7,400 HP = 16 knots | Supplies: 890 tons of coal

Armour: nil

Armament: 10-228.6/13.9 MLR, 6-177.8/15.9 MLR

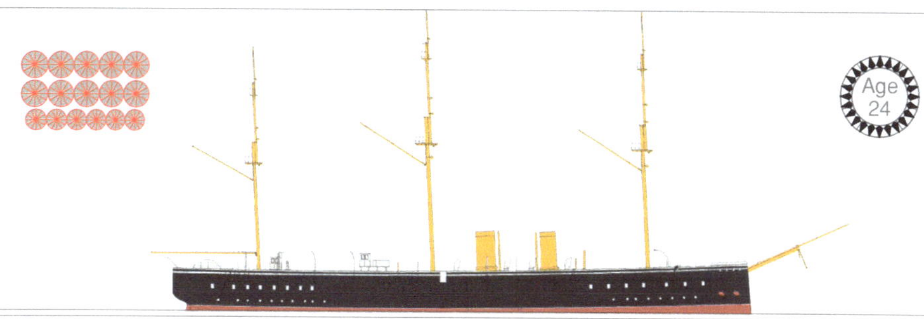

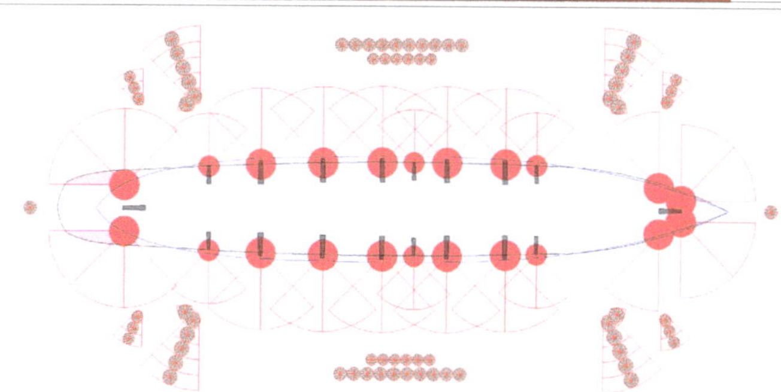

Inconstant (1866)

1 Penn horizontal single expansion trunk E | 7,360 HP = 16,2 knots |
1866 11 01 - 1868 11 12 - 1869 08 14 | Pembroke DYd, Pembroke

2 | "Iris" type (1875)
Unprotected cruisers, Britain

Displacement: 3,730 tons at draft: 6.71 m | Length: 101.45 m oa | Beam: 14.02 m | Crew: 275 men

Machinery: 8 oval boilers, 4 cylindrical boilers | 2 * 3,000 HP steam EE * 2 screws | 6,000 HP = 17 knots | Supplies: 780 tons of coal

Armour: nil

Rearmed (1886): 13-127/26, 4-47/43, 4-356 TT (^)

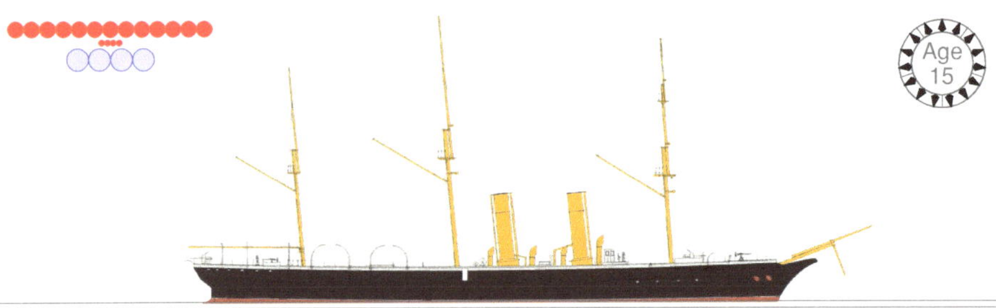

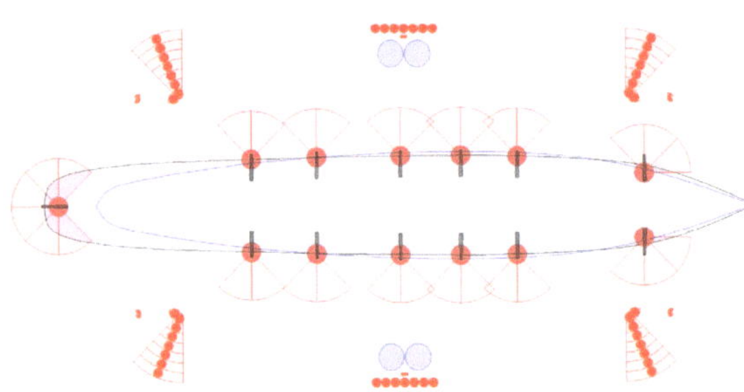

Iris (1875)
2 Maudslay horizontal direct acting compound EE | 7,330 HP = 17.89 knots |
1875 11 10 - 1877 04 12 - 1879 04 01 | Pembroke DYd, Pembroke

Mercury (1876) | Length: 96.01 m oa |
2 Maudslay horizontal direct acting compound EE | 7,735 HP = 18.75 knots |
1876 03 16 - 1878 04 17 - 1879 09 01 | Pembroke DYd, Pembroke

2 | "Iris" type (1875)
Unprotected cruisers, Britain (continued)

Mercury

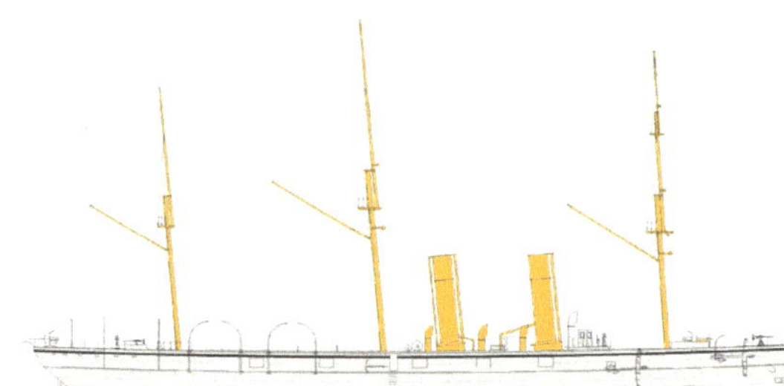

3 | "Boadicea" type (1873)
Unprotected cruisers, Britain (continued)

Euryalus

3 | "Boadicea" type (1873)
Unprotected cruisers, Britain

Displacement: 4,000 tons at draft: 7.2 m | Length: 85.34 m pp | Beam: 13.86 m | Crew: 375 men
Machinery: 10 cylindrical boilers | 1 steam E * 1 screw | 5,000 HP = 15 knots | Supplies: 570 tons of coal
Armour: nil
Armament: 14-177.8/15.9 MLR; 2-160/16 MLR

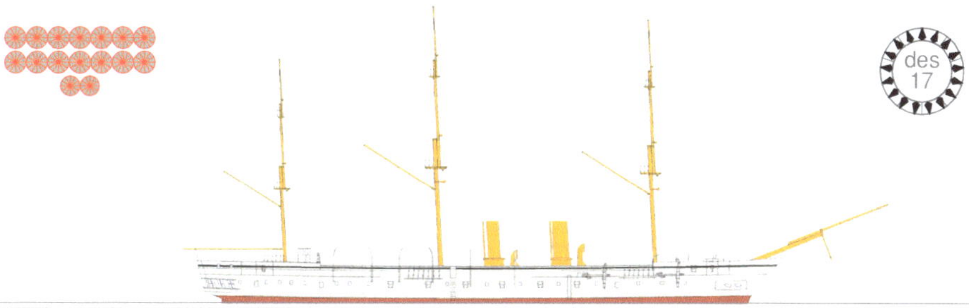

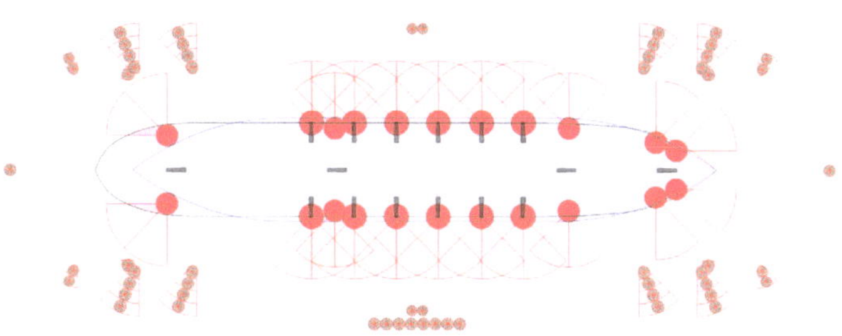

Bacchante (1873) | 4,070 tons
1 Rennie horizontal compound expansion E | 5,420 HP = 15.06 knots |
1873 03 15 - 1876 10 19 - 1879 07 01 | Portsmouth DYd, Portsmouth

Boadicea (1873) | 3,913 tons
1 Rennie horizontal compound expansion E | 5,130 HP = 14,7 knots |
1873 01 30 - 1875 10 16 - 1877 05 01 | Portsmouth DYd, Portsmouth

3 | "Boadicea" type (1873)
Unprotected cruisers, Britain (continued)

Bacchante

Armament: 14-177.8/15.9 MLR; 2-356 TT

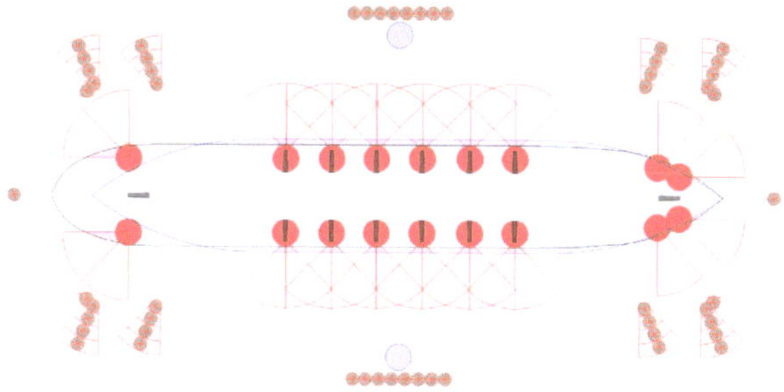

Euryalus (1873) | 3,932 tons
1 Ravenhill horizontal compound expansion E | 5,110 HP = 14.72 knots |
1873 11 01 - 1877 01 31 - 1878 06 01 | Chatham DYd, Chatham

1 | "Rover" type (1872)
Unprotected cruiser, Britain

Displacement: 3,460 tons at draft: 7.01 m | Length: 85.34 m pp | Beam: 13.25 m | Crew: 340 men

Machinery: 10 rectangular boilers | 1 steam E * 1 screw | 4,500 HP = 14.5 knots | Supplies: 410 tons of coal

Armour: nil

Armament: 2-177.8/15.9 MLR, 16-160/15.5 MLR

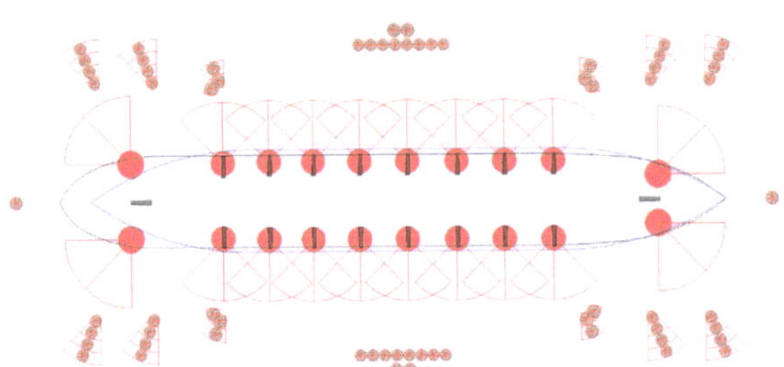

Rover (1872)
1 Ravenhill horizontal compound expansion E | 4,960 HP = 14.53 knots |
1872 01 01 - 1874 08 12 - 1875 09 21 | Thames IW, Blackwall

2 | "Active" (1867) type
Unprotected cruisers, Britain

Displacement: 3,080 tons at draft: 6.7 m | Length: 82.29 m pp | Beam: 12.80 m | Crew: 340 men

Machinery: 5 rectangular boilers | 1 steam E * 1 screw | 4,500 HP = 14.5 knots | Supplies: 410 tons of coal

Armour: nil

Armament: 6-177.8/15.9 MLR, 4-160/15.5 MLR

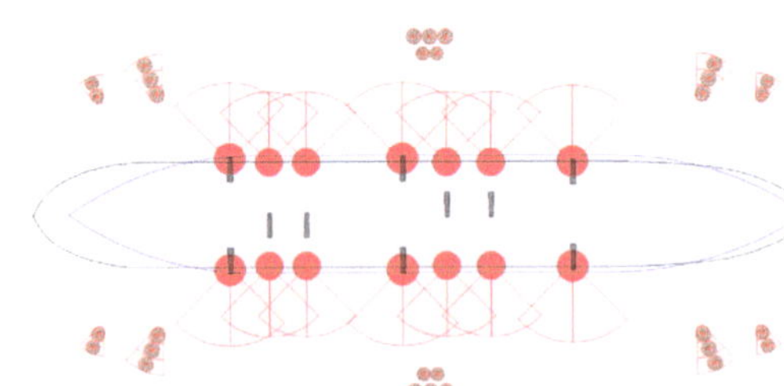

Active (1867)
1 Humphrys horizontal single expansion E | 4,130 HP = 15 knots |
1867 09 01 - 1869 03 13 - 1871 03 01 | Thames IW, Blackwall

Volage (1867)
1 Penn horizontal single expansion E | 4,530 HP = 15.3 knots |
1867 09 01 - 1869 02 27 - 1870 03 01 | Thames IW, Blackwall

2 | "Alacrity" type (1884)
Unprotected cruisers, Britain

Displacement: 1,650 tons at draft 3.96 m | Length: 76.2 m pp | Beam: 9.91 m | Crew: 93 men

Machinery: 4 cylindrical boilers | 2 * 1,000 HPn / 1,500 HPf steam EE * 2 screws | 2,000 HPn = 16 knots / 3,000 HPf = 17 knots | Supplies: 400 tons of coal

Armament: 4-127/25, 4-57/43

Age 6

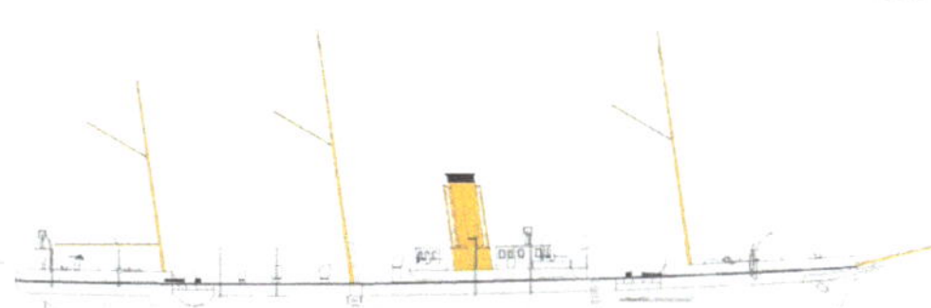

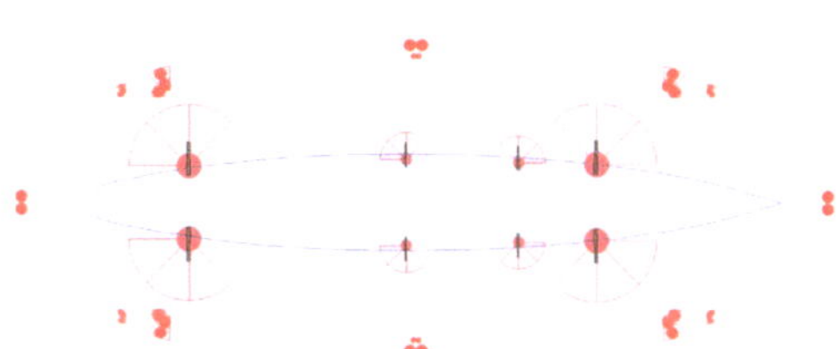

Alacrity (1884)
2 builder's HDA compound EE | 3,180 HPf = 18 knots |
1884 02 14 - 1885 02 28 - 1886 06 22 | Palmer, Jarrow

Surprise (1884)
2 builder's HDA compound EE | 3,018 HPf = 17.85 knots |
1884 02 14 - 1885 01 17 - 1886 07 20 | Palmer, Jarrow

2 | "Alacrity" type (1884)
Unprotected cruisers, Britain (continued)

Surprise

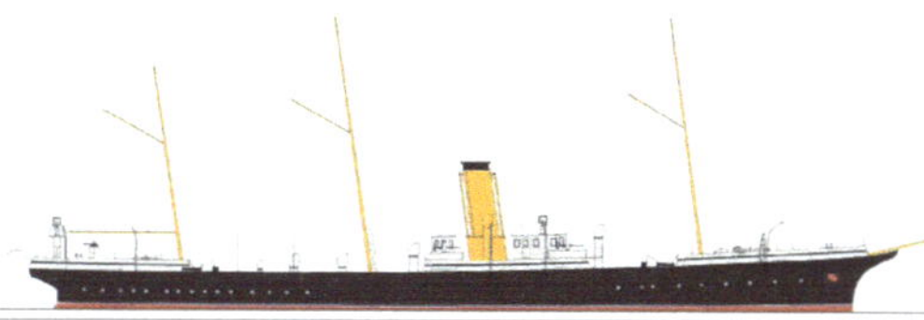

HDAC = horizontal direct acting compound engine

WOODEN CRUISERS

6 | "Emerald" type (1874)
Wooden screw corvettes, Britain

Displacement: 2,120 tons at draft: 5.48 m | Length: 67.05 m pp | Beam: 12.19 m | Crew: 230 men

Machinery: 6 cylindrical boilers | 1 steam E * 1 screw | 2,000 HP = 13 knots |
Supplies: 260 tons of coal

Armament: 12-160/16 MLR

Age 17

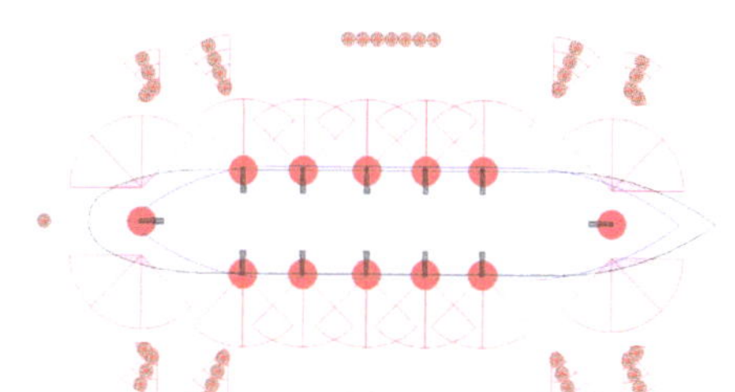

Emerald (1874)
1 Thompson horizontal compound exp. E | 2,170 HP = 13.9 knots
1874 07 29 - 1876 08 18 - 1878 07 02 | Pembroke DYd

Garnet (1875)
1 Leslie horizontal compound expansion E | 2,000 HP = 13.2 knots
1875 03 16 - 1877 06 30 - 1878 10 31 | Chatham DYd

6 | "Emerald" type (1874)
Wooden screw corvettes, Britain (continued)

Opal (1873)
1 Napier horizontal compound expansion E | 2,100 HP = 12.5 knots
1873 10 13 - 1875 03 09 - 1876 01 01 | Doxford, Sunderland

Ruby (1874)
1 Earle horizontal compound expansion E | 2,018 HP = 12.3 knots
1874 07 08 - 1876 08 09 - 1877 06 14 | Earle, Hull

Tourmaline (1874)
1 Leslie horizontal compound expansion E | 1,972 HP = 12.6 knots
1874 07 17 - 1875 10 30 - 1876 10 25 | Raylton Dixon, Tees

Turquoise (1874)
1 Earle horizontal compound expansion E | 1,994 HP = 12.3 knots
1874 07 08 - 1876 04 22 - 1877 09 13 | Earle, Hull

Opal

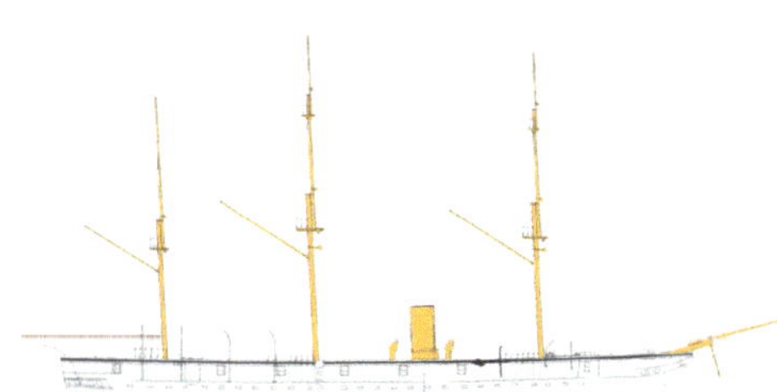

1 | "Amethyst" type (1871)
Wooden screw corvette, Britain

Displacement: 1,970 tons at draft: 5.48 m | Length: 67.05 m oa | Beam: 11.27 m | Crew: 225 men

Machinery: 6 cylindrical boilers | 1 steam E * 1 screw | 2,200 HP = 13 knots | Supplies: 230 tons of coal

Armament: 14-160/16 MLR

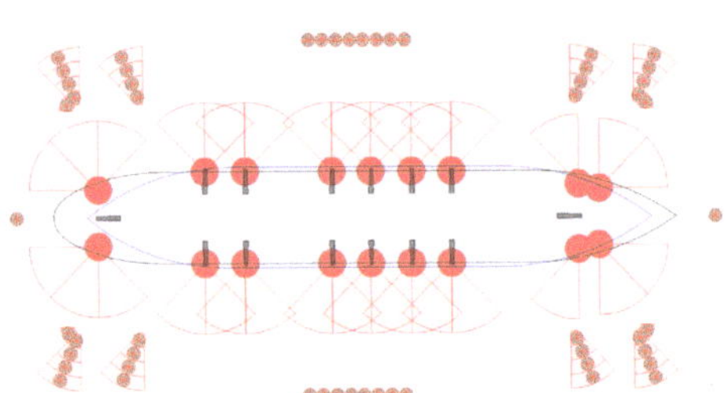

Sapphire (1873)
1 Leslie horizontal compoud expansion E | 2,364 HP = 13.58 knots
1873 06 17 - 1874 09 24 - 1875 08 01 | Devonport DYd

1 | "Juno" type (1866)
Wooden screw corvette, Britain

Displacement: 2,240 tons at draft: 5.05 m | Length: 60.96 m oa | Beam: 12.92 m | Crew: 200 men

Machinery: 4 cylindrical boilers | 1 steam E * 1 screw | 1,500 HP = 11 knots | Supplies: 270 tons of coal

Armament (1886): disarmed

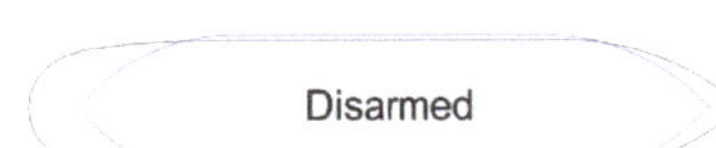

Thalia (1866)
1 Napier horizontal single expansion E | 1,597 HP = 11.13 knots
1866 01 01 - 1869 07 14 - 1870 03 01 | Woolwich DYd

2 | "Eclipse" type (1866)
Wooden screw corvettes, Britain

Displacement: 1,755 tons at draft: 4.97 m | Length: 64.61 m oa | Beam: 10.97 m | Crew: 180 men

Machinery: 4 cylindrical boilers | 1 steam E * 1 screw | 1,500 HP = 12 knots | Supplies: 259 tons of coal

Armament: 2-177.8/15.9 MLR, 4-160/16 MLR

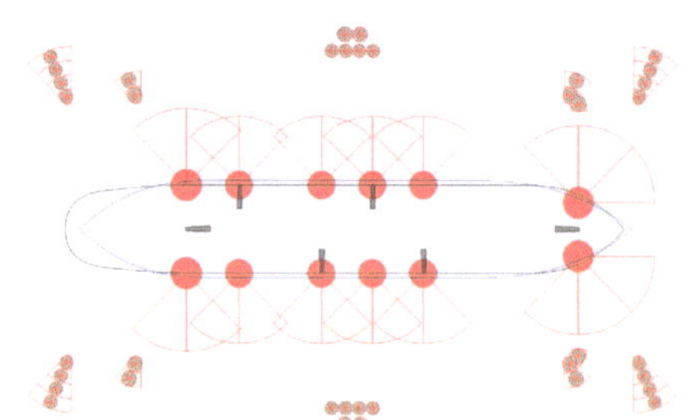

Danae (1866)
1 Napier horizontal single expansion E | 2,089 HP = 13.17 knots |
1866 01 01 - 1867 05 21 - 1868 11 01 | Portsmouth DYd

Eclipse (1866)
1 Penn horizontal single expansion E | 1,594 HP = 12.24 knots |
1866 01 01 - 1867 11 14 - 1868 06 01 | Sheerness DYd

TORPEDO SHIPS

0 +1 | "Antelope" type (1889)
Torpedo gunboat, Britain

BUILDING

0Y:3M
Antelope

Displacement: 810 tons at draft: 3.66 m | Length: 70.1 m pp | Beam: 8,23 m | Crew: 91 men

Machinery: 4 locomotive boilers | 2 * 1,500 HPn / 1,750 HPf steam EE * 2 screws | 2,500 HPn = 16.5 knots / 3,500 HPf = 18.7 knots | Supplies: 100 tons of coal

Armament: 2-120/40, 4-47/43, 1-11.5 Mg, 5-356 TT (2*2 trn, 1 >)

Age 1

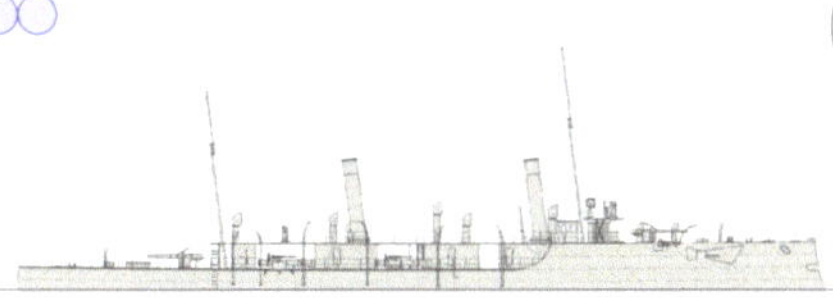

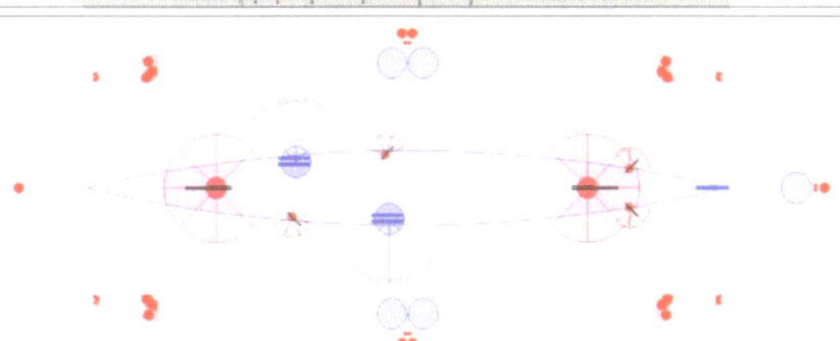

Antelope (1889)
2 Yarrow vertical triple expansion EE |
1889 10 21 - ***** - ***** | Devonport DYd

1 +8 | "Sharpshooter" type (1888)
Torpedo gunboats, Britain

BUILDING

0Y:12M	0Y:12M	1Y:6M	1Y:6M
Gleaner	Gossamer	Sheldrake	Skipjack
1Y:9M	1Y:9M	1Y:9M	1Y:9M
Seagull	Salamander	Speedwell	Spanker

Displacement: 735 tons at draft: 3.2 m | Length: 70.1 m pp | Beam: 8.23 m | Crew: 91 men

Machinery: 4 locomotive boilers | 2 * 1,500 HPn / 1,750 HPf steam EE * 2 screws | 2,500 HPn = 16.5 knots / 3,500 HPf = 19 knots | Supplies: 100 tons of coal

Armament: 2-120/30, 4-47/43, 1-11.5 Mg, 5-356 TT (2*2 trn, 1 >)

Age 2

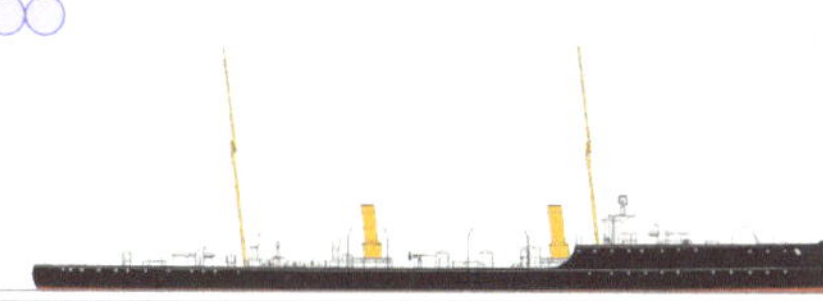

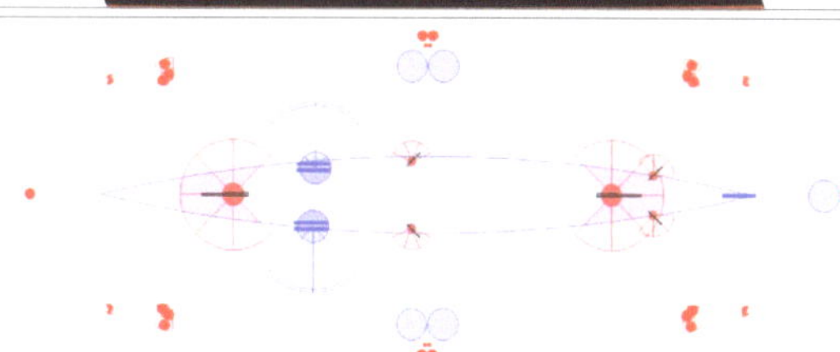

Gleaner (1889)
2 builder's vertical triple expansion EE |
1889 01 21 - ***** - ***** | Sheerness DYd

Gossamer (1889)
2 builder's vertical triple expansion EE |
1889 01 21 - ***** - ***** | Sheerness DYd

Salamander (1888)
2 Maudslay vertical triple expansion EE |
1888 04 23 - 1889 05 31 - ***** | Chatham DYd

Seagull (1888)
2 Maudslay vertical triple expansion EE |
1888 04 23 - 1889 05 31 - ***** | Chatham DYd

Sharpshooter (1888)
2 Belliss vertical triple expansion EE |
1888 01 13 - 1888 11 30 - 1889 08 01 | Devonport DYd

1 +8 | "Sharpshooter" type (1888)
Torpedo gunboats, Britain (continued)

Unidentified

Sheldrake (1888)
2 Madden vertical triple expansion EE |
1888 07 04 - 1889 03 30 - ***** | Chatham DYd

Skipjack (1888)
2 Laird vertical triple expansion EE |
1888 07 04 - 1889 04 30 - ***** | Chatham DYd

Spanker (1888)
2 Maudslay vertical triple expansion EE |
1888 04 12 - 1889 02 27 - ***** | Devonport DYd

Speedwell (1888)
2 Maudslay vertical triple expansion EE |
1888 04 18 - 1889 03 15 - ***** | Devonport DYd

Sister ships can be found in the Australian and Indian sections

3 | "Grasshopper" type (1886)
Torpedo gunboats, Britain

Displacement: 526 tons at draft: 3.15 m | Length: 60.96 m. pp | Beam: 7.01 m | Crew: 66 men

Machinery: 2 locomotive boilers | 2 * 800 HPn / 1,350 HPf steam EE * 2 screws | 1,600 HPn = 16.5 knots / 2,700 HPf = 19.5 knots | Supplies: 100 tons of coal

Armament: 1-127/25, 6-47/43, 4-356 TT (1 <, 2*1 trainable, 1 >)

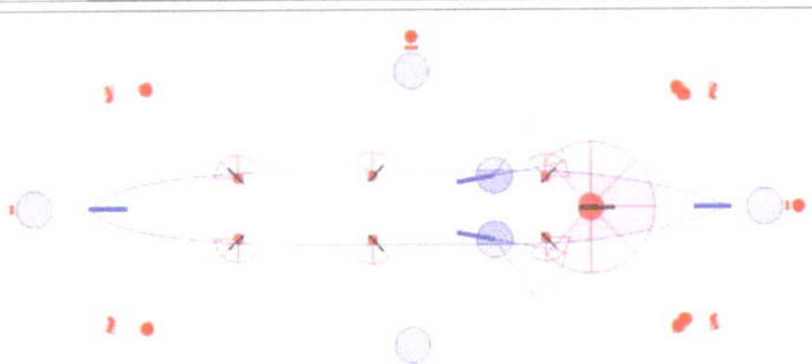

Grasshopper (1886)
2 Maudslay vertical triple expansion EE |
1886 04 27 - 1887 08 30 - 1888 01 01 | Sheerness DYd

Sandfly (1886)
2 Maudslay vertical triple expansion EE |
1886 04 19 - 1887 09 20 - 1888 07 01 | Devonport DYd

Spider (1886)
2 Maudslay vertical triple expansion EE |
1886 06 09 - 1887 10 17 - 1888 12 01 | Devonport DYd

1 | "Rattlesnake" type (1885)
Torpedo gunboat, Britain

Displacement: 550 tons at draft: 3.15 m | Length: 60.96 m pp | Beam: 7.01 m | Crew: 66 men

Machinery: 2 locomotive boilers | 2 * 800 HPn / 1,350 HPf steam EE * 2 screws | 1,600 HP = 16.5 knots / 2,700 HPf = 19.5 knots | Supplies: 100 tons of coal

Armament: 1-127/25, 6-47/43, 4-356 TT (1 <, 2*1 trainable, 1 >)

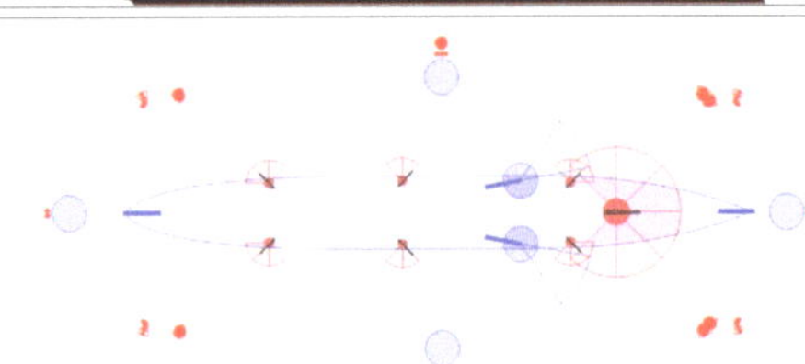

Rattlesnake (1885)
2 Laird vertical triple expansion EE |
1885 11 16 - 1886 09 11 - 1887 05 01 | Laird, Birkenhead

1 | "Polyphemus" type (1878)
Torpedo ram, Britain

Displacement: 2,640 tons at draft: 6.25 m | Length: 73.15 m pp | Beam: 12.19 m | Crew: 146 men

Machinery: 10 locomotive boilers | 2 * 2,750 HPn / 3,500 HPf steam EE * 2 screws | 5,500 HP = 16 knots / 7,000 HPf = 18 knots | Supplies: 270 tons of coal

Armament: 6-25,4/40 Mg, 5-356 TT (2*2 ||, 1 >)

Polyphemus (1878)
2 Humphrys compound expansion EE |
1878 09 21 - 1881 06 15 - 1882 09 01 | Chatham DYd

1 | "Vesuvius" type (1873)
Torpedo vessel, Britain

Displacement: 245 tons at draft: 2.59 m | Length: 27.4 m pp | Beam: 6.71 m | Crew: 15 men

Machinery: 2 cylindrical boilers | 2 * 175 HP compound expansion EE * 2 screws | 350 HP = 9.5 knots | Supplies: 25 tons of coal

Armament: 1-406,4 TT (>)

Age 17

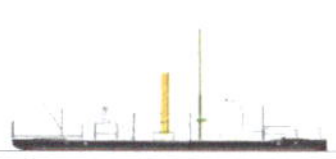

Vesuvius (1873)
1873 03 16 - 1874 03 24 - 1874 09 11 | Pembroke DYd

6 | "82" type (1887)
Torpedo boats 1 class, Britain

Displacement: 85 tons at draft: 1.07 m | Length: 40 m oa, 39.67 m pp | Beam: 4.14 m | Crew: 19 men

Machinery: 2 Yarrow boilers | 1 vertical compound expansion E * 1 screw | 1,800 HP = 23.5 knots | Supplies: 26.5 tons of coal

Armament: 3-47/43, 3-356 TT (1*2 trainable, 1 >)

Age 3

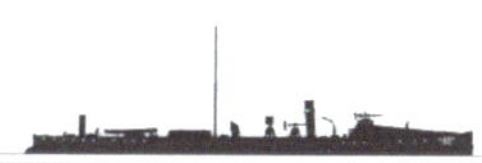

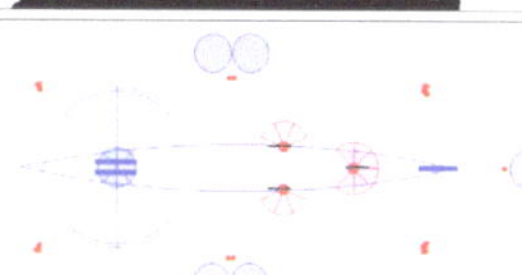

82 (1887)
1887 01 01 - 1887 06 06 - 1888 01 01 | Yarrow, Poplar

83 (1887)
1887 01 01 - 1887 06 06 - 1888 01 01 | Yarrow, Poplar

84 (1887)
1887 01 01 - 1887 06 06 - 1888 01 01 | Yarrow, Poplar

85 (1887)
1887 01 01 - 1887 06 06 - 1888 01 01 | Yarrow, Poplar

86 (1887)
1887 01 01 - 1887 06 06 - 1888 01 01 | Yarrow, Poplar

87 (1887)
1887 01 01 - 1887 06 06 - 1888 01 01 | Yarrow, Poplar

1 | "80" type (1886)
Torpedo boat 1 class, Britain

Displacement: 105 tons at draft: 1.63 m | Length: 41.15 m pp | Beam: 4.27 m | Crew: 21 men

Machinery: 1 locomotive boiler | 1 vertical compound expansion E * 1 screw | 1,500 HP = 23 knots | Supplies: 32 tons of coal

Armament: 4-47/43, 3-356 TT (1*2 trainable, 1 >)

Age 4

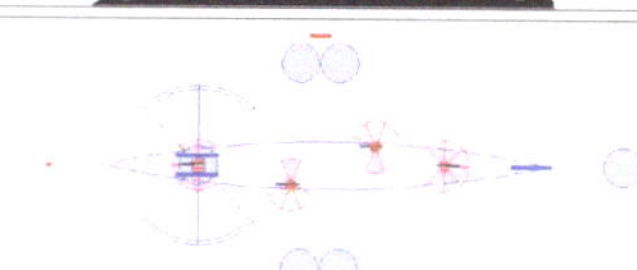

80 (1886)
1886 01 01 - 1886 06 06 - 1887 01 01 | Yarrow, Poplar

1 | "79" type (1886)
Torpedo boat 1 class, Britain

Displacement: 75 tons at draft: 1.68 m | Length: 39.22 m oa | 39.01 m wl | 36,62 m pp | Beam: 3.96 m | Crew: 16 men

Machinery: 1 locomotive boiler | 1 vertical triple expansion E * 1 screw | 1,000 HP = 22.25 knots | Supplies: 15 tons of coal

Armament: 3-25,4/40 Mg, 5-356 TT (2*2 trainable, 1 >)

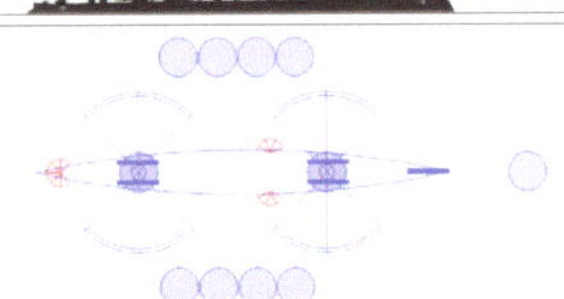

79 (1886)
1886 01 01 - 1886 06 06 - 1887 01 01 | Yarrow, Poplar

1 | "81" type (1886)
Torpedo boat 1 class, Britain

Displacement: 137 tons at draft 2.9 m | Length: 46.85 m oa, 45.72 m pp | Beam: 5.35 m | Crew: 25 men

Machinery: 1 locomotive boiler | 1 vertical compound expansion E * 1 screw | 1,330 HP = 23.75 knots | Supplies: 35 tons of coal

Armament: 4-47/43, 3-356 TT (1*2 trainable, 1 >)

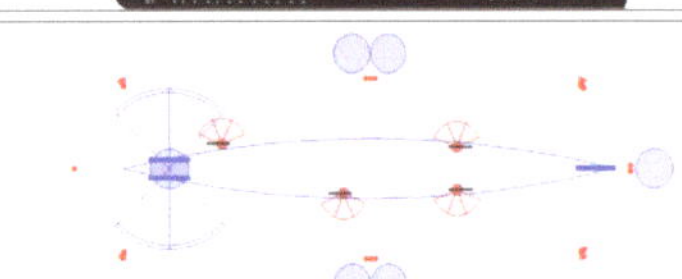

81 (1886)
1886 01 01 - 1886 06 06 - 1886 12 12 | White, Cowes

5 | "34" type (1885)
Torpedo boats 1 class, Britain

Displacement: 60 tons at draft: 1.32 m, 66 tons full | Length: 38.1 m pp | Beam: 4.42 m | Crew: 16 men

Machinery: 1 locomotive boiler | 1 vertical compound expansion E * 1 screw | 950 HP = 18 knots | Supplies: 20 tons of coal

Armament: 4-25.4/40 Mg, 5-356 TT (2*2 trainable, 1 >)

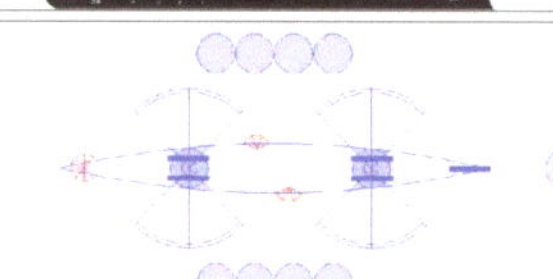

34 (1885)
1885 01 01 - 1885 06 06 - 1886 01 01 | White, Cowes

35 (1885)
1885 01 01 - 1886 01 01 - 1886 06 06 | White, Cowes

36 (1885)
1885 01 01 - 1886 01 01 - 1886 06 06 | White, Cowes

37 (1886)
1886 01 01 - 1887 01 01 - 1888 01 01 | White, Cowes

38 (1886)
1886 01 01 - 1887 01 01 - 1888 01 01 | White, Cowes

20 | "41" type (1885)
Torpedo boats 1 class, Britain

Displacement: 70 tons at draft: 1.82 m, 93 tons full; Length: 39.09 m oa | 38.86 m wl | Beam: 3.81 m | Crew: 16 men

Machinery: 1 locomotive boiler, 1 vertical compound expansion E * 1 screw | 700 HP = 20 knots | Supplies: 20 tons of coal

Armament: 4-25,4/40 Mg, 4-356 TT (2*2 trainable)

Age 5

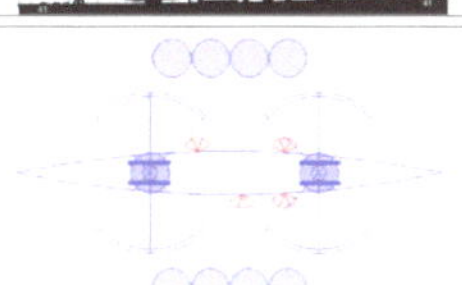

41 (1885)
1885 01 01 - 1885 06 06 - 1886 01 01 | Thornycroft, Woolston

42 (1885)
1885 01 01 - 1885 06 06 - 1886 01 01 | Thornycroft, Woolston

43 (1885)
1885 01 01 - 1885 06 06 - 1886 01 01 | Thornycroft, Woolston

44 (1885)
1885 01 01 - 1885 06 06 - 1886 01 01 | Thornycroft, Woolston

45 (1885)
1885 01 01 - 1885 06 06 - 1886 01 01 | Thornycroft, Woolston

46 (1885)
1885 01 01 - 1885 06 06 - 1886 01 01 | Thornycroft, Woolston

47 (1885)
1885 01 01 - 1885 06 06 - 1886 01 01 | Thornycroft, Woolston

48 (1885)
1885 01 01 - 1885 06 06 - 1886 01 01 | Thornycroft, Woolston

49 (1885)
1885 01 01 - 1886 01 01 - 1886 06 06 | Thornycroft, Woolston

50 (1885)
1885 01 01 - 1886 01 01 - 1886 06 06 | Thornycroft, Woolston

20 | "41" type (1885)
Torpedo boats 1 class, Britain (continued)

51 (1885)
1885 01 01 - 1886 01 01 - 1886 06 06 | Thornycroft, Woolston

52 (1885)
1885 01 01 - 1886 01 01 - 1886 06 06 | Thornycroft, Woolston

53 (1885)
1885 01 01 - 1886 01 01 - 1886 06 06 | Thornycroft, Woolston

54 (1885)
1885 01 01 - 1886 01 01 - 1886 06 06 | Thornycroft, Woolston

55 (1885)
1885 01 01 - 1886 01 01 - 1886 06 06 | Thornycroft, Woolston

56 (1885)
1885 01 01 - 1886 01 01 - 1886 06 06 | Thornycroft, Woolston

57 (1885)
1885 01 01 - 1886 01 01 - 1887 01 01 | Thornycroft, Woolston

58 (1885)
1885 01 01 - 1886 01 01 - 1887 01 01 | Thornycroft, Woolston

59 (1885)
1885 01 01 - 1886 01 01 - 1887 01 01 | Thornycroft, Woolston

60 (1885)
1885 01 01 - 1886 01 01 - 1887 01 01 | Thornycroft, Woolston

22 | "30" type (1885)
Torpedo boats 1 class, Britain

Displacement: 60 tons at draft: 1.68 m | Length: 38.18 m oa | Beam: 3.96 m | Crew: 16 men

Machinery: 1 locomotive boiler | 1 vertical compound expansion E * 1 screw | 670 HP = 19.5 knots | Supplies: 20 tons of coal

Armament: 4-25,4/40 Mg, 5-356 TT (2*2 trainable, 1 >)

Age 5

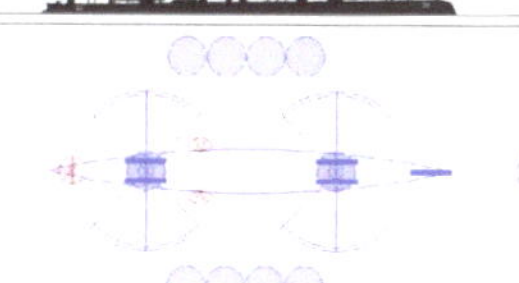

30 (1885)
1885 01 01 - 1885 06 06 - 1886 01 01 | Yarrow, Poplar

31 (1885)
1885 01 01 - 1885 06 06 - 1886 01 01 | Yarrow, Poplar

32 (1885)
1885 01 01 - 1885 06 06 - 1886 01 01 | Yarrow, Poplar

33 (1885)
1885 01 01 - 1885 06 06 - 1886 01 01 | Yarrow, Poplar

61 (1885)
1885 01 01 - 1885 06 06 - 1886 01 01 | Yarrow, Poplar

62 (1885)
1885 01 01 - 1885 06 06 - 1886 01 01 | Yarrow, Poplar

63 (1885)
1885 01 01 - 1885 06 06 - 1886 01 01 | Yarrow, Poplar

64 (1885)
1885 01 01 - 1885 06 06 - 1886 01 01 | Yarrow, Poplar

65 (1886)
1886 01 01 - 1886 06 06 - 1887 01 01 | Yarrow, Poplar

65 (1886)
1886 01 01 - 1886 06 06 - 1887 01 01 | Yarrow, Poplar

22 | "30" type (1885)
Torpedo boats 1 class, Britain (continued)

65 (1886)
1886 01 01 - 1886 06 06 - 1887 01 01 | Yarrow, Poplar

66 (1886)
1886 01 01 - 1886 06 06 - 1887 01 01 | Yarrow, Poplar

67 (1886)
1886 01 01 - 1886 06 06 - 1887 01 01 | Yarrow, Poplar

68 (1886)
1886 01 01 - 1886 06 06 - 1887 01 01 | Yarrow, Poplar

69 (1886)
1886 01 01 - 1886 06 06 - 1887 01 01 | Yarrow, Poplar

70 (1886)
1886 01 01 - 1886 06 06 - 1887 01 01 | Yarrow, Poplar

71 (1886)
1886 01 01 - 1886 06 06 - 1887 01 01 | Yarrow, Poplar

72 (1886)
1886 01 01 - 1886 06 06 - 1887 01 01 | Yarrow, Poplar

73 (1886)
1886 01 01 - 1886 06 06 - 1887 01 01 | Yarrow, Poplar

74 (1886)
1886 01 01 - 1886 06 06 - 1887 01 01 | Yarrow, Poplar

75 (1886)
1886 01 01 - 1886 06 06 - 1887 01 01 | Yarrow, Poplar

76 (1886)
1886 01 01 - 1886 06 06 - 1887 01 01 | Yarrow, Poplar

77 (1886)
1886 01 01 - 1886 06 06 - 1887 01 01 | Yarrow, Poplar

78 (1886)
1886 01 01 - 1886 06 06 - 1887 01 01 | Yarrow, Poplar

5 | "25" type (1885)
Torpedo boats 1 class, Britain

Displacement: 60 tons at draft: 1.82 m | Length: 39.09 m oa | 38.86 m wl | Beam: 3.81 m | Crew: 16 men

Machinery: 1 locomotive boiler | 1 vertical compound expansion E * 1 screw | 700 HP = 20.75 knots | Supplies: 20 tons of coal

Armament: 4-25,4/40 Mg, 5-356 TT (2*2 trainable, 1 >)

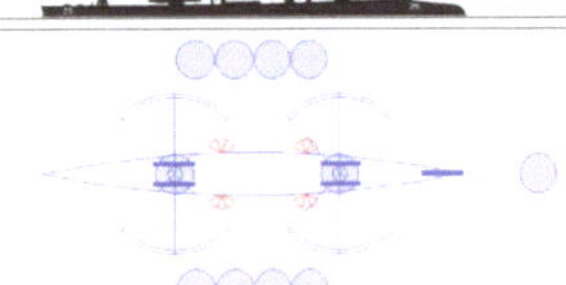

25 (1885)
1885 01 01 - 1885 06 06 - 1886 01 01 | Thornycroft, Woolston

26 (1885)
1885 01 01 - 1885 06 06 - 1886 01 01 | Thornycroft, Woolston

27 (1885)
1885 01 01 - 1885 06 06 - 1886 01 01 | Thornycroft, Woolston

28 (1885)
1885 01 01 - 1885 06 06 - 1886 01 01 | Thornycroft, Woolston

29 (1885)
1885 01 01 - 1885 06 06 - 1886 01 01 | Thornycroft, Woolston

2 | "Suhum"/"21" type (1882)
Torpedo boats 1 class, Russia

NO PHOTO AVAILABLE

Displacement: 64 tons at draft: 0.61 m | Length: 34.59 m oa | 34.04 m pp | Beam: 3.81 m | Crew: 14 men

Machinery: 1 locomotive boiler | 1 vertical compound expansion E * 1 screw | 700 HP = 20 knots | Supplies: 12 tons of coal

Armament: 1-25,4/40 Mg, 3-356 TT (1 trainable, 2 >)

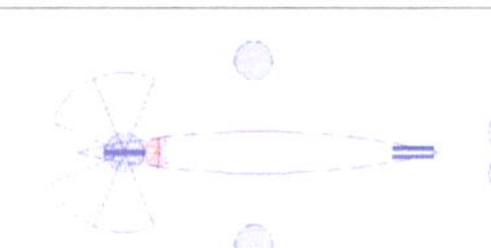

21 (1884)
1884 01 01 - 1885 01 01 - 1886 01 01 | Thornycroft, Woolston

22 (1884)
1884 01 01 - 1885 01 01 - 1886 01 01 | Thornycroft, Woolston

Sister ships are in the Australian, Russian, Swedish sections

2 | "Suhum"/"23" type (1882)
Torpedo boats 1 class, Russia

Displacement: 67 tons at draft: 2.06 m | Length: 34.59 m oa | Beam: 3.84 m | Crew: 14 men

Machinery: 1 locomotive boiler | 1 vertical compound expansion E * 1 screw | 600 HP = 18.75 knots | Supplies: 13 tons of coal

Armament: 3-25,4/40 Mg, 3-356 TT (1 trainable, 2 >)

Age 8

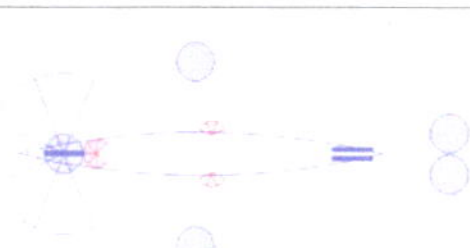

23 (1884)
1884 01 01 - 1885 01 01 - 1886 01 01 | Yarrow, Poplar

24 (1884)
1884 01 01 - 1885 01 01 - 1886 01 01 | Yarrow, Poplar

Sister ships are in the Australian, Russian, Swedish sections

2 | "Batum"/"39" type (1879)
Torpedo boats 1 class, Russia

Displacement: 40 tons at draft: 1.22 m | Length: 30.48 m pp | Beam: 3.81 m | Crew: 15 men

Machinery: 1 locomotive boiler | 1 vertical compound expansion E * 1 screw | 500 HP = 20 knots | Supplies: 15 tons of coal

Armament: 2-37/23 Mg, 1-356 TT (1 >)

Age 11

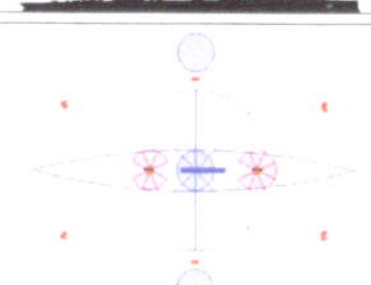

39 (1884)
1884 01 01 - 1885 01 01 - 1885 06 06 | Yarrow, Poplar

40 (1884)
1884 01 01 - 1885 01 01 - 1885 06 06 | Yarrow, Poplar

Sister ships can be found in the Argentinean, Austria-Hungarian, Brazilean, Chilean, Dutch, Greek, Japanese, and Russian sections

1 | "20" type (1879)
Torpedo boat, Britain

Displacement: 28 tons at draft: 1.22 m | Length: 26.52 m oa | Beam: 3.05 m | Crew: 15 men

Machinery: 1 locomotive boiler | 1 vertical compound expansion E * 1 screw | 360 HP = 16.75 knots | Supplies: 3 tons of coal

Armament: 2-356 TT (>)

20 (1879)
1879 01 01 - 1880 01 01 - 1881 01 01 | Rennie, London

1 | "13" type (1878)
Torpedo boat, Britain

Displacement: 28 tons at draft: 1.22 m | Length: 26.52 m oa | 24.99 m pp | Beam: 3.28 m | Crew: 15 men

Machinery: 1 locomotive boiler | 1 vertical compound expansion E * 1 screw | 450 HP = 20 knots | Supplies: 6 tons of coal

Armament: 2-356 TT (>)

13 (1878)
1878 01 01 - 1878 06 06 - 1880 01 01 | Maudsley, London

1 | "15" type (1878)
Torpedo boat, Britain

Displacement: 28 tons at draft: 1.7 m | Length: 26.52 m oa | Beam: 3.12 m | Crew: 15 men

Machinery: 1 locomotive boiler | 1 vertical compound expansion E * 1 screw | 450 HP = 21 knots | Supplies: 3 tons of coal

Armament: 2-356 TT (>)

15 (1878)
1878 01 01 - 1879 01 01 - 1880 01 01 | Hanna, D. & W., Glasgow

1 | "19" type (1877)
Torpedo boat, Britain

Displacement: 28 tons at draft: 1.22 m | Length: 28.35 m oa | Beam: 3.28 m | Crew: 15 men

Machinery: 1 locomotive boiler | 1 vertical compound expansion E * 1 screw | 460 HP = 21 knots | Supplies: 5 tons of coal

Armament: 2-356 TT (>)

19 (1877)
1877 01 01 - 1878 01 01 - 1878 06 06 | White, Cowes

2 | "unnamed"/"17" type (1878)
Torpedo boats, Russia

Displacement: 33 tons at draft: 1.37 m | Length: 26.21 m oa | 24.6 m pp | Beam: 3.35 m | Crew: 15 men

Machinery: 1 locomotive boiler | 1 vertical compound expansion E * 1 screw | 450 HP = 21 knots | Supplies: 5 tons of coal

Armament: 1-406 TT (>)

Age 12

17 (1878)
1878 01 01 - 1879 01 01 - 1880 01 01 | Yarrow, Poplar

18 (1878)
1878 01 01 - 1879 01 01 - 1880 01 01 | Yarrow, Poplar

1 | "14" type (1877)
Torpedo boat, Britain

Displacement: 33 tons at draft: 0.55 m | Length: 26.21 m oa | 25.91 m wl; 25.6 m pp | Beam: 3.35 m | Crew: 15 men

Machinery: 1 locomotive boiler | 1 vertical compound expansion E * 1 screw | 550 HP = 21.75 knots | Suppl.: 5 tons of coal

Armament: 1-406 TT (>)

Age 13

14 (1877)
1877 01 01 - 1877 06 06 - 1878 01 01 | Yarrow, Poplar

1 | "10" type (1877)
Torpedo boat, Britain

NO PHOTO AVAILABLE

Displacement: 28 tons at draft: 1.6 m | Length: 27.69 m wl | 25.76 m pp | Beam: 3.12 m | Crew: 15 men

Machinery: 1 locomotive boiler | 1 vertical compound expansion E * 1 screw | 460 HP = 20 knots | Supplies: 6 tons of coal

Armament: 1-406 TT (>)

Age 13

10 (1877)
1877 01 01 - 1879 01 01 - 1879 06 06 | Thornycroft, Woolston

10 | "2" type (1877)
Torpedo boats, Britain

Displacement: 26.52 tons at draft: 1.6 m | Length: 26.31 m wl, 25.76 m pp | Beam: 3.12 m | Crew: 15 men

Machinery: 1 locomotive boiler | 1 vertical compound expansion E * 1 screw | 460 HP = 20 knots | Supplies: 6 tons of coal

Armament: 1-406 TT (>)

Age 13

2 (1877)
1877 01 01 - 1878 01 01 - 1878 06 06 | Thornycroft, Woolston

3 (1877)
1877 01 01 - 1878 01 01 - 1878 06 06 | Thornycroft, Woolston

4 (1877)
1877 01 01 - 1878 01 01 - 1878 06 06 | Thornycroft, Woolston

5 (1877)
1877 01 01 - 1878 01 01 - 1878 06 06 | Thornycroft, Woolston

6 (1877)
1877 01 01 - 1878 01 01 - 1878 06 06 | Thornycroft, Woolston

7 (1877)
1877 01 01 - 1878 01 01 - 1878 06 06 | Thornycroft, Woolston

8 (1877)
1877 01 01 - 1878 01 01 - 1878 06 06 | Thornycroft, Woolston

9 (1877)
1877 01 01 - 1878 01 01 - 1878 06 06 | Thornycroft, Woolston

11 (1878)
1878 01 01 - 1879 01 01 - 1879 06 06 | Thornycroft, Woolston

12 (1878)
1878 01 01 - 1879 01 01 - 1879 06 06 | Thornycroft, Woolston

1 | "1" Lightning type (1876)
Torpedo boat, Britain

Displacement: 27 tons at draft: 1.6 m | Length: 26.52 m oa, 25.76 m pp | Beam: 3.28 m | Crew: 15 men

Machinery: 1 locomotive boiler | 1 vertical compound expansion E * 1 screw | 390 HP = 18 knots | Supplies: 6 tons of coal

Armament: 1-406,4 TT (>)

Age 14

Lightning, 1 (1876)
1876 01 01 - 1877 01 01 - 1877 06 06 | Thornycroft, Woolston

0 +10 | "39" type (1888)
Torpedo boats 2 class, Britain

BUILDING

1Y:12M 39 · 1Y:12M 40 · 1Y:12M 41 · 1Y:12M 42 · 1Y:12M 43
1Y:12M 44 · 1Y:12M 45 · 1Y:12M 46 · 1Y:12M 47 · 1Y:12M 48

Displacement: 16.5 tons at draft: 0.91 m | Length: 18.29 m oa | Beam: 2.82 m | Crew: 9 men

Machinery: 1 locomotive boiler | 1 vertical compound expansion E * 1 screw | 240 HP = 16.5 knots | Supplies: app. 1 ton of coal

Armament: 1-11.2 Mg, 2-356 TT (Djevetskiy)

Age 2

39 (1888)
1888 01 01 - 1889 01 01 - ***** | Yarrow, Poplar

40 (1888)
1888 01 01 - 1889 01 01 - ***** | Yarrow, Poplar

41 (1888)
1888 01 01 - 1889 01 01 - ***** | Yarrow, Poplar

42 (1888)
1888 01 01 - 1889 01 01 - ***** | Yarrow, Poplar

43 (1888)
1888 01 01 - 1889 01 01 - ***** | Yarrow, Poplar

44 (1888)
1888 01 01 - 1889 01 01 - ***** | Yarrow, Poplar

45 (1888)
1888 01 01 - 1889 01 01 - ***** | Yarrow, Poplar

46 (1888)
1888 01 01 - 1889 01 01 - ***** | Yarrow, Poplar

47 (1888)
1888 01 01 - 1889 01 01 - ***** | Yarrow, Poplar

48 (1888)
1888 01 01 - 1889 01 01 - ***** | Yarrow, Poplar

2 | "49" (1888) type
Torpedo boats 2 class, Britain

Displacement: 15 tons at draft: 0.73 m | Length: 18.14 m oa | Beam: 2.59 m | Crew: 9 men

Machinery: 1 locomotive boiler | 1 vertical compound expansion E * 1 screw | 200 HP = 16.5 knots | Suppl.: app.1 ton of coal

Armament: 1-25.4/40 Mg, 1-356 TT (1 trainable)

Age 2

49 (1888)
1888 01 01 - 1888 06 06 - 1889 01 01 | Yarrow, Poplar

50 (1888)
1888 01 01 - 1888 06 06 - 1889 01 01 | Yarrow, Poplar

Sister ships can be found in the Argentnean and Russian sections

2 | "99" (1884) type
Torpedo boats 2 class, Britain

Displacement: 12 tons at draft: 0.84 m | Length: 20.22 m oa | 18.8 m pp | Beam: 2.29 m | Crew: 7 men

Machinery: 1 boiler | 1 vertical triple expansion E * 1 screw | 190 HP = 16.5 knots | Supplies: 1 ton of coal

Armament: 2-356 TT (2 >)

Age 6

99 (1884) | locomotive boiler
1884 01 01 - 1885 01 01 - 1886 01 01 | Thornycroft, Woolston

100 (1884) | Thornycroft boiler
1884 01 01 - 1885 01 01 - 1886 01 01 | Thornycroft, Woolston

12 | "1" (1883) type
Torpedo boats 2 class, Britain

NO PHOTO AVAILABLE

Displacement: 10.5 tons at draft: 1.45 m, 14 tons full | Length: 17.07 m oa | Beam: 2.82 m | Crew: 9 men

Machinery: 1 locomotive boiler | 1 vertical compound expansion E * 1 screw | 140 HP = 15.5 knots | Suppl.: app. 1 ton of coal

Armament: 1-11.2 Mg, 2-356 TT (2*1 Djevetskiy)

Age 6

1 (1883)
1883 01 02 - 1884 01 01 - 1885 01 01 | White, Cowes

2 (1883)
1883 01 01 - 1884 01 01 - 1885 01 01 | White, Cowes

3 (1883)
1883 01 01 - 1884 01 01 - 1885 01 01 | White, Cowes

4 (1884)
1884 01 01 - 1885 01 01 - 1886 01 01 | White, Cowes

5 (1884)
1884 01 01 - 1885 01 01 - 1886 01 01 | White, Cowes

6 (1884)
1884 01 01 - 1885 01 01 - 1886 01 01 | White, Cowes

7 (1885)
1885 01 01 - 1886 01 01 - 1886 06 06 | White, Cowes

8 (1885)
1885 01 01 - 1886 01 01 - 1886 06 06 | White, Cowes

9 (1885)
1885 01 01 - 1886 01 01 - 1886 06 06 | White, Cowes

10 (1886)
1886 01 01 - 1887 01 01 - 1888 01 01 | White, Cowes

12 | "1" type (1883)
Torpedo boats 2 class, Britain (continued)

11 (1886)
1886 01 01 - 1887 01 01 - 1888 01 01 | White, Cowes

12 (1886)
1886 01 01 - 1887 01 01 - 1888 01 01 | White, Cowes

1 | "98" (1882) type
Torpedo boat 2 class, Britain

Displacement: 14.5 tons at draft: 0.84 m | Length: 20.22 m oa, 18.8 m pp | Beam: 2.29 m | Crew: 7 men

Machinery: 1 locomotive boiler | 1 Ruther turbine * 1 screw | 167 HP = 11.75 knots | Supplies: 1 ton of coal

Armament: 2-356 TT (2 >)

Age 8

98 (1882)
1882 01 01 - 1883 01 01 - 1884 01 01 | Thornycroft, Woolston

20 | "76" type (1879)
Torpedo boats 2 class, Britain

Italian TB2cl "Euterpe"

Displacement: 12.89 tons at draft: 0.76 m | Length: 19.2 m oa, 18.8 m pp | Beam: 2.29 m | Crew: 7 men

Machinery: 1 locomotive boiler | 1 vertical compound expansion E * 1 screw | 170 HP = 16.5 knots | Supplies: 1 ton of coal

Armament: 2-356 TT (2 >)

Age 11

76 (1879)
1879 01 01 - 1880 01 01 - 1881 01 01 | Thornycroft, Woolston

77 (1879)
1879 01 01 - 1880 01 01 - 1881 01 01 | Thornycroft, Woolston

78 (1879)
1879 01 01 - 1880 01 01 - 1881 01 01 | Thornycroft, Woolston

79 (1879)
1879 01 01 - 1880 01 01 - 1881 01 01 | Thornycroft, Woolston

80 (1879)
1879 01 01 - 1880 01 01 - 1881 01 01 | Thornycroft, Woolston

81 (1879)
1879 01 01 - 1880 01 01 - 1881 01 01 | Thornycroft, Woolston

82 (1879)
1879 01 01 - 1880 01 01 - 1881 01 01 | Thornycroft, Woolston

83 (1880)
1880 01 01 - 1881 01 01 - 1881 06 06 | Thornycroft, Woolston

84 (1880)
1880 01 01 - 1881 01 01 - 1882 01 01 | Thornycroft, Woolston

85 (1880)
1880 01 01 - 1881 01 01 - 1882 01 01 | Thornycroft, Woolston

20 | "76" type (1879)
Torpedo boats 2 class, Britain (continued)

86 (1880)
1880 01 01 - 1881 01 01 - 1882 01 01 | Thornycroft, Woolston

87 (1880)
1880 01 01 - 1881 01 01 - 1882 01 01 | Thornycroft, Woolston

88 (1880)
1880 01 01 - 1881 01 01 - 1882 01 01 | Thornycroft, Woolston

89 (1881)
1881 01 01 - 1882 01 01 - 1883 01 01 | Thornycroft, Woolston

90 (1881)
1881 01 01 - 1882 01 01 - 1883 01 01 | Thornycroft, Woolston

91 (1881)
1881 01 01 - 1882 01 01 - 1883 01 01 | Thornycroft, Woolston

92 (1881)
1881 01 01 - 1882 01 01 - 1883 01 01 | Thornycroft, Woolston

93 (1881)
1881 01 01 - 1882 01 01 - 1883 01 01 | Thornycroft, Woolston

94 (1881)
1881 01 01 - 1882 01 01 - 1883 01 01 | Thornycroft, Woolston

95 (1881)
1881 01 01 - 1882 01 01 - 1883 01 01 | Thornycroft, Woolston

Sister ships can be found in the Danish & Italian sections

4 | "74" type (1879)
Torpedo boats 2 class, Britain

Displacement: 12.5 tons at draft 0.69 m | Length: 19.2 m oa, 18.29 m pp | Beam: 2.36 m | Crew: 7 men

Machinery: 1 locomotive boiler, 1 vertical compound expansion E * 1 screw | 200 HP = 16 knots | Supplies: app. 1 ton of coal

Armament: 2-356 TT (2 >)

74 (1879)
1879 01 01 - 1881 01 01 - 1881 06 06 | Yarrow, Poplar

75 (1880)
1880 01 01 - 1881 06 06 - 1881 12 12 | Yarrow, Poplar

96 (1882)
1882 01 01 - 1883 01 01 - 1884 01 01 | Yarrow, Poplar

97 (1882)
1882 01 01 - 1883 01 01 - 1884 01 01 | Yarrow, Poplar

10 | "64" type (1879)
Torpedo boats 2 class, Britain

NO PHOTO AVAILABLE

Displacement: 10.9 tons at draft: 0.3 m | Length: 19.2 m oa, 18.44 m pp | Beam: 2.25 m | Crew: 7 men

Machinery: 1 locomotive boiler | 1 vertical compound expansion E * 1 screw | 110 HP = 15 knots | Supplies: 1 ton of coal

Armament: 2-356 TT (2 >)

Age 11

64 (1879)
1879 01 01 - 1880 01 01 - 1881 01 01 | Thornycroft, Woolston

65 (1879)
1879 01 01 - 1880 01 01 - 1881 01 01 | Thornycroft, Woolston

66 (1879)
1879 01 01 - 1880 01 01 - 1881 01 01 | Thornycroft, Woolston

67 (1879)
1879 01 01 - 1880 01 01 - 1881 01 01 | Thornycroft, Woolston

68 (1879)
1879 01 01 - 1880 01 01 - 1881 01 01 | Thornycroft, Woolston

69 (1879)
1879 01 01 - 1880 01 01 - 1881 01 01 | Thornycroft, Woolston

70 (1879)
1879 01 01 - 1880 01 01 - 1881 01 01 | Thornycroft, Woolston

71 (1880)
1880 01 01 - 1881 01 01 - 1881 06 06 | Thornycroft, Woolston

72 (1880)
1880 01 01 - 1881 01 01 - 1881 06 06 | Thornycroft, Woolston

73 (1880)
1880 01 01 - 1881 01 01 - 1881 06 06 | Thornycroft, Woolston

1 | "63" type (1878)
Spar torpedo boat 2 class, Britain

Displacement: 15 tons at draft: 0.5 m | Length: 18.29 m oa | Beam: 2.29 m | Crew: 7 men

Machinery: 1 locomotive boiler | 1 vertical compound expansion E * 1 screw | ??? HP = 16 knots | Supplies: 3 tons of coal

Armament: 2 spar torpedoes

Age 12

63 (1878)
1878 01 01 - 1878 06 06 - 1878 12 12 | Herreschoff, Bristol

Not active. Used for experimental purposes

Sister ship was aquired by the Russian navy

12 | "51" type (1877)
Torpedo boats 2 class, Britain

NO PHOTO AVAILABLE

Displacement: 10.75 tons at draft: 0.28 m | Length: 17.37 m oa | Beam: 2.13 m | Crew: 7 men

Machinery: 1 locomotive boiler | 1 vertical compound expansion E * 1 screw | 120 HP = 12 knots | Supplies: 1 tons of coal

Armament: 2-356 TT (Djevetskiy)

Age 13

51 (1877)
1877 01 01 - 1878 01 01 - 1878 06 06 | Thornycroft, Woolston

52 (1877)
1877 01 01 - 1878 01 01 - 1878 06 06 | Thornycroft, Woolston

53 (1877)
1877 01 01 - 1878 01 01 - 1878 06 06 | Thornycroft, Woolston

54 (1877)
1877 01 01 - 1878 01 01 - 1878 06 06 | Thornycroft, Woolston

55 (1877)
1877 01 01 - 1878 01 01 - 1878 06 06 | Thornycroft, Woolston

56 (1877)
1877 01 01 - 1878 01 01 - 1878 06 06 | Thornycroft, Woolston

57 (1878)
1878 01 01 - 1879 01 01 - 1880 01 01 | Thornycroft, Woolston

58 (1878)
1878 01 01 - 1879 01 01 - 1880 01 01 | Thornycroft, Woolston

59 (1878)
1878 01 01 - 1879 01 01 - 1880 01 01 | Thornycroft, Woolston

60 (1878)
1878 01 01 - 1879 01 01 - 1880 01 01 | Thornycroft, Woolston

12 | "51" type (1877)
Torpedo boats 2 class, Britain (continued)

61 (1878)
1878 01 01 - 1879 01 01 - 1880 01 01 | Thornycroft, Woolston

62 (1878)
1878 01 01 - 1879 01 01 - 1880 01 01 | Thornycroft, Woolston

2 | "Beagle" type (1888)
Steel sloops, Britain

Displacement: 1,170 tons at draft: 3.81 m | Length: 59.44 m pp | Beam: 8.53 m | Crew: 138 men

Machinery: 2 locomotive boilers | 2 * 1,000 HP steam EE * 2 screws | 2,000 HP = 14.5 knots | Supplies: 180 tons of coal

Armament: 8-127/25, 8-25.4/40 Mg

Age 2

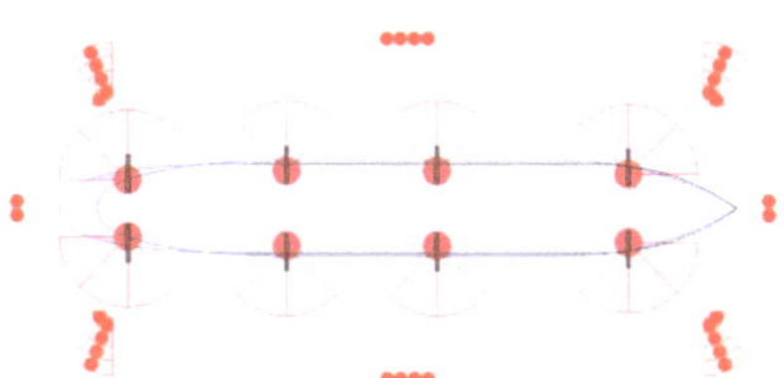

Basilisk (1888)
2 Rennie horizontal triple expansion EE |
1888 05 01 - 1889 04 06 - 1889 12 01 | Sheerness DYd

Beagle (1888)
2 Rennie horizontal triple expansion EE |
1888 05 14 - 1889 02 28 - 1889 09 01 | Portsmouth DYd

4 | "Swallow" type (1885)
Composite sloops, Britain

Displacement: 1,140 tons at draft: 3.81 m | Length: 59.43 m pp | Beam: 8.53 m | Crew: 138 men

Machinery: 2 locomotive boilers | 2 * 750 HP steam EE * 2 screws | 1,570 HP = 13.5 knots | Supplies: app. 180 tons of coal

Armament: 8-127/25, 8-25.4/40 Mg

Age 5

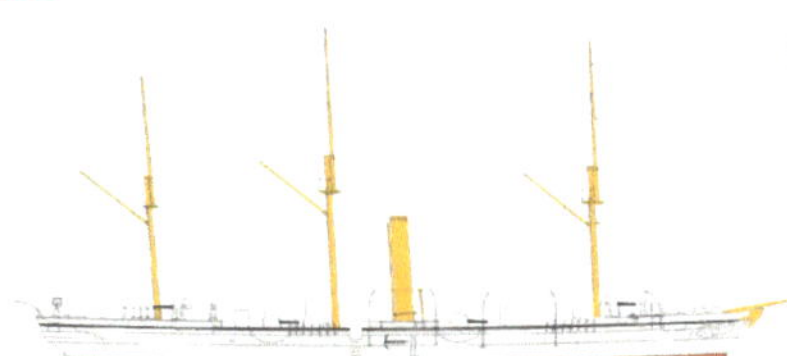

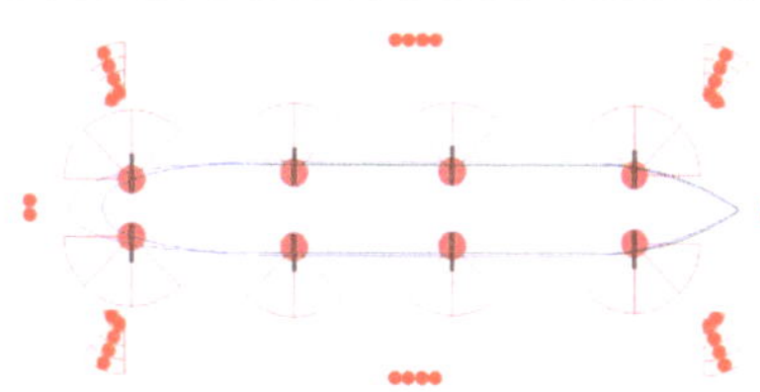

Buzzard (1886)
2 Barrow horizontal triple expansion EE |
1886 05 01 - 1887 05 10 - 1888 05 01 | Sheerness DYd

Daphne (1887)
2 Greenock foundry horizontal triple expansion EE |
1887 06 20 - 1888 05 29 - 1889 05 01 | Sheerness DYd

Nymphe (1887)
2 Greenock foundry horizontal triple expansion EE |
1887 07 05 - 1888 05 01 - 1889 02 01 | Portsmouth DYd

Swallow (1885)
2 Rennie horizontal triple expansion EE |
1885 01 01 - 1885 10 27 - 1886 10 01 | Sheerness DYd

4 | "Swallow" type (1885)
Wooden screw sloops, Britain (continued)

Nymphe

5 | "Mariner" type (1883)
Composite screw sloops, Britain (continued)

Melita

Reindeer

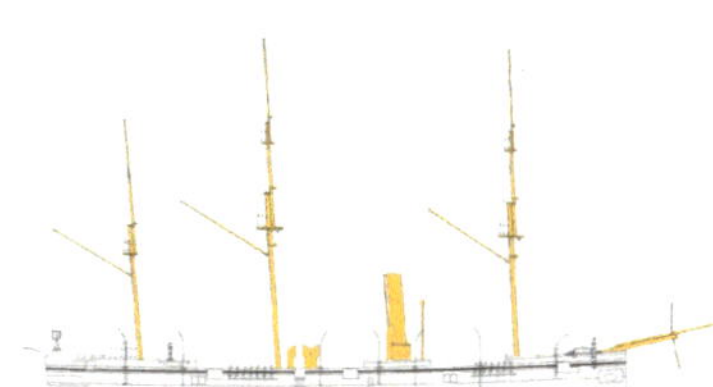

Reindeer (1883)
1 Leslie horizontal compound expansion E | 1,030 HP = 11.5 knots |
1883 01 01 - 1883 11 14 - 1884 01 01 | Devonport DYd

5 | "Mariner" type (1883)
Composite screw sloops, Britain

Displacement: 970 tons at draft: 4.26 m | Length: 50.9 m pp | Beam: 9.75 m | Crew: 126 men

Machinery: 2 locomotive boilers | 1 steam E * 1 screw | 1,200 HP = 12.5 knots | Supplies: app. 150 tons

Armament: 8-127/25, 1-95.3/26, 8-25.4/40 Mg

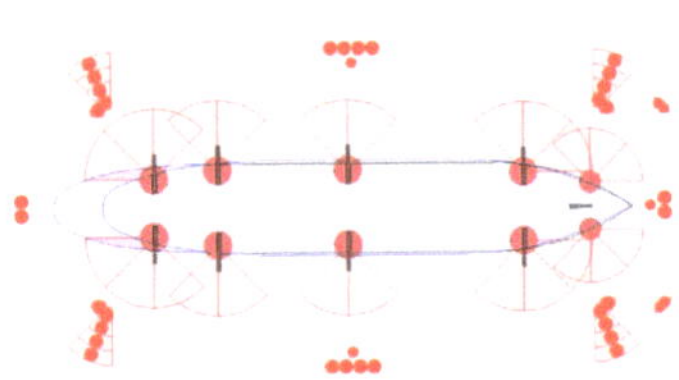

Icarus (1884)
1 Barrow horizontal compound expansion E | 1,230 HP = 12.5 knots
1884 01 01 - 1885 07 27 - 1886 01 01 | Devonport DYd

Melita (1887)
1 Malta DYd horizontal comp. expansion E | 1,200 HP = 12.5 kts
1887 01 01 - 1888 03 20 - 1889 01 01 | Malta DYd

Mariner (1883)
1 Leslie horizontal compound expansion E | 970 HP = 11.5 knots
1883 01 01 - 1884 06 23 - 1885 01 01 | Devonport DYd

Racer (1883)
1 Leslie horizontal compound expansion E | 920 HP = 11.5 knots
1883 01 01 - 1884 08 06 - 1885 01 01 | Devonport DYd

2 | "Dolphin" type (1881)
Composite screw sloops, Britain

Displacement: 925 tons at draft: 4.26 m | Length: 47.85 m pp | Beam: 9.75 m | Crew: 115 men

Machinery: 2 locomotive boilers | 1 steam E * 1 screw | 720 HP = 11.3 knots | Supplies: 135 tons of coal

Armament: 2-152.4/26, 2-127/25, 1-95.3/21, 3-25.4/40 Mg

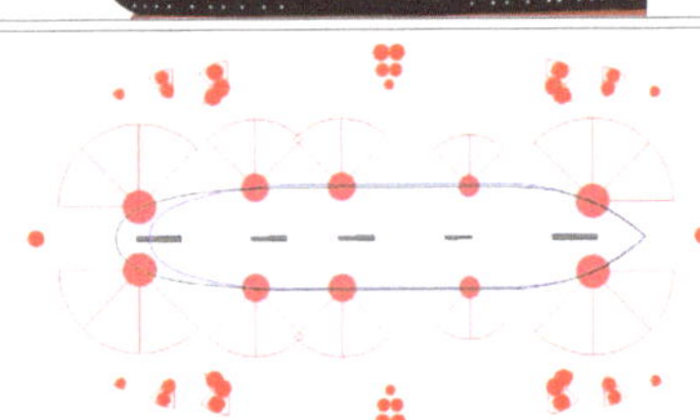

Dolphin (1881)
1 Leslie horizontal compound expansion E | 720 HP = 11.3 knots
1881 01 01 - 1882 12 09 - 1883 01 01 | Raylton Dixon, Tees

Wanderer (1882)
1 Leslie horizontal compound expansion E | 750 HP = 11.3 knots
1882 01 01 - 1883 02 08 - 1884 01 01 | Raylton Dixon, Tees

Wanderer

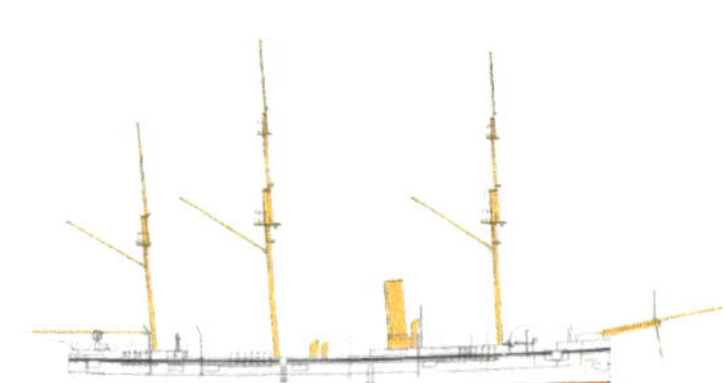

7 | "Satellite" type (1880)
Composite screw sloops, Britain

Displacement: 1,420 tons at draft: 4.8 m | Length: 60.96 m pp; Beam: 11.58 m | Crew: 200 men

Machinery: 2 locomotive boilers | 1 steam E * 1 screw | 1,500 HP = 13 knots | Supplies: app. 300 tons of coal

Armament: 8-152.4/25, 1-95.3/21, 2-25.4/40 Mg

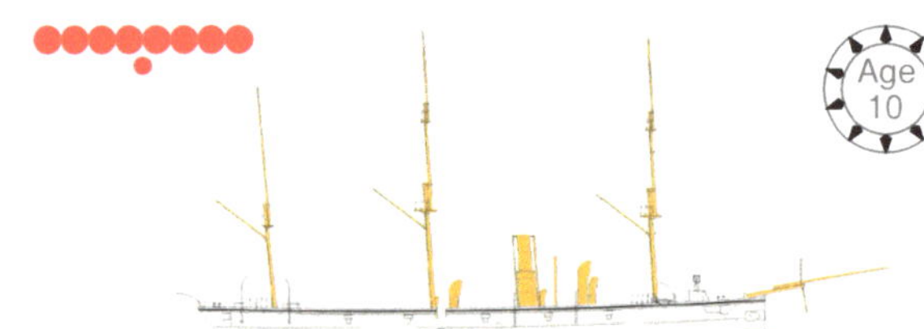

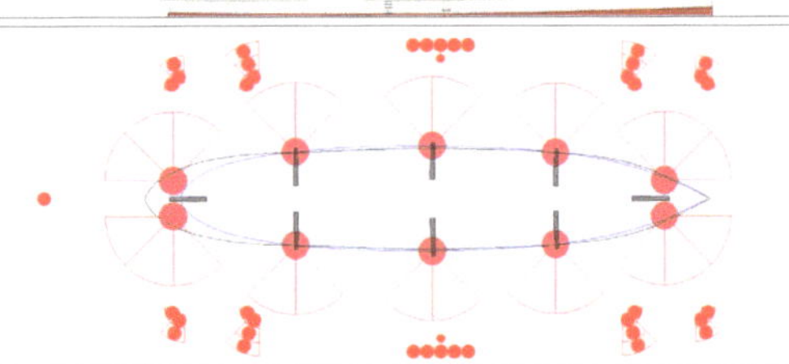

Caroline (1881)
1 Maudslay horizontal compound expansion E | 1,400 HP = 13 kts
1881 10 01 - 1882 11 25 - 1885 12 01 | Sheerness DYd

Heroine (1880)
1 Humphrys horizontal comp. expansion E | 1,470 HP = 13.1 knots
1880 08 01 - 1881 12 03 - 1883 06 01 | Devonport DYd

Hyacinth (1880)
1 Humphrys horizontal comp. expansion E | 1,470 HP = 13.1 knots
1880 08 01 - 1881 12 20 - 1883 07 01 | Devonport DYd

Pylades (1883)
1 Laird horizontal compound expansion E | 1,640 HP = 13 knots
1883 01 01 - 1884 11 05 - 1886 08 01 | Sheerness DYd

7 | "Satellite" type (1880)
Composite screw sloops, Britain (continued)

Rapid (1881)
1 Maudslay horizontal compound expansion E | 1,400 HP = 13 kts
1881 04 01 - 1883 03 21 - 1884 07 01 | Devonport DYd

Royalist (1881)
1 Maudslay horizontal compound expansion E | 1,510 HP = 13 kts
1881 04 01 - 1883 03 07 - 1884 11 01 | Devonport DYd

Satellite (1880)
1 Humphrys horizontal compound expansion E | 1,400 HP = 13 kts
1880 10 01 - 1881 08 13 - 1882 12 01 | Sheerness DYd

Heroine

Royalist

7 | "Doterel" type (1877)
Composite screw sloops, Britain (continued)

Espeigle

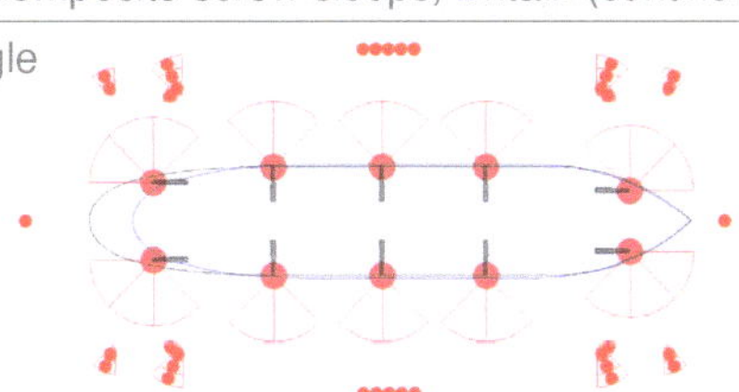

7 | "Doterel" type (1877)
Composite screw sloops, Britain

Displacement: 1,130 tons at draft: 4.8 m | Length: 58.81 m pp | Beam: 10.97 m | Crew: 140 men
Machinery: 2 locomotive boilers | 1 steam E * 1 screw | 1,000 HP = 11 knots | Supplies: 150 tons of coal
Armament: 2-177.8/15.9 MLR, 4-160/16 MLR

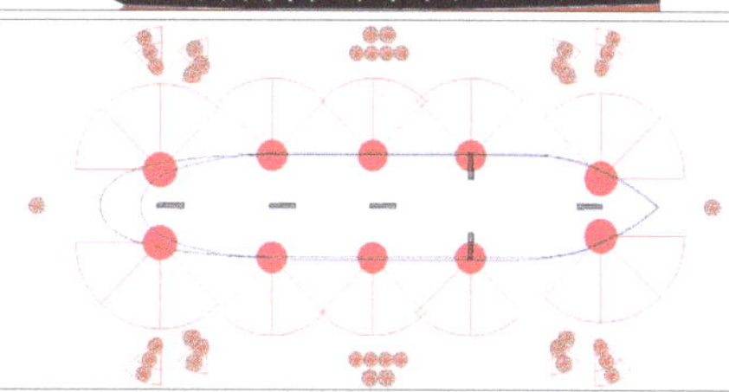

Dragon (1877)
1 Maudslay horizontal comp. expansion E | 1,056 HP = 11.52 knots
1877 04 26 - 1878 05 30 - 1879 02 19 | Devonport DYd

Espeigle (1879)
1 Maudslay horizontal compound expansion E | 1,140 HP = 11.5 kts
1879 09 23 - 1880 08 03 - 1881 10 11 | Devonport DYd

Gannet (1877)
1 Humphrys horizontal comp. expansion E | 1,107 HP = 11.53 knots
1877 01 01 - 1878 08 30 - 1879 04 17 | Sheerness DYd

Kingfisher (1878)
1 Maudslay horizontal comp. expansion E | 1,090 HP = 11.64 knots
1878 09 23 - 1879 12 16 - 1880 08 17 | Sheerness DYd

7 | "Doterel" type (1877)
Composite screw sloops (continued)

Miranda (1878)
1 Napier horizontal compound expansion E | 1,020 HP = 11 knots
1878 07 08 - 1879 09 30 - 1880 07 22 | Devonport DYd

Mutine (1879)
1 Maudslay horizontal comp. expansion E | 1,120 HP = 11.64 knots
1879 06 07 - 1880 07 20 - 1881 05 10 | Devonport DYd

Pegasus (1877)
1 Laird horizontal compound expansion E | 972 HP = 11.47 knots
1877 05 09 - 1878 06 13 - 1879 03 05 | Devonport DYd

Gannet

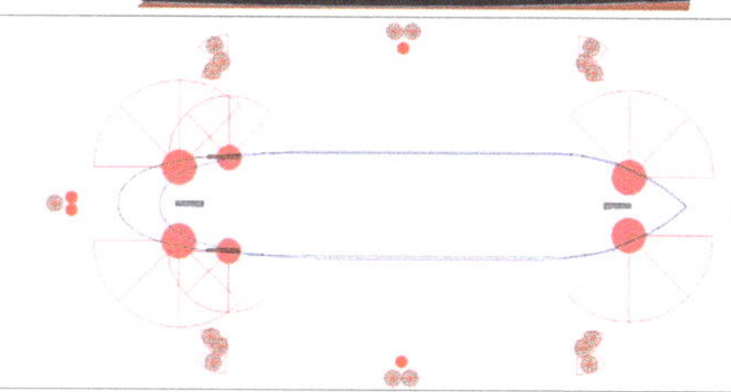

Pegasus

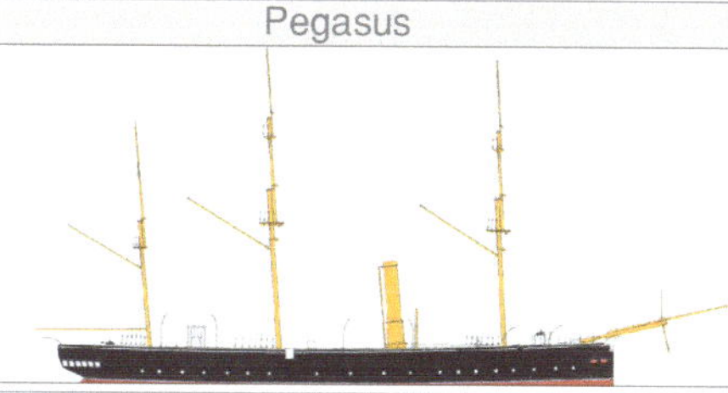

5 | "Penguin" type (1874)
Composite screw sloops (continued)

Wild Swan (1874)
1 Leslie horizontal compound expansion E | 797 HP = 10,35 knots
1874 09 14 - 1876 01 28 - 1876 08 23 | Napier, Govan

Pelican
Wild Swan

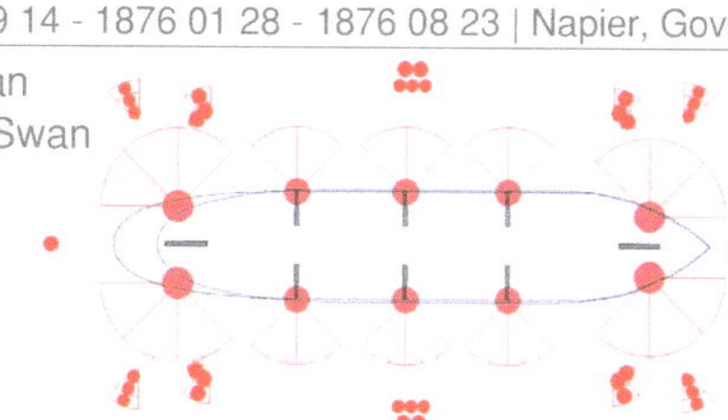

5 | "Penguin" type (1874)
Composite screw sloops, Britain

Displacement: 1,130 tons at draft: 4.8 m | Length: 58.81 m pp | Beam: 10.97 m | Crew: 140 men

Machinery: 2 locomotive boilers | 1 steam E * 1 screw | 1,000 HP = 11 knots | Supplies: 150 tons of coal

Armament: 2-177.8/15.9 MLR, 4-160/16 MLR

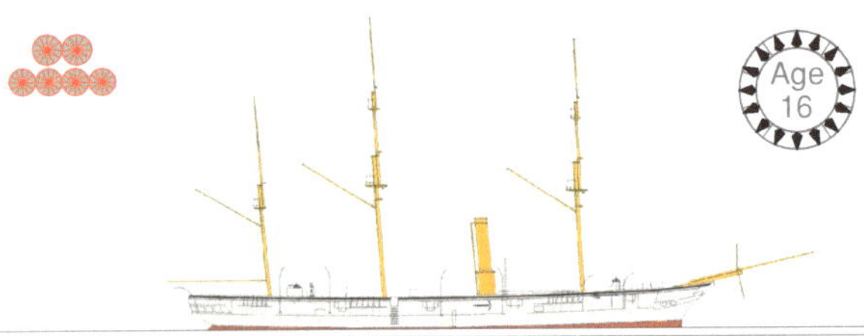

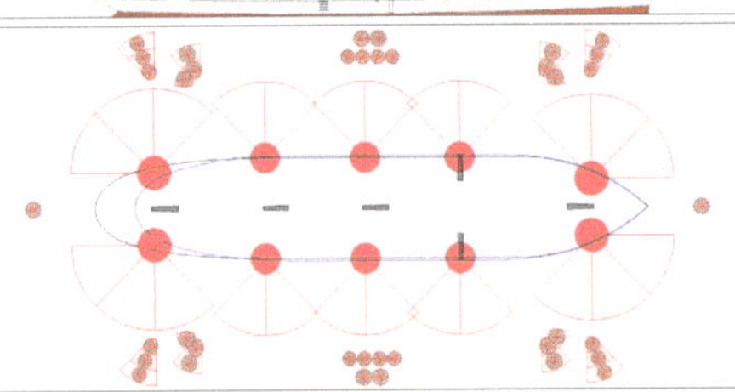

Cormorant (1875)
1 Humphrys horizontal compound expansion E | 951 HP = 11.31 kts
1875 01 01 - 1877 09 12 - 1878 07 28 | Chatham DYd

Osprey (1875)
1 Humphrys horizontal compound expansion E | 946 HP = 11.38 kts
1875 01 01 - 1876 08 05 - 1877 04 19 | Sheerness DYd

Pelican (1875)
1 Humphrys horizontal comp. expansion E | 1,056 HP = 12.24 knots
1875 03 08 - 1877 04 26 - 1877 11 29 | Devonport DYd

Penguin (1874)
1 Leslie horizontal compound expansion E | 666 HP = 9.87 knots
1874 07 14 - 1876 03 25 - 1877 08 23 | Napier, Govan

1 | "Fantome" type (1872)
Composite screw sloop, Britain

Displacement: 940 tons at draft: 3.81 m | Length: 48.76 m pp | Beam: 9.44 m | Crew: 125 men

Machinery: 3 cylindrical boilers | 1 steam E * 1 screw | 1,000 HP = 11 knots | Supplies: 100 tons of coal

Armament: 2-177.8/15.9 MLR, 2-160/16 MLR

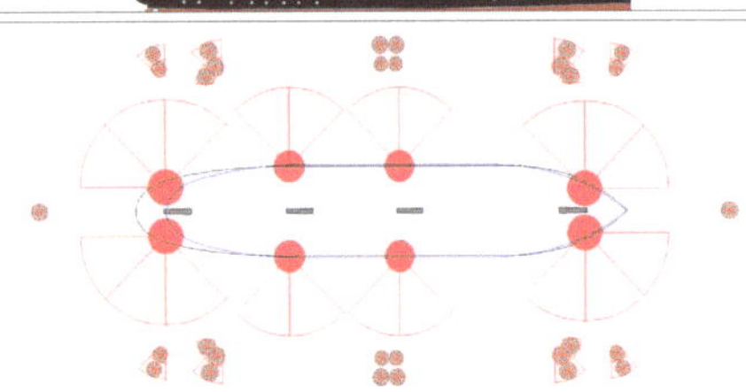

Egeria (1872)
1 Humphrys horizontal compound expansion E | 1,011 HP = 11.3 kts
1872 12 30 - 1873 11 01 - 1874 11 01 | Pembroke DYd

Magpie

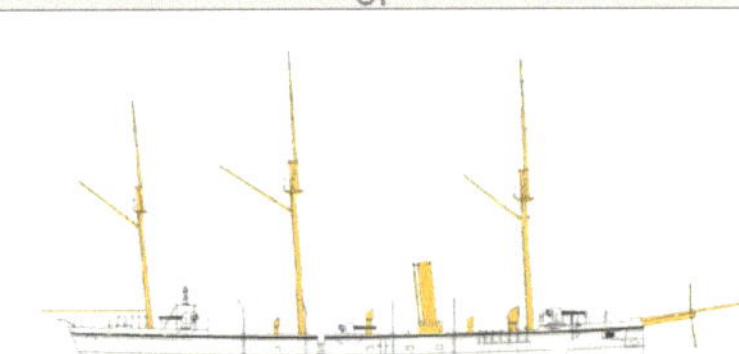

5 +4 | "Redbreast" type (1888)
Composite gunboats, Britain

BUILDING

1Y:12M	1Y:12M	1Y:12M	1Y:12M
Redpoll	Sparrow	Thrush	Widgeon

Displacement: 805 tons at draft: 3.35 m | Length: 50.29 m pp | Beam: 9.45 m | Crew: 76 men

Machinery: 2 cylindrical boilers | 1 steam E * 1 screw | 1,200 HP = 13 knots | Supplies: 105 tons of coal

Armament: 6-101.6/40, 2-47/43

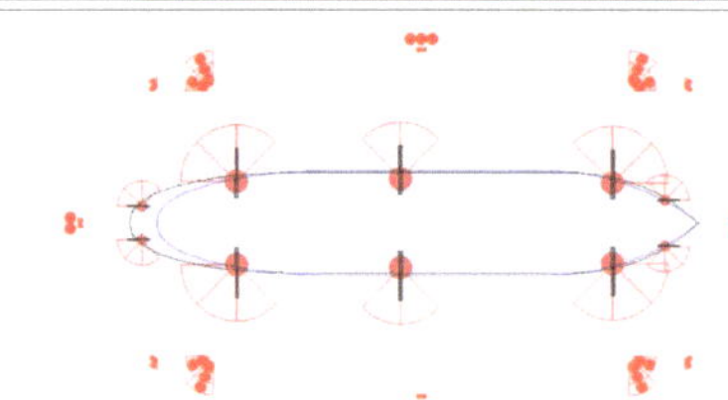

Goldfinch (1888)
1 builder's vertical triple expansion E |
1888 01 01 - 1889 05 18 - 1889 11 11 | Sheerness DYd

Lapwing (1888)
1 builder's vertical triple expansion E |
1888 05 01 - 1889 04 12 - 1889 10 10 | Devonport DYd

Magpie (1888)
1 Earle vertical triple expansion E |
1888 01 01 - 1889 03 15 - 1889 09 09 | Pembroke DYd

Redbreast (1888)
1 Earle vertical triple expansion E |
1888 01 01 - 1889 04 25 - 1889 11 11 | Pembroke DYd

5 + 4 | "Redbreast" type (1888)
Composite gunboats, Britain (continued)

Redbreast

Ringdove (1888)
1 builder's vertical triple expansion E |
1888 06 01 - 1889 04 30 - 1889 11 11 | Devonport DYd

Redpoll (1888)
1 Earle vertical triple expansion E |
1888 01 01 - 1889 06 13 - ***** | Pembroke DYd

Sparrow (1888)
1 builder's vertical triple expansion E |
1888 01 01 - 1889 09 26 - ***** | Scotts SB, Greenock

Thrush (1888)
1 builder's vertical triple expansion E |
1888 01 01 - 1889 06 22 - ***** | Scotts SB, Greenock

Widgeon (1888)
1 Rennie vertical triple expansion E |
1888 01 01 - 1889 08 09 - ***** | Pembroke DYd

6 | "Pigmy" type (1887)
Composite gunboats, Britain

Displacement: 755 tons at draft = 3.43 m | Length: 50.29 m. pp | Beam: 9.14 m | Crew: 76 men

Machinery: 2 cylindrical boilers | 1 steam E * 1 screw | 1,200 HP = 13 knots | Supplies: 105 tons of coal

Armament: 6-101.6/27, 4-11.4 Mg

Age 3

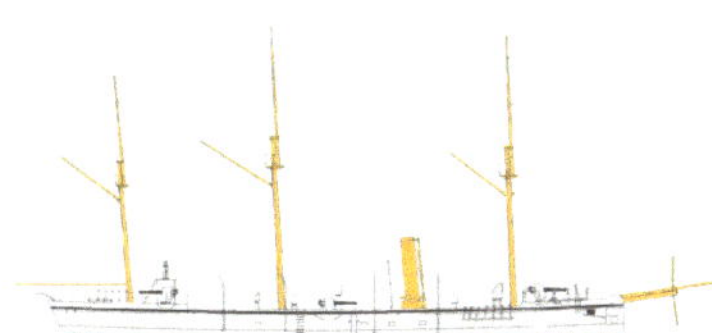

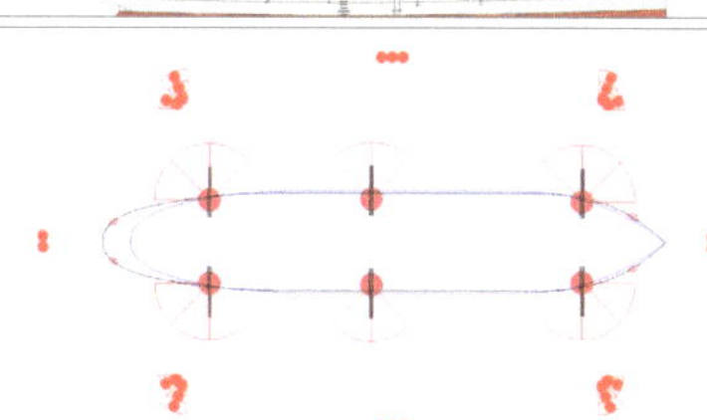

6 | "Pigmy" type (1887)
Composite gunboats, Britain (continued)

Pigmy (1887)
1 Barrow vertical triple expansion E |
1887 01 01 - 1888 07 27 - 1889 07 01 | Sheerness DYd

Plover (1887)
1 Barrow vertical triple expansion E |
1887 01 01 - 1888 10 18 - 1889 07 01 | Pembroke DYd

Partridge (1887)
1 builder's vertical triple expansion E |
1887 06 06 - 1888 05 10 - 1888 12 01 | Devonport DYd

Peacock (1887)
1 Barrow vertical triple expansion E |
1887 01 01 - 1888 06 22 - 1888 11 01 | Pembroke DYd

Pheasant (1887)
1 builder's vertical triple expansion E |
1887 06 06 - 1888 04 10 - 1888 10 01 | Devonport DYd

Pigeon (1887)
1 Barrow vertical triple expansion E |
1887 01 01 - 1888 09 05 - 1889 03 01 | Pembroke DYd

Partridge

3 | "Bramble" type (1885)
Composite gunboats, Britain

Displacement: 715 tons at draft: 3.35 m, 810 tons full | Length: 50,29 m pp | Beam: 8.84 m | Crew: 76 men

Machinery: 2 cylindrical boilers | 1 steam E * 1 screw | 1,000 HP = 13 knots | Supplies: 105 tons of coal

Armament: 6-101.6/27, 4 Mg

Age 5

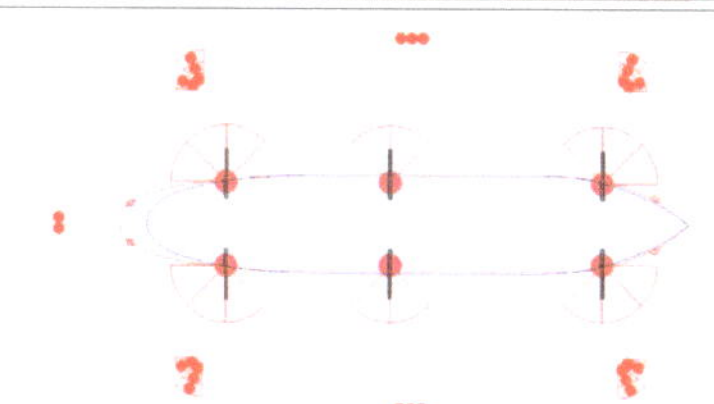

Bramble (1885)
1 builder's vertical triple expansion E |
1885 08 01 - 1886 12 11 - 1887 10 01 | Harland, Belfast

Lizard (1885)
1 builder's vertical triple expansion E |
1885 01 01 - 1886 11 27 - 1887 10 01 | Harland, Belfast

Rattler (1885)
1 Leslie vertical triple expansion E |
1885 08 15 - 1886 08 04 - 1886 12 04 | Armstrong, Elswick

2 | "Curlew" type (1885)
Gunboats, Britain

Displacement: 950 tons at draft: 3.2 m | Length: 59.44 m pp | Beam: 8.53 m | Crew: 46 men

Machinery: 4 cylindrical boilers | 2 * 750 HP steam EE * 2 screws | 1,500 HP = 14 knots | Supplies: 250 tons of coal

Armament: 1-152.4/25, 3-127/25, 7-11.4 Mg, 3-356 TT (2*1 trainable side sets, 1 >)

Age 5

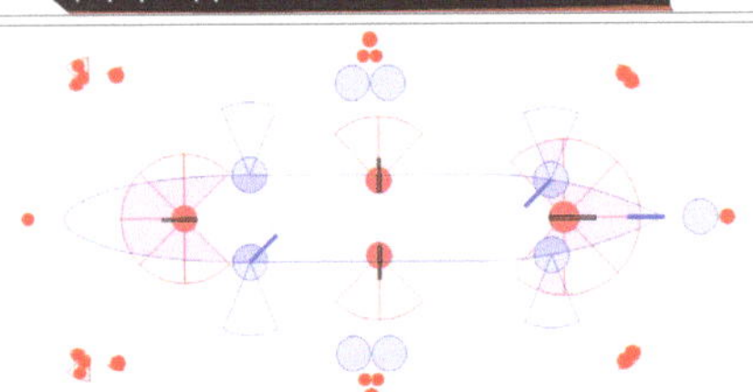

Curlew (1885)
1 Penn horizontal direct acting compound E |
1885 01 05 - 1885 10 23 - 1886 06 22 | Devonport DYd

Landrall (1885)
1 Penn horizontal direct acting compound E |
1885 01 05 - 1886 01 19 - 1887 03 10 | Devonport DYd

3 | "Albacore" type (1881)
Composite gunboats, Britain

Displacement: 560 tons at draft: 2.13 m | Length: 41.15 m pp | Beam: 7.92 m | Crew: 60 men

Machinery: 2 cylindrical boilers | 1 steam E * 1 screw | 650 HP = 10.7 knots | Supplies: 85 tons of coal

Armament: 2-127/25, 2-101.6/27

Age 9

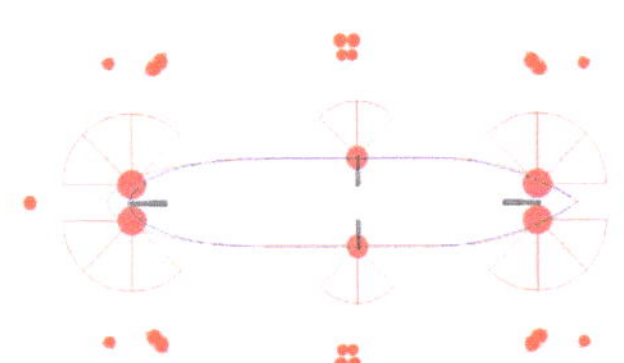

Albacore (1881)
1 builder's horizontal compound E |
1881 01 01 - 1883 01 13 - 1884 02 01 | Laird, Birkenhead

Misletoe (1881)
1 builder's horizontal compound E |
1881 01 01 - 1883 02 07 - 1884 08 01 | Laird, Birkenhead

Watchful (1881)
1 builder's horizontal compound E |
1881 01 01 - 1883 02 13 - 1884 08 01 | Laird, Birkenhead

2 | "Algerine" type (1879)
Composite gunboats, Britain

Displacement: 835 tons at draft: 4.11 m | Length: 47.85 m pp | Beam: 8.99 m | Crew: 100 men

Machinery: 2 cylindrical boilers | 1 return compound E * 1 screw | 810 HP = 10.5 knots | Supplies: 110 tons of coal

Armament: 1-177.8/15.9 MLR, 2-160/16 MLR

Age 11

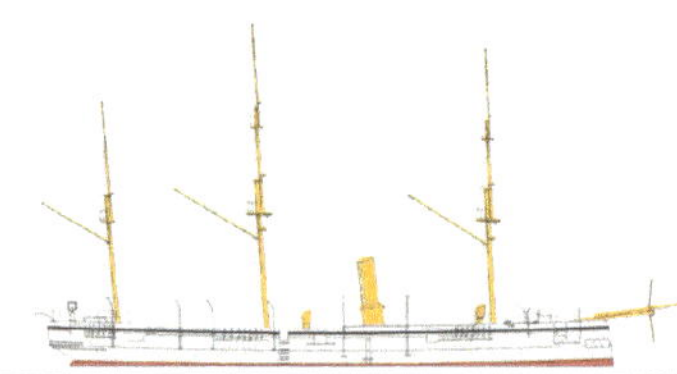

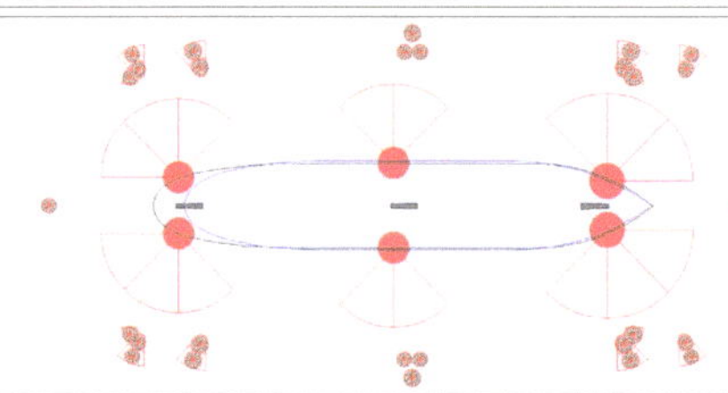

Algerine (1879)
1879 01 01 - 1880 11 06 - 1881 12 12 | Harland, Belfast

Ranger (1879)
1879 01 01 - 1880 02 12 - 1881 01 01 | Elder, Glasgow

Ranger

2 | "Linnet" type (1878)
Composite gunboats, Britain

Displacement: 756 tons at draft: 3,12 m, 788 tons at draft: 3.3 m | Length: 50.29 m pp | Beam: 8,84 m | Crew: 92 men

Machinery: 2 cylindrical boilers | 1 horizontal compound E * 1 screw | 870 HP = 11.8 knots | Supplies: 110 tons of coal

Armament: 1-177.8/15.9 MLR, 3-95.3/21

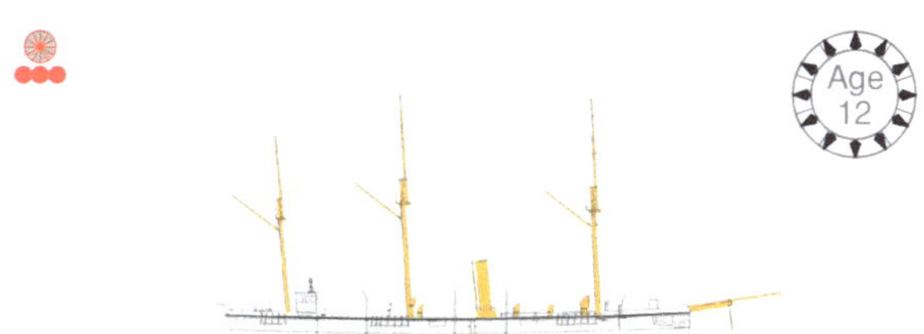

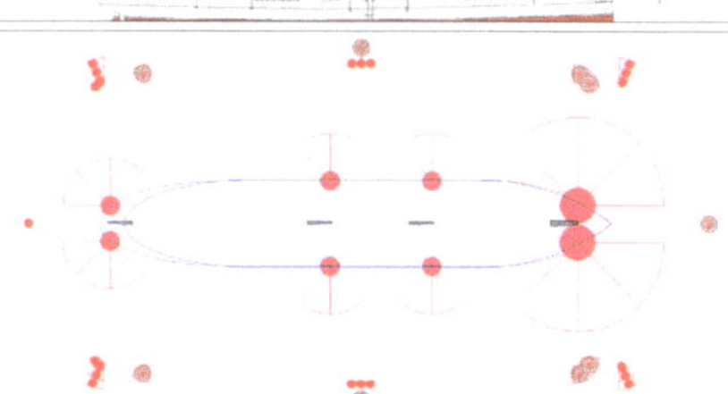

Linnet (1878) | 1 Rennie engine
1878 12 12 - 1880 01 30 - 1881 01 01 | Thames IW, Blackwall

Swift (1878) | 1 Rennie engine
1878 12 12 - 1879 11 29 - 1881 01 01 | Thames IW, Blackwall

3 | "Condor" type (1875)
Composite gunboats, Britain

Displacement: 774 tons at draft: 3,96 m, 780 tons full; Length: 47.85 m pp | Beam: 8.99 m | Crew: 100 men

Machinery: 2 cylindrical boilers | 1 compound recyprocating E * 1 screw | 721 HP = 11.5 knots | Supplies: app. 90 tons of coal

Armament: 1-177.8/15.9 MLR, 2-160/15.9 MLR

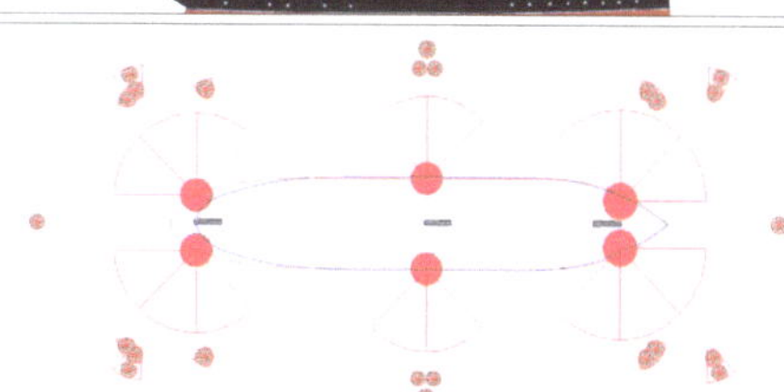

Falcon (1876)
1876 01 01 - 1877 01 04 - 1878 01 01 | Laird, Birkenhead

Griffon (1876)
1876 01 01 - 1876 12 16 - 1877 09 01 | Laird, Birkenhead

Flamingo (1875)
1875 12 15 - 1876 12 13 - 1877 06 01 | Devonport DYd

3 | "Frolic" type (1871)
Composite gunboats, Britain

Displacement: 610 tons at draft: 2.9 m | Length: 47.2 m pp | Beam: 7.62 m | Crew: 80 men

Machinery: 2 cylindrical boilers | 1 horizontal single expansion E * 1 screw | 896 HP = 10.7 knots | Supplies: 80 tons of coal

Armament: 1-177.8/15.9 MLR, 1-160/15.9 MLR, 2-95.3/21

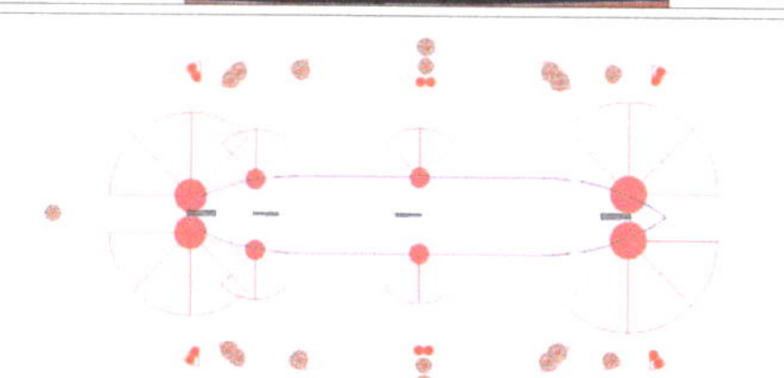

Frolic (1871)
1871 01 01 - 1872 02 29 - 1873 01 01 | Chatham DYd

Ready (1871)
1871 01 01 - 1872 09 24 - 1874 01 01 | Chatham DYd

Rifleman (1871)
1871 01 01 - 1872 11 20 - 1873 12 12 | Chatham DYd

3 | "Beacon" type (1867)
Composite gunboats, Britain

Displacement: 603 tons at draft: 2.9 m | Length: 47.24 m pp | Beam: 7.62 m | Crew: 80 men

Machinery: 2 cylindrical boilers | 1 horizontal direct acting recyprocating E * 1 screw | 506 HP = 9.6 knots | Supplies: 90 tons of coal

Armament: 1-177.8/15.9 MLR, 1-160/16 MLR, 2-95.3/21

Age 24

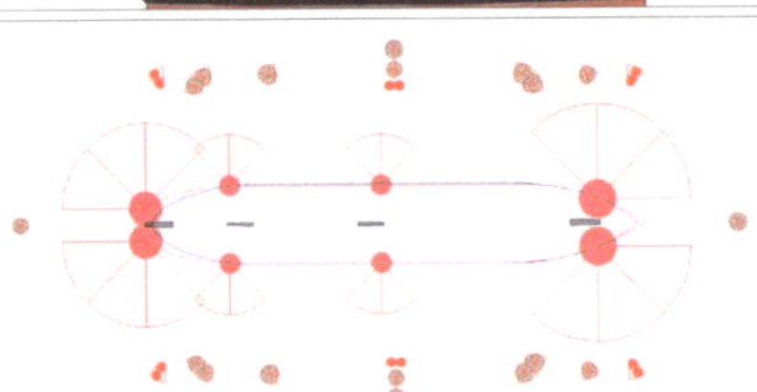

Avon (1866)
1866 01 01 - 1867 10 02 - 1869 11 01 | Portsmouth DYd

Elk (1867)
1867 01 01 - 1868 01 10 - 1869 01 01 | Portsmouth DYd

Midge (1867)
1867 01 01 - 1868 05 21 - 1869 01 01 | Elder, Glasgow

Midge

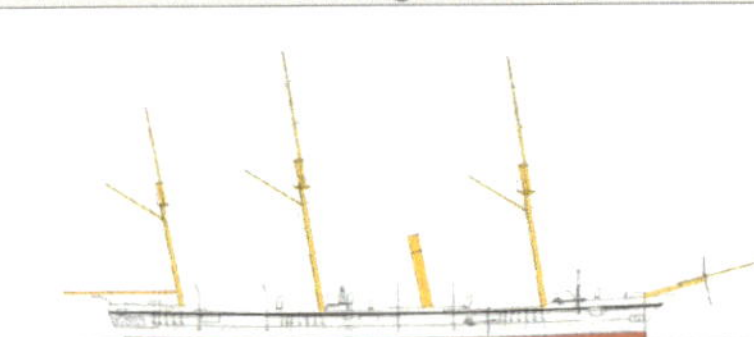

2 | "Vixen" type (1864)
Armoured gunboats, Britain

Displacement: 1,230 tons at draft: 3.55 m | Length: 48.77 m pp | Beam: 9.75 m | Crew: 80 men

Machinery: 2 cylindrical boilers | 2 horizontal E * 2 screws | 700 HP = 9.6 knots | Supplies: app. 100 tons of coal

Armour: iron | Battery: 114 | WLBelt: 114

Armament: 2-177.8/15.9 MLR, 2-95.3/21

Age 26

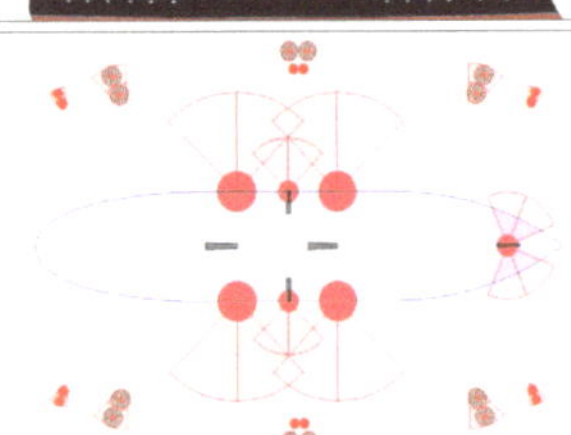

Viper (1864)
1864 01 01 - 1865 12 21 - 1867 01 01 | Dudgeon, Poplar

Vixen (1864)
1864 01 01 - 1865 11 18 - 1866 01 01 | Lungley, Deptford

1 | "Waterwitch" type (1864)
Armoured gunboat, Britain

Displacement: 1280 tons at draft: 3.63 m | Length: 49.38 m pp | Beam: 9.78 m | Crew: 80 men

Machinery: 2 cylindrical boilers | 2 Ruthven hydraulic turbines * 2 screws | 780 HP = 8.9 knots | Supplies: app. 100 tons of coal

Armour: iron | Battery: 114, WLBelt: 114

Armament: 2-177.8/15.9 MLR, 2-95.3/21

Age 26

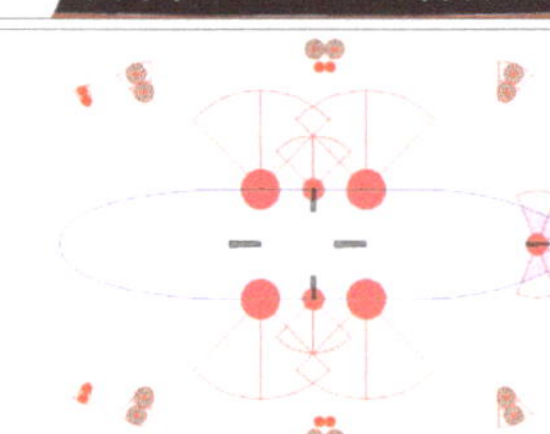

Waterwitch (1864)
1864 01 01 - 1865 06 28 - 1867 01 01 | Thames IW, Blackwall

1 | "Ranger" type (1858)
Wooden gunboat, Britain

Displacement: 570 tons at draft: 3.35 m | Length: 44.2 m pp | Beam: 7.72 m | Crew: 60 men

Machinery: 2 cylindrical boilers | 1 return connecting rod E * 1 screw | 334 HP = 9.9 knots | Supplies: app. 50 tons of coal

Armament: 1-177.8/15.9 MLR, 2-120/8, 2-95.3/21

Age 31

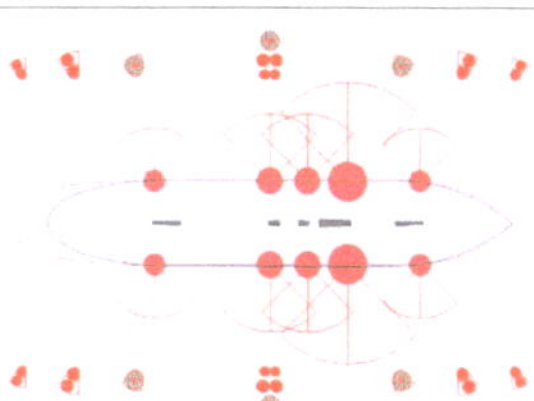

Nimble (1859)
1859 10 03 - 1860 09 15 - 1861 04 08 | Pembroke DYd

1 | "Drudge" type (1887)
Gunboat, Britain

NO PHOTO AVAILABLE

Displacement: 890 tons at draft: 3.8 m | Length: 38.1 m pp | Beam: 10.7 m | Crew: app. 80 men

Machinery: 2 cylindrical boilers | 2 * 155 HP vertical triple expansion EE * 2 screws | 310 HP = 8.5 knots | Supplies: 110 tons of coal

Armament: 1-343/30, 1-120/40

Age 3

Drudge (1887)
1887 01 12 - 1887 06 15 - 1887 08 23 | Armstrong, Elswick

1 | "Handy" type (1882)
Gunboat, Britain

Displacement: 508 tons at draft: 2.75 m | Length: 35.1 m pp | Beam: 11.3 m | Crew: app. 80 men

Machinery: 2 cylindrical boilers | 2 * 155 HP vertical compound EE * 2 screws | 310 HP = 8.5 knots | Supplies: 110 tons of coal

Armament: 1-343/30

Age 8

Handy (1882)
1882 08 01 - 1882 12 30 - 1883 07 12 | Armstrong, Elswick

2 | "Bouncer" type (1880)
Gunboats, Britain

Displacement: 265 tons at draft: 1.83 m | Length: 26.63 m. pp | Beam: 7.96 m | Crew: 30 men

Machinery: 1 cylindrical boiler | 2 * 130 HP horizontal reciprocating EE * 2 screws | 268 HP = 8.5 knots | Supplies: 35 tons of coal

Armament: 1-254/14.8 MLR

Bouncer (1880)
1880 01 01 - 1881 01 01 - 1882 01 01 | Pembroke DYd

Insolent (1880)
1880 01 01 - 1881 01 01 - 1882 01 01 | Pembroke DYd

10 | "Banterer" type (1880)
Gunboats, Britain

Displacement: 465 tons at draft: 3.05 m | Length: 38.1 m pp | Beam: 7.16 m | Crew: 92 men

Machinery: 2 cylindrical boilers | 1 horizontal reciprocating E * 1 screw | 440 HP = 9.5 knots | Supplies: 40 tons of coal

Armament: 2-160/15.9 MLR, 2-95.3/21, 2-25.4/40 Mg

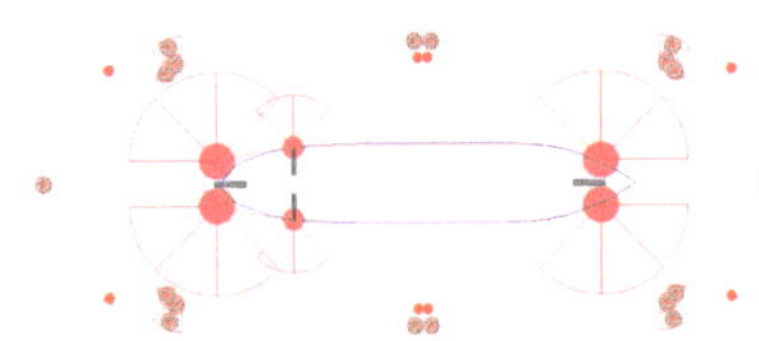

Banterer (1880)
1880 01 01 - 1880 11 02 - 1882 10 01 | Barrow SB, Barrow

Bullfrog (1880)
1880 01 01 - 1881 02 03 - 1882 01 01 | Pembroke DYd

Cockchafer (1880)
1880 01 01 - 1881 02 19 - 1882 01 01 | Pembroke DYd

Espoir (1880)
1880 01 01 - 1880 11 02 - 1882 01 01 | Barrow SB, Barrow

Grappler (1880)
1880 01 01 - 1880 10 05 - 1882 01 01 | Barrow SB, Barrow

Raven (1881)
1881 01 01 - 1882 05 18 - 1883 04 01 | Samuda, Poplar

Redwing (1880)
1880 01 01 - 1880 05 25 - 1880 12 12 | Pembroke DYd

10 | "Banterer" type (1880)
Gunboats, Britain (continued)

Starling (1881)
1881 01 01 - 1882 04 19 - 1882 08 01 | Samuda, Poplar

Stork (1881)
1881 01 01 - 1882 05 18 - 1883 01 01 | Samuda, Poplar

Wrangler (1880)
1880 01 01 - 1880 10 05 - 1881 10 01 | Barrow SB, Barrow

4 | "Gadfly" type (1878)
Gunboats, Britain

Displacement: 254 tons at draft: 1.83 m | Length: 25.91 m pp | Beam: 7.96 m | Crew: 30 men

Machinery: 1 cylindrical boiler | 2 * 130 HP horizontal compound reciprocating EE * 2 screws | 260 HP = 8.5 knots | Supplies: 35 tons of coal

Armament: 1-254/14.8 MLR

Gadfly (1878)
1878 01 01 - 1879 05 05 - 1880 05 05 | Pembroke DYd

Griper (1879)
1879 01 01 - 1879 09 15 - 1880 09 09 | Pembroke DYd

Pincher (1878)
1878 01 01 - 1879 05 05 - 1880 05 05 | Pembroke DYd

Tickler (1879)
1879 01 01 - 1879 09 15 - 1880 09 09 | Pembroke DYd

7 | "Sheldrake" type (1874) Gunboats, Britain

Foxhound

Displacement: 455 tons at draft: 2.59 m | Length: 38.1 m pp | Beam: 7.16 m | Crew: 60 men

Machinery: 2 cylindrical boilers | 1 horizontal compound reciprocating E * 1 screw | 387 HP = 10 knots | Supplies: app. 50 tons of coal

Armament: 2-160/16 MLR, 2-95.3/21

Age 14

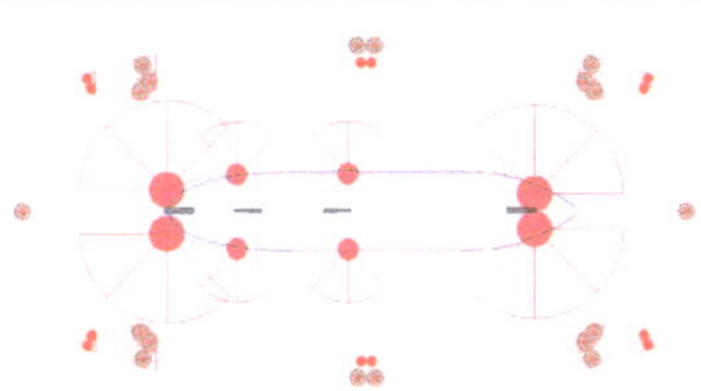

Firebrand (1876)
1876 01 01 - 1877 04 30 - 1878 01 01 | Thompson, Clydebank

Firefly (1877)
1877 01 01 - 1877 06 28 - 1878 01 01 | Thompson, Clydebank

Firm (1876)
1876 01 01 - 1877 02 14 - 1878 01 01 | Earle, Hull

Forester (1876)
1876 01 01 - 1877 02 26 - 1878 01 01 | Earle, Hull

Forward (1876)
1876 01 01 - 1877 01 29 - 1878 01 01 | Barrow SB, Barrow

Foxhound (1876)
1876 01 01 - 1877 01 29 - 1878 01 01 | Barrow SB, Barrow

Sheldrake (1874)
1874 01 01 - 1875 07 03 - 1876 01 01 | Napier, Govan

7 | "Sheldrake" type (1874) Gunboats, Britain (continued)

Firebrand

12 | "Medina" type (1875) Gunboats, Britain (continued)

Esk, Tweed

Spey (1875)
1875 10 10 - 1876 10 05 - 1877 10 10 | Palmer, Jarrow

Tay (1875)
1875 10 10 - 1876 10 19 - 1877 10 10 | Palmer, Jarrow

Tees (1875)
1875 10 10 - 1876 10 19 - 1877 10 10 | Palmer, Jarrow

Trent (1876)
1876 08 08 - 1877 08 23 - 1878 08 08 | Palmer, Jarrow

Tweed (1876)
1876 08 08 - 1877 08 23 - 1878 08 08 | Palmer, Jarrow

12 | "Medina" type (1875) Gunboats, Britain

Displacement: 386 tons at draft: 1.7 m | Length: 33.53 m pp | Beam: 10.36 m | Crew: 51 men

Machinery: 2 cylindrical boilers | 1 horizontal compound reciprocating E * 1 screw | 310 HP = 9.5 knots | Supplies: app. 50 tons of coal

Armament: 3-160/16 MLR

Age 15

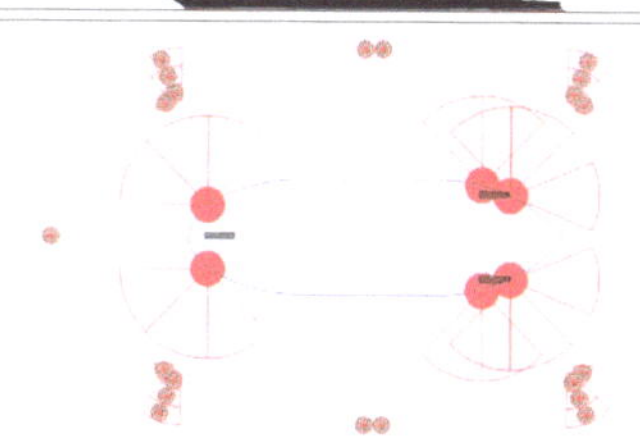

Dee (1876)
1876 04 04 - 1877 04 04 - 1878 04 04 | Palmer, Jarrow

Don (1876)
1876 04 04 - 1877 04 14 - 1878 04 04 | Palmer, Jarrow

Esk (1876)
1876 04 04 - 1877 04 28 - 1878 04 04 | Palmer, Jarrow

Medina (1875)
1875 08 08 - 1876 08 03 - 1877 08 08 | Palmer, Jarrow

Medway (1875)
1875 08 08 - 1876 10 03 - 1877 08 08 | Palmer, Jarrow

Sabrina (1875)
1875 10 10 - 1876 10 03 - 1877 10 10 | Palmer, Jarrow

Slaney (1876)
1876 04 04 - 1877 04 28 - 1878 04 04 | Palmer, Jarrow

20 | "Arrow" type (1870)
Gunboats, Britain

Displacement: 254 tons at draft 1.83 m | Length: 25.91 m pp | Beam: 7.96 m | Crew: 30 men

Machinery: 1 cylindrical boiler | 2 * 130 HP horizontal compound reciprocating EE * 2 screws | 260 HP = 8.5 knots | Supplies: app. 8 tons of coal

Armament: 1-254/14.5 MLR

Age 20

Ant (1872)
1872 08 08 - 1873 08 14 - 1874 08 08 | Laird, Birkenhead

Arrow (1870)
1870 04 04 - 1871 04 22 - 1872 04 04 | Rennie, London

Badger (1871)
1871 03 03 - 1872 03 13 - 1873 03 03 | Chatham DYd

Blazer (1870)
1870 01 01 - 1870 12 07 - 1872 01 01 | Portsmouth DYd

Bloodhound (1870)
1870 09 01 - 1871 04 22 - 1871 04 25 | Armstrong, Low Walker

Bonetta (1870)
1870 05 05 - 1871 05 20 - 1872 05 05 | Rennie, London

Bulldog (1871)
1871 09 09 - 1872 09 17 - 1873 09 09 | Campbell, North Woolwich

20 | "Arrow" type (1870)
Gunboats, Britain (continued)

Bustard (1870)
1870 01 01 - 1871 01 07 - 1872 01 01 | Napier, Govan

Comet (1870)
1870 01 01 - 1870 12 08 - 1872 01 01 | Portsmouth DYd

Cuckoo (1872)
1872 08 08 - 1873 08 14 - 1874 08 08 | Laird, Birkenhead

Fidget (1871)
1871 03 03 - 1872 03 13 - 1873 03 03 | Chatham DYd

Hyaena (1872)
1872 08 08 - 1873 08 30 - 1874 08 08 | Laird, Birkenhead

Kite (1870)
1870 01 01 - 1870 12 08 - 1872 01 01 | Napier, Govan

Mastiff (1870)
1870 09 01 - 1871 04 04 - 1871 04 06 | Armstrong, Low Walker

Pickle (1871)
1871 11 11 - 1872 11 15 - 1873 11 11 | Campbell, North Woolwich

Pike (1871)
1871 10 10 - 1872 10 16 - 1873 10 10 | Campbell, North Woolwich

Scourge (1870)
1870 03 03 - 1871 03 25 - 1872 03 03 | Chatham DYd

Snake (1870)
1870 03 03 - 1871 03 25 - 1872 03 03 | Chatham DYd

Snap (1871)
1871 12 12 - 1872 12 11 - 1873 12 12 | Campbell, North Woolwich

Weazel (1872)
1872 09 09 - 1873 09 04 - 1874 09 09 | Laird, Birkenhead

3 | "Ariel" type (1872)
Gunboats, Britain

Displacement: 430 tons at draft: 3,2 m | Length: 38.1 m pp | Beam: 6.85 m | Crew: 60 men

Machinery: 2 cylindrical boilers | 1 horizontal compound reciprocating E * 1 screw | 400 HP = 9.5 knots | Supplies: 40 tons of coal

Armament: 2-160/15.9 MLR, 2-95.3/21

Age 19

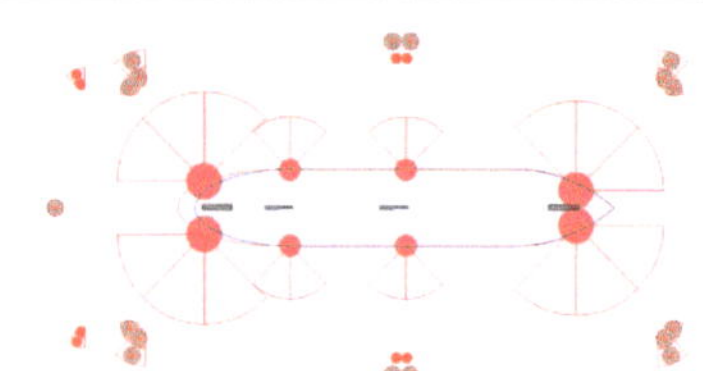

Goshawk (1871)
1871 01 01 - 1872 01 23 - 1873 09 01 | Pembroke DYd

Merlin (1871)
1871 01 01 - 1871 12 09 - 1873 11 01 | Pembroke DYd

Swinger (1871)
1871 01 01 - 1872 02 07 - 1873 09 01 | Pembroke DYd

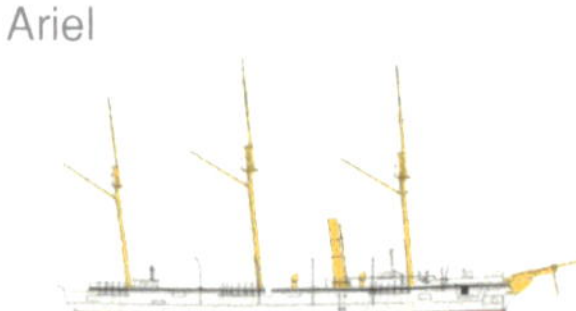

1 | "Plucky" type (1869)
Wooden gunboat, Britain

NO PHOTO AVAILABLE

Displacement: 212 tons at draft: 1.87 m | Length: 23.38 m pp | Beam: 7.66 m | Crew: 31 men

Machinery: 1 cylindrical boilers | 2 * 110 HP steam EE * 2 screws | 220 HP = 7 knots | Supplies: app. 20 tons of coal

Armament: 1-228,6/13,9 MLR

Age 19

Plucky (1871)
2 horizontal single expansion engines | 224 HP = 7.7 knots | 1871 01 01 - 1872 02 07 - 1873 09 01 | Pembroke DYd

1 | "Staunch" type (1867)
Gunboat, Britain

Displacement: 164 tons at draft: 1.98 m, 200 tons full | Length: 22.9 m pp | Beam: 7.62 m | Crew: 31 men

Machinery: 1 cylindrical boiler | 2 * 75 HP steam EE * 2 screws | designed 150 HP = 7 knots | Supplies: app. 8 tons of coal

Armament: 1-228.6/14.5 MLR

Age 23

Staunch (1867)
2 Stephenson horizontal single expansion | 134 HP = 7.7 knots | 1867 06 17 - 1867 11 04 - 1867 12 04 | Armstrong, Low Walker

4 | "Britomart" type (1859)
Wooden gunboats, Britain

Displacement: 330 tons at draft: 2.74 m | Length: 36.6 m pp | Beam: 6.71 m | Crew: 36 men

Machinery: 1 rectangular boiler | 1 reciprocating single expansion E * 1 screw | 260 HP = 9 knots | Supplies: app. 35 tons of coal

Armament: 2-206/14 MLR

Age 31

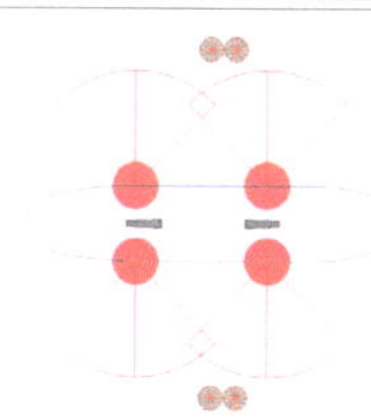

Britomart (1859)
1859 01 01 - 1860 01 01 - 1861 01 01 | Smith's, Middlesborough

Cherub (1860)
1860 12 03 - 1865 03 29 - 1866 04 14 | Portsmouth DYd

Minstrel (1866)
1866 01 01 - 1867 01 01 - 1868 01 01 | Portsmouth DYd

Tyrian (1865)
1865 01 01 - 1866 01 01 - 1867 01 01 | Courtenay, ?

Her Majesty's Australian Ships

Besides of the Australian ships there were British fleet ships assigned to the Australian detachment consisted on the 1890-01-01 of 1 belted cruiser, 2 screw sloops and 2 gunboats, also to replace 1 2nd class protected cruiser and 1 screw corvette went from Australia to Britain there was 1 2nd class protected cruiser sent to the Australian detachment.

1 | "Cerberus" type (1867)
Monitor, Australia

MONITOR

Displacement: 3,344 tons at draft: 4.67 m | Length: 68.58 m pp | Beam: 13.72 m | Crew: 155 men

Machinery: 4 rectangular boilers | 2 * 700 HP steam EE * 2 screws | 1,400 HP = 11 knots | Supplies: 120 tons of coal

Armour: iron | CT: 229-203; Turrets: 254-229; Breastwork: 229-203; WLBelt: 152(^38) 203(^38) 152(^38);

Armament: 4-254/14 MLR, 8-47/43

Age 23

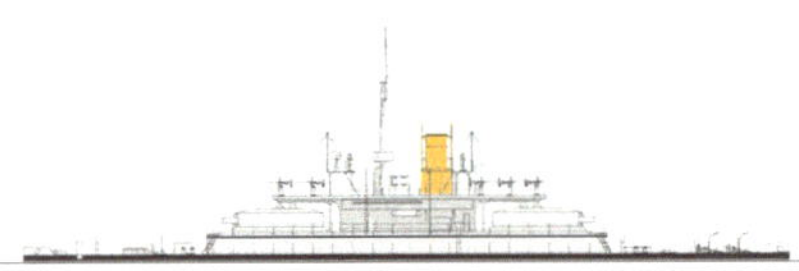

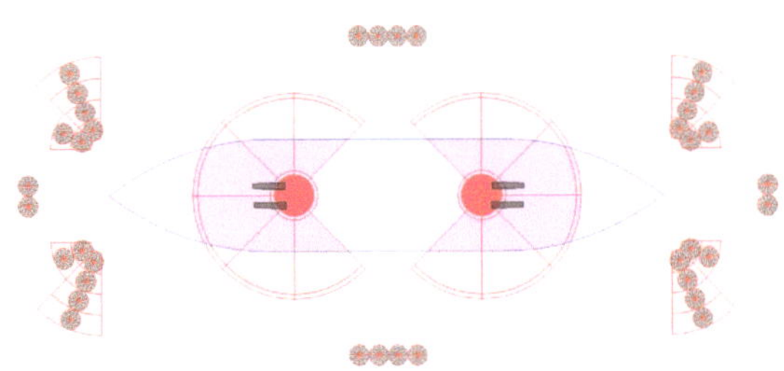

Cerberus (1867)
2 Maudslay horizontal return connecting rod EE | 1,369 HP = 9.75 knots |
1867 09 01 - 1868 12 02 - 1870 09 01 | Palmer, Jarrow

Sister ship can be found in the Indian section

1 | "Protector" type (1882)
Protected cruiser, Australia

PROTECTED CRUISERS

Displacement: 921 tons at draft: 3.8 m | Length: 65.3 m oa, 55 m pp | Beam: 9.1 m | Crew: 90 men

Machinery: 4 cylindrical boilers | 2 * 750 HP steam EE * 2 screws | 1500 HP = 14 knots | Supplies: 150 / +25 tons of coal

Armour: steel | CT: 114; Breastwork: 25 | WLBelt: 25

Armament: 1-203.2/28, 5-152.4/26, 4-25.4/40 Mg

Age 8

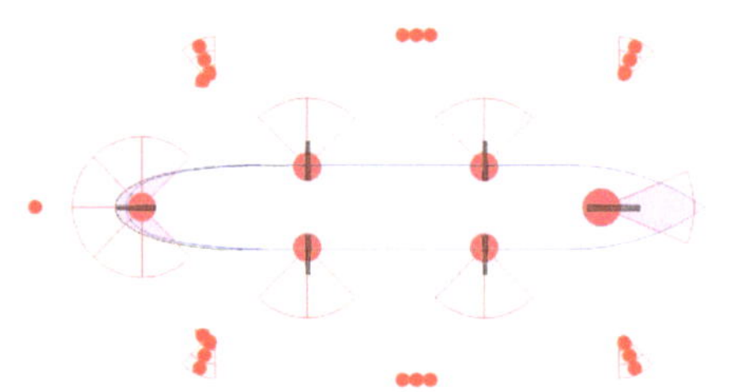

Protector (1882)
2 Leslie horizontal direct acting compound EE | 1,645 HP = 14.15 knots |
1882 08 02 - 1883 12 27 - 1884 06 19 | Armstrong, Low Walker

0 +5 | "Phoenix" type (1888)
Protected cruisers 3 class, Australia

BUILDING IN BRITAIN

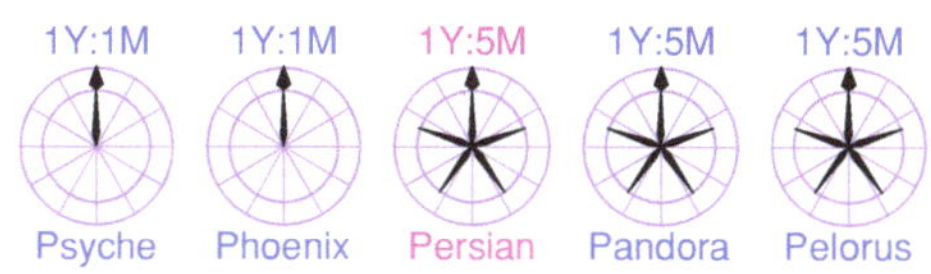

Displacement: 2,575 tons at draft: 4.72 m | Length: 84.73 m | Beam: 12.5 m | Crew: 217 men

Machinery: 4 cylindrical boilers | 2 * 2,000 HPn / 3,750 HPf steam EE * 2 screws | 4,000 HPn = 17 knots / 7,500 HPf = 19 knots | Supplies: 300 / +142 tons of coal

Armour: steel | CT: 76(^25); Gun shields: 51 |
WLDeck: 0(/51^25)

Armament: 8-120/40, 8-47/43, 2-25.4/40 Mg, 2-356 TT (^)

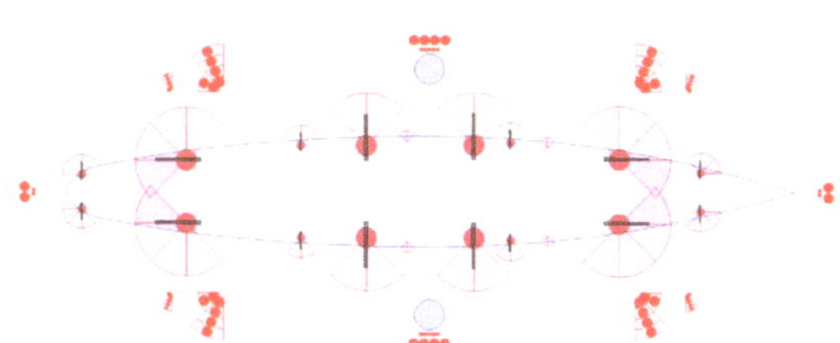

Persian (1888)
2 Leslie vertical triple expansion EE |
1888 08 16 - ***** - ***** |/ Armstrong, Elswick

Pandora (1888)
2 Leslie vertical triple expansion EE |
1888 08 15 - 1889 08 27 - ***** | Armstrong, Elswick

0 +5 | "Phoenix" type (1888)
Protected cruisers 3 class, Australia (continued)

Pelorus (1888)
2 Leslie vertical triple expansion EE |
1888 08 15 - 1889 11 25 - ***** | Armstrong, Elswick

Phoenix (1888)
2 builder's vertical triple expansion EE |
1888 12 01 - 1889 10 28 - ***** | Thompson, Clydebank

Psyche (1888)
2 builder's vertical triple expansion EE |
1888 12 06 - 1889 12 10 - ***** | Thompson, Clydebank

4 more cruisers are built for the British navy and can be found in the British section

WOODEN CRUISERS

1 | "Jason" type (1858)
Training ship (ex-screw corvette), Britain

Displacement: 2,424 tons at draft: 5.43 m | Length: 68.58 m oa | Beam: 12.39 m | Crew: 240 men

Machinery: 4 boilers | 1 steam E * 1 screw | 1,339 HP = 10.21 knots | Supplies: 259 tons of coal

Armament: 20-203.2/14.8 MLR, 1-177.8/15.8

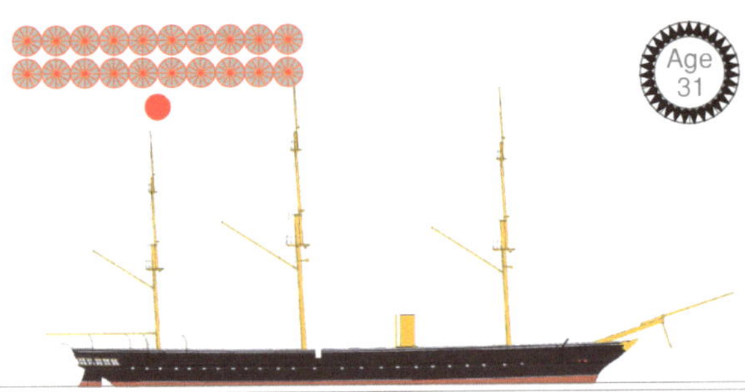

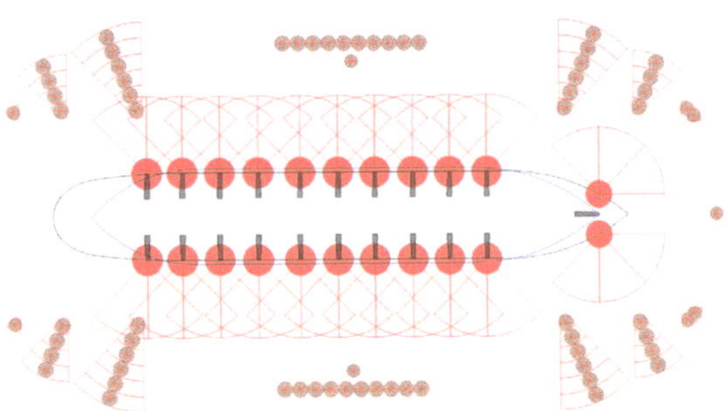

Wolverine (1859)
1 Ravenhill horizontal single expansion E | 1,339 HP = 10.2 knots |
1859 04 01 - 1863 08 29 - 1864 04 01 | Woolwich DYd

1 | "Nelson" type (1809)
Training station (ex-sail battleship), Britain

Dimensioins: 4,096 tons at draft: 7.32 m | Length: 74.37 m. oa | Beam: 16 46 m | Crew: 131 men

Machnery: ?? rectangular boilers | 1 steam E * 1 screw | 2,000 HP = 11 knots | Supplies: app. 400 tons of coal

Rearmed: 2-177.8/14 MLR, 14-160/14.8 MLR

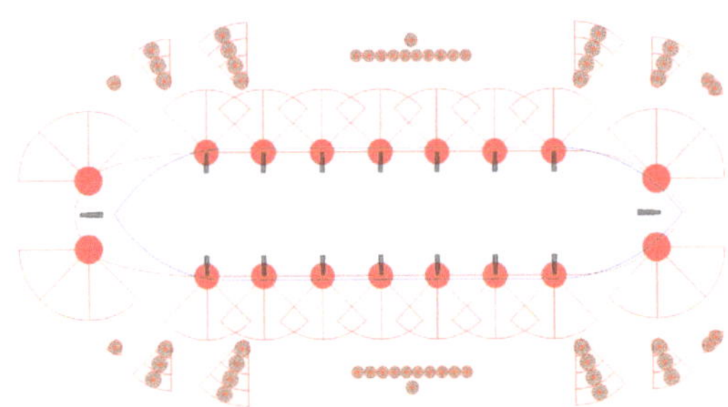

Nelson (1809)
1 Ravenhill horizontal single expansion E | 2,102 HP = 11.53 knots
1809 12 01 - 1814 07 08 - 1814 08 17 | Woolwich DYd

TORPEDO SHIPS

0 +2 | "Sharpshooter" type (1888)
Torpedo gunboats, Britain

BUILDING IN BRITAIN

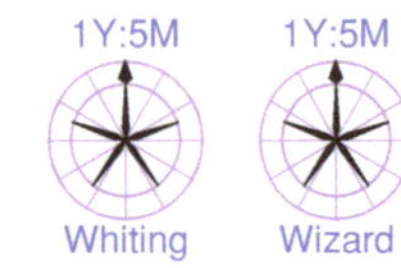

Displacement: 735 tons at draft: 3.2 m | Length: 70.1 m pp | Beam: 8.23 m | Crew: 91 men

Machinery: 4 locomotive boilers | 2 * 1,250 HPn / 1,750 HPf steam EE * 2 screws | 2,500 HP = 16.5 knots / 3,500 HPf = 19 knots | Supplies: 100 tons of coal

Armament: 2-120/30, 4-47/43, 1-11.5 Mg, 5-356 TT (2*2 trn, 1 >)

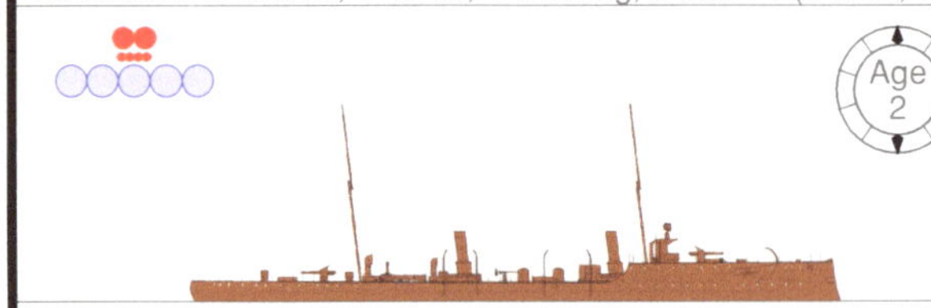

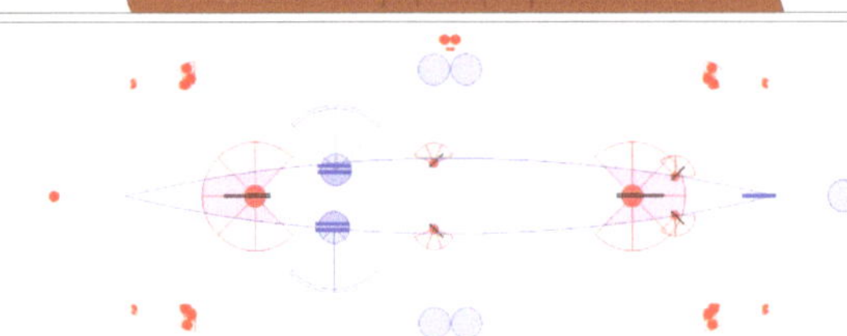

Whiting (1888)
2 Belliss vertical triple expansion EE |
1888 08 17 - 1889 07 24 - ***** | Armstrong, Elswick

Wizard (1888)
2 Belliss vertical triple expansion EE |
1888 08 17 - 1889 08 27 - ***** | Armstrong, Elswick

11 more ships are built for the British and Indian navies and can be found in the respective sections

0 +1 | "Countess of Hopetown" (1889) type
Torpedo boat 1 class, Ausiralia

BUILDING IN BRITAIN

0Y:12M

Countess of Hopetown

Displacement: 82 tons at draft: 1.68 m | Length: 39.22 m oa | 36.62 m pp | Beam: 4.11 m | Crew: 19 men

Machinery: 1 locomotive boiler | 1 vertical triple expansion E * 1 screw | 1,150 HP = 23 knots | Supplies: 15 tons of coal

Armament: 3-47/43, 3-356 TT (1 >, 1*2 trainable)

Age 1

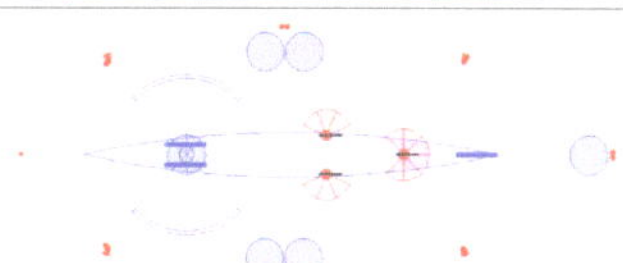

Countess of Hopetown (1889)
1889 01 01 - 1889 12 12 - ***** | Yarrow, Poplar

1 | "Suhum"/"Childers" type (1882)
Torpedo boat 1 class, Russia

Displacement: 65 tons at draft: 1.68 m | Length: 34.44 m oa | 34.04 m pp | Beam: 3.81 m | Crew: 14 men

Machinery: 1 locomotive boiler | 1 vertical compound expansion E * 1 screw | 670 HP = 19 knots | Supplies: 12 tons of coal

Armament: 1-37/23R, 2-381 TT (>)

Age 8

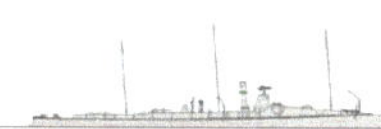

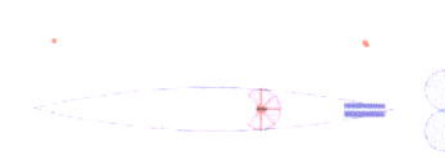

Childers (1882)
1882 12 12 - 1883 06 06 - 1884 01 01 | Thornycroft, Woolston

Sister ships can be found in the British, Russian, Swedish sections

2 | "Acheron" type (1878)
Spar torpedo boats, Ausiralia

Avernus

Displacement: 16 tons at draft: 0.91 m | Length: 23.77 m oa | Beam: 3.05 m | Crew: 9 men

Machinery: 1 locomotive boiler | 1 vertical compound E * 1 screw | 200 HP = 16 knots | Supplies: app. 2 tons of coal

Armament: 1 spar torpedo

Age 12

Acheron (1878)
1878 01 01 - 1879 01 01 - 1880 01 01 | Atlas, Sydney

Avernus (1878)
1878 01 01 - 1879 01 01 - 1880 01 01 | Atlas, Sydney

1 | "Gordon" type (1886)
Torpedo boat 2 class, Ausiralia

Displacement: 10.5 tons at draft: 1.45 m, 12 ton full | Length: 17.08 m oa | Beam: 2.9 m | Crew: 11 men

Machinery: 1 locomotive boiler | 1 compound E * 1 screw | 150 HP = 15 knots | Supplies: app. 1 ton of coal

Armament: 1-25.4/40 Mg, 2-356 TT (2 *1 Djevetskiy)

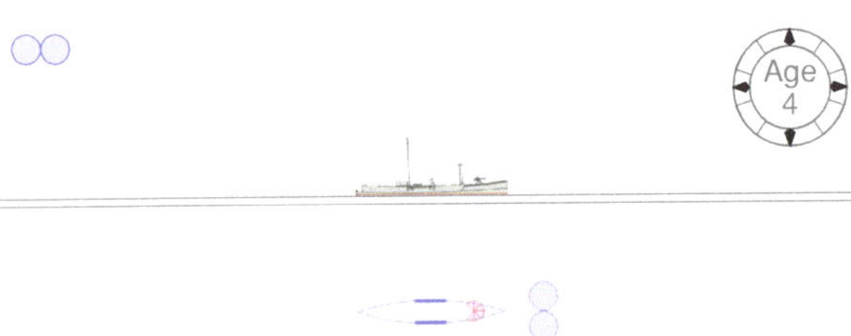

Gordon (1886)
1886 01 01 - 1886 06 06 - 1887 01 01 | White, Cowes

Sister ships can be found in the New Zealandian section

4 | "Mosquito" type (1883)
Torpedo boats 2 class, Ausiralia

Displacement: 10.9 tons at draft: 0.3 m | Length: 19.2 m oa | 18.44 m pp | Beam: 2.25 m | Crew: 7 men

Machinery: 1 locomotive boiler | 1 vertical compound E * 1 screw | 110 HP = 15 knots | Supplies: 1 ton of coal

Armament: 2 spar TT

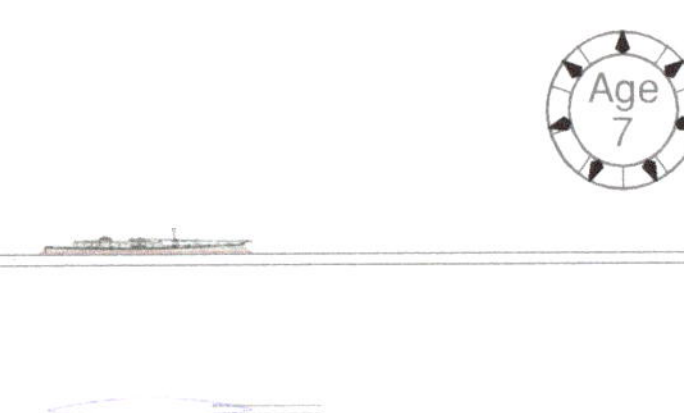

Mosquito (1883)
1883 01 01 - 1884 01 01 - 1885 01 01 | Thornycroft, Woolston

Lonsdale (1883)
1883 01 01 - 1884 01 01 - 1885 01 01 | Thornycroft, Woolston

Nepean (1883)
1883 01 01 - 1884 01 01 - 1885 01 01 | Thornycroft, Woolston

Tasmania

191 (1883)
1883 01 03 - 1883 06 06 - 1883 12 12 | Thornycroft, Woolston

SLOOPS & GUNBOATS

2 | "Gayundah" type (1883)
Gunboats, Ausiralia

Displacement: 385 tons at draft: 2.9 m | Length: 36.6 m oa, 35.1 m pp | Beam: 7.92 m | Crew: 55 men

Machinery: 1 cylindrical boiler | 2 * 190 HP vertical compound EE * 2 screws | 385 HP = 10.57 knots | Supplies: 58 tons of coal

Armament: 1-203.2/28, 1-152.4/26, 2-25.4/40 Mg

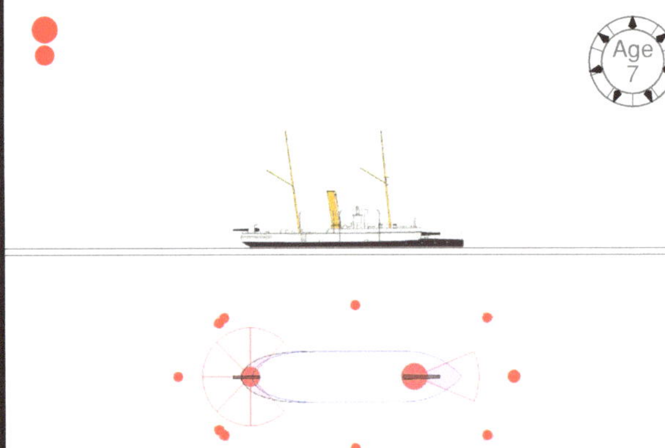

Gayundah (1883)
1883 06 26 - 1884 05 06 - 1884 09 26 | Armstrong, Low Walker

Paluma (1883)
1883 06 26 - 1884 05 06 - 1884 09 26 | Armstrong, Low Walker

5 | "Bonito" type (1883)
Gunboats (ex-barges), Australia

NO PHOTO AVAILABLE

Displacement: 450 tons at draft: 2.74 m | Length: 44.2 m oa | Beam: 8.53 | Crew: ?? men

Machinery: ?? boilers | 1 horizontal direct acting E * 1 screw | 400 HP = 10 knots | Supplies: app. 50 tons

Armament: 1-127/26

Age 7

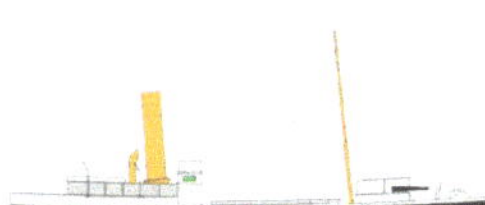

Bonito (1883)
1883 01 01 - 1884 01 01 - 1884 12 12 | Walkers, Maryborough

Bream (1883)
1883 01 01 - 1884 01 01 - 1885 01 01 | Walkers, Maryborough

Dolphin (1884)
1884 01 01 - 1885 01 01 - 1886 01 01 | Walkers, Maryborough

Pumba (1884)
1884 01 01 - 1885 01 01 - 1886 01 01 | Walkers, Maryborough

Stingaree (1884)
1884 01 01 - 1885 01 01 - 1886 01 01 | Walkers, Maryborough

1 | "Victoria" type (1882)
Gunboat, Ausiralia

Displacement: 544 tons at draft: 3.35 m | Length: 42.67 m pp | Beam: 8.23 m | Crew: 53 men

Machinery: 1 cylindrical boiler | 2 * 400 HP horizontal comp. EE * 2 screws | 807 HP = 12.58 knots | Supplies: 70 / +15 tons of coal

Armament: 1-254/30, 1-152.4/26, 2-76.2/30, 2-25.4/40 Mg

Age 8

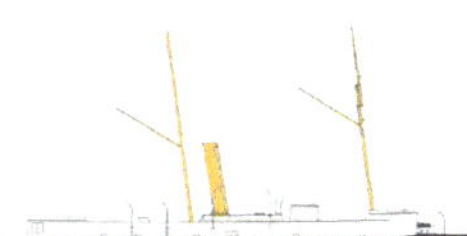

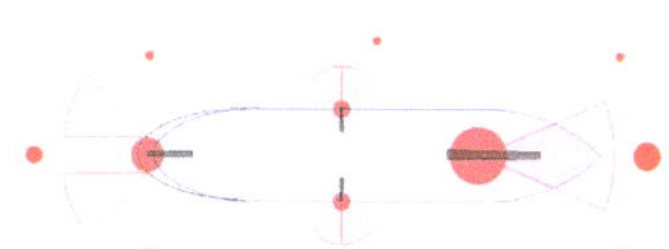

Victoria (1882)
1882 09 30 - 1883 06 25 - 1884 01 25 | Armstrong, Low Walker

1 | "Albert" type (1882)
Gunboat, Ausiralia

Displacement: 350 tons at draft: 2.9 m | Length: 36.6 m oa, 35.1 m pp | Beam: 7.62 m | Crew: 55 men

Machinery: 1 cylindrical boiler | 2 * 200 HP vertical compound EE * 2 screws | 400 HP = 10.5 knots | Supplies: 60 tons of coal

Armament: 1-203.2/28, 1-152.4/26, 2-95.3/26, 2-25.4/40 Mg

Age 8

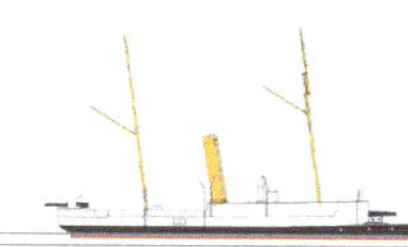

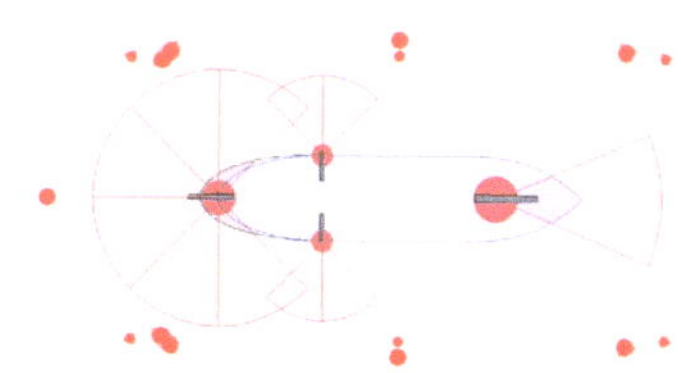

Albert (1882)
1882 09 20 - 1883 06 06 - 1884 01 05 | Armstrong, Low Walker

1 | "Gannet" type (1882)
Gunboat (ex-steamer), Australia

NO PHOTO AVAILABLE

Displacement: 346 tons at draft: 3.35 m | Length: 40.84 m oa | Beam: 6.71 m | Crew: 25 men

Machinery: ?? boilers, 1 horizontal direct acting E * paddles | 540 HP = 12 knots | Supplies: app 40 tons of coal

Armament: 1-152.4/26

Age 8

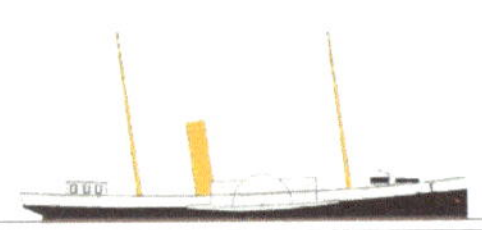

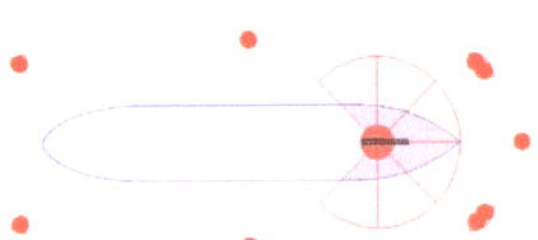

Gannet (1882)
1882 01 01 - 1883 01 01 - 1884 01 01 | Barrow SB, Barrow

1 | "Otter" type (1882)
Gunboat (ex-barge), Australia

Dimensions: 220 tons at draft: ?? m | L: ?? m | B: ?? m | Crew: ?? men

Machinery: | 12 knots |
Supplies: app. 15 tons of coal

Armament: 1-127/26

Age 8

NO PLAN AVAILABLE

Otter (1882)
1882 01 01 - 1883 01 01 - 1884 01 01 | Ramage & Ferg., Leith

2 | "Batman" type (1883)
Gunboats (ex-dredger), Australia

Displacement: 388 tons at drafi: 3.05 m | Length: 46.94 m oa | Beam: 7.93 m | Crew: 25 men

Machinery: ?? boilers | 1 horizontal direct acting E * 1 screw | 350 HP = ?? knots | Supplies: app. 40 tons of coal

Armament: 1-152.4/26

Age 7

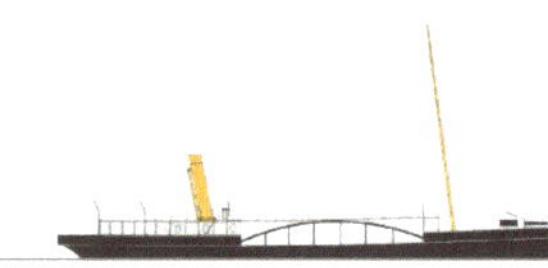

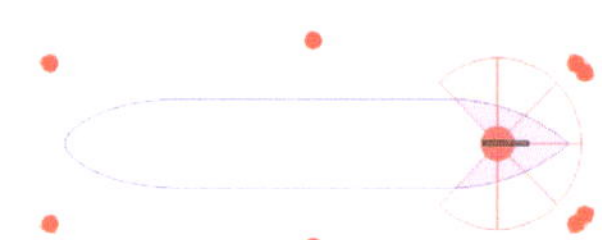

Batman (1883)
1883 01 01 - 1883 12 12 - 1884 06 06 | Simons, Renfrew

Fawkner (1883)
1883 01 01 - 1883 12 12 - 1884 06 06 | Portsmouth DYd

Her Majesty's New Zealand ships

4 | "Mosquito"/"1" type (1883)
Torpedo boats 2 class, Australia

TORPEDO SHIPS

Displacement: 10.9 tons at draft: 0.3 m | Length: 19.2 m oa | 18.44 m pp | Beam: 2.25 m | Crew: 7 men

Machinery: 1 locomotive boiler | 1 vertical compound E * 1 screw | 110 HP = 15 knots | Supplies: 1 ton of coal

Armament: 2 spar TT

Age 7

1 (1883)
1883 01 01 - 1884 01 01 - 1885 01 01 | Thornycroft, Woolston

2 (1883)
1883 01 01 - 1884 01 01 - 1885 01 01 | Thornycroft, Woolston

3 (1883)
1883 01 01 - 1884 01 01 - 1885 01 01 | Thornycroft, Woolston

4 (1883)
1883 01 01 - 1884 01 01 - 1885 01 01 | Thornycroft, Woolston

Sister ships can be found in the Australian section

Her Majesty's Indian ships

Besides of the Indian ships there were ships British fleet ships assigned to the East Indian detachment consisted on the 1890-Jan-01 of 1 2nd class protected cruiser, 2 3d class protected cruisers, 1 unprotected cruiser, 1 screw corvette, 3 screw sloops and 2 gunboats.

MONITORS

1 | "Abyssinia" type (1868)
Monitor, India

Displacement: 2,901 tons at draft: 4.45 m | Length: 68.58 m pp | Beam: 12,8 m | Crew: 100 men

Machinery: 4 boilers | 2 * 600 HP steam EE * 2 screws | 1,200 HP = 10 knots | Supplies: 92 tons of coal

Armour: iron | CT: 76; Turret: 254-203; UPBelt: 203-178 | WLBelt: 152(^38-25) 178(^38-25) 152(^38-25)

Armament: 4-254/14 MLR

Age 22

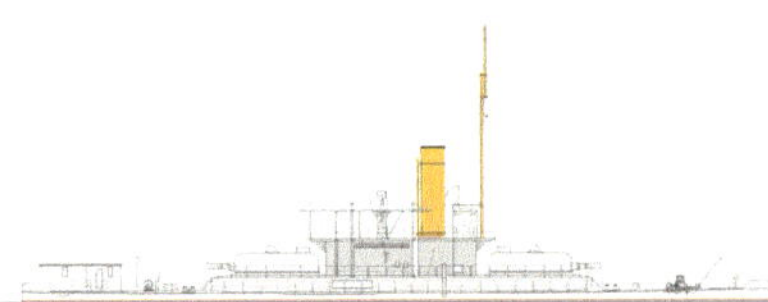

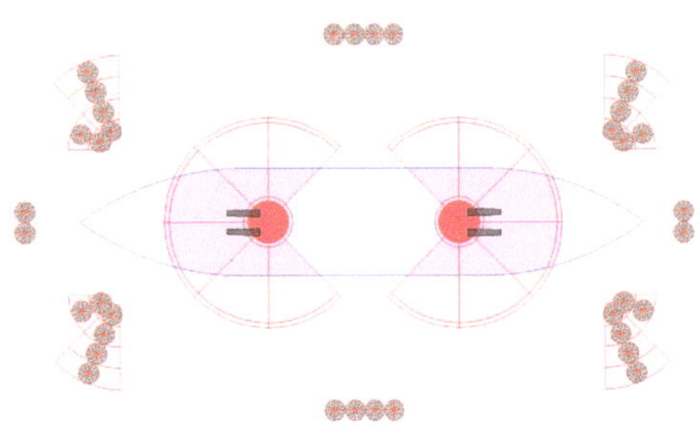

Abyssinia (1868)
2 Dudgeon horizontal return connecting rod EE | 1,200 HP = 9.59 knots |
1868 07 23 - 1870 02 19 - 1870 10 01 | Dudgeon, Poplar

1 | "Cerberus" type (1867)
Monitor, Australia

Displacement: 3,344 tons at draft: 4.67 m | Length: 68.58 m pp | Beam: 13.72 m | Crew: 155 men

Machinery: 4 rectangular boilers | 2 * 700 HP steam EE * 2 screws | 1,400 HP = 11 knots | Suppl.: 120 tons of coal

Armour: iron | CT: 229-203; Turrets: 254-229; Breastwork: 229-203 | WLBelt: 152(^38) 203(^38) 152(^38)

Armament: 4-254/14 MLR

Age 22

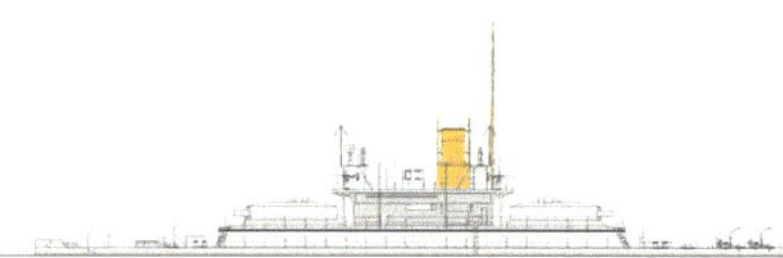

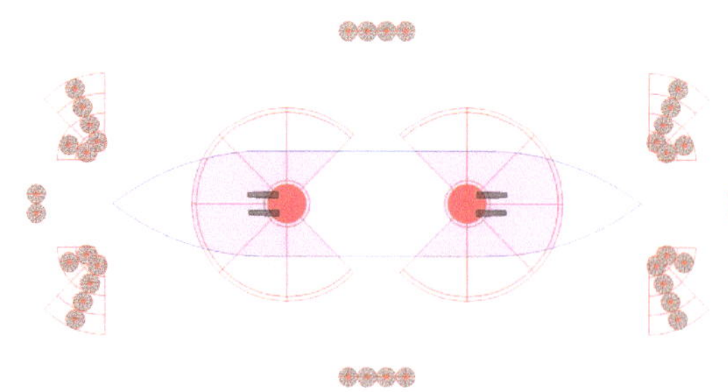

Magdala (1868)
2 Ravenhill horizontal return connecting rod EE | 1,436 HP = 10.6 knots |
1868 10 06 - 1870 03 02 - 1870 11 01 | Thames IW, Blackwall

TORPEDO SHIPS

0 +2 | "Sharpshooter"/"Assaye" type (1888) Torpedo gunboats, Britain

BUILDING IN BRITAIN

1Y:2M Assaye | 1Y:2M Plassey

Displacement: 735 tons at draft: 3.2 m | Length: 70.1 m pp | Beam: 8.23 m | Crew: 91 men

Machinery: 4 locomotive boilers | 2 * 1,250 HPn / 1,750 HPf steam EE * 2 screws | 2,500 HPn = 16.5 knots / 3,500 HPf = 19 knots | Supplies: 100 tons of coal

Armament: 2-120/30, 4-47/43, 1-11.5 Mg, 5-356 TT (2*2 trn, 1 >)

Age 2

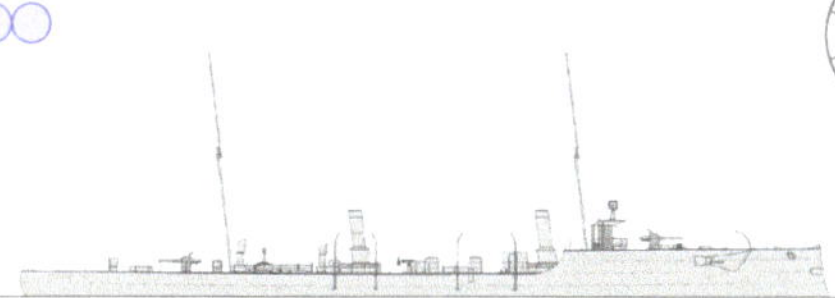

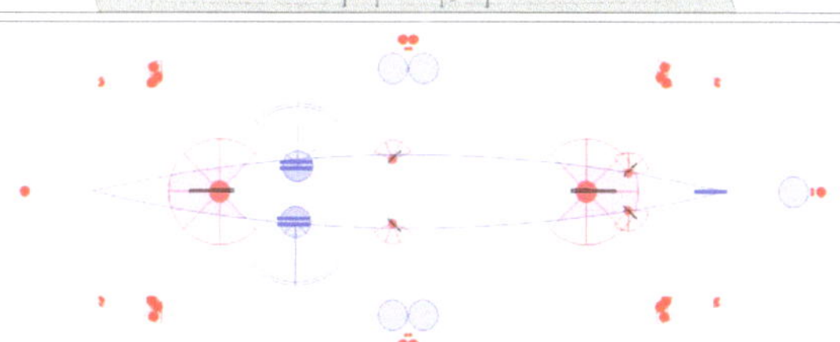

Assaye (1888)
2 Maudslay vertical triple expansion EE |
1888 11 19 - ***** - ***** | Armstrong, Elswick

Plassey (1888)
2 Maudslay vertical triple expansion EE |
1888 11 19 - ***** - ***** | Armstrong, Elswick

11 more ships are built for the British and Australian navies and can be found in the respective sections

0 +3 | "Rajput" type (1888) Torpedo boats 1 class, India

BUILDING

1Y:12M Rajput | 1Y:12M Sikh | 1Y:12M Mahratta

Displacement: 65 tons at draft: 2.29 m | Length: 41.15 m oa | 39.62 m pp | Beam: 4.42 m | Crew: 18 men

Machinery: 2 locomotive boilers | 1 horizontal expansion E * 1 screw | 1,000 HP = 22 knots | Supplies: app. 10 tons of coal

Armament: 4-25,4/40 Mg, 5-356 TT (2*2 trainable, 1 >)

Age 2

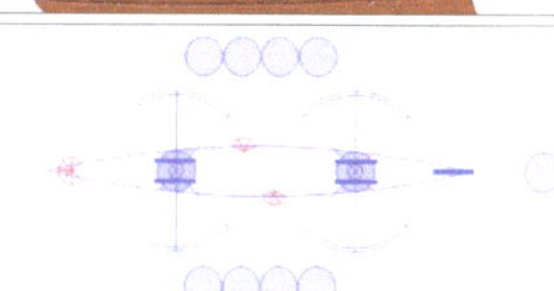

Mahratta (1888)
1888 01 01 - 1889 01 01 - ***** | White, Cowes

Rajput (1888)
1888 01 01 - 1889 01 01 - ***** | White, Cowes

Sikh (1888)
1888 01 01 - 1889 01 01 - ***** | White, Cowes

3 | "Baluchi" type (1887) Torpedo boats 1 class, India

NO PHOTO AVAILABLE

Displacement: 96 tons at draft: 2.13 m | Length: 41.03 m oa | 40.73 m pp | Beam: 4.5 m | Crew: 18 men

Machinery: 2 Thornycroft boilers | 1 vertical compound E * 1 screw | 1,260 HP = 23 knots | Supplies: app. 10 tons of coal

Armament: 2-25,4/40 Mg, 5-356 TT (2*2 trainable, 1 >)

Age 3

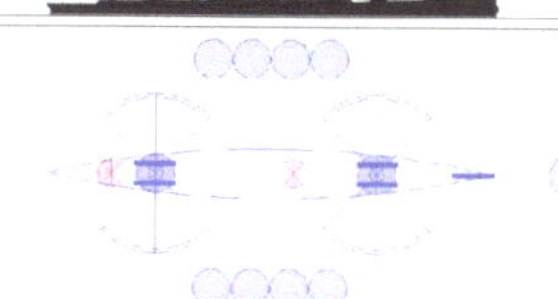

Baluchi (1887)
1887 01 01 - 1888 01 01 - 1889 01 01 | Thornycroft, Woolston

Karen (1887)
1887 01 01 - 1888 01 01 - 1889 01 01 | Thornycroft, Woolston

Pathan (1887)
1887 01 01 - 1888 01 01 - 1889 01 01 | Thornycroft, Woolston

Stay in Britain

1 | "Gurkha" type (1886)
Torpedo boat 1 class, India

Displacement: 92 tons at draft: 2.13 m | Length: 41.15 m oa | 39.62 m pp | Beam: 4.27 m | Crew: 18 men

Machinery: 1 locomotive boiler | 1 vertical compound E * 1 screw | 1,000 HP = 21.5 knots | Supplies: app. 10 tons of coal

Armament: 4-25,4/40 Mg, 5-356 TT (2*2 trainable, 1 >)

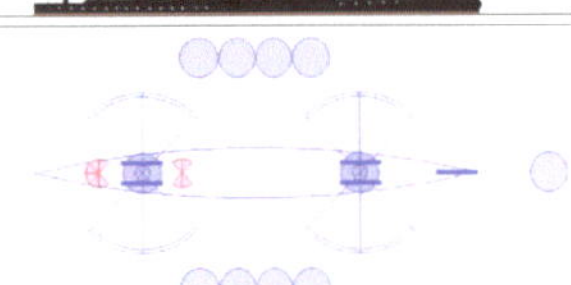

Gurkha (1886)
1886 01 01 - 1887 01 01 - 1888 01 01 | Hanna, Glasgow

Stays in Britain

ITALIAN FLEET

National flag

Fleet age 4

Ensign

Jack

Admiral

Months derivation +44

Warships built from 1858 01 01 to 1890 01 01

212 warships = 241,056 tons = 1,137 tons / average ship

Fleet age - 1886 05 18

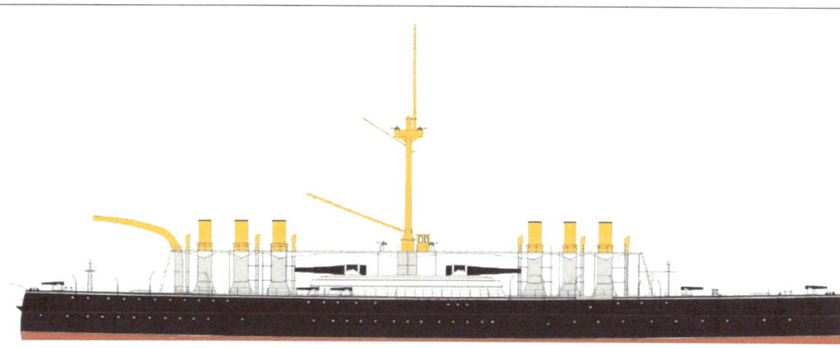
1889 fleet largest ship: Italia (1876) protected cruiser

Italian naval construction

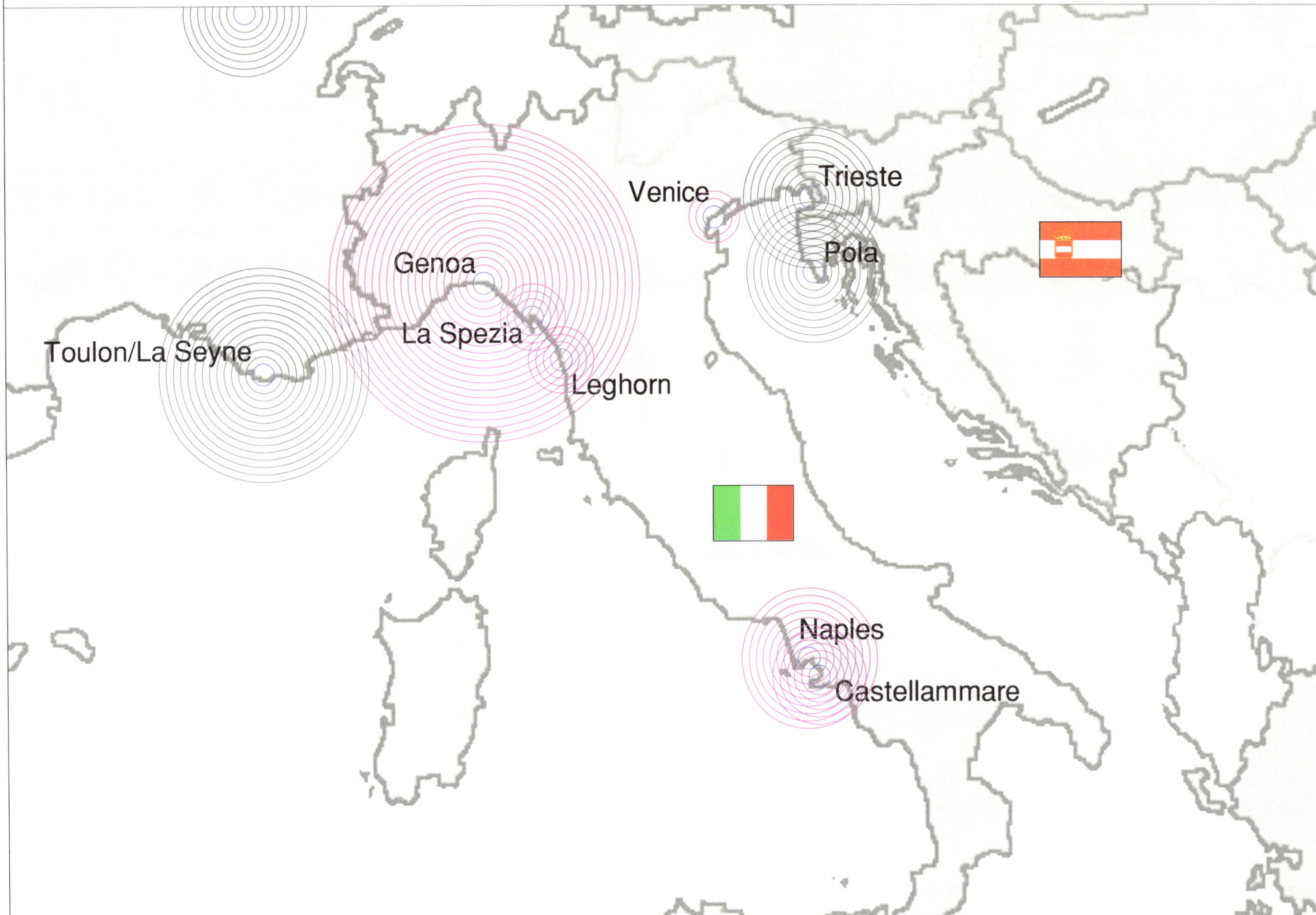

Ansaldo

11.2 tons / average ship / month

1890-Jan-01
6 ships building

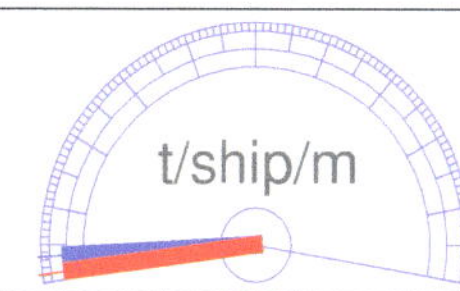

3,378 t.
tons / month = 67.0

t = 78 t/m = 3.3	124 S
t = 78 t/m = 3.3	125 S
t = 78 t/m = 3.3	126 S
t = 78 t/m = 3.3	127 S
t = 821 t/m = 19.3	Minerva
t = 2,245 t/m = 34.6	Liguria

Genoa, Italy

17 years of naval construction
15 ships | 5,368 tons

The plant's largest ship:
Liguria (1889) protected cruiser
2,245 tons

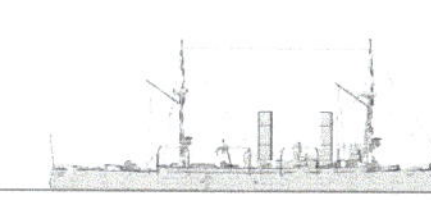

The plant's most effectively built ship:
Liguria (1889) protected cruiser
34.6 tons/month

Specialisation in:
Protected cruiser | 1 ship | 2,245 tons
Gunboat | 1 ship | 1,366 tons
Torpedo boats | 12 ships | 936 tons

Castellammare dockyard

52.5 tons / average ship / month

1890-Jan-01
5 ships building

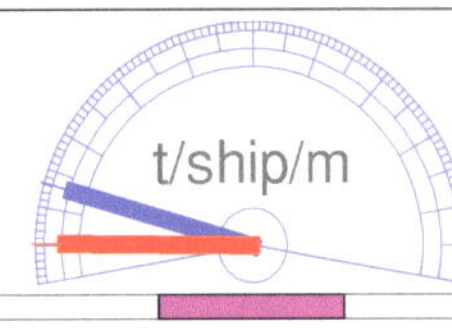

18,574 t.
tons / month = 262.5

Re Umberto	Partenope	Lombardia	Euridice	Iride
t = 13,673	t = 821	t = 2,245	t = 904	t = 931
t/m = 132.7	t/m = 30.4	t/m = 44.2	t/m = 34.2	t/m = 44.2

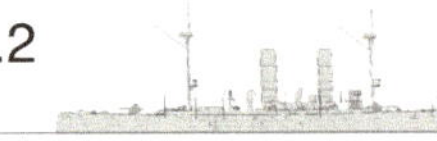

Castellammare, Italy

29 years of naval construction
21 ships | 77,305 tons

The plant's largest ship:
Italia (1876) cruiser-battleship
13,678 tons

The plant's most effectively built ship:
Re Umberto (1884) battleship
132.7 tons/month

Specialisation in:
Battleships | 3 ships | 34,521 tons
Protected cruisers | 3 ships | 19,461 tons
Battery ironclads | 3 ships | 11,450 tons

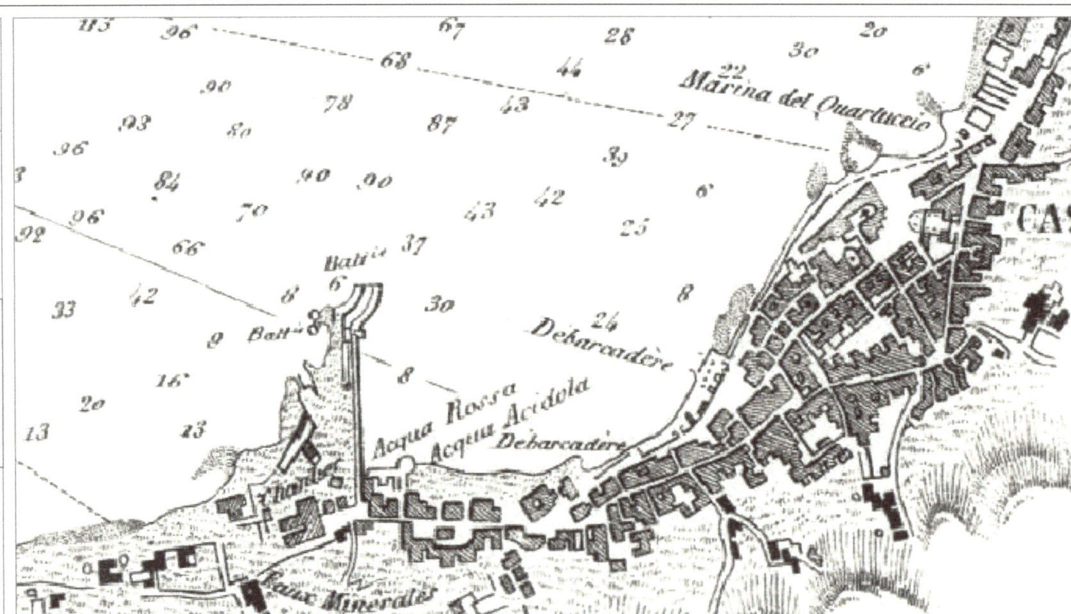

Gravero

3.3 tons / average ship / month

1890-Jan-01
4 ships building

t/ship/m

312 t.
tons / month = 13.0

t = 78 t/m = 3.3	116 S
t = 78 t/m = 3.3	117 S
t = 78 t/m = 3.3	118 S
t = 78 t/m = 3.3	119 S

Genoa, Italy

4 years of naval construction
9 ships | 1,204 tons

The plant's largest ship:
68 S (1886) torpedo boat
78 tons

The plant's most effectively built ship:
68 S (1886) torpedo boat
3.3 tons/month

Specialisation in:
Torpedo boats | 8 ships | 624 tons
Harbourgh vessel | 1 ship | 580 tons

La Spezia dockyard

78.3 tons / average ship / month

1890-Jan-01
3 ships building

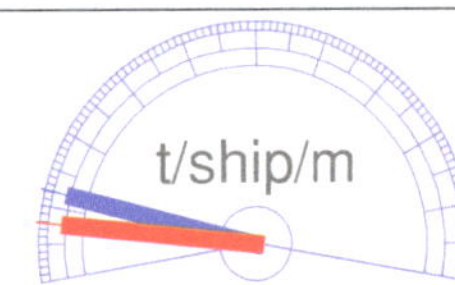

24,283 t.
tons / month = 234.8

t = 9,886 Andrea Doria
t/m =
88.3

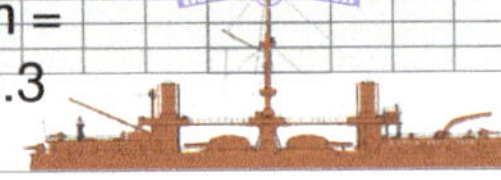

t = 13,641 Sardegna
t/m =
122.3

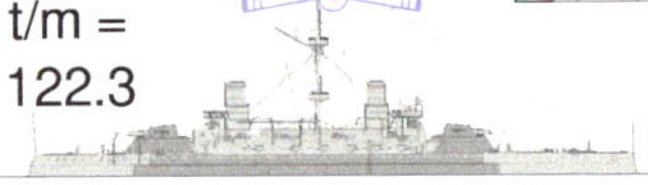

t = 756 Confienza
t/m =
24.2

La Spezia, Italy

25 years of naval construction
9 ships | 43,248 tons

The plant's largest ship:
Sardegna (1885) battleship
13,641 tons

The plant's most effectively built ship:
Sardegna (1885) battleship
122.3 tons/month

Specialisation in:
Battleships | 3 ships | 34,552 tons
Battery ironclad | 1 ship | 5,761 tons
Topedo-gunboats | 3 ships | 2,413 tons

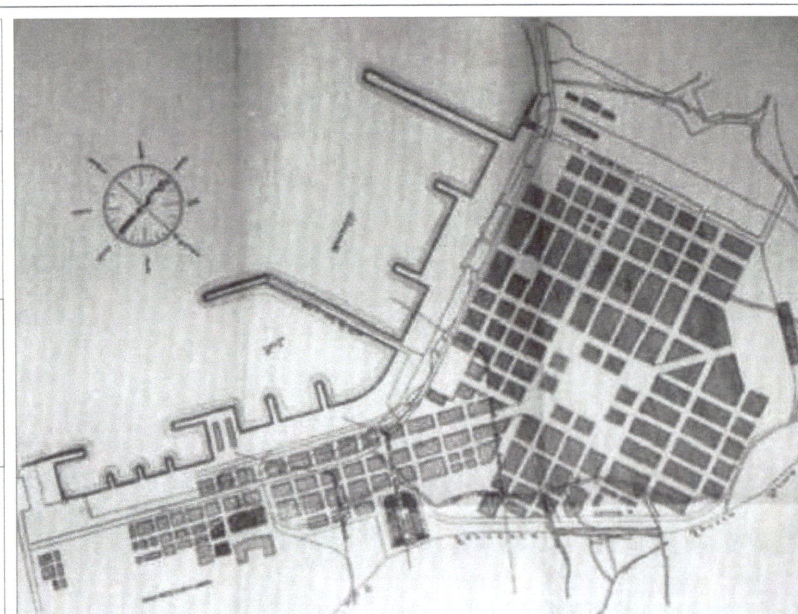

Odero

4.9 tons / average ship / month

1890-Jan-01
9 ships building

t/ship/m

1,555 t.
tons / month = 43.6

t = 78 t/m = 3.3 120 S	t = 78 t/m = 3.3 121 S	t = 78 t/m = 3.3 122 S	t = 78 t/m = 3.3 123 S	t = 821 t/m = 19.3 Urania
	t = 78 t/m = 3.3 120 S	t = 78 t/m = 3.3 121 S	t = 78 t/m = 3.3 122 S	t = 78 t/m = 3.3 123 S

Sestri Ponente, Italy

7 years of naval construction
27 ships | 3,687 tons

The plant's largest ship:
Messagiero (1883) aviso
1,005 tons

NO PLAN AVAILABLE

The plant's most effectively built ship:
Urania (1889) torpedo gunboat
17.6 tons/month

Specialisation in:
Torpedo boats | 27 ships | 1,751 tons
Patrol vessel | 1 ship | 1,005 tons
Topcdo gunboat | 1 ship | 931 tons

Orlando

30.1 tons / average ship / month

1890-Jan-01
3 ships building

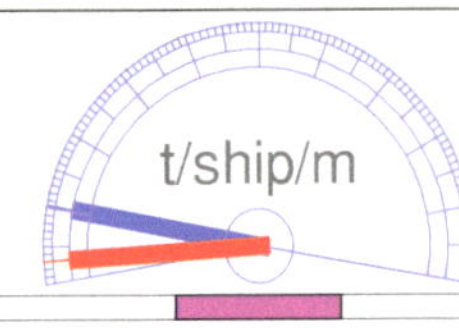

5,311 t.
tons / month = 90.4

t = 2,245
t/m =
33.8

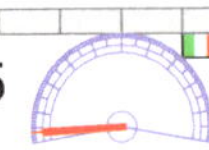

Umbria

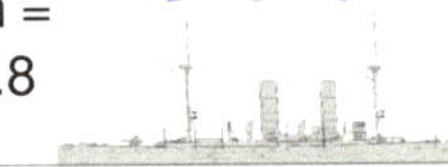

t = 2,245
t/m =
35.5

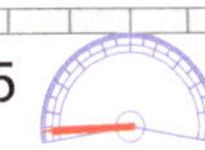

Etruria

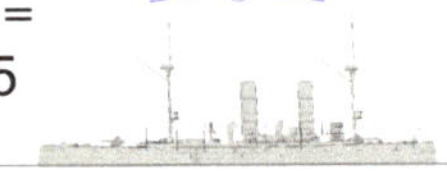

t = 821
t/m =
21.1

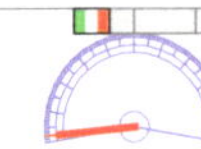

Arethusa

Leghorn, Italy

24 years of naval construction
19 ships | 29,973 tons

The plant's largest ship:
Lepanto (1876) cruiser-battleship
13,336 tons

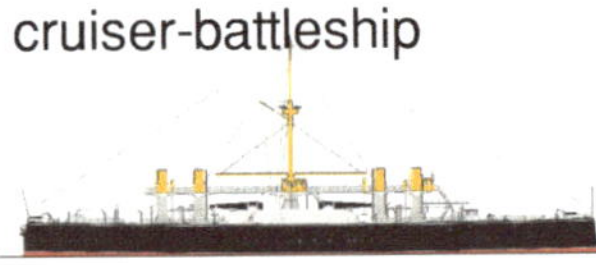

The plant's most effectively built ship:
Lepanto (1876) cruiser-battleship
103.3 tons/month

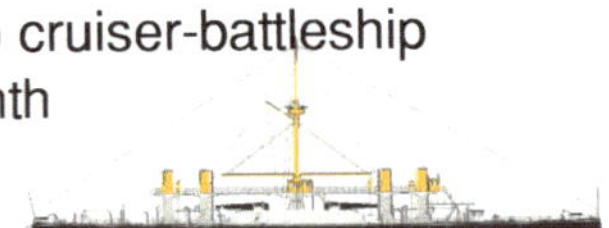

Specialisation in:
Protected cruisers | 5 ships | 24,902 tons
Gunboats | 5 ships | 3,946 tons
Topedo-gunboat | 1 ship | 821 tons

Pattison

3.3 tons / average ship / month

1890-Jan-01
8 ships building

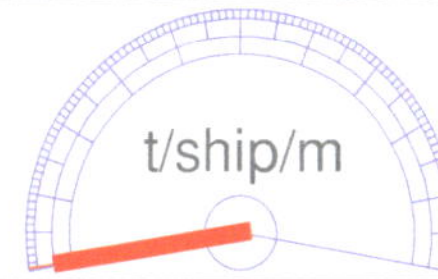

624 t.
tons / month = 26.0

t = 78 t/m = 3.3 112 S	t = 78 t/m = 3.3 113 S	t = 78 t/m = 3.3 114 S	t = 78 t/m = 3.3 115 S	t = 78 t/m = 3.3 132 S
	t = 78 t/m = 3.3 133 S	t = 78 t/m = 3.3 134 S	t = 78 t/m = 3.3 135 S	

Naples, Italy

7 years of naval construction
20 ships | 1,323 tons

The plant's largest ship:
112 S (1888) torpedo boat
78 tons

The plant's most effectively built ship:
134 S (1890) torpedo boat
3.3 tons/month

Specialisation in:
Torpedo boats | 20 ships | 1,323 tons

Venice dockyard

66.4 tons / average ship / month

1890-Jan-01
2 ships building

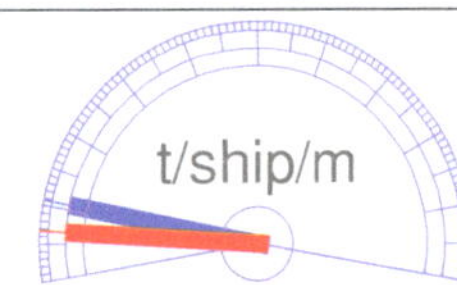

14,097 t.
tons / month = 132.8

t = 13,058	Sicilia
t/m = 103.9	

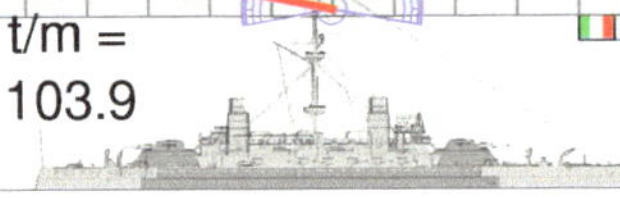

t = 1,039	Curtatone
t/m = 28.9	

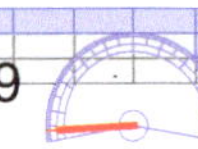

Venice, Italy

23 years of naval construction
15 ships | 38,722 tons

The plant's largest ship:
Sicillia (1884) battleship
13,058 tons

The plant's most effectively built ship:
Francesco Morosini (1881) battleship
107.0 tons/month

Specialisation in:
Battleships | 2 ships | 22,944 tons
Screw sailing cruisers | 2 ships | 4,001 tons
Protected cruiser | 1 ship | 3,538 tons

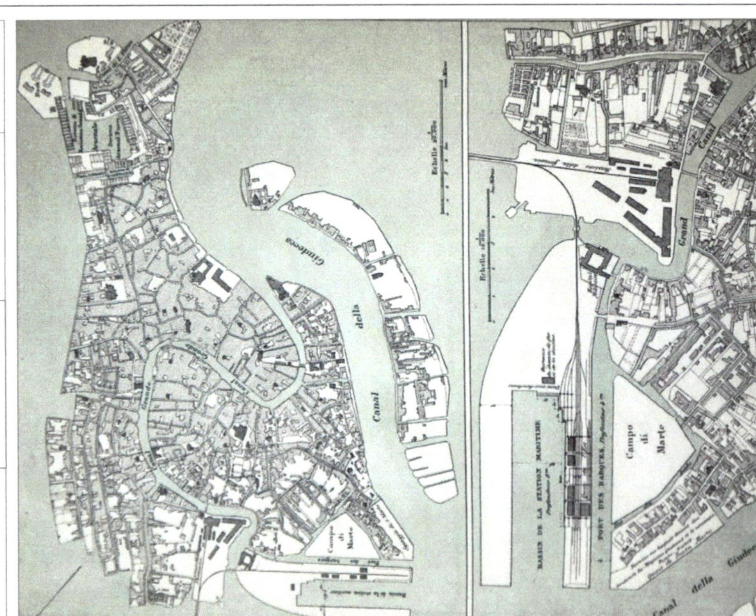

Italian fleet effective list

5 battleships

Francesco Morosini

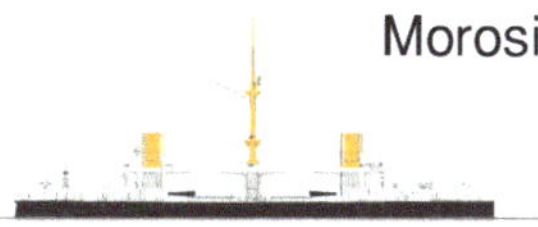

Ruggiero di Lauria

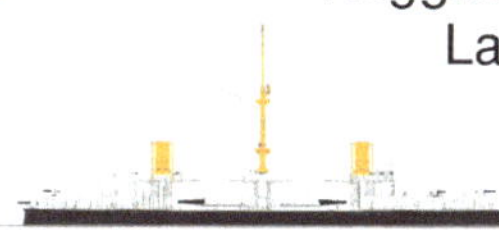

Dandolo

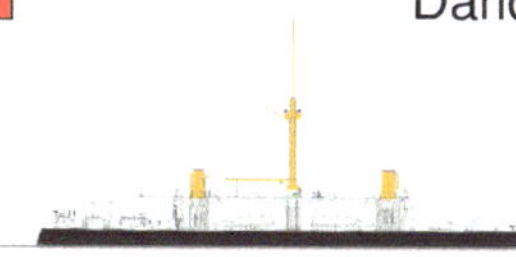

Duilio

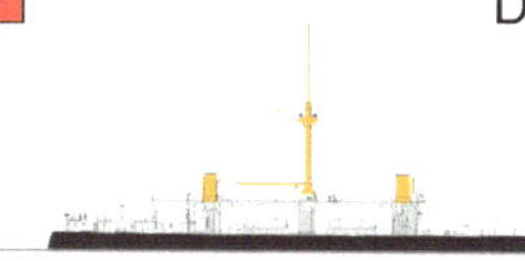

Affondatore

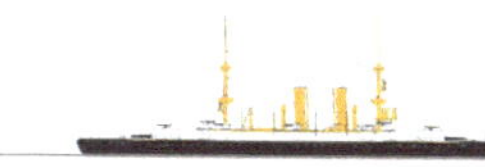

9 boardside ironclads

Palestro

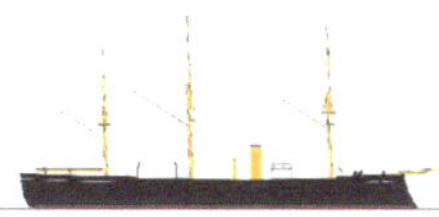

Roma

Training ship

Venezia

Regina Maria Pia

San Martino

Ancona

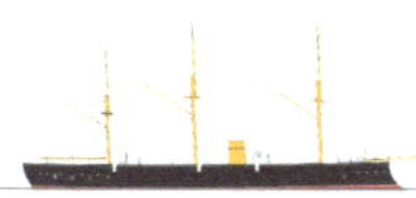

Castelfidardo

Formidabile

Training ship

Terribile

Training ship

9 protected cruisers

Piemonte

Dogali

Ettore Fieramosca

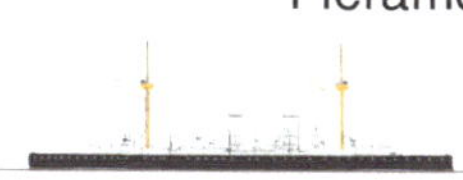

Etna

Vesuvio

Stromboli

Giovanni Bausan

Italia

Lepanto

5 unprotected & wooden cruisers

Amerigo Vespucci

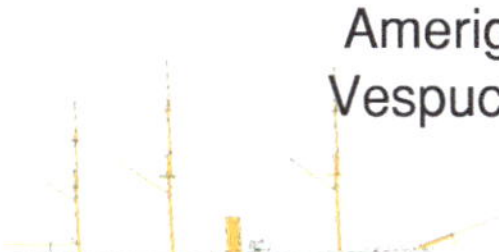

Flavio Gioia

Cristoforo Colombo

Vettor Pisani

Training ship

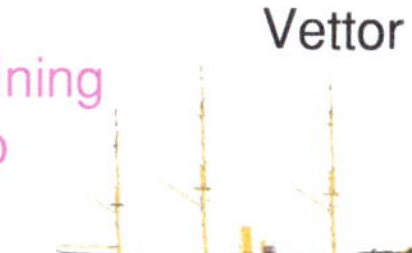

Caracciolo

Training ship

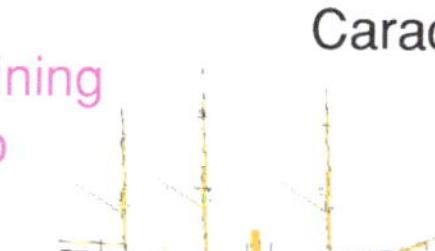

7 torpedo gunboats

Saetta	Damaged, in reserve Folglore	Goito	Montebello	Monzambano
	Tripoli	Pietro Micca		

121 torpedo boats

Aquila	Avvoltoio	Falco	Nibbio	Sparviero	Fatum	103 S	104 S	105 S	106 S
107 S	108 S	109 S	110 S	111 S	56 S	57 S	58 S	59 S	60 S
61 S	62 S	63 S	64 S	65 S	66 S	67 S	68 S	69 S	70 S
71 S	72 S	73 S	74 S	75 S	78 S	79 S	84 S	85 S	86 S
87 S	88 S	89 S	90 S	91 S	92 S	93 S	94 S	95 S	96 S
97 S	98 S	99 S	100 S	101 S	102 S	76 YA	77 YA	78 YA	79 YA
26 T	27 T	28 T	29 T	30 T	31 T	32 T	33 T	34 T	35 T
36 T	37 T	38 T	39 T	40 T	41 T	42 T	43 T	44 T	45 T
46 T	47 T	48 T	49 T	50 T	51 T	52 T	53 T	54 T	55 T
80 T	81 T	82 T	83 T	11 T	22 Y	25 Y	23 T	24 T	12 T
13 T	14 T	15 T	3 T	4 T	5 T	6 T	7 T	8 T	9 T
10 T	16 T	17 T	18 T	19 T	20 T	21 T	2 Y	1 T	

7 sloops & avisos

Archimede

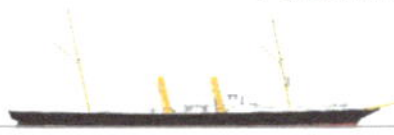

Galileo Galilei

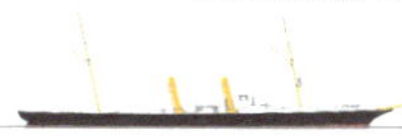

Agostino Barbarigo

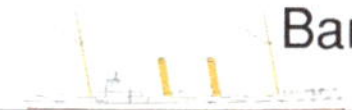

Marcantonio Colonna

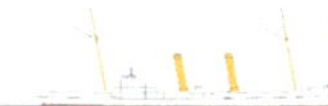

Rapido

Staffetta

Vedetta

9 gunboats

Castore

Polluce

Volturno

Andrea Provana

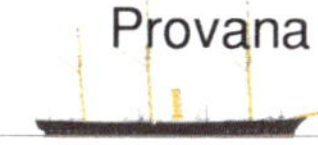

Sebastiano Veniero

Carridi

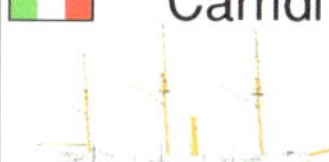

Scilla

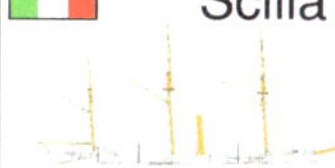

Guardi-ano

Sentinella

Fleet auxiliaries

Headquater ship

Principe Amedeo

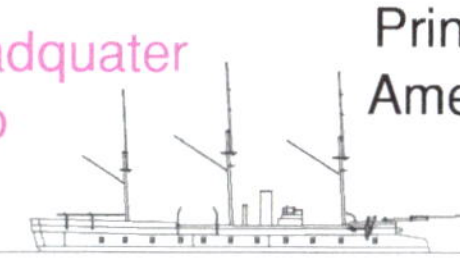

Depot ship

Vareze

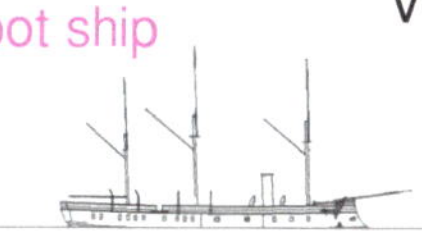

Messagiero

NO PLAN AVAILABLE

Esploratore

NO PLAN AVAILABLE

Sesia

NO PLAN AVAILABLE

Italian section ship names glossary

1 T (TB2)
10 T (TB2)
100 S (TB1)
101 S (TB1)
102 S (TB1)
103 S (TB1)
104 S (TB1)
105 S (TB1)
106 S (TB1)
107 S (TB1)
108 S (TB1)
109 S (TB1)

11 T (TB1)
110 S (TB1)
111 S (TB1)
112 S (TB1)
113 S (TB1)
114 S (TB1)
115 S (TB1)
116 S (TB1)
117 S (TB1)
118 S (TB1)
119 S (TB1)

12 T (TB2)
120 S (TB1)
121 S (TB1)
122 S (TB1)
123 S (TB1)
124 S (TB1)
125 S (TB1)
126 S (TB1)
127 S (TB1)
128 S (TB1)
129 S (TB1)

13 T (TB2)
130 S (TB1)
131 S (TB1)
132 S (TB1)
133 S (TB1)
134 S (TB1)
135 S (TB1)
14 T (TB2)
15 T (TB2)
16 T (TB2)
17 T (TB2)
18 T (TB2)
19 T (TB2)

2 Y (TB2)
20 T (TB2)
21 T (TB2)
22 Y (TB1)
23 T (TB1)
24 T (TB1)
25 Y (TB1)
26 T (TB1)
27 T (TB1)
28 T (TB1)
29 T (TB1)

3 T (TB2)
30 T (TB1)
31 T (TB1)
32 T (TB1)
33 T (TB1)
34 T (TB1)
35 T (TB1)
36 T (TB1)
37 T (TB1)
38 T (TB1)
39 T (TB1)

4 T (TB2)
40 T (TB1)
41 T (TB1)
42 T (TB1)
43 T (TB1)
44 T (TB1)
45 T (TB1)
46 T (TB1)
47 T (TB1)
48 T (TB1)
49 T (TB1)

5 T (TB2)
50 T (TB1)
51 T (TB1)
52 T (TB1)
53 T (TB1)
54 T (TB1)
55 T (TB1)
56 S (TB1)
57 S (TB1)
58 S (TB1)
59 S (TB1)

6 T (TB2)
60 S (TB1)
61 S (TB1)
62 S (TB1)
63 S (TB1)
64 S (TB1)
65 S (TB1)
66 S (TB1)
67 S (TB1)
68 S (TB1)
69 S (TB1)

7 T (TB2)
70 S (TB1)
71 S (TB1)
72 S (TB1)
73 S (TB1)
74 S (TB1)
75 S (TB1)
76 YA (TB1)
77 YA (TB1)
78 YA (TB1)
79 YA (TB1)

8 T (TB2)
80 T (TB1)
81 T (TB1)
82 T (TB1)
83 T (TB1)
84 S (TB1)
85 S (TB1)
86 S (TB1)
87 S (TB1)
88 S (TB1)
89 S (TB1)

9 T (TB2)
90 S (TB1)
91 S (TB1)
92 S (TB1)
93 S (TB1)
94 S (TB1)
95 S (TB1)
96 S (TB1)
97 S (TB1)
98 S (TB1)
99 S (TB1)

A
Affondatore (CDB)
Agostino Barbarigo (ZCD)
Amerigo Vespucci (CS)
Ancona (BO)
Andrea Doria (BB)
Andrea Provana (GB)
Aquila (TB1)
Archimede (ZCD)
Arethusa (DC)
Avvoltoio (TB1)

B
Betta '13 (harbV)
Caracciolo (trailS)
Carridi (GB)
Castelfidardo (BO)
Castore (GB)
Confienza (DC)
Cristoforo Colombo (CW)
Curtatone (GB)
Dandolo (BB)
Dogali (CP)
Duilio (BB)

E
Esploratore (CZ)
Etna (CP)
Etruria (CP)
Ettore Fieramosca (CP)
Euridice (DC)
Falco (TB1)
Fatum (TB1)
Flavio Giola (CS)
Folglore (DC)
Formidabile (trailS)
Francesco Morosini (BB)

G
Galileo Galilei (ZCD)
Giovanni Bausan (CP)
Goito (DC)
Guardiano (GB)

I
Iride (DC)
Italia (CP)
Lepanto (CP)
Liguria (CP)
Lombardia (CP)

M
Marcantonio Colonna (ZCD)
Messagiero (ZCD)
Minerva (DC)
Montebello (DC)
Monzambano (DC)

N
Nibbio (TB1)

P
Palestro (commandS)
Partenope (DC)
Piemonte (CP)
Pietro Micca (DC)
Polluce (GB)

R
Rapido (CZ)
Re Umberto (BB)
Regina Maria Pia (BO)
Roma (BO)
Ruggiero di Lauria (BB)

S
Saetta (DC)
San Martino (BO)
Sardegna (BB)
Scilla (GB)
Sebastiano Veniero (GB)
Sentinella (GB)
Sesia (PS)
Sicilia (BB)
Sparviero (TB1)
Staffetta (CZ)
Stromboli (CP)

T
Terribile (trailS)
Tripoli (DC)

U
Umbria (CP)
Urania (DC)

V
Vareze (tendr)
Vedetta (CZ)
Venezia (trailS)
Vesuvio (CP)
Vettor Pisani (trailS)
Volturno (GB)

Italian fleet naval guns

Calibre (mm)	Mark	Calibre (in)/pdr	Gun type	First Introduced	Weight (kg)	Gun lenght (in/cal)	Bore lenght (cal)	Charge weight (lbs)	Muzzle energy (ft/t)	Projectile weight (lbs)	Muzzle velocity (ft/s)	Wrought iron perforation (in)
450.0	Armstrong	17.72"	BLR	1883	100 000	392.4 / 22.15	20.50	630.0	40 060	2 000.0	1 700	28.5
431.8	new	17"	BLR	1883	104 300	489.0 / 28.77	27.00	600.0	55 030	2 000.0	1 992	33.7
431.8	1882	17"	BLR	1883	101 500	468.0 / 27.53	26.00	480.0	51 930	2 000.0	1 935	32.8
400.0	Krupp	15.75"	BLR	1884	119 000	536.0 / 34.00	32.00	615.0	46 061	1 980.0	2 017	29.2
342.9	I, II - III, IV	13.5"	BLR	1883	68 075 - 70 107	433.0 / 32.08	30.00	630.0	35 230	1 250.0	2 016	30.1
254.0	II, III, IV	10"	BLR	1887	29 465	342.4 / 34.24	32.00	252.0	14 430	500.0	1 914	22.3
228.6	V	9"	MLR	1865	12 193	156.0 / 17.33	13.89	50.0	3 681	256.0	1 440	11.4
210.0	Krupp	8.24"	BLR	-	12 300	247.3 / 29.91	27.10	103.6	5 876	308.6	1 657	15.6
160.0	III	64 cwt	MLR	1864	3 250	118.0 / 18.73	15.47	10.0	865	64.5	1 390	-
152.4	-	6"	BLR	-	6 600	249.2 / 41.54	40.00	18.3	4 334	100.0	2 500	19.5
152.4	Armstrong	6"	BLR	1887	6 200		32.25			100.0	1 980	
152.4	VI	6"	BLR	1873	5 080	173.5 / 28.92	26.00	45.0	2 424	100.0	1 870	11.9
150.0	Germany	5.9"	BLR	1878	4 040	176.0 / 29.80	27.20	33.1	112	112.4	1 624	11.0
150.0	Krupp	5.9"	BLR	1865	3 940	151.6 / 25.67	23.00	20.9	1 447	69.4	1 558	9.1
120.0	II, L	4.7"	BLR	1884	2 083	194.1 / 41.09	40.00	12.0	486	45.0	2150	8.5
120.0	-	4.7"	BLR	-	1 690	156.0 / 33.02	35.00	12.0	-	36.2	-	-
120.0	Krupp	4.7"	BLR	1859	1 380	115.2 / 25.00	-	9.9	680	36.1	1 493	7.0
120.0	II, L	4.7"	MLR	1865	1 778	120.0 / 25.53	22.00	6.8	312	38.8	1 425	-
76.2	-	3"	BLR	-	610	123.6 / 41.20	40.00	1.9	423	12.5	2 210	8.1
57.0	I & II	6 pr	BLR	1876	406	96.5 / 43.00	40.00	1.2	345	6.0	1 818	3.7
47.0	I & II	3 pr	BLR	1876	254	79.6 / 43.00	40.00	1.8	321	3.3	1 873	3.1
37.0	Hotchkiss R	1 pr	BLR	1873	209	33.5 / 23.0	29.13	0.2	-	1.1	1 450	-
37.0	Hotchkiss	1 pr	BLR	1884	72	33.5 / 23.0	33.15	0.2	-	1.1	1 450	-

WARSHIPS SECTION

Ship classes	Building (exports including)	Active
Battleships	4	4
Boardside ironclads		9
Armoured cruisers		
Coast defence armoured ships		1
Protected cruisers	4	9
Unprotected cruisers		2
Torpedo gunboats	7	7
Torpedo boats 1 class	24	97
Torpedo boats 2 class		20
Submarines		
Wooden screw cruisers		3
Wooden screw sloops		4
Gunboats	1	7
Gunboats coastal		2

BATTLESHIPS

0 +3 | "Re Umberto" type (1884) Battleships, Italy

BUILDING

4Y:3M

Sardegna

Displacement: 13,641 tons at draft: 8.84 m, 15,426 tons full | Length: 130.73 m oa | Beam: 23.44 m | Crew: 794 men

Machinery: 18 cylindrical boilers | 2 * 7,600 HPn / 11,400 HPf triple expansion EE * 2 screws | 15,200 HPn = 19 knots / 22,800 HPf = 21.2 knots | Supplies: 1,650 / +350 tons of coal

Armour: compound | CT: 305; Barbettes: 350; Hoist: 305; Gun shields: 51; Battery: 102; UPBelt: 0 =70= 102 =70= 0 | WLBelt: 0/102(^76) 114/102(^76) 0/102(^76)

Armament: 4-343/30, 8-152.4/40, 16-120/40, 20-57/43, 10-37/23, 5-450 TT (3 ^, 2 v)

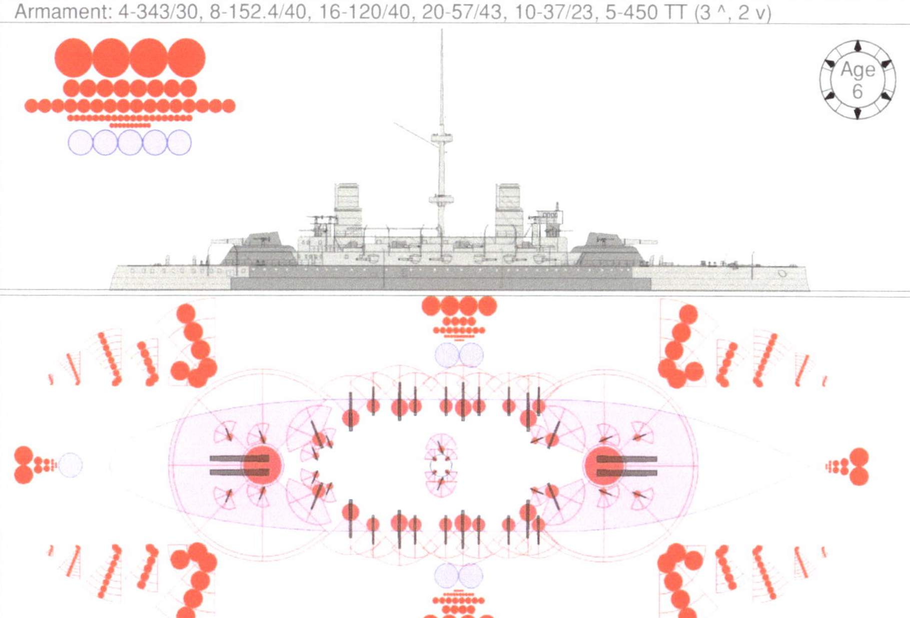

Sardegna (1885)

1885 10 24 - ***** - ***** | La Spezia DYd, La Spezia

0 +3 | "Re Umberto" type (1884) Battleships, Italy (continued)

BUILDING

5Y:2M 5Y:6M

Sicillia Re Umberto

Displacement: 13,673 tons at draft: 9.29 m, 15,454 tons full | Length: 127.6 m oa | Beam: 23.44 m | Crew: 733 men

Machinery: 18 cylindrical boilers | 2 * 7,600 HPn / 9,800 HPf vertical compound EE * 2 screws | 15,200 HPn = 19 knots / 19,500 HPf = 20 knots | Supplies: 1,650 / +350 tons of coal

Armour: compound | CT: 305; Barbettes: 350; Hoist: 305; Gun shields: 51; Battery: 102; UPBelt: 0 =70= 102 =70= 0 | WLBelt: 0/102(^76) 114/102(^76) 0/102(^76)

Armament: 4-343/30, 8-152.4/40, 16-120/40, 20-57/43, 10-37/23, 5-450 TT (3 ^, 2 v)

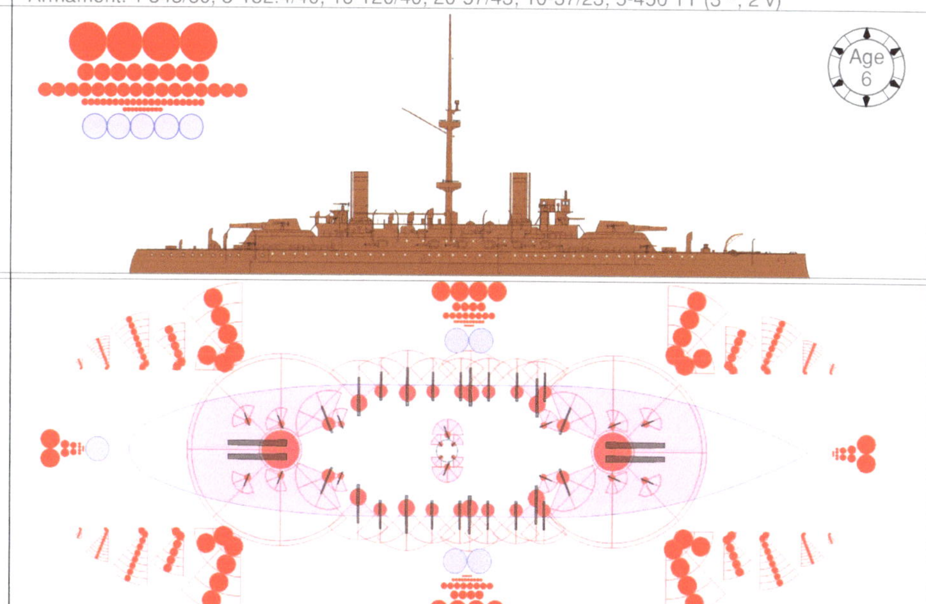

Re Umberto (1884)

1884 07 10 - 1888 10 17 - ***** | Castellammare DYd, Castellammare

Sicillia (1884)

1884 11 03 - ***** - ***** | Venice DYd, Venice

2 +1 | "Ruggiero di Lauria" type (1881)
Battleships, Italy

BUILDING

6Y:12M

Andrea Doria

Displacement: 9,886 tons at draft: 8.37 m, 10,997 tons full | Length: 105.9 m oa | Beam: 19.84 m | Crew: 507 men

Machinery: 8 oval boilers | 2 * 5,000 HP compound EE * 2 screws | 10,000 HP = 16 knots | Supplies: 850 tons of coal

Armour: steel | CT: 249: Citadel: 361; Barbettes (Cupolas): 361(^51) | WLBelt: 0/102(^76) =361= 451/102(^76) =361= 0/102(^76)

Armament: 4-431.8/27 MLR, 2-152.4/32, 4-120/32, 2-356 TT (^)

Age 9

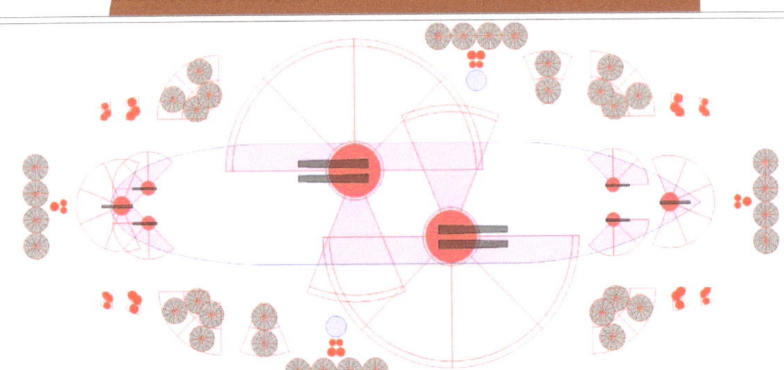

Andrea Doria (1882)
1882 01 07 - 1885 11 21 - ***** | La Spezia DYd, La Spezia

Francesco Morosini (1881)
1881 12 04 - 1885 07 30 - 1889 08 21 | Venice DYd, Venice

Ruggiero di Lauria (1881)
1881 08 03 - 1884 08 09 - 1888 02 01 | Castellammare DYd, Castellammare

2 +1 | "Ruggiero di Lauria" type (1881)
Battleships, Italy (continued)

Ruggiero di Lauria

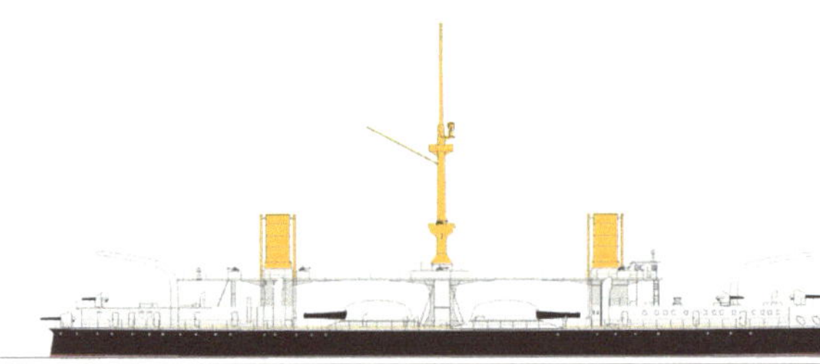

Francesco Morosini

2 | "Duilio" type (1873)
Battleships, Italy

Displacement: 10,962 tons at draft: 8.31 m, 12,071 tons full | Length: 109.16 m oa | Beam: 19.74 m | Crew: 420 men

Machinery: 8 rectangular boilers | 2 * 3,850 HP vertical reciprocating EE * 2 screws | 7,711 HP = 15.04 knots | Supplies: 732 / +268 tons of coal

Armour: steel | Turrets: 432; Citadel: 432 |
WLBelt: 0/51(^51) =400= 546(^38) =400= 0/51(^51)

Armament: 4-420/20 MLR, 3-356 TT (v), 1-2nd class torpedo boat

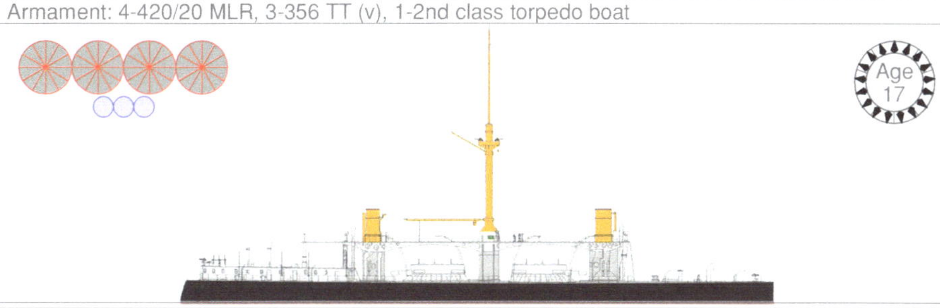

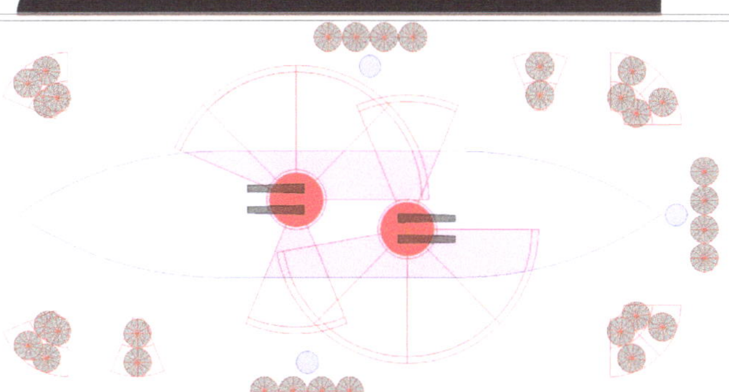

Dandolo (1873)
1873 01 06 - 1878 07 10 - 1882 04 11 | La Spezia DYd, La Spezia

Duilio (1873)
1873 01 06 - 1876 05 08 - 1880 01 06 | Castellammare DYd, Castellammare

1 | "Affondatore" type (1863)
Coast defence battleship, Italy

COAST DEFENCE BATTLESHIP

Displacement: 4,006 tons at draft: 6.3 m, 4,307 tons full | Length: 93.8 m oa | Beam: 12.2 m | Crew: 309 men

Machinery: 8 rectangular boilers | 1 single expansion E * 1 screw | 3,240 HP = 13 knots | Supplies: 566 / +80 tons of coal

Armour: iron | CT: 152; Turrets: 127 |
WLBelt: 114(^51) 127(^51) 114(^51)

Armament: 2-254/30, 6-120/40, 8-57/43, 1-75/25, 4-37/23

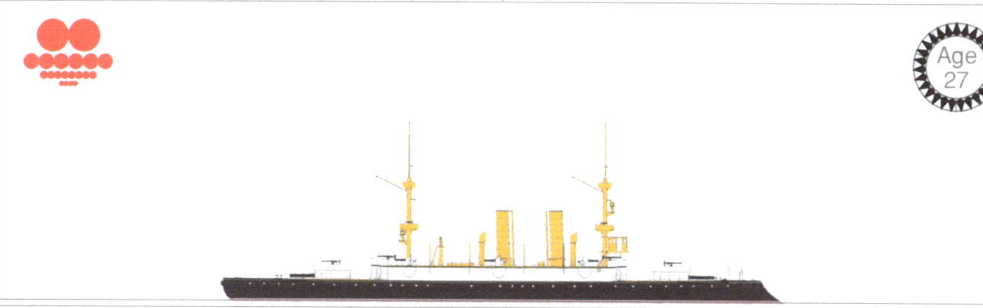

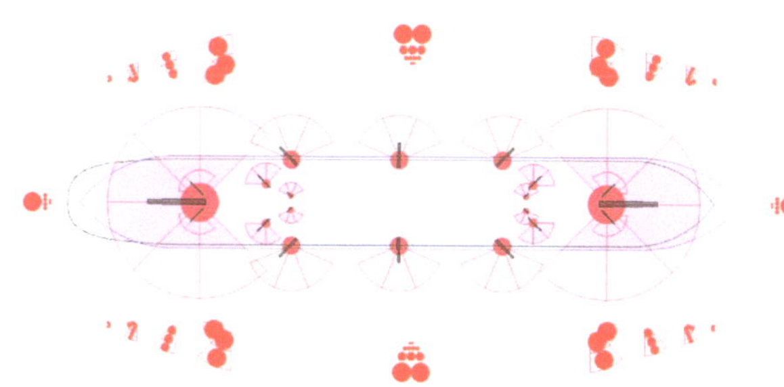

Affondatore (1863)
1863 04 11 - 1865 03 11 - 1870 01 01 | Harrison, Millwall

BATTERY IRONCLADS

1 | "Principe Amedeo" type (1865)
Headquater ship (ex-battery ironclad), Italy

Displacement: 5,761 tons at draft: 8 m | Length: 78.82 m pp | Beam: 17.3 m | Crew: 548 men

Machinery: 6 cylindrical boilers | 1 single expansion E * 1 screw | 6,117 HP = 12.85 knots | Supplies: app. 600 tons of coal

Armour: iron | Battery: 109 | WLBelt: 109

Armament: 1-278.4/12 MLR, 6-254/14.5 MLR

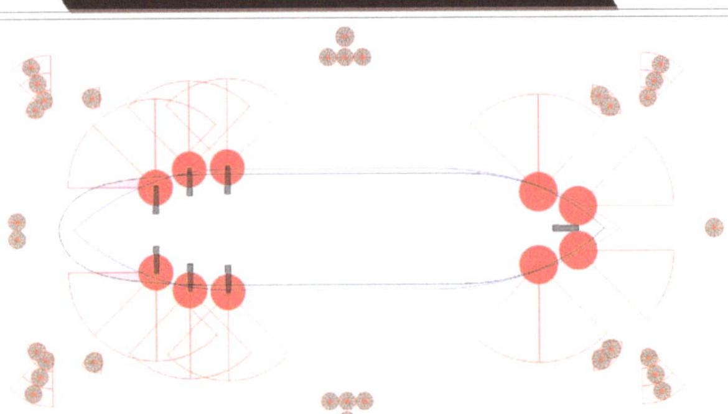

Palestro (1865)
1865 08 01 - 1871 10 02 - 1875 07 11 | La Spezia DYd, La Spezia

2 | "Roma" type (1863)
Battery ironclads, Italy

Displacement: 5,698 tons at draft: 7.57 m, 6,151 tons full | Length: 79.67 m pp | Beam: 17.33 m | Crew: 549 men

Machinery: 6 cylindrical boilers | 1 single expansion E * 1 screw | 3,670 HP = 13 knots | Supplies: 621 tons of coal

Armour: iron | WLBelt: 150

Rearmed (1886): 11-210/25

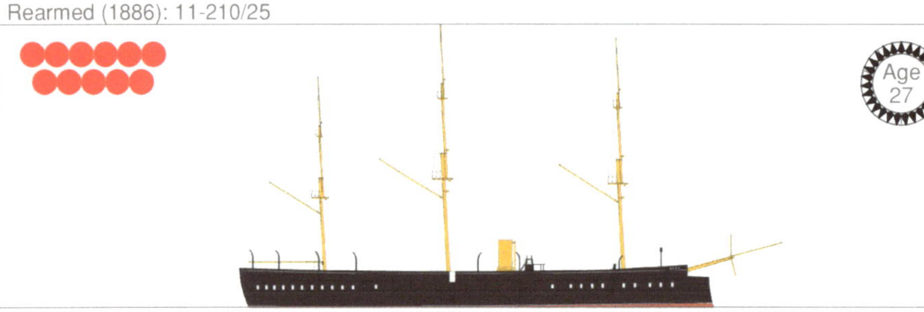

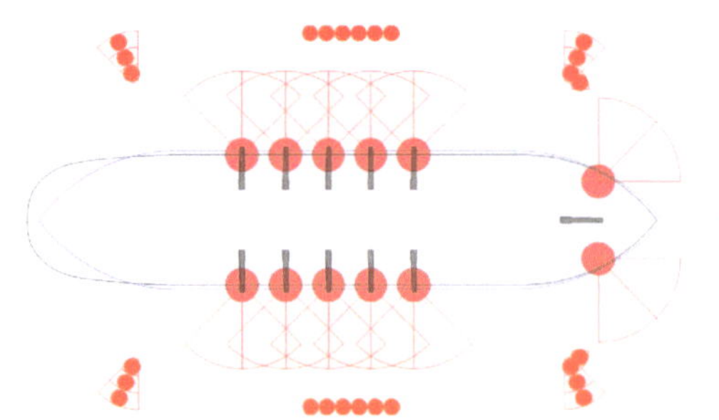

Roma (1863)
1863 02 01 - 1865 12 18 - 1869 05 01 | La Foce, Genoa

The sister ship "Venezia" is used as a torpedo school

4 | "Castelfidardo" type (1862)
Battery ironclads, Italy

Displacement: 4,192 tons at draft: 6.35 m, 4,527 tons full | Length: 81.8 m oa | Beam: 15.6 m | Crew: 480 men

Machinery: 8 rectangular boilers | 1 recyprocating E * 1 screw | 2,125 HP = 12.1 knots | Supplies: 485 tons of coal

Armour: iron | Battery: 109 |
WLBelt: 109

Rearmed (1884): 8-152.4/32, 6-120/27, 4-57/43, 8-37/23R

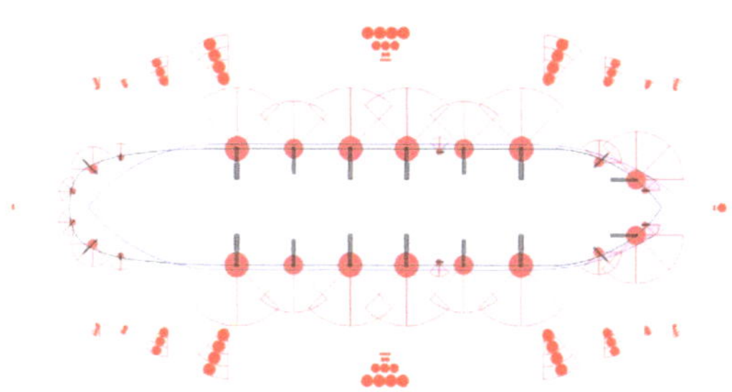

Ancona (1862)
1862 08 11 - 1864 10 17 - 1866 04 01 | Arman, Bordeaux

Castelfidardo (1862)
1862 07 27 - 1863 08 01 - 1864 05 01 | Goüin et Guibert, Nantes

Regina Maria Pia (1862)
1862 07 22 - 1863 04 28 - 1864 04 27 | Forges et Chantiers de la Méditerranée, La Seyne

4 | "Castelfidardo" type (1862)
Battery ironclads, Italy (continued)

San Martino (1862)
1862 07 22 - 1863 09 21 - 1864 11 19 | Forges et Chantiers de la Méditerranée, La Seyne

San Martino

Rearmed (1885): 8-152.4/32, 5-120/27, 4-57/43, 8-37/23R, 3-356 TT

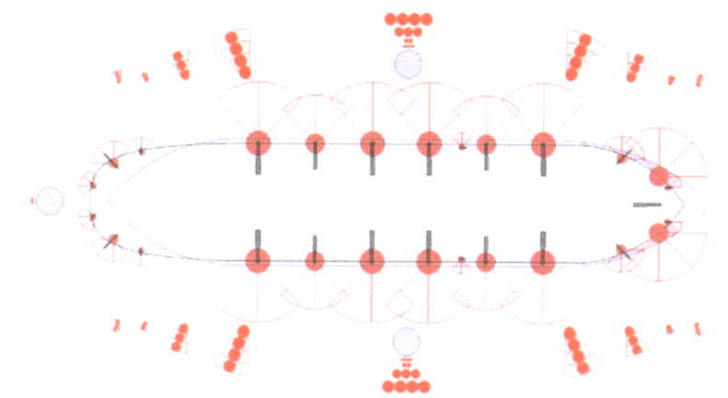

Ancona

Rearmed (1884): 8-152.4/32, 6-120/27, 4-57/43, 2-37/23R, 3-356 TT

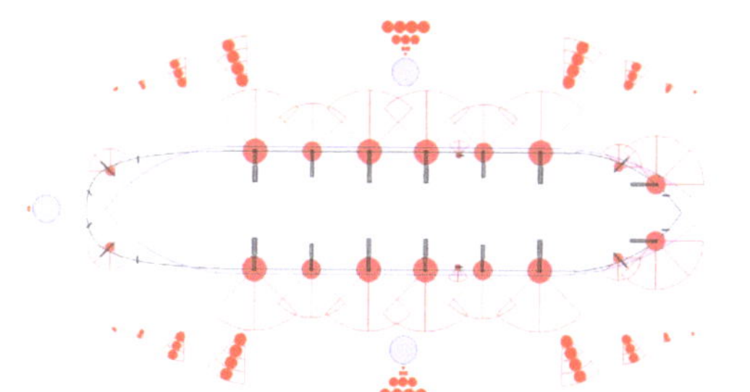

Regina Maria Pia

PROTECTED CRUISERS

2 | "Italia" type (1876)
Protected cruisers 1 class, Italy

Displacement: 13,678 tons at draft: 8.75 m, 15,407 tons full | Length: 124.7 m oa | Beam: 22.54 m | Crew: 669 men

Machinery: 8 oval, 16 cylindrical boilers | 4 * 3,000 HP vertical compound EEEE * 4 screws | 11,986 HP = 17.8 knots | Supplies: 1,650 / +350 tons of coal

Armour: steel | CT: 102; Citadel: 483; Funnel bases: 406; Hoist: 456 | WLDeck: 0/102(^76)

Armament: 1-431.8/27 MLR, 3-431.8/26 MLR, 7-150/25, 4-120/23, 4-356 TT (^)

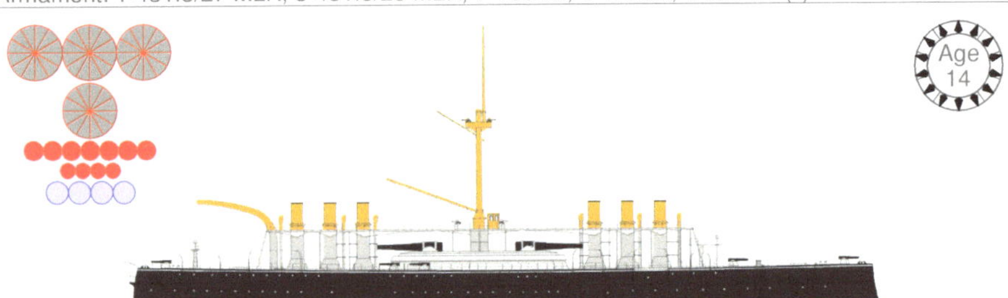

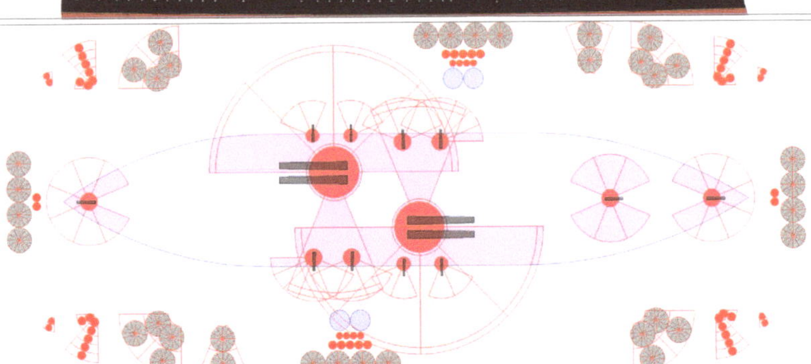

Italia (1876)
1876 01 03 - 1880 09 29 - 1885 10 16 | Castellammare DYd, Castellammare

Lepanto (1876)
1876 11 04 - 1883 03 17 - 1887 08 16 | Orlando, Leghorn

2 | "Italia" type (1876)
Protected cruisers 1 class, Italy (continued) (Data below is for "Lepanto")

Displacement: 13,336 tons at draft: 9.39 m, 15,649 tons full | Length: 124.7 m oa | Beam: 22.34 m | Crew: 669 men

Machinery: 8 oval, 16 cylindrical boilers | 4 * 4,000 HP vertical compound EEEE * 4 screws | 15,797 HP = 18.4 knots | Supplies: 1,650 / +350 tons of coal

Armour: steel | CT: 102: Citadel: 483; Funnel bases: 406; Hoists: 456 | WLDeck: 0/102(^76)

Armament: 4-431.8/27 MLR, 8-152.4/32, 4-120/32, 4-356 TT (^)

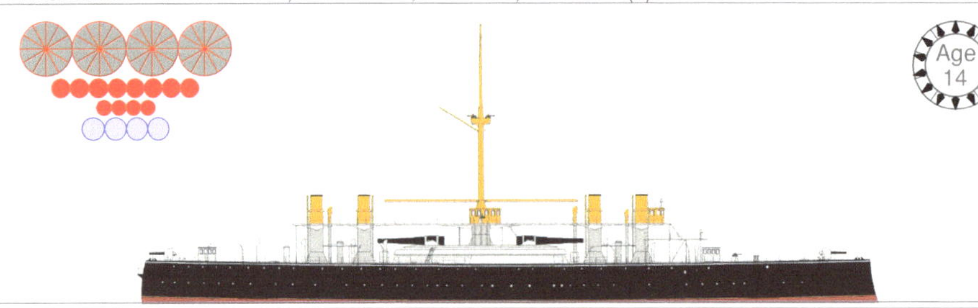

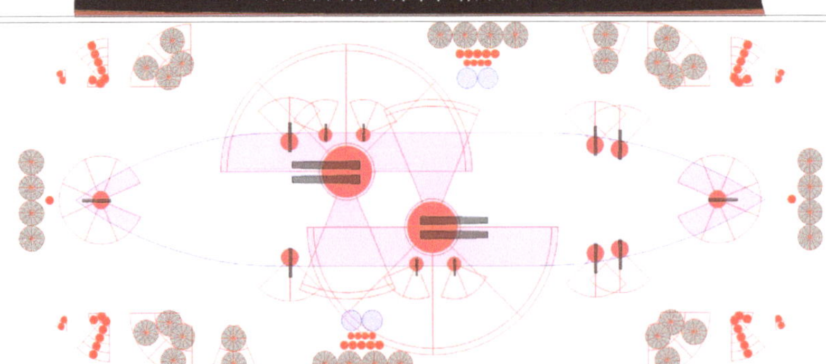

0 +4 | "Lombardia" type (1888)
Protected cruisers 2 class, Italy

BUILDING

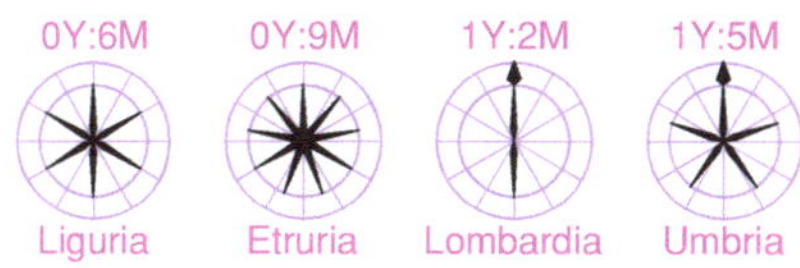

Displacement: 2,245 tons at draft: 4.87 m, 2,750 tons full | Length: 84.8 m oa | Beam: 12.03 m | Crew: 213 men

Machinery: 4 cylindrical boilers | 2 * 3,750 HP horizontal triple expansion EE * 2 screws | 7,500 HP = 18.5 knots | Supplies: 430 tons of coal

Armour: steel | CT: 51(^25) |
WLDeck: 0(/51^51)

Armament: 4-152.4/40, 6-120/40, 10-57/43, 2-450 TT (^)

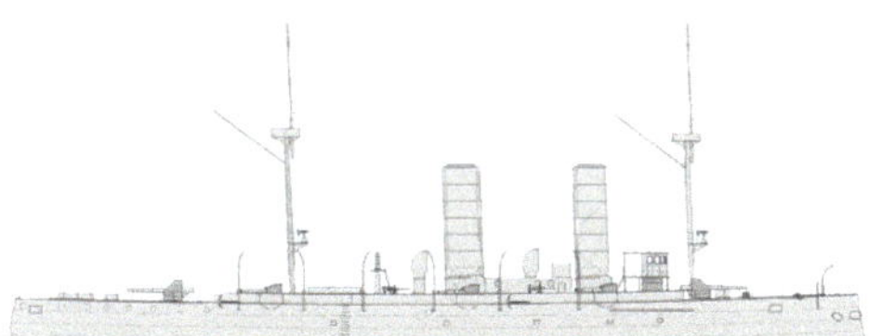

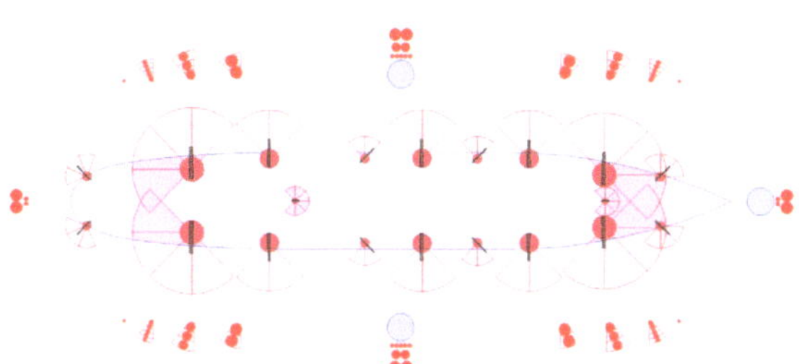

Etruria (1889)
1889 04 01 - ***** - ***** | Orlando, Leghorn

Liguria (1889)
1889 07 01 - ***** - ***** | Ansaldo, Genoa

Lombardia (1888)
1888 11 19 - ***** - ***** | Castellammare DYd, Castellammare

0 +4 | "Lombardia" type (1888)
Protected cruisers 2 class, Italy (continued)

Umbria (1888)
1888 08 01 - ***** - ***** | Orlando, Leghorn

Liguria

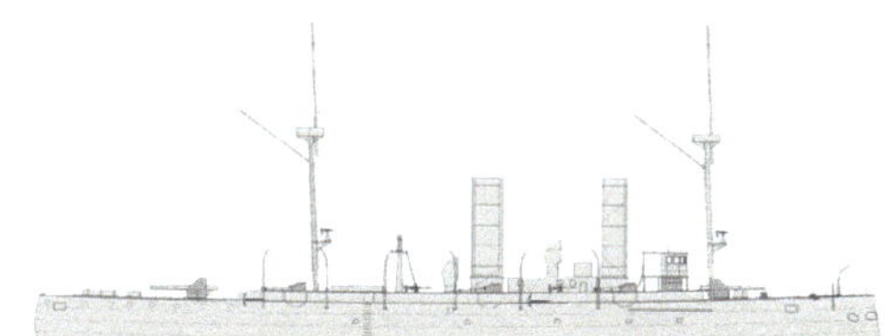

1 | "Piemonte" type (1887)
Protected cruisers 2 class, Italy

Displacement: 2,443 tons at draft: 7.86 m, 2,900 tons full | Length: 97.83 m oa, 92.73 m pp | Beam: 11.62 m | Crew: 298 men

Machinery: 4 cylindrical boilers | 2 * 3,550 HPn / 5,750 HPf vertical double expansion EE * 2 screws | 7,100 HPn = 20.5 knots / 11,500 HPf = 21.5 knots | Supplies: 200 / +400 tons of coal

Armour: steel | CT: 76 | WLDeck: 0(/76-51^25)

Armament: 6-152.4/40, 6-120/40, 10-57/43, 6-37/23, 4-25.4/40 Mg, 3-356 TT (^)

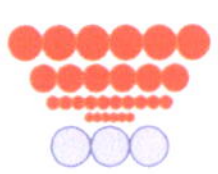

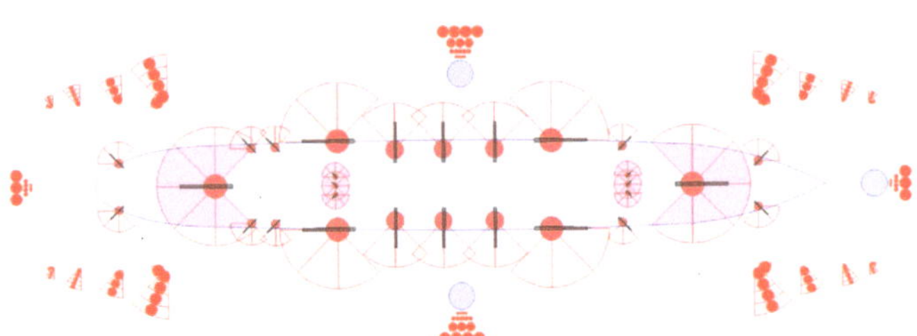

Piemonte (1887)

2 Humphrys vertical triple expantion engines | 7,100 HP = 20.44 knots / 12,980 HPf = 22.4 knots

1887 01 01 - 1888 08 23 - 1889 09 08 | Armstrong, Elswick

1 | "Dogali" type (1885)
Protected cruiser 2 class, Italy

Displacement: 2,050 tons at draft: 4.42 m | Length: 76.2 m. pp | Beam: 11.28 m | Crew: 224 men

Machinery: 4 cylindrical boilers | 2 * 2,500 HPn / 3,750 HPf vertical triple expansion EE * 2 screws | 5,000 HP = 17.75 knots / 7,500 HPf = 19.5 knots | Supplies: 200 / +280 tons of coal

Armour: steel | CT: 51; Shields: 102 | WLDeck: 0(/51^25)

Armament: 6-152.4/32, 9-57/43, 6-25.4/40 Mg, 4-356 TT (^)

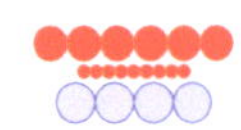

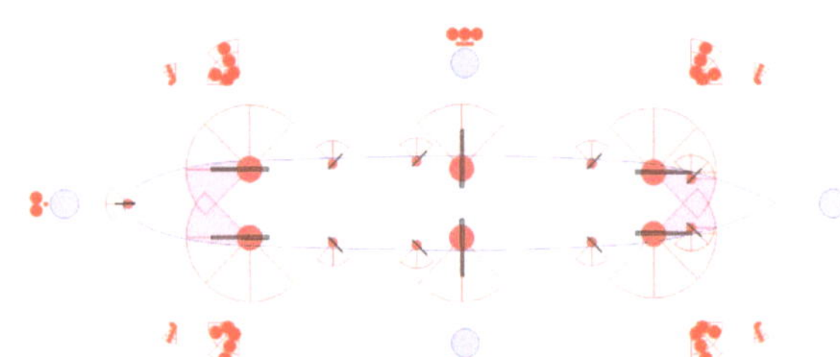

Dogali (1885)

2 Leslie horizontal triple expantion engines | 7,179 HPf = 19.66 knots

1885 02 13 - 1885 12 23 - 1887 04 28 | Armstrong, Elswick

5 | "Giovanni Bausan" type (1882)
Protected cruisers 2 class, Italy

Displacement: 3,079 tons at draft: 5.98 m | Length: 89.32 m oa | Beam: 12.85 m | Crew: 295 men

Machinery: 4 cylindrical boilers | 2 * 3,200 HPf double expansion * 2 screws | 6,470 HPf = 17.4 knots | Supplies: 400 / +200 tons of coal

Armour: steel | CT: 51; Hosts: 25 |
WLDeck: 0/38(^13)

Armament: 4-254/30, 6-152.4/26, 4-57/43, 6-37/23R, 3-37/23, 3-356 TT (^)

Age 8

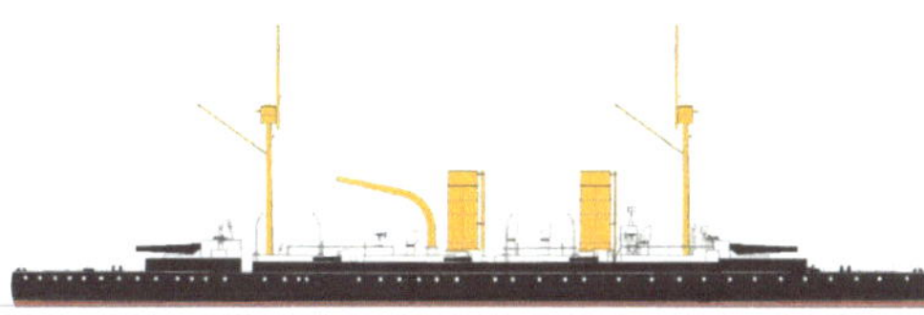

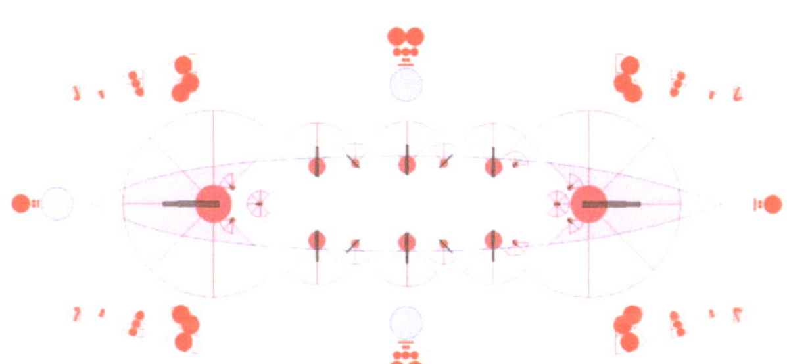

Giovanni Bausan (1882)
1882 08 21 - 1883 12 15 - 1885 05 09 | Armstrong, Elswick

Etna (1883)
1883 01 19 - 1885 09 26 - 1887 12 03 | Castellammare DYd, Castellammare

Ettore Fieramosca (1885)
1885 12 31 - 1888 08 30 - 1889 11 16 | Orlando, Leghorn

5 | "Giovanni Bausan" type (1882)
Protected cruisers 2 class (continued)

Stromboli (1883)
1883 09 27 - 1886 02 04 - 1888 03 21 | Venice DYd, Venice

Vesuvio (1883)
1883 07 10 - 1886 03 21 - 1888 03 16 | Orlando, Leghorn

Etna (and other italian built ships)

Displacement: 3,538 tons at draft: 5.83 m | Length: 91.4 m oa | Beam: 13.22 m | Crew: 308 men

Machinery: 4 cylindrical boilers | 2 * 3,200 HP double expansion EE * 2 screws | 6,252 HP = 16.6 knots | Supplies: 400 / +200 tons of coal

Armour: steel | CT: 51; Hosts: 25 |
WLDeck: 0/38(^13)

Armament: 4-254/30, 6-152.4/26, 4-57/43, 6-37/23R, 3-37/23, 3-356 TT (^)

Ettore Fieramosca

UNPROTECTED CRUISERS

2 | "Flavio Giola" type (1879)
Unprotected cruisers, Italy

Displacement: 2,493 tons at draft: 5.19 m | Length: 78 m pp | Beam: 12.78 m | Crew: 268 men
Machinery: 8 cylindrical boilers | 1 horizontal reciprocating E * 1 screw | 4,156 HP = 14 knots | Supplies: app. 400 tons of coal
Armament: 8-150/26, 3-75/24, 4-25.4/40

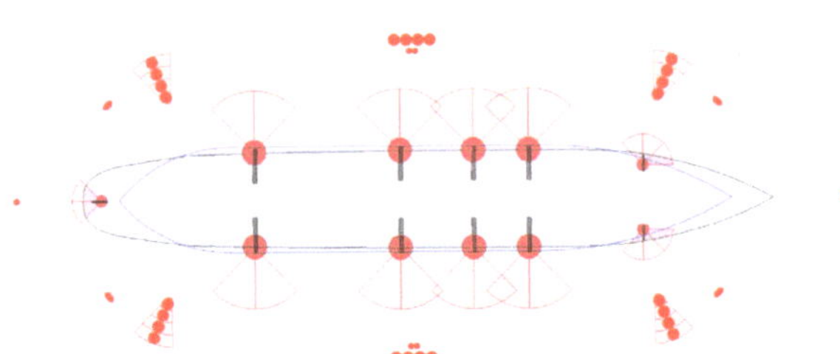

Amerigo Vespucci (1879)
1879 12 09 - 1882 07 31 - 1884 09 01 | Venice DYd, Venice

Flavio Giola (1879)
1879 06 26 - 1881 06 17 - 1883 01 26 | Castellammare DYd, Castellammare

1 | "Cristoforo Colombo" type (1873)
Wooden screw corvette, Italy

Displacement: 2,325 tons at draft: 5.25 m | Length: 75.72 m pp | Beam: 11.3 m | Crew: 207 men
Machinery: 6 cylindrical boilers | 1 recyprocating E * 1 screw | 3,782 HP = 16 knots | Supplies: app. 400 tons of coal
Armament: 8-120/24

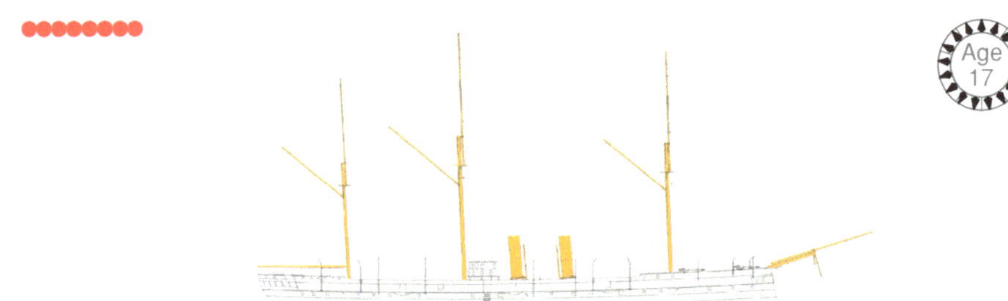

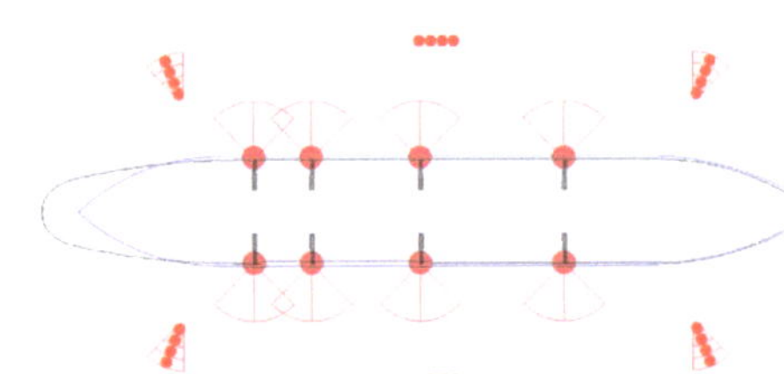

Cristoforo Colombo (1873)
1873 02 01 - 1875 09 17 - 1876 11 16 | Venice DYd, Venice

WOODEN CRUISERS

1 | "Vettor Pisani" type (1867)
Wooden screw corvette, Italy

Displacement: 1,676 tons at draft: 5.28 m | Length: 65.1 m pp | Beam: 11.84 m | Crew: 226 men
Machinery: 2 cylindrical boilers | 1 recyprocating E * 1 screw | 1,004 HP = 9.76 knots | Supplies: app. 200 tons of coal
Armament: 6-120/24, 2-75/24

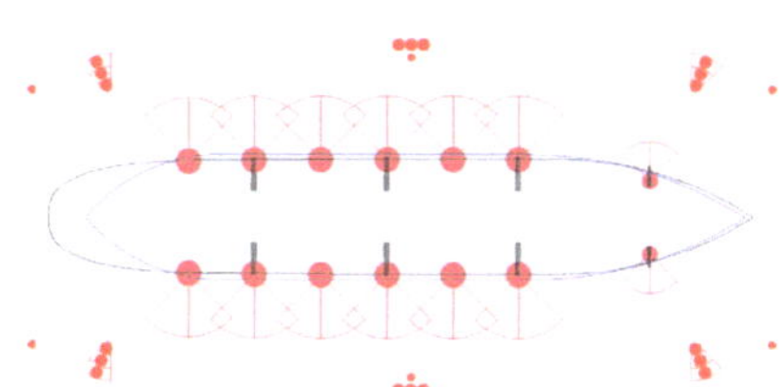

Vettor Pisani (1867)
1867 05 11 - 1869 07 22 - 1871 04 10 | Venice DYd, Venice

1 | "Caracciolo" type (1865)
Wooden screw corvette, Italy

Displacement: 1,553 tons at draft: 4.97 m | Length: 64.3 m pp | Beam: 10.94 m | Crew: 247 men
Machinery: 4 cylindrical boilers | 1 recyprocating E * 1 screw | 973 HP = 9.2 knots | Supplies: app. 300 tons of coal
Armament: 6-160/17 MLR

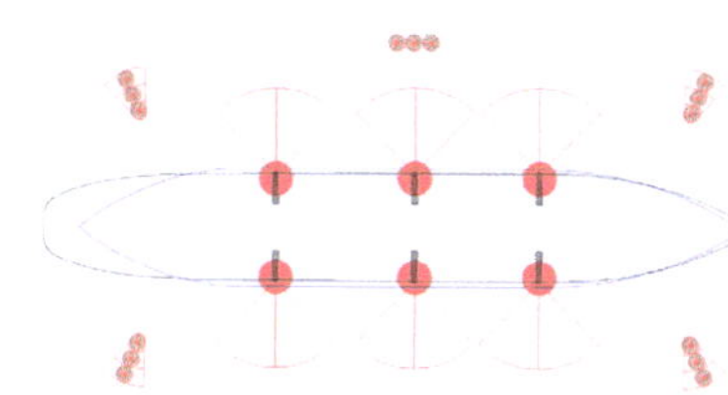

Caracciolo (1865)
1865 10 01 - 1869 01 18 - 1870 07 20 | Castellammare DYd, Castellammare

2 | "Archimede" type (1885)
Despatch vessels, Italy

Galileo Galilei

Displacement: 772 tons at draft: 3.75 m | Length: 70 m pp | Beam: 8.03 m | Crew: 73 men

Machinery: 4 cylindrical boilers | 1 vertical triple expansion E * 1 screw | 1,411 HP = 12 knots | Supplies: 210 tons of coal

Armament: 4-120/25

Age 5

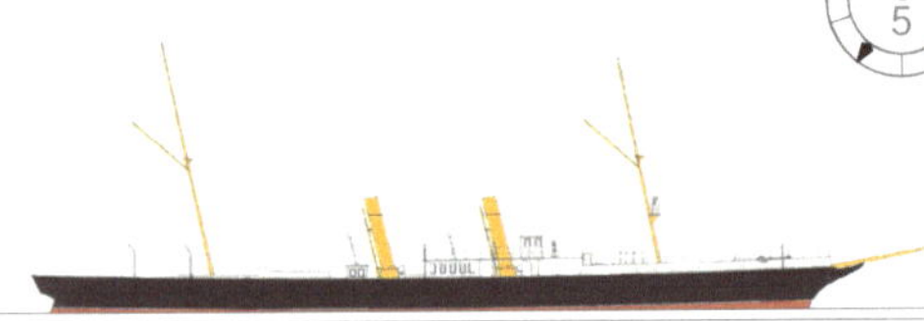

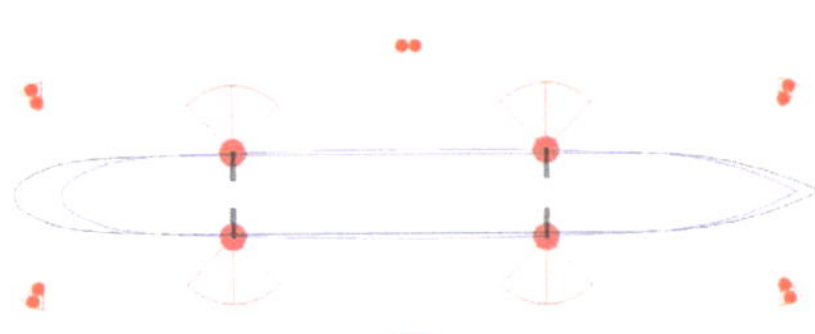

Archimede (1885)
1885 08 01 - 1887 03 08 - 1888 02 01 | Venice DYd

Galileo Galilei (1886)
1886 09 01 - 1887 05 03 - 1888 10 01 | Venice DYd

1 | "Messagiero" type (1883)
Paddle despatch vessel, Italy

Displacement: 1,005 tons at draft: 3.16 m | Length: 72 m oa | Beam: 9.14 m | Crew: 104 men

Machinery: 4 rectangular boilers | 1 recyprocating E * 1 screw | 1,797 HP = 14 knots | Supplies: app. 600 tons of coal

Armament: 4-75/21

Age 7

NO PLAN AVAILABLE

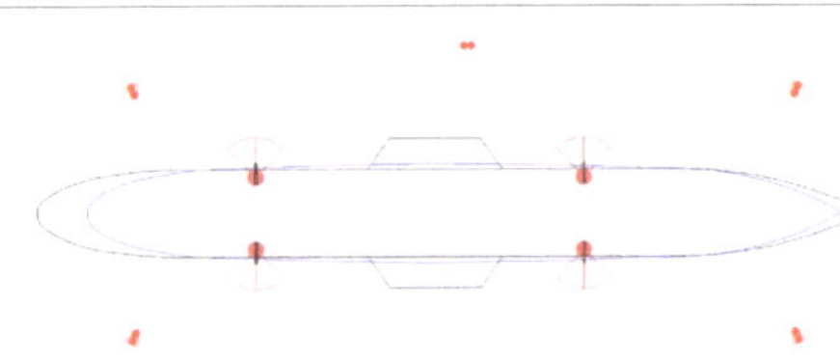

Messagiero (1883)
1883 01 01 - 1885 07 13 - 1888 10 11 | Odero, Sestri Ponente

2 | "Agostino Barbarigo" type (1876)
Despatch vessels, Italy

Displacement: 614 tons at draft: 3.43 m | Length: 66 m pp | Beam: 7.35 m | Crew: 104 men

Machinery: 4 cylindrical boilers | 1 vertical triple expansion E * 1 screw | 1,760 HP = 15 knots | Supplies: 200 tons of coal

Armament: 2-75/21, 1-356 TT

Age 14

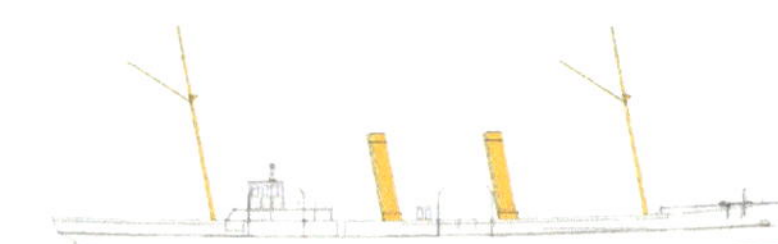

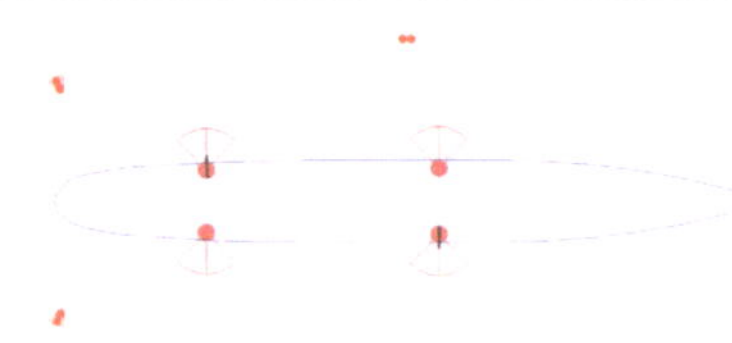

Agostino Barbarigo (1876)
1876 08 19 - 1879 01 23 - 1879 07 11 | Venice DYd

Marcantonio Colonna (1877)
1877 01 07 - 1879 08 06 - 1880 05 01 | Venice DYd

TORPEDO SHIPS

0 +6 | "Partenope" type (1888)
Torpedo gunboats, Italy

BUILDING

0Y:7M Arethusa | 0Y:11M Iride | 0Y:11M Urania | 0Y:11M Euridice | 0Y:11M Minerva | 1Y:7M Partenope

Displacement: 821 tons at draft: 3.48 m | Length: 73.09 m oa | Beam: 8.22 m | Crew: 96 men

Machinery: 4 locomotive boilers | 2 * 1,900 HP triple expansion EE * 2 screws | 3,884 HP = 18.1 knots | Supplies: app. 80 tons of coal

Armour: steel | CT: 30 | WLDeck: 0(^30)

Armament: 1-120/40, 6-57/43, 2-37/23, 5-450 TT (4*1 trainable, 1 >)

Age 1

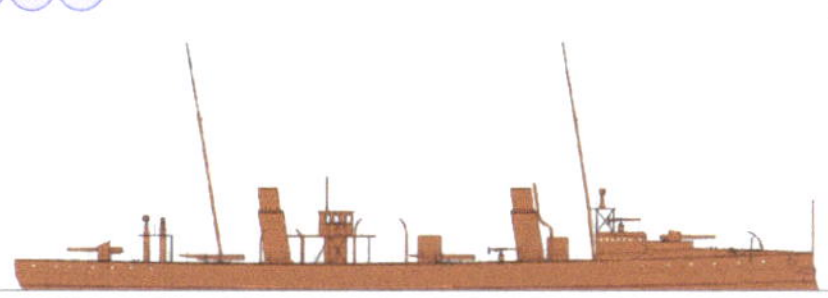

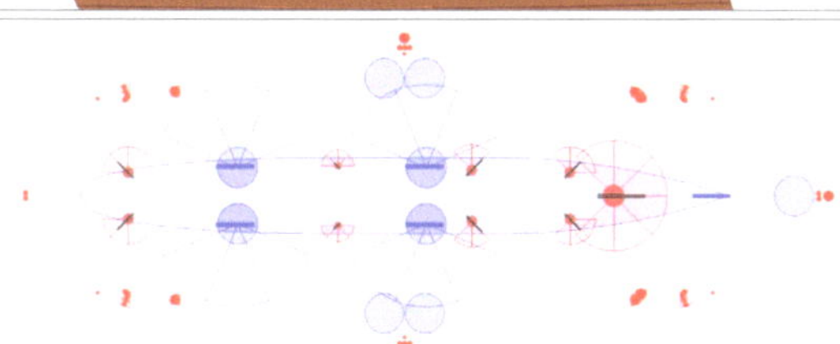

Arethusa (1889)
1889 06 01 - ***** - ***** | Orlando, Leghorn

Euridice (1889)
1889 02 14 - ***** - ***** | Castellammare DYd

Iride (1889)
1889 02 21 - ***** - ***** | Castellammare DYd

Minerva (1889)
1889 02 01 - ***** - ***** | Ansaldo, Genoa

Partenope (1888)
1888 06 08 - 1889 12 23 - ***** | Castellammare DYd

Urania (1889)
1889 02 16 - ***** - ***** | Odero, Sestri Ponente

0 +1 | "Confienza" type (1887)
Torpedo gunboat, Italy

BUILDING

2Y:4M Confienza

Displacement: 756 tons at draft: 3.04 m, 969 tons full | Length: 73.4 m oa | Beam: 8.05 m | Crew: 105 men

Machinery: 4 locomotive boilers | 2 * 1,000 HP double expansion EE * 2 screws | 1,962 HP = 17 knots | Suppl.: app. 100 tons of coal

Armour: steel | CT: 40 | WLDeck: 0(^40)

Armament: 1-120/32, 6-57/43, 2-37/23, 4-356 TT (3*1 trainable, 1 >)

Age 3

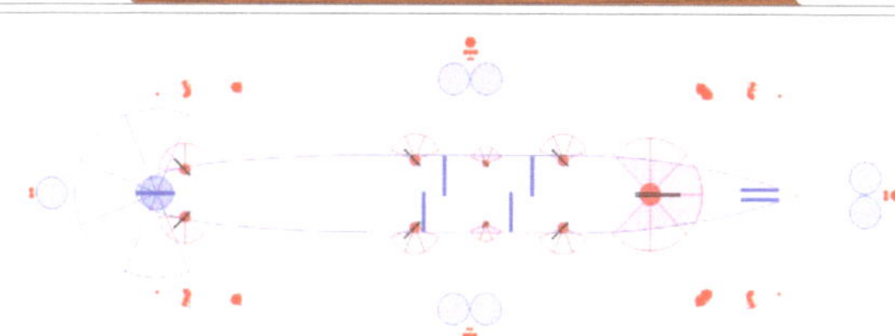

Confienza (1887)
1887 09 01 - 1889 07 28 - ***** | La Spezia DYd

2 | "Folglore" type (1886)
Torpedo gunboats, Italy

Displacement: 364 tons at draft: 2.15 m | Length: 56.7 m oa | Beam: 6.31 m | Crew: 57 men

Machinery: 4 locomotive boilers | 2 * 1,100 HP double expansion EE * 2 screws | 2,150 HP = 17 knots | Suppl.: app. 40 tons of coal

Armour: nil

Armament: 2-57/43, 4-37/23, 3-356 TT (1*1 trainable, 2 >>)

Age 4

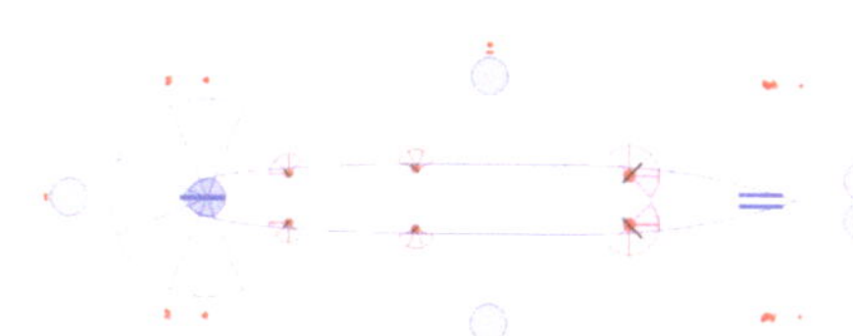

Folglore (1886) *
1886 01 01 - 1886 09 29 - 1887 02 16 | Castellammare DYd

Saetta (1886)
1886 01 01 - 1887 05 30 - 1888 02 16 | Castellammare DYd

* "Folglore" is laid up due to the heavy dammage

3 | "Goito" type (1885)
Torpedo gunboats, Italy

Displacement: 829 tons at draft: 3.6 m | Length: 73.4 m oa | Beam: 7.88 m | Crew: 105 men

Machinery: 6 locomotive boilers | 3 * 900 HP double expansion EEE * 3 screws | 2,700 HP = 18 knots | Suppl.: app. 100 tons of coal

Armour: steel | CT: 40 | WLDeck: 0(^40)

Armament: 4-57/43, 2-37/23, 3-37/23R, 5-356 TT (4*1 trainable, 1 >)

Age 5

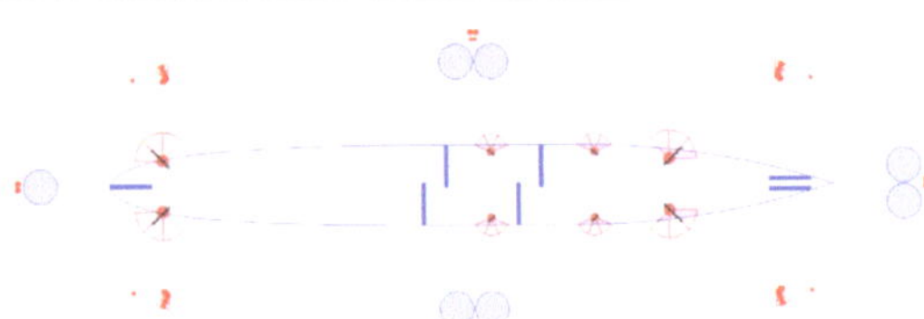

Goito (1885)
1885 09 01 - 1887 07 06 - 1888 02 16 | Castellammare DYd

Montebello (1885)
1885 09 25 - 1888 03 14 - 1889 01 21 | La Spezia DYd

Monzambano (1885)
1885 08 25 - 1888 03 14 - 1889 08 11 | La Spezia DYd

Tripoli (1885)
1885 06 10 - 1886 08 25 - 1886 12 12 | Castellammare DYd

3 | "Goito" type (1885)
Torpedo gunboats, Italy (continued)

Montebello

Displacement: 801 tons at draft: 3.31 m, 955 tons full | Length: 73.4 m oa | Beam: 7.88 m | Crew: 105 men

Machinery: 6 locomotive boilers | 3 * 900 HP triple expansion EEE * 3 screws | 2,700 HP = 18 knots | Supplies: app. 100 tons of coal

Armour: steel | CT: 40 | WLDeck: 0(^40)

Armament: 6-57/43, 2-37/23, 4-356 TT (2*1 trainable, 2 >>)

Age 5

Tripoli

1 | "Pietro Micca" type (1875)
Torpedo gunboat, Italy

Displacement: 526.5 tons at draft: 3.62 m, 598 tons full | Length: 61.87 m pp | Beam: 5.97 m | Crew: 73 men

Machinery: 4 locomotive boilers | 1 vertical single expansion E * 1 screw | 571 HP = 12.88 knots | Supplies: app. 40 tons of coal

Armour: steel | WLDeck: 0(^18)

Armament: 2-25,4/30, 1-356 TT (1 >)

Age 15

Pietro Micca (1875)
1875 02 15 - 1876 08 01 - 1877 07 03 | Venice DYd

5 | "Aquila" type (1887)
Torpedo boats, Italy

Displacement: 125 tons at draft: 1.12 m | Length: 47.61 m oa | Beam: 5.1 m | Crew: 24 men

Machinery: 2 loc. boilers | 2 * 1,000 HP vertical double expansion EE * 2 screws | 2,000 HP = 26 knots | Supplies: 30 tons of coal

Armament: 1-37/23, 1-37/23R, 3-356 TT (2*1 trainable, 1 >)

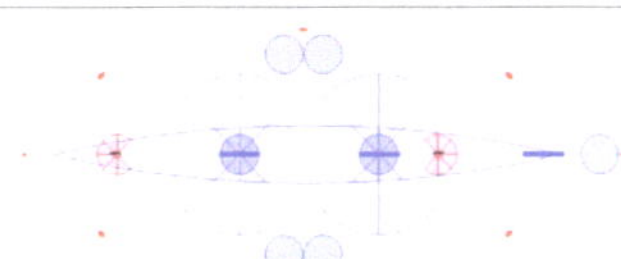

Aquila (1887)
1887 01 01 - 1888 01 01 - 1889 01 01 | Schichau, Elbing

Avvoltoio (1887)
1887 01 01 - 1888 01 01 - 1889 01 01 | Schichau, Elbing

Falco (1887)
1887 01 01 - 1888 01 01 - 1889 01 01 | Schichau, Elbing

Nibbio (1887)
1887 01 01 - 1888 01 01 - 1889 01 01 | Schichau, Elbing

Sparviero (1887)
1887 01 01 - 1888 01 01 - 1889 01 01 | Schichau, Elbing

Sister ship can be found in the Russian section

1 | "Fatum" type (1887)
Torpedo boat, Italy

Displacement: 42 tons at draft: 1.53 m | Length: 30.84 m oa | Beam: 3.51 m | Crew: 11 men

Machinery: 1 loc. boiler | 2 *250 HP vertical compound reciprocating EE * 2 screws | 500 HP = 19 knots | Supplies: 5 tons of coal

Armament: 2-25.4/40 Mg, 2-356 TT (2 >>)

Fatum (1887)
1887 01 01 - 1888 01 01 - 1889 01 01 | Orlando, Leghorn

10 +24 | "103 S" type (1886)
Torpedo boats, Italy

BUILDING

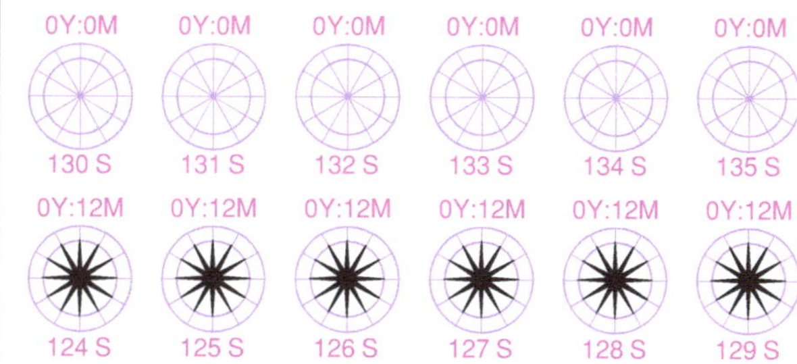

Displacement: 83 tons at draft: 1.12 m, 103 tons full | Length: 39.9 m oa | Beam: 4.8 m | Crew: 17 men

Machinery: 1 locomotive boiler | 1 vertical double expansion E * 1 screw | 900 HP = 19 knots | Supplies: 24 tons of coal

Armament: 2-37/23R, 2-356 TT (1*1 trainable, 1 >)

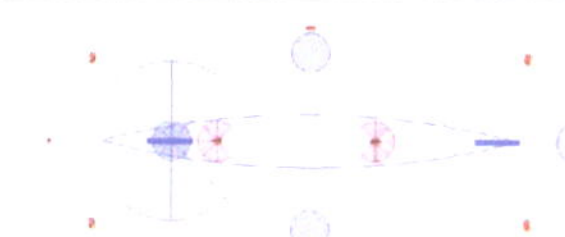

56 S (1886)
1886 12 12 - 1887 12 12 - 1888 12 12 | Ansaldo, Genoa

103 S (1886)
1886 01 01 - 1886 09 09 - 1887 06 06 | Schichau, Elbing

104 S (1886)
1886 01 01 - 1886 09 09 - 1887 06 06 | Schichau, Elbing

105 S (1886)
1886 01 01 - 1886 09 09 - 1887 06 06 | Schichau, Elbing

106 S (1886)
1886 01 01 - 1886 09 09 - 1887 06 06 | Schichau, Elbing

107 S (1886)
1886 01 01 - 1886 09 09 - 1887 06 06 | Schichau, Elbing

108 S (1886)
1886 01 01 - 1886 09 09 - 1887 06 06 | Schichau, Elbing

109 S (1886)
1886 01 01 - 1886 09 09 - 1887 06 06 | Schichau, Elbing

110 S (1886)
1886 01 01 - 1886 09 09 - 1887 06 06 | Schichau, Elbing

111 S (1886)
1886 01 01 - 1886 09 09 - 1887 06 06 | Schichau, Elbing

10 +24 | "103 S" type (1886)
Torpedo boats, Italy (continued)

1Y:12M	1Y:12M	1Y:12M	1Y:12M	1Y:12M	1Y:12M
112 S	113 S	114 S	115 S	116 S	117 S
1Y:12M	1Y:12M	1Y:12M	1Y:12M	1Y:12M	1Y:12M
118 S	119 S	120 S	121 S	122 S	123 S

112 S (1888)
1888 01 01 - 1889 01 01 - ***** | Pattison, Naples

113 S (1888)
1888 01 01 - 1889 01 01 - ***** | Pattison, Naples

114 S (1888)
1888 01 01 - 1889 01 01 - ***** | Pattison, Naples

115 S (1888)
1888 01 01 - 1889 01 01 - ***** | Pattison, Naples

116 S (1888)
1888 01 01 - 1889 01 01 - ***** | Gravero, Genoa

117 S (1888)
1888 01 01 - 1889 01 01 - ***** | Gravero, Genoa

118 S (1888)
1888 01 01 - 1889 01 01 - ***** | Gravero, Genoa

119 S (1888)
1888 01 01 - 1889 01 01 - ***** | Gravero, Genoa

120 S (1888)
1888 01 01 - 1889 01 01 - ***** | Odero, Sestri Ponente

121 S (1888)
1888 01 01 - 1889 01 01 - ***** | Odero, Sestri Ponente

122 S (1888)
1888 01 01 - 1889 01 01 - ***** | Odero, Sestri Ponente

123 S (1888)
1888 01 01 - 1889 01 01 - ***** | Odero, Sestri Ponente

124 S (1889)
1889 01 01 - ***** - ***** | Ansaldo, Genoa

125 S (1889)
1889 01 01 - ***** - ***** | Ansaldo, Genoa

126 S (1889)
1889 01 01 - ***** - ***** | Ansaldo, Genoa

127 S (1889)
1889 01 01 - ***** - ***** | Ansaldo, Genoa

128 S (1889)
1889 01 01 - ***** - ***** | Odero, Sestri Ponente

129 S (1889)
1889 01 01 - ***** - ***** | Odero, Sestri Ponente

10 +24 | "103 S" type (1886)
Torpedo boats, Italy (continued)

130 S (1890)
1890 01 01 - ***** - ***** | Odero, Sestri Ponente

131 S (1890)
1890 01 01 - ***** - ***** | Odero, Sestri Ponente

132 S (1890)
1890 01 01 - ***** - ***** | Pattison, Naples

133 S (1890)
1890 01 01 - ***** - ***** | Pattison, Naples

134 S (1890)
1890 01 01 - ***** - ***** | Pattison, Naples

135 S (1890)
1890 01 01 - ***** - ***** | Pattison, Naples

4 | "76 YA" type (1886)
Torpedo boats, Italy

Displacement: 108.5 tons at draft: 1.54 m | Length: 41.18 m oa | Beam: 4.27 m | Crew: 20 men

Machinery: 2 loc. boilers | 2 * 800 HP vertical double expansion EE * 2 screws | 1,600 HP = 26 knots | Suppl.: app. 20 tons of coal

Armament: 2-37/23R, 4-356 TT (1*2 trainable, 2 >>)

Age 4

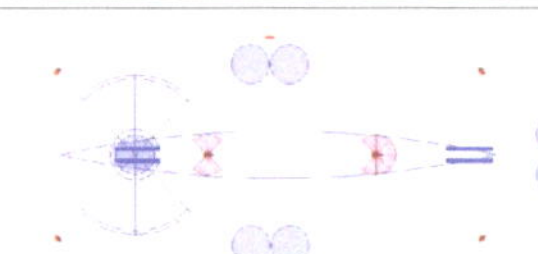

76 YA (1886)
1886 01 01 - 1886 06 06 - 1887 01 01 | Yarrow, Poplar

77 YA (1886)
1886 01 01 - 1886 06 06 - 1887 01 01 | Yarrow, Poplar

78 YA (1886)
1886 01 01 - 1887 01 01 - 1888 01 01 | Venice DYd

79 YA (1886)
1886 01 01 - 1887 01 01 - 1888 01 01 | Venice DYd

38 | "Habicht"/"56 S" type (1884)
Torpedo boats, Austria-Hungary

Displacement: 78 tons at draft: 1.12 m | Length: 39.9 m oa | Beam: 4.8 m | Crew: 17 men

Machinery: 1 locomotive boilers | 1 vertical double expansion E * 1 screw | 728 HP = 19 knots | Supplies: 24 tons of coal

Armament: 2-37/23R, 2-356 TT (2 >>)

Age 5

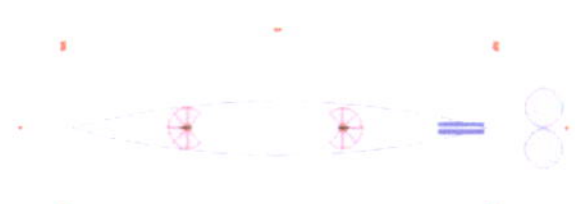

57 S (1885)
1885 01 01 - 1885 09 09 - 1887 01 01 | Schichau, Elbing

58 S (1885)
1885 01 01 - 1885 09 09 - 1887 01 01 | Schichau, Elbing

59 S (1885)
1885 01 01 - 1885 09 09 - 1887 01 01 | Schichau, Elbing

60 S (1887)
1887 01 01 - 1888 01 01 - 1889 01 01 | Pattison, Naples

61 S (1887)
1887 01 01 - 1888 01 01 - 1889 01 01 | Pattison, Naples

62 S (1887)
1887 01 01 - 1888 01 01 - 1889 01 01 | Pattison, Naples

63 S (1887)
1887 01 01 - 1888 01 01 - 1889 01 01 | Pattison, Naples

64 S (1887)
1887 01 01 - 1888 01 01 - 1889 01 01 | Pattison, Naples

65 S (1887)
1887 01 01 - 1888 01 01 - 1889 01 01 | Pattison, Naples

66 S (1887)
1887 01 01 - 1888 01 01 - 1889 01 01 | Guppy, Naples

38 | "Habicht"/"56 S" type (1884)
Torpedo boats, Austria-Hungary (continued)

67 S (1887)
1887 01 01 - 1888 01 01 - 1889 01 01 | Guppy, Naples

68 S (1886)
1886 01 01 - 1887 01 01 - 1888 01 01 | Gravero, Genoa

69 S (1886)
1886 01 01 - 1887 01 01 - 1888 01 01 | Gravero, Genoa

70 S (1886)
1886 01 01 - 1887 01 01 - 1888 01 01 | Gravero, Genoa

71 S (1886)
1886 01 01 - 1887 01 01 - 1888 01 01 | Gravero, Genoa

72 S (1886)
1886 01 01 - 1887 01 01 - 1888 01 01 | Odero, Sestri Ponente

73 S (1886)
1886 01 01 - 1887 01 01 - 1888 01 01 | Odero, Sestri Ponente

74 S (1886)
1886 01 01 - 1887 01 01 - 1888 01 01 | Odero, Sestri Ponente

75 S (1886)
1886 01 01 - 1887 01 01 - 1888 01 01 | Odero, Sestri Ponente

84 S (1885)
1885 01 01 - 1885 09 09 - 1886 06 06 | Schichau, Elbing

85 S (1885)
1885 01 01 - 1885 09 09 - 1886 06 06 | Schichau, Elbing

86 S (1887)
1887 01 01 - 1888 01 01 - 1889 01 01 | Ansaldo, Genoa

87 S (1887)
1887 01 01 - 1888 01 01 - 1889 01 01 | Ansaldo, Genoa

88 S (1887)
1887 01 01 - 1888 01 01 - 1889 01 01 | Ansaldo, Genoa

89 S (1887)
1887 01 01 - 1888 01 01 - 1889 01 01 | Ansaldo, Genoa

90 S (1887)
1887 01 01 - 1888 01 01 - 1889 01 01 | Ansaldo, Genoa

91 S (1887)
1887 01 01 - 1888 01 01 - 1889 01 01 | Ansaldo, Genoa

92 S (1887)
1887 01 01 - 1888 01 01 - 1889 01 01 | Ansaldo, Genoa

93 S (1887)
1887 01 01 - 1888 01 01 - 1889 01 01 | Odero, Sestri Ponente

94 S (1887)
1887 01 01 - 1888 01 01 - 1889 01 01 | Odero, Sestri Ponente

95 S (1887)
1887 01 01 - 1888 01 01 - 1889 01 01 | Odero, Sestri Ponente

96 S (1887)
1887 01 01 - 1888 01 01 - 1889 01 01 | Odero, Sestri Ponente

97 S (1887)
1887 01 01 - 1888 01 01 - 1889 01 01 | Odero, Sestri Ponente

98 S (1887)
1887 01 01 - 1888 01 01 - 1889 01 01 | Odero, Sestri Ponente

99 S (1886)
1886 01 01 - 1886 09 09 - 1887 06 06 | Schichau, Elbing

38 | "Habicht"/"56 S" type (1884)
Torpedo boats, Austria-Hungary (continued)

100 S (1886)
1886 01 01 - 1886 09 09 - 1887 06 06 | Schichau, Elbing

101 S (1886)
1886 01 01 - 1886 09 09 - 1887 06 06 | Schichau, Elbing

102 S (1886)
1886 01 01 - 1886 09 09 - 1887 06 06 | Schichau, Elbing

Sister ships can be found in the Austria-Hungarian, German, Russian and Turquish sections

34 | "Aldebaran" type (1882)
Torpedo boats, Italy

Displacement: 38.5 tons at draft: 1.67 m | Length: 30.58 m oa | Beam: 3.58 m | Crew: 11 men

Machinery: 1 locomotive boilers | 1 vertical double expansion E * 1 screw | 430 HP = 21 knots | Supplies: app. 8 tons of coal

Armament: 1-25.4/40, 2-356 TT (2 >>)

Age 8

26 T (1882)
1882 01 01 - 1883 01 01 - 1883 06 06 | Thornycroft, Woolston

27 T (1882)
1882 01 01 - 1883 01 01 - 1883 06 06 | Thornycroft, Woolston

28 T (1882)
1882 01 01 - 1883 01 01 - 1883 06 06 | Thornycroft, Woolston

29 T (1882)
1882 01 01 - 1883 01 01 - 1883 06 06 | Thornycroft, Woolston

30 T (1882)
1882 01 01 - 1883 01 01 - 1883 06 06 | Thornycroft, Woolston

31 T (1882)
1882 01 01 - 1883 01 01 - 1883 06 06 | Thornycroft, Woolston

32 T (1883)
1883 01 01 - 1883 06 06 - 1884 01 01 | Thornycroft, Woolston

33 T (1883)
1883 01 01 - 1883 06 06 - 1884 01 01 | Thornycroft, Woolston

34 T (1883)
1883 01 01 - 1883 06 06 - 1884 01 01 | Thornycroft, Woolston

35 T (1883)
1883 01 01 - 1883 06 06 - 1884 01 01 | Thornycroft, Woolston

35 | "26 T" type (1882)
Torpedo boats, Italy (continued)

36 T (1882)
1882 01 01 - 1883 01 01 - 1884 01 01 | Orlando, Leghorn

37 T (1883)
1883 01 01 - 1884 01 01 - 1885 01 01 | Pattison, Naples

38 T (1883)
1883 01 01 - 1884 01 01 - 1885 01 01 | Odero, Sestri Ponente

39 T (1883)
1883 01 01 - 1884 01 01 - 1885 01 01 | Pattison, Naples

40 T (1882)
1882 01 01 - 1883 01 01 - 1883 06 06 | Thornycroft, Woolston

41 T (1883)
1883 01 01 - 1884 01 01 - 1885 01 01 | Orlando, Leghorn

42 T (1884)
1884 01 01 - 1885 01 01 - 1886 01 01 | Guppy, Naples

43 T (1884)
1884 01 01 - 1885 01 01 - 1886 01 01 | Guppy, Naples

44 T (1884)
1884 01 01 - 1885 01 01 - 1886 01 01 | Orlando, Leghorn

45 T (1884)
1884 01 01 - 1885 01 01 - 1886 01 01 | Orlando, Leghorn

46 T (1882)
1882 01 01 - 1883 01 01 - 1884 01 01 | Odero, Sestri Ponente

47 T (1882)
1882 01 01 - 1883 01 01 - 1884 01 01 | Odero, Sestri Ponente

48 T (1884)
1884 01 01 - 1885 01 01 - 1886 01 01 | Pattison, Naples

49 T (1884)
1884 01 01 - 1885 01 01 - 1886 01 01 | Pattison, Naples

50 T (1885)
1885 01 01 - 1886 01 01 - 1887 01 01 | Odero, Sestri Ponente

51 T (1885)
1885 01 01 - 1886 01 01 - 1887 01 01 | Odero, Sestri Ponente

52 T (1885)
1885 01 01 - 1886 01 01 - 1887 01 01 | Orlando, Leghorn

53 T (1885)
1885 01 01 - 1886 01 01 - 1887 01 01 | Orlando, Leghorn

54 T (1885)
1885 01 01 - 1886 01 01 - 1887 01 01 | Pattison, Naples

55 T (1885)
1885 01 01 - 1886 01 01 - 1887 01 01 | Pattison, Naples

80 T (1886)
1886 01 01 - 1887 01 01 - 1888 01 01 | Odero, Sestri Ponente

81 T (1886)
1886 01 01 - 1887 01 01 - 1888 01 01 | Odero, Sestri Ponente

82 T (1886)
1886 01 01 - 1887 01 01 - 1888 01 01 | Odero, Sestri Ponente

83 T (1886)
1886 01 01 - 1887 01 01 - 1888 01 01 | Odero, Sestri Ponente

Sister ships can be found in the Swedish section

1 | "Clio" type (1881)
Torpedo boat, Italy

Displacement: 30.5 tons at draft: 1.53 m | Length: 23.35 m oa | Beam: 3 m | Crew: 10 men

Machinery: 1 locomotive boiler | 1 triple expansion reciprocating E * 1 screw | 250 HP = 18 knots | Supplies: app. 5 tons of coal

Armament: 2-356 TT (2 >>)

Age 9

11 T (1881)
1881 01 01 - 1882 01 01 - 1883 01 01 | Orlando, Leghorn

2 | "Batum"/"Sparviero" type (1879)
Torpedo boats, Russia

Displacement: 39.5 tons at draft: 1.49 m | Length: 30.43 m oa | Beam: 3.81 m | Crew: 11 men

Machinery: 1 locomotive boiler | 1 vertical reciprocating E * 1 screw | 620 HP = 22.5 knots | Supplies: 8.3 tons of coal

Armament: 1-25.4/40, 2-356 TT (2 >>)

22 Y (1880)
1880 06 06 - 1881 01 01 - 1881 05 11 | Yarrow, Poplar

25 Y (1880)
1880 01 01 - 1881 01 01 - 1881 06 06 | Yarrow, Poplar

Sister ships can be found in the Argentinean, Austria-Hungarian, Brazilean, British, Chilean, Dutch, Greek, Japanese and Russian sections

2 | "Gabbiano" type (1880)
Torpedo boats, Italy

Displacement: 34.5 tons at draft: 1.47 m | Length: 29.18 m oa | Beam: 3.28 m | Crew: 11 men

Machinery: 1 locomotive boiler | 1 vertical double expansion E * 1 screw | 470 HP = 20 knots | Supplies: app. 6 tons of coal

Armament: 1-25.4/40 Mg, 2-356 TT (2 >>)

23 T (1880)
1880 01 01 - 1881 01 01 - 1881 06 06 | Thornycroft, Woolston

24 T (1880)
1880 01 01 - 1881 01 01 - 1881 06 06 | Thornycroft, Woolston

4 | "Mosca" type (1883)
Torpedo boats 2 class, Italy

Displacement: 16 tons at draft: 1.34 m | Length: 20.12 m oa | Beam: 2.44 m | Crew: 10 men

Machinery: 1 locomotive boiler | 1 vertical reciprocating E * 1 screw | 250 HP = 20 knots | Supplies: app. 2 tons of coal

Armament: 2-356 TT (2 >>)

12 T (1883)
1883 01 01 - 1883 06 06 - 1884 01 01 | Thornycroft, Woolston

13 T (1883)
1883 01 01 - 1883 06 06 - 1884 01 01 | Thornycroft, Woolston

14 T (1883)
1883 01 01 - 1883 06 06 - 1884 01 01 | Thornycroft, Woolston

15 T (1883)
1883 01 01 - 1883 06 06 - 1884 01 01 | Thornycroft, Woolston

14 | "Euterpe" type (1882)
Torpedo boats 2 class, Italy

Displacement: 13.5 tons at draft: 1.14 m | Length: 19.2 m oa | Beam: 2.29 m | Crew: 10 men

Machinery: 1 locomotive boiler | 1 vertical triple expansion E * 1 screw | 170 HP = 17.3 knots | Supplies: app. 1 tons of coal

Armament: 2-356 TT (2 >>)

Age 8

3 T (1883)
1883 01 01 - 1885 01 01 - 1885 06 06 | Thornycroft, Woolston

4 T (1884)
1884 01 01 - 1885 01 01 - 1885 06 06 | Thornycroft, Woolston

5 T (1884)
1884 01 01 - 1885 01 01 - 1885 06 06 | Thornycroft, Woolston

6 T (1884)
1884 01 01 - 1885 01 01 - 1885 06 06 | Thornycroft, Woolston

7 T (1883)
1883 01 01 - 1884 01 01 - 1884 06 06 | Thornycroft, Woolston

8 T (1883)
1883 01 01 - 1883 06 06 - 1884 01 01 | Thornycroft, Woolston

9 T (1883)
1883 01 01 - 1884 01 01 - 1884 06 06 | Thornycroft, Woolston

10 T (1883)
1883 01 01 - 1884 01 01 - 1884 06 06 | Thornycroft, Woolston

16 T (1883)
1883 01 01 - 1883 06 06 - 1884 01 01 | Thornycroft, Woolston

17 T (1883)
1883 01 01 - 1883 06 06 - 1884 01 01 | Thornycroft, Woolston

1 | "Avvoltoio" type (1878)
Torpedo boat 2 class, Italy

Displacement: 25 tons at draft: 0.95 m | Length: 26.6 m oa | Beam: 3.3 m | Crew: 10 men

Machinery: 1 locomotive boiler | 1 vertical double expansion E * 1 screw | 400 HP = 21 knots | Supplies: app. 4 tons of coal

Armament: 2-356 TT (2 >>)

Age 12

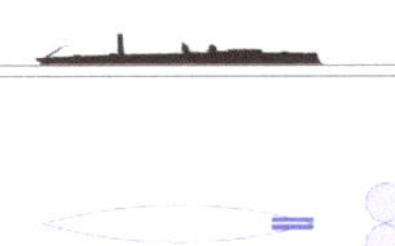

2 Y (1878)
1878 01 01 - 1879 01 01 - 1881 01 01 | Yarrow, Poplar

14 | "Euterpe" type (1882)
Torpedo boats 2 class, Italy (continued)

18 T (1883)
1883 01 01 - 1883 06 06 - 1884 01 01 | Thornycroft, Woolston

19 T (1883)
1883 01 01 - 1883 06 06 - 1884 01 01 | Thornycroft, Woolston

20 T (1882)
1882 01 01 - 1882 06 06 - 1883 01 01 | Thornycroft, Woolston

21 T (1882)
1882 01 01 - 1882 06 06 - 1883 01 01 | Thornycroft, Woolston

Sister ships can be found in the British, Danish, German sections

1 | "II"/"Nibbio" type (1876)
Spar torpedo boat, Netherlands

NO PHOTO AVAILABLE

Displacement: 25.5 tons at draft: 1.05 m | Length: 24.38 m oa | Beam: 3.06 m | Crew: 10 men

Machinery: 1 locomotive boiler | 1 compound reciprocating E * 1 screw | 250 HP = 18 knots | Supplies: app. 3 tons of coal

Armament: 2 spar torpedoes

Age 13

1 T (1877)
1877 01 01 - 1878 01 01 - 1879 01 01 | Thornycroft, Woolston

Sister ship can be found in the Dutch section

AUXILIARIES

1 | "Roma" type (1863)
Training ship (ex-battery ironclad), Italy

Displacement: 5,772 tons at draft: 7.6 m | Length: 79.65 m pp | Beam: 17.48 m | Crew: 551 men
Machinery: 6 cylindrical boilers | 1 single expansion E * 1 screw | 3,670 HP = 13 knots | Supplies: 621 tons of coal
Armour: iron | Battery: 114 | WLBelt: 150
Rearmed (1881): 2-75/28, 2-57/23

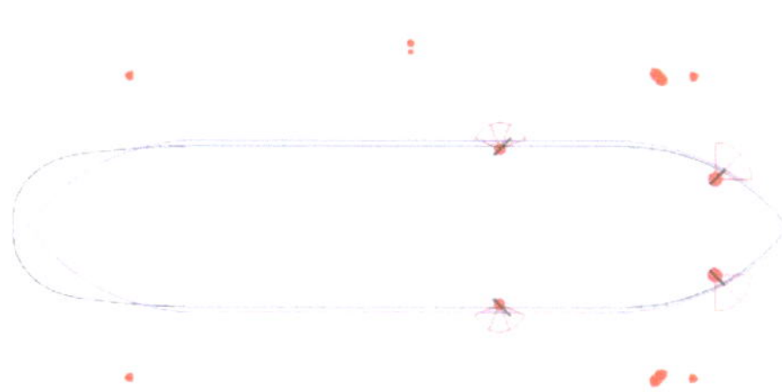

Venezia (1863)
1863 02 01 - 1869 01 21 - 1873 04 01 | La Foce, Genoa

2 | "Formidabile" type (1860)
Training stations (ex-battery ironclads), Italy

Displacement: 2,682 tons at draft: 5.45 m | Length: 65.8 m oa | Beam: 14.44 m | Crew: 371 men
Machinery: 6 rectangular boilers | 1 single expansion E * 1 screw | 1,100 HP = 10 knots | Supplies: 310 tons of coal
Armour: iron | Battery: 100 | WLBelt: 110
Rearmed (1885): 2-152,4/25, 2-150/30, 2-120/27, 2-356 TT (^)

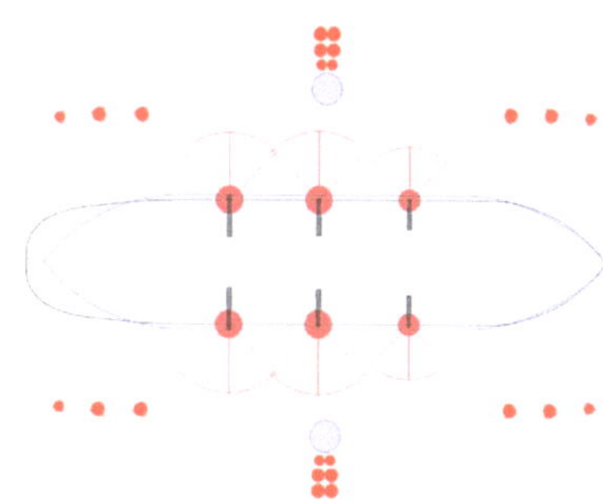

Terribile (1860)
1860 06 01 - 1861 02 16 - 1861 09 01 | F.et Ch.de la M., La Seyne

2 | "Formidabile" type (1860)
Training stations, Italy (comtinued)

Displacement: 2,682 tons at draft: 5.45 m | Length: 65.8 m oa | Beam: 14.44 m | Crew: 371 men
Machinery: 6 rectangular boilers | 1 single expansion E * 1 screw | 1,100 HP = 10 knots | Supplies: 310 tons of coal
Armour: iron | Battery: 100 | WLBelt: 110
Rearmed (1887): 6-120/27

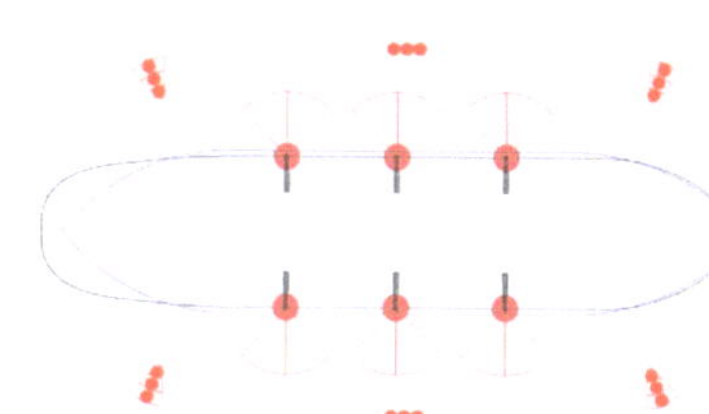

Formidabile (1860)
1860 12 01 - 1861 10 01 - 1862 05 01 | F.et Ch.de la M., La Seyne

SLOOPS & GUNBOATS

1 | "Rapido" type (1873)
Aviso, Italy

Displacement: 1,433 tons at draft: 3.81 m | Length: 78.16 m pp | Beam: 9.27 m | Crew: 144 men

Machinery: 4 cylindrical boilers | 1 vertical double expansion E * 1 screw | 1,920 HP = 13.5 knots | Supplies: app. 110 tons of coal

Armament: 2-75/21, 1-356 TT

Age 17

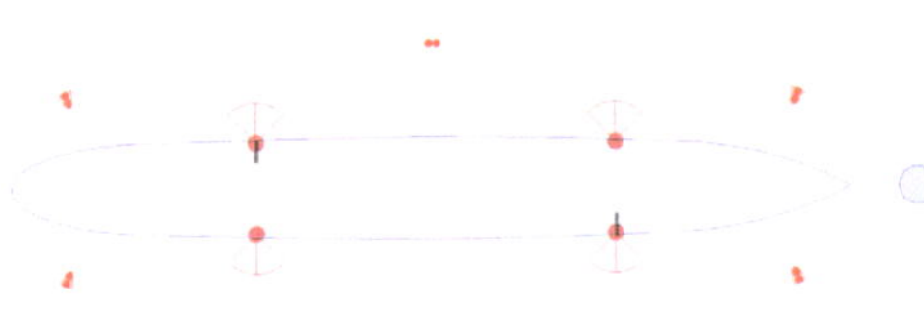

Rapido (1873)
1873 10 01 - 1876 11 16 - 1877 05 01 | Orlando, Leghorn

1 | "Staffetta" type (1873)
Aviso, Italy

Displacement: 1,366 tons at draft: 3.97 m | Length: 77.08 m pp | Beam: 9.43 m | Crew: 144 men

Machinery: 2 cylindrical boilers | 1 double expansion E * 1 screw | 1,700 HP = 12.5 knots | Supplies: 210 tons of coal

Armament: 4-120/21, 1-356 TT (^)

Age 17

Staffetta (1873)
1873 08 01 - 1876 06 24 - 1877 12 01 | Ansaldo, Genoa

1 | "Esploratore" type (1862)
Wooden paddle aviso, Italy

Displacement: 981 tons at draft: 3.62 m | Length: 71.77 m pp | Beam: 9.14 m | Crew: 125 men

Machinery: 4 rectangular boilers | 1 recyprocating compound E * 1 screw | 1,681 HP = 17 knots | Supplies: app. 90 tons of coal

Armament: 2-120/21

Age 28

NO PLAN AVAILABLE

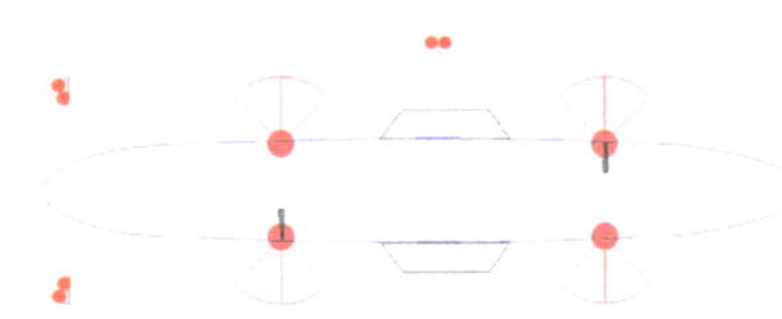

Esploratore (1862)
1862 07 05 - 1863 02 01 - 1864 08 01 | Wigram, Blackwall

1 | "Vedetta" type (1862)
Sloop, Italy

Displacement: 814 tons at draft: 3.56 m | Length: 56.25 m oa | Beam: 8.24 m | Crew: 95 men

Machinery: 2 rectangular boilers | 1 horizontal single expansion E * 1 screw | 670 HP = 11 knots | Supplies: 140 tons of coal

Armament: 2-75/21

Age 28

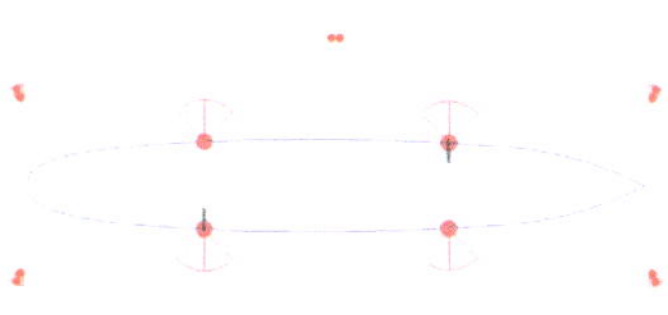

Vedetta (1862)
1862 01 01 - 1866 10 24 - 1869 04 16 | La Foce, Genoa

2 | "Castore" type (1887)
Gunboats, Italy

Displacement: 640 tons at draft: 2.45 m | Length: 30.07 m oa | Beam: 11.28 m | Crew: 40 men

Machinery: 2 cyl. boilers | 2 * 180 HP vertical triple expansion EE * 2 screws | 364 HP = 8 knots | Supplies: 200 / +400 tons of coal

Armament: 1-400/32

Age 3

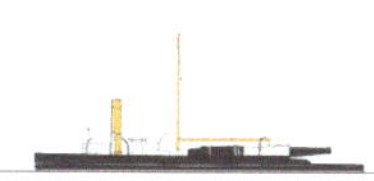

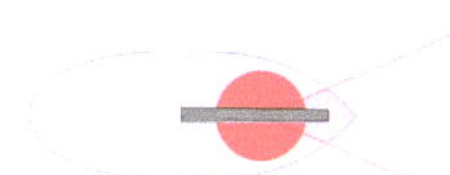

Castore (1887)
1887 02 22 - 1888 09 01 - 1889 09 09 | Armstrong, Elswick

Polluce (1887)
1887 02 22 - 1888 10 01 - 1889 09 09 | Armstrong, Elswick

2 | "Curtatone" type (1886)
Gunboats, Italy

Displacement: 1,039 tons at draft: 4.53 m | Length: 54.08 m pp | Beam: 9.89 m | Crew: 121 men

Machinery: 3 cylindrical boilers | 1 compound E * 1 screw | 1,043 HP = 10.71 knots | Supplies: app. 90 tons of coal

Armament: 4-120/32, 4-57/43, 2-37/23, 2-25.4/40 Mg, 1-356 TT (^)

Age 4

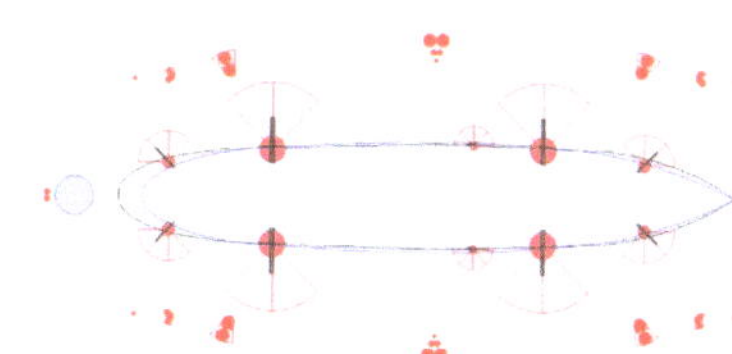

Curtatone (1887)
1887 01 01 - 1888 08 14 - 1889 12 12 | Venice DYd, Venice

Volturno (1886)
1886 01 01 - 1887 12 23 - 1889 01 01 | Venice DYd, Venice

2 | "Andrea Provana" type (1883) Gunboats, Italy

Displacement: 632 tons at draft: 3.4 m | Length: 51.5 m oa | Beam: 8.01 m | Crew: 95 men

Machinery: 2 cylindrical boilers | 1 vertical triple expansion E * 1 screw | 1,087 HP = 13.6 knots | Supplies: app. 50 tons of coal

Armament: 6-120/21, 2-25.4/40 Mg

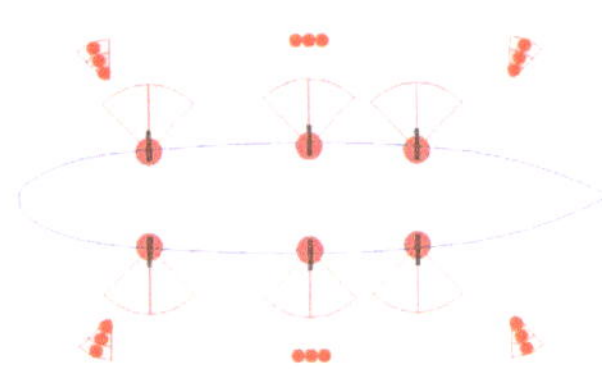

Andrea Provana (1883)
1883 01 01 - 1884 05 25 - 1886 01 01 | Orlando, Leghorn

Sebastiano Veniero (1883)
1883 01 01 - 1884 05 25 - 1886 01 01 | Orlando, Leghorn

2 | "Carridi" type (1873) Gunboats, Italy

Displacement: 1,059 tons at draft: 4.25 m | Length: 57.65 m wl | Beam: 8.76 m | Crew: 139 men

Machinery: 2 cylindrical boilers | 1 horizontal triple expansion E * 1 screw | 956 HP = 10.9 knots | Supplies: app. 90 tons of coal

Armament: 1-165/32, 2-120/21, 2-75/21, 2-25.4/40 Mg

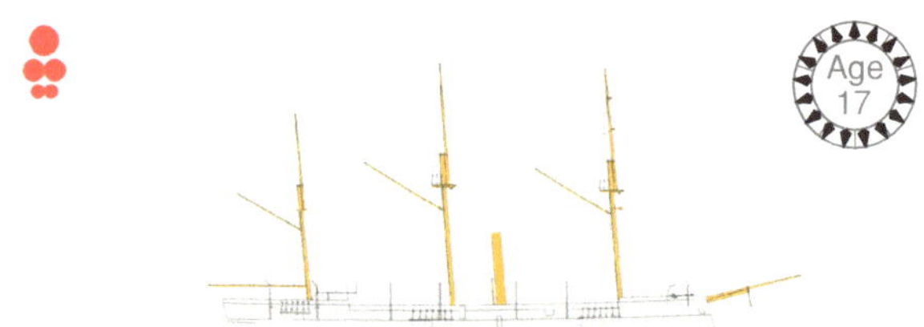

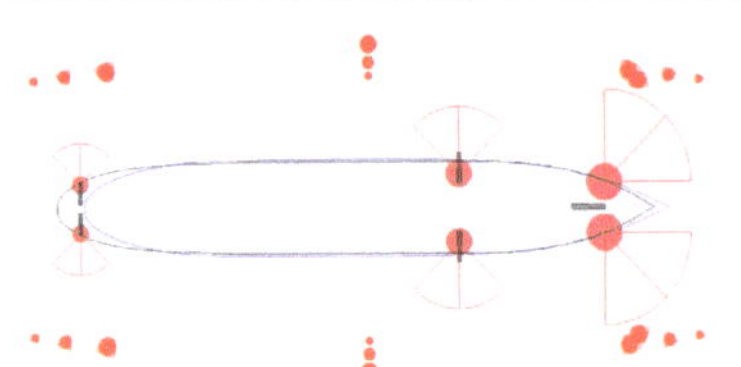

Carridi (1873)
1873 01 01 - 1875 03 22 - 1876 01 01 | Castellammare DYd

Scilla (1873)
1873 01 01 - 1874 10 10 - 1876 01 01 | Castellammare DYd

2 | "Guardiano" type (1873) Gunboats, Italy

NO PHOTO AVAILABLE

Displacement: 261 tons at draft: 1.98 m | Length: 30.5 m oa | Beam: 8.31 m | Crew: 48 men

Machinery: 4 cylindrical boilers | 2 * 120 HP horizontal compound EE * 2 screws | 231 HP = 8.7 knots | Supplies: 200 tons of coal

Armament: 1-228.6/13.9 MLR

Guardiano (1873)
1873 01 01 - 1874 05 01 - 1876 01 01 | La Spezia DYd

Sentinella (1873)
1873 01 01 - 1874 12 31 - 1876 01 01 | La Spezia DYd

TURKISH FLEET

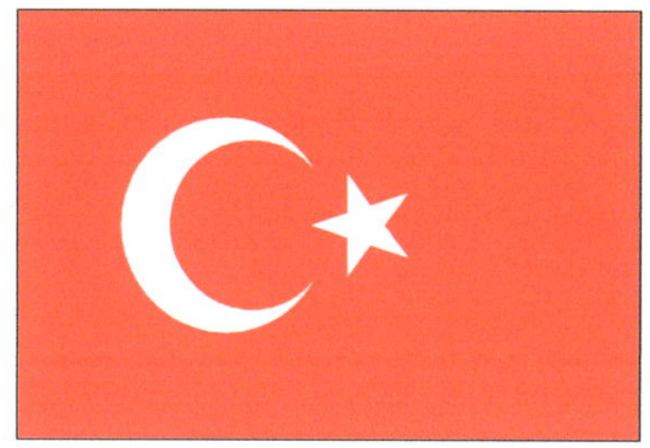

National flag

Fleet age 15

Ensign

Jack

Admiral

Months derivation -82

Warships built from 1858 01 01 to 1890 01 01

50 warships = 29,625 tons = 593 tons / average ship

Fleet age - 1875 11 03

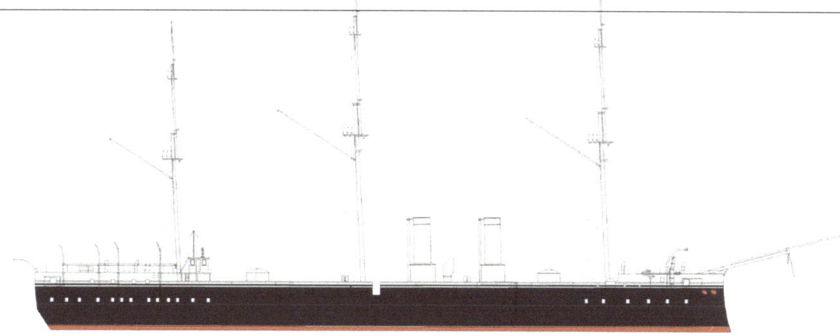

1889 fleet largest ship: Messudieh (1873) battery ironclad

Turkish naval construction

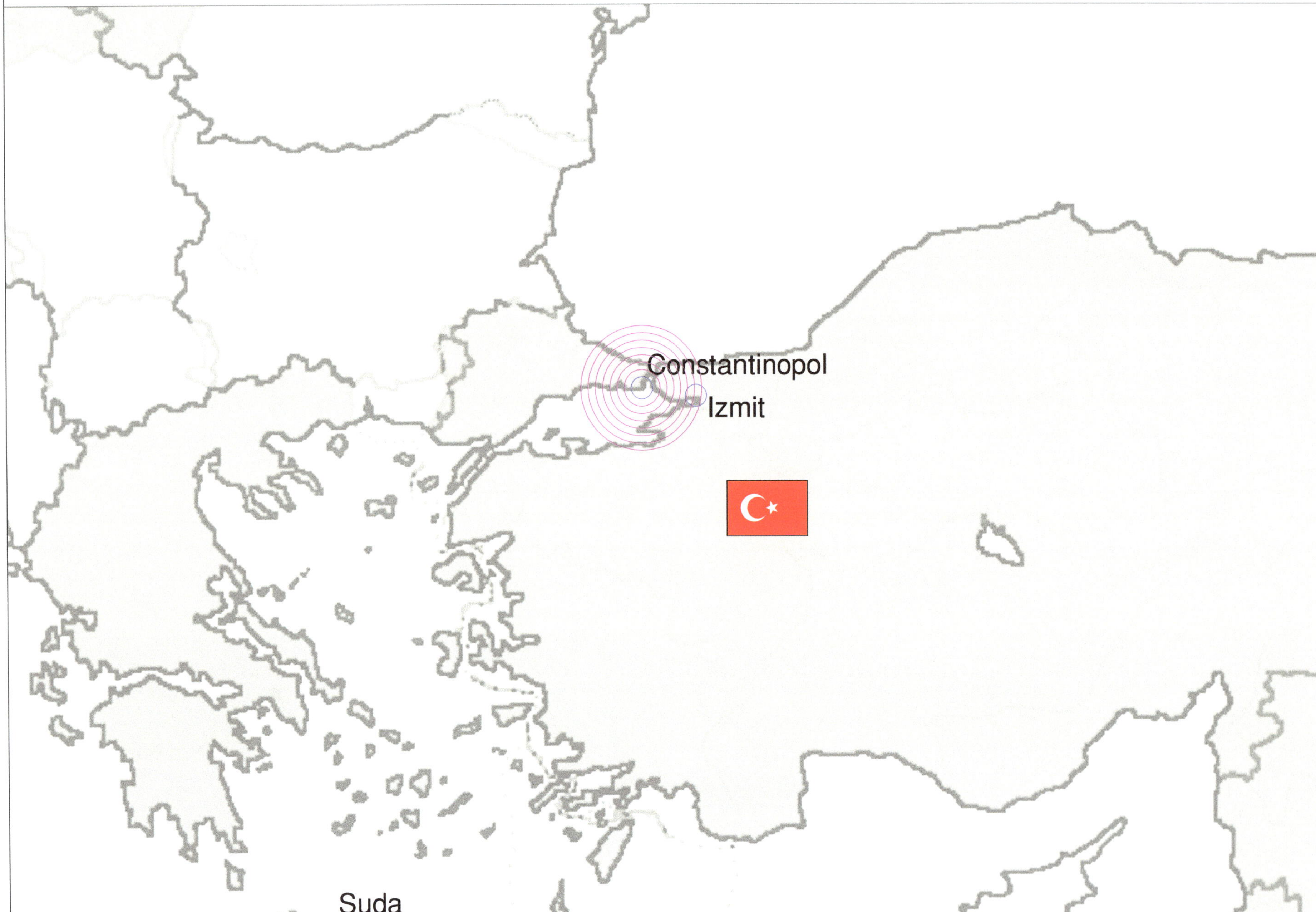

Constantinopol dockyard

7.3 tons / average ship / month

1890-Jan-01
7 ships building

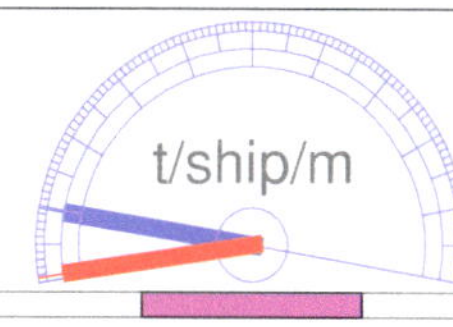

9,918 t.
tons / month = 51.2

Ship	t	t/m
Hamidieh	6,594	28.9
Heibetnuma	1,463	10.2
Lufti Humayun	1,292	9.0
Assari Teraki	42	0.8
Nimet	42	0.5
Shan Aver	42	0.2
Sahin-I Deria	443	3.3

Constantinopol, Turkey

32+ years of naval construction
30 ships | 25,984 tons

The plant's largest ship:
Hamidieh (1874) battery ironclad
6,594 tons

The plant's most effectively built ship:
Selimiye (1866) wooden screw frigate
96.9 tons/month

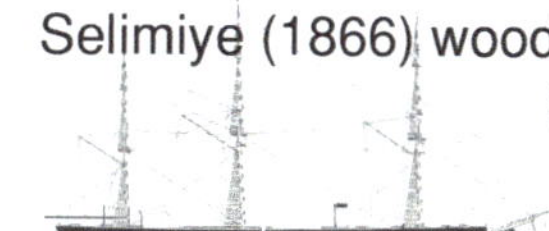

Specialisation in:
Battery ironclads | 2 ships | 9,355 tons
Screw sailing cruisers | 3 ships | 7,397 tons
Gunboats | 16 ships | 4,592 tons

Turkish fleet effective list

13 boardside ironclads

- Messudieh
- Mukaddami Khair — in reserve
- Muin-i-Zaffer — in reserve
- Avni Illah — in reserve
- Fethi Bulend
- Idjalieh — in reserve
- Assari Shevket — in reserve
- Nijmi Shevket — in reserve
- Assari Tewfik — in reserve
- Orkanieh — in reserve
- Mahmudieh — in reserve
- Abdul Aziz — in reserve
- Osmanieh — in reserve

4 coast defence armoured ships

- Hifzi Rahman
- Hizber
- Fethul Islam
- Memduhiye

2 wooden cruisers

- Mehemet Selim
- Ertogrul

11 screw sloops &15 gunboats

- Mansurah
- Muzaffer
- Beyrut
- Iskende-rieh
- Meric
- Sed-el-Bahr
- Utarit
- Zuhaf
- Bursa
- Edirneh
- Sinub
- Firat — NO PLAN AVAILABLE
- Sat — NO PLAN AVAILABLE
- Istankoi
- Rodos
- Amaly Kavak — NO PLAN AVAILABLE
- Yali Kiosk — NO PLAN AVAILABLE
- Sahir — NO PLAN AVAILABLE
- Shahiddin — NO PLAN AVAILABLE
- Mujde Resan

>>> screw sloops & gunboats (continued)

Seyr-i-Deria

Mosul — NO PLAN AVAILABLE

Seyar — NO PLAN AVAILABLE

Akka — NO PLAN AVAILABLE

Necmi-fesan — NO PLAN AVAILABLE

Shevket Numa — NO PLAN AVAILABLE

2 submarines

Abdul Hamid

Abdul Medjid

14 torpedo boats

Giljom

Saiki

Teiki Bahri

Tini Za-ffer

Vesirehi Nussret

Muhabet — NO PLAN AVAILABLE

Satvet — NO PLAN AVAILABLE

Timsah — NO PLAN AVAILABLE

Pervins

Seham

Burhan ed Din

Mecidieh

Tevfik

Semsir-i-Huchum

17 paddle gunboats

Is ed Din

Arkadi — NO PLAN AVAILABLE

Sureya — NO PLAN AVAILABLE

Ismail — NO PLAN AVAILABLE

Fuad — NO PLAN AVAILABLE

Ha Nieh — NO PLAN AVAILABLE

Taliah — NO PLAN AVAILABLE

Kandiah — NO PLAN AVAILABLE

Sultaniye — NO PLAN AVAILABLE

Serifiye — NO PLAN AVAILABLE

Ihsaniye — NO PLAN AVAILABLE

Asir — NO PLAN AVAILABLE

Taif — NO PLAN AVAILABLE

Eser-I Nussret — NO PLAN AVAILABLE

Medar-I Zaffer — NO PLAN AVAILABLE

Istanbul — NO PLAN AVAILABLE

Resmo — NO PLAN AVAILABLE

Fleet auxiliaries

ex-screw battleships

1 training station: Fethiye

1 laid up: Peyk-I Zafer

1 depot ship: Sadrye

ex-screw frigates

3 training stations: Rehberi Tewfik, Selimieh, Hudavendigar

3 depot ships: Peyk-I Nusret, Mubir-i-Surur, Mukad.-I Seref

ex-paddle corvettes

1 Depot ship: Mecidiyeh

1 Laid up: Eser-I Cedid

Turkish section ship names glossary

A
Abdul Hamid (SS)
Abdul Medjid (SS)
Akka (GB)
Arkadi (CZP)
Asir (YA)
Assari Shevket (BO)
Assari Teraki (TB1)
Assari Tewfik (BO)
Avni Illah (BO)
Aynali Kavak (GB)
Azizieh (BO)

B
Beyrut (CZ)
Burhan ed Din (TB1)
Bursa (CZ)

E
Edirneh (CZ)
Ertogrul (CW)
Eser-I Cedid (deptS)
Eser-I Nussret (YA)

F
Fatih (TB1)
Fethi Bulend (BO)
Fethiye (drilS)
Fethul Islam (MR)
Firat (GB)
Fuad (CZP)

G
Giljom (TB1)

H
Ha Nieh (CZP)
Hamidieh (BO)
Heibetnuma (CW)
Hifzi Rahman (MM)
Hizber (MR)
Hudavendigar (drilS)

I
Idjalieh (BO)
Ihsaniye (YA)
Is ed Din (CZP)
Iskenderieh (CZ)
Ismail (CZP)
Istanbul (YA)
Istankoi (GB)

K
Kandiah (CZP)

L
Lufti Humayun (CW)

M
Mahmudieh (BO)
Mansurah (CZ)
Mecidieh (TB1)
Mecidiyeh (deptS)
Medar-I Zaffer (YA)
Mehemet Selim (trailS)
Memduhiye (MR)
Meric (CZ)
Messudieh (BO)
Mosul (GB)
Mubir-i-Surur (deptS)
Muhabet (TB1)
Muin-i-Zaffer (BO)
Mujde Resan (GB)
Mukaddami Khair (BO)
Mukaddeme-I Seref (deptS)
Muzaffer (CZ)

N
Namet (DC)
Nasir (TB1)
Necmifesan (GB)
Nijmi Shevket (BO)
Nimet (TB1)

O
Orkanieh (BO)
Osmanieh (BO)

P
Peyk-I Nusret (deptS)
Peyk-I Zafer (BW)

R
Rehberi Tewfik (drilS)
Resmo (YA)
Rodos (GB)

S
Sadrye (deptS)
Sahin-I Deria (DC)
Sahir (GB)
Saiki (TB1)
Sat (GB)
Satvet (TB1)
Sed-el-Bahr (CZ)
Selimiye (drilS)
Semshur I-Hucum (TB2)
Serifiye (YA)
Seyar (GB)
Seyf-i Bahri (TB1)
Shahiddin (GB)
Shan Aver (TB1)
Shevket Numa (GB)
Sinub (CZ)
Sultaniye (YA)
Sureya (CZP)

T
Taif (YA)
Taliah (CZP)
Tevfik (TB1)
Timsah (TB1)
Tir-i Zaffer (TB1)

U
Utarit (CZ)

V
Vesile-I Nusret (TB1)

Y
Yali Kiosk (GB)

Z
Ziver-i-Deria (GB)
Zuhaf (CZ)

Y1889 Turkish fleet lost ship

Peyk-I Meserret (Depot ship), +++++ in accident

Turkish fleet naval guns

Calibre (mm)	Mark	Calibre (in)/pdr	Gun type	First Introduced	Weight (kg)	Gun lenght (in/cal)	Bore lenght (cal)	Charge weight (lbs)	Muzzle energy (ft/t)	Projectile weight (lbs)	Muzzle velocity (ft/s)	Wrought iron perforation (in)
254.0	II	10"	MLR	1868	18 289	180.0 / 18.00	14.55	70.0	5 408	410.0	1 379	13.1
240.0	Krupp	9.45"	BLR	1865	-	330.7 / 35.00	35.00	-	11 910	-	1 903	21.0
240.0	Krupp	9.45"	BLR	1865	13 780	185.4 / 19.62	16.80	50.7	4 736	261.5	1 391	13.0
228.6	V	9"	MLR	1865	12 193	156.0 / 17.33	13.89	50.0	3 681	256.0	1 440	11.4
203.2	III	8"	MLR	1862	9 144	144.0 / 18.00	14.75	35.0	2 398	179.0	1 390	9.7
177.8	III	7"	MLR	1865	6 604	133.0 / 19.00	15.86	30.0	1 848	114.6	1 525	9.2
177.8	-	7"	BLR	1864	4 166	120.0 / 17.14	14.21	11.0	909	91.3	1 100	-
170.0	Krupp	6.80"	BLR		5 510	167.3 / 25.00	21.90	30.9	2 112	112.9	1 654	10.3
152.4	VI	6"	BLR	1872	5 080	173.5 / 28.92	26.00	45.0	2 424	100.0	1 870	11.9
150.0	Krupp	5.9"	BLR	-	4 700	205.2 / 34.75	32.20	41.9	2 784	112.4	1 890	12.8
150.0	Germany	5.9"	BLR	1878	4 040	176.0 / 29.80	27.20	33.1	1 115	112.4	1 624	11.0
150.0	Krupp	5.9"	BLR	1865	3 940	151.6 / 25.67	23.00	20.9	1 447	69.4	1 558	9.1
150.0	Germany	5.9"	BLR	1864	3 440	128.7 / 21.80	19.10	14.3	1 418	65.0	1 555	8.0
120.0	Krupp	4.7"	BLR	1859	1 380	115.2 / 25.00	-	9.9	680	36.1	1 493	7.0
120.0	II, L	4.7"	MLR	1865	1 778	-	22.00	6.8	546	40.7	1 425	-
120.0	-	4.7"	BLR	1865	1 778	-	22.39	5.0	380	38.8	1 160	-
105.0	Krupp	3.96"	BLR	1865	1 150	145.0 / 35.08	35.00	8.8	-	39.7	1 526	-
101.6	II, III - IV, V, VI	4"	BLR	1859	1 169 - 1 321	120.0 / 30.00	27.00	12.0	625	25.0	1 900	7.2
101.6	18 cwt	4"	MLR	1867	914	120.0 / 30.00	22.00	4.5	315	25.0	1 350	-
95.3	I	20 pr	BLR	1867	610	107.5 / 28.65	29.00	6.0	380	19.5	1 677	-
90.0	-	3.54"	BLR	1867	100	70.8 / 20.00	18.75	-	-	6.6	1 545	-
76.2	I	12 pr	BLR	1865	356	92.4 / 30.80	28.00	4.0	249	12.5	1 530	-
60.0	Krupp	2.36"	BLR	1865	100	49.2 / 20.83	-	0.9	-	6.6	1 545	-
57.0	I & II	6 pr	BLR	1876	406	96.5 / 43.00	40.00	1.2	345	6.0	1 818	3.7
47.0	I & II	3 pr	BLR	1876	254	79.6 / 43.00	40.00	1.8	321	3.3	1 873	3.1
37.0	Hotchkiss R	1 pr	BLR	1873	209	33.5 / 23.0	19.13	0.2	-	1.1	1 450	-
37.0	Hotchkiss	1 pr	BLR	1884	72	33.5 / 23.0	19.15	0.2	-	1.1	1 450	-

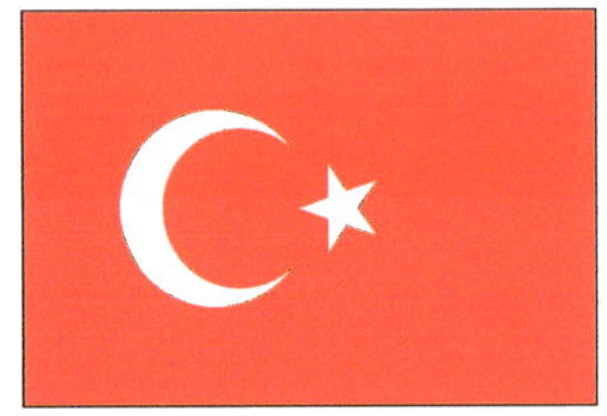

WARSHIPS SECTION

Ship classes	Building (exports including)	Active
Battleships		
Boardside ironclads	1	13
Armoured cruisers		
Coast defence armoured ships		1
Protected cruisers		
Unprotected cruisers		
Torpedo gunboats	2	
Torpedo boats 1 class	3	13
Torpedo boats 2 class		1
Submarines		2
Wooden screw cruisers	2	1
Wooden screw sloops		11
Gunboats		15
Gunboats coastal		17

BATTERY IRONCLADS

0 +1 | "Hamidieh" type (1874)
Battery ironclad, Turkey

BUILDING

15Y:1M

Hamidieh

Displacement: 6,594 tons at draft: 7.57 m | Length: 89 m pp | Beam: 16.99 m | Crew: 350 men

Machinery: 4 rectangular boilers | 1 horizontal single expansion E * 1 screw | 6,800 HP = 13 knots | Supplies: 600 tons of coal

Armour: iron | CT: 178; Battery: 178(^13) | WLBelt: 127(^13) 229(^13) 127(^13)

Amament: 4-240/30, 10-150/35, 6-37/23

Age 16

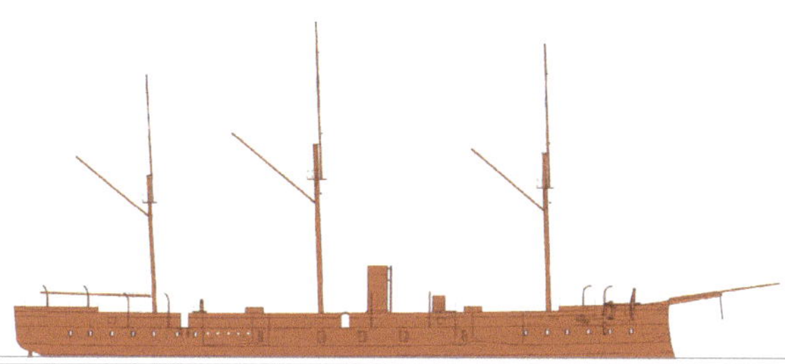

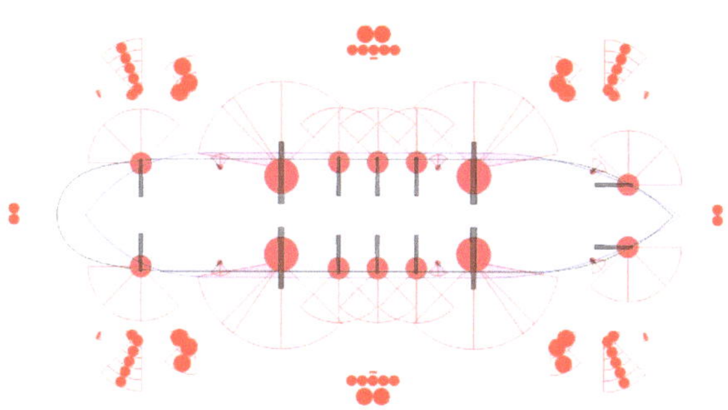

Hamidieh (1874)

1874 12 01 - 1885 02 01 - 1894 01 01 | Constantinopol DYd, Constantinopol

1 | "Messudieh" (1873) type
Battery ironclad, Turkey

Displacement: 9,120 tons at draft: 7.77 m, 9,710 tons = 7.9 m full | Length: 101.3 m pp | Beam: 17.98 m | Crew: 654 men

Machinery: 9 rectangular boilers | 1 horizontal direct acting E * 1 screw | 6,580 HP = 13.2 knots | Supplies: 600 tons of coal

Armour: iron | CT: 203; Battery: 305-254(^38) | WLBelt: 178(^38) 305 178(^38)

Armament: 12-254/14.5 MLR, 3-177.8/15.9 MLR, 6-101.6/29

Age 17

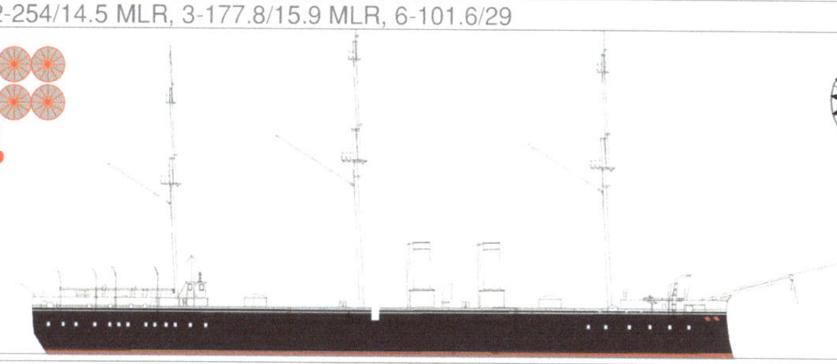

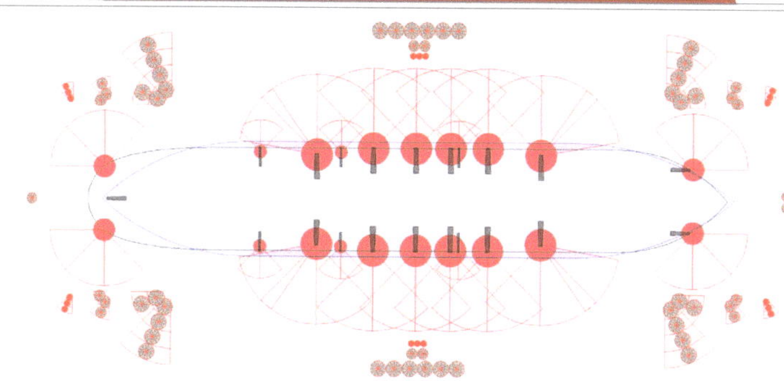

Messudieh (1873)

1873 01 01 - 1874 10 28 - 1875 12 01 | Thames Iron Works, Blackwall

Sister ship can be found in the British section

2 | "Fethi Bulend" type (1868)
Battery ironclads, Turkey

Displacement: 2,761 tons at draft: 5.51 m | Length: 72.01 m pp | Beam: 11.99 m | Crew: 180 men
Machinery: 4 rectangular boilers | 1 horizontal compound E * 1 screw | 3,250 HP = 13 knots | Supplies: 300 tons of coal
Armour: iron | Casemates: 229-152 |
WLBelt: 152(^76) 229(^76) 152(^76)
Rearmed (1882): 4-228.6/13.9 MLR, 1-170/25, 2-90/24

Age 22

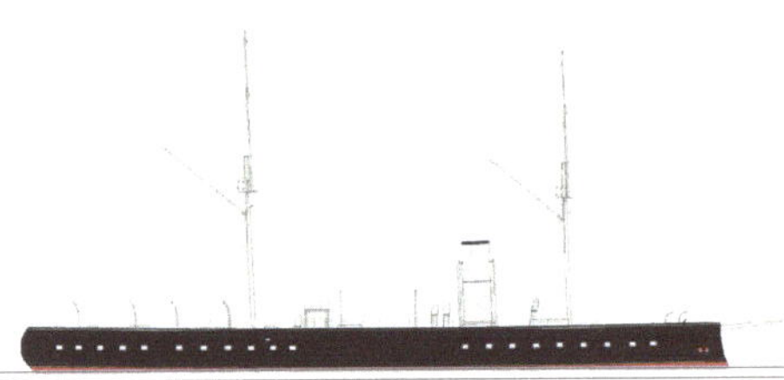

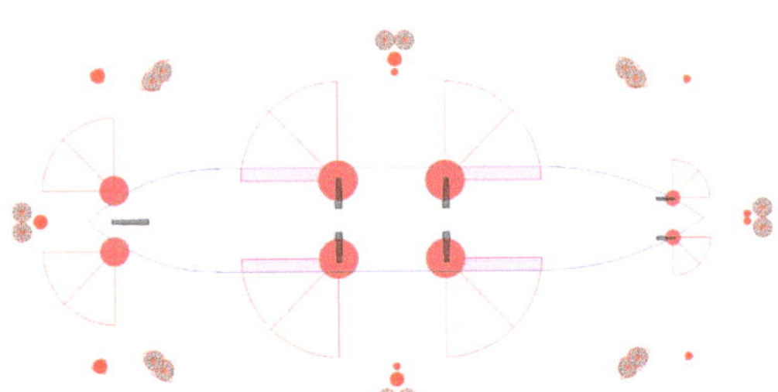

Fethi Bulend (1868)
1868 05 01 - 1870 01 01 - 1870 12 12 | Thames Iron Works, Blackwall

Mukaddami Khair (1870)
1870 01 01 - 1873 01 01 - 1876 01 01 | Constantinopol DYd, Constantinopol

Note: Mukaddami Khair /// G (1888): 4-228,6/13,9 MLR, 2-90/24, 2-65/35, 2-37/23R, 1-25,4/40 Mg

1 | "unnamed"/"Lufti Djelil" type (1868)
Monitor, Egypt

Displacement: 2,540 tons at draft: 4.42 m | Length: 62.18 m pp | Beam: 14 m | Crew: 130 men
Machinery: 2 locomotive boilers | 2 * 1,000 HP horizontal compound EE * 2 screws | 2,000 HP = 12 knots | Supplies: 300 tons of coal
Armour: iron | Turret: 140 |
WLBelt: 117(^38) 140(^38) 117(^38)
Armament: 2-228.6 /13.9 MLR, 2-177.8/15.9 MLR, 1-120/23

Age 23

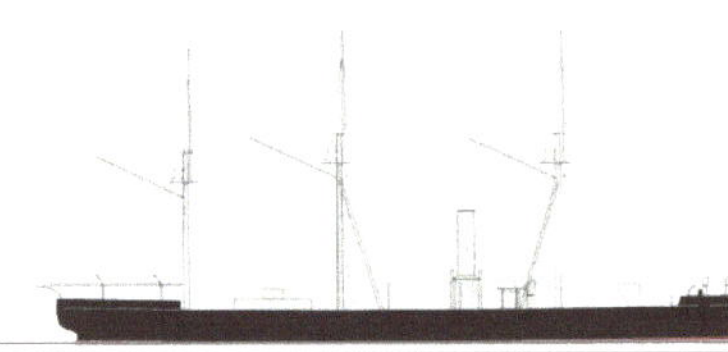

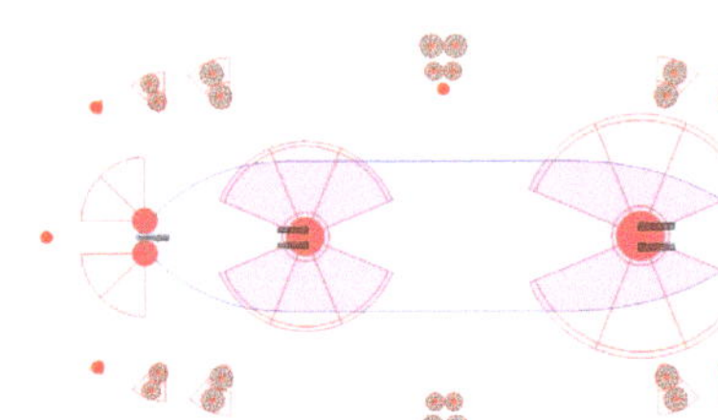

Hifzi Rahman (1867)
1868 01 01 - 1869 01 01 - 1870 03 01 | Arman, Bordeaux

1 | "Iclalieh" type (1868)
Battery ironclad, Turkey

Displacement: 3,266 tons at draft: 5.28 m | Length: 65 m pp | Beam: 12.98 m | Crew: 180 men
Machinery: 2 rectangular boilers | 1 horizontal compound E * 1 screw | 1,800 HP = 12 knots |
Supplies: 300 tons of coal
Armour: iron | Battery: 114; Barbette: 127 |
WLBelt: 102 114 152 114 102
Rearmed (1885): 2-240/22, 1-150/22, 2-90/24, 2-70/25, 2-37/23, 2-25.4/40 Mg

Age 22

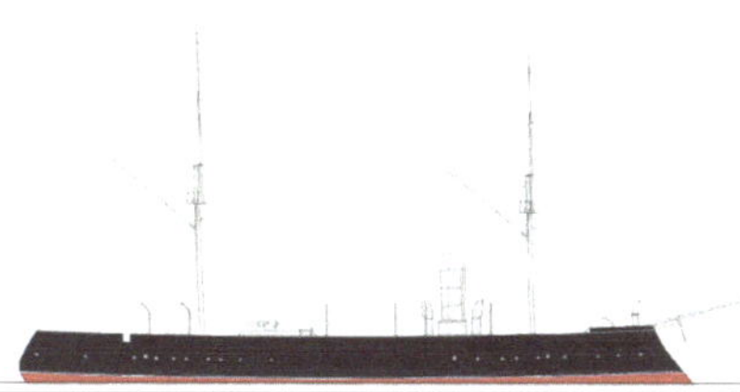

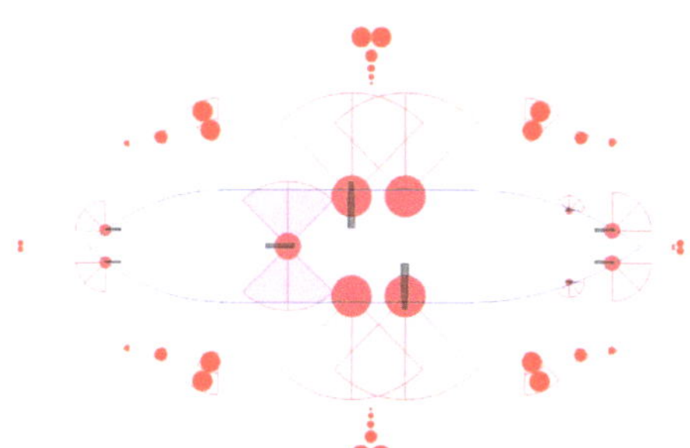

Iclalieh (1868)
1868 05 01 - 1870 01 01 - 1871 02 01 | STT, Trieste

2 | "Avni Illah" type (1868)
Battery ironclad, Turkey

Displacement: 2,362 tons at draft: 5 m | Length: 68.99 m oa | Beam: 10.97 m | Crew: 140 men
Machinery: 4 rectangular boilers | 1 horizontal compound E * 1 screw | 2,200 HP = 12 knots |
Supplies: 220 tons of coal
Armour: iron | Battery: 127 |
WLBelt: 76 127 152 127 76
Rearmed (1882): 4-228.6/13.9 MLR, 2-90/24

Age 22

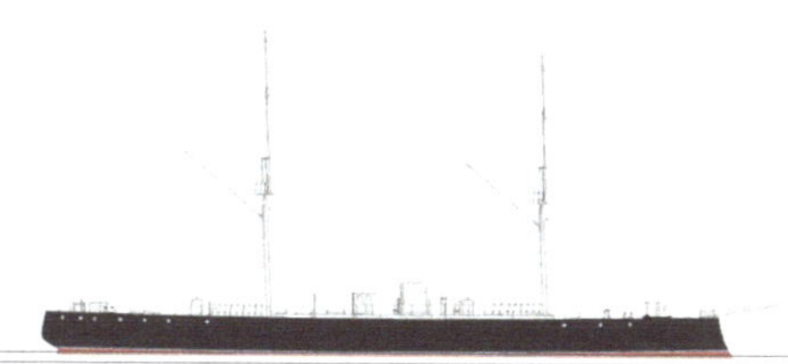

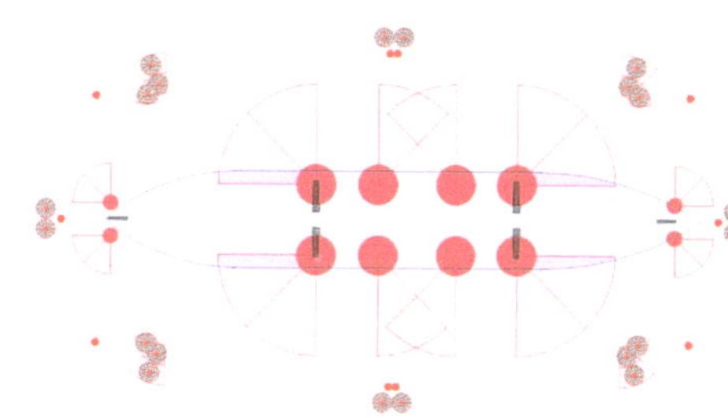

Avni Illah (1868)
1868 01 01 - 1869 04 21 - 1870 01 01 | Thames Iron Works, Blackwall

Muin-i-Zaffer (1868)
1868 01 01 - 1869 06 01 - 1870 01 01 | Samuda, Poplar

2 | "Kahira"/"Assari Shevket" type (1867)
Battery ironclads, Egypt

Displacement: 2,047 tons at draft: 5 m | Length: 62 m pp | Beam: 12.98 m | Crew: 170 men

Machinery: 2 locomotive boilers | 2 * 870 HP horizontal compound EE * 2 screws | 1,750 HP = 12 knots | Supplies: 300 tons of coal

Armour: iron | Battery: 114; Barbette: 114 |
WLBelt: 114(^25) 152(^25) 114(^25)

Armament: 1-228.6/13.9 MLR, 4-177.8/15.9 MLR

Age 23

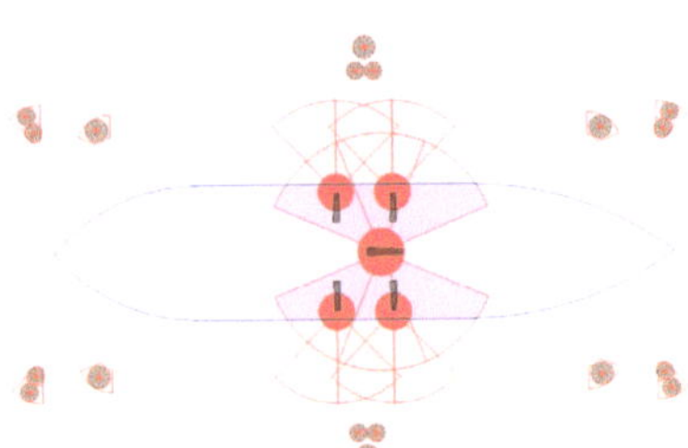

Assari Shevket (1867)
1867 01 01 - 1868 01 01 - 1870 03 03 | Forges et Chantiers de la Méditerranée, La Seyne

Nijmi Shevket (1867)
1867 01 01 - 1868 01 01 - 1870 03 03 | Forges et Chantiers de la Méditerranée, La Seyne

1 | "Ibrahimieh"/"Assari Tewfik" type (1867)
Battery ironclad, Egypt

Displacement: 4,687 tons at draft: 6.5 m | Length: 83.01 m pp | Beam: 16 m | Crew: 320 men

Machinery: 6 rectangular boilers | 1 compound E * 1 screw | 3,560 HP = 13 knots | Supplies: 400 tons of coal

Armour: iron | Battery: 152; Barbette: 127 |
WLBelt: 76(^25) 203(^25) 76(^25)

Armament: 8-228.6/13.9 MLR

Age 24

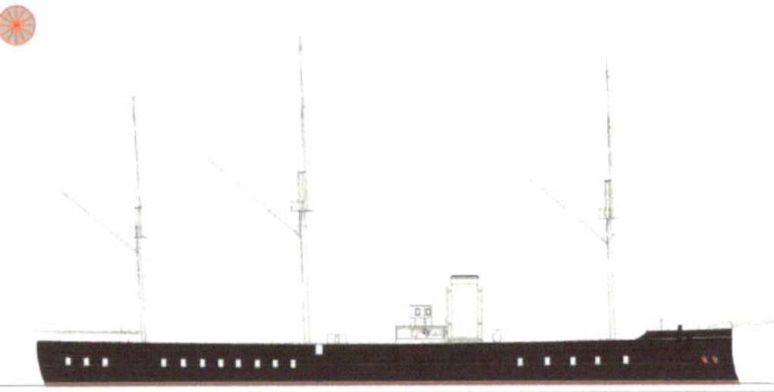

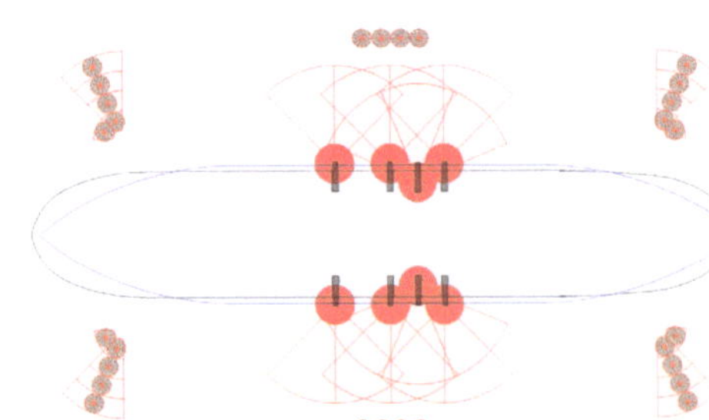

Assari Tewfik (1867)
1867 01 01 - 1868 01 01 - 1870 01 01 | Forges et Chantiers de la Méditerranée, La Seyne

4 | "Osmanieh" type (1863)
Battery ironclads, Turkey

Displacement: 6,400 tons at draft: 7.8 m | Length: 89.31 m pp | Beam: 16.99 m | Crew: 340 men

Machinery: 6 rectangular boilers | 1 horizontal compound E * 1 screw | 3,735 HP = 12 knots | Supplies: 546 / +204 tons of coal

Armour: iron | Battery: 114 127 114 | WLBelt: 76(^25) 140(^25) 76(^25)

Rearmed (1884): 1-228.6/13.9 MLR, 14-203.3/14.8 MLR, 4-47/43, 2-25.4/40 Mg

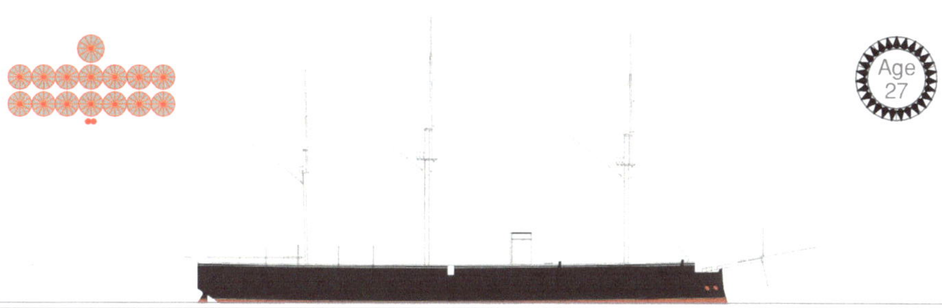

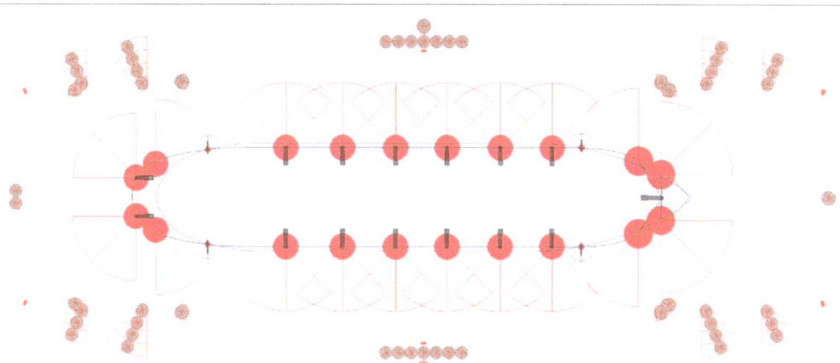

Azizieh (1863)
1863 05 01 - 1864 12 01 - 1865 08 01 | Napier, Govan

Mahmudieh (1864)
1864 01 01 - 1864 12 13 - 1866 01 01 | Thames Iron Works, Blackwall

Orkanieh (1863)
1863 06 06 - 1865 06 26 - 1866 06 06 | Napier, Govan

4 | "Osmanieh" type (1863)
Battery ironclads, Turkey (continued)

Osmanieh (1863)
1863 03 01 - 1864 09 02 - 1865 11 01 | Napier, Govan

Orkanieh

Mahmudieh

Abdul Aziz

WOODEN CRUISERS

0 +1 | "Lufti Humayun" type (1882)
Composite screw corvette, Turkey

BUILDING

7Y:12M

Lufti Humayun

Displacement: 1,313 tons at draft: 2.96 m | Length: 64.01 m pp | Beam: 9.14 m | Crew: 150 men
Machinery: 2 cylindrical boilers | 1 vertical triple expansion E * 1 screw | 3,500 HP = 14 knots |
Supplies: 200 tons of coal
Armament: 2-170/25, 6-120/25, 4-47/43, 6-37/23R, 2-381 TT (^)

NO PLAN AVAILABLE
Author's impression

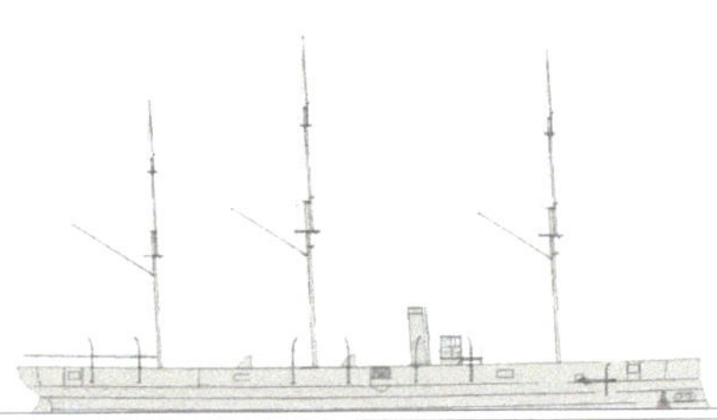

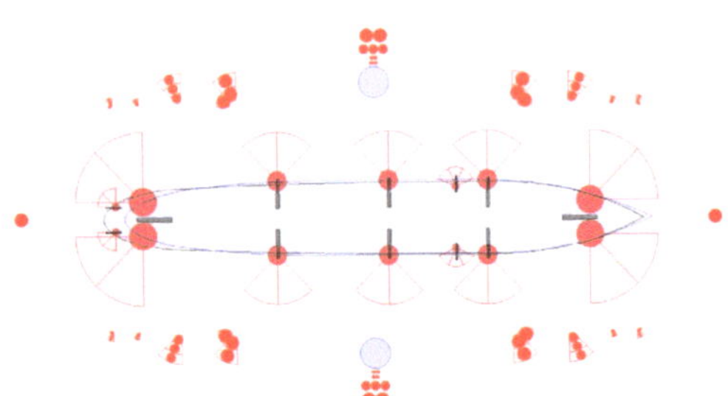

Lufti Humayun (1882)
1882 01 01 - ***** - ***** | Constantinopol DYd, Constantinopol

0 +1 | "Heibetnuma" type (1881)
Composite screw corvette, Turkey

BUILDING

8Y:12M

Heibetnuma

Displacement: 1,958 tons at draft: 4.7 m | Length: 68.88 m pp | Beam: 11.28 m | Crew: 200 men
Machinery: 6 rectangular boilers | 1 vertical triple expansion E * 1 screw | 2,786 HP = 12 knots |
Supplies: 280 tons of coal
Armament: 3-170/25, 6-120/25, 6-47/43, 4-37/23R, 2-381 TT (^)

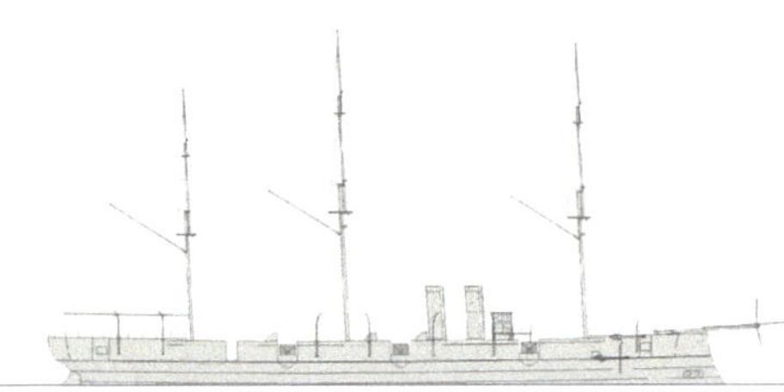

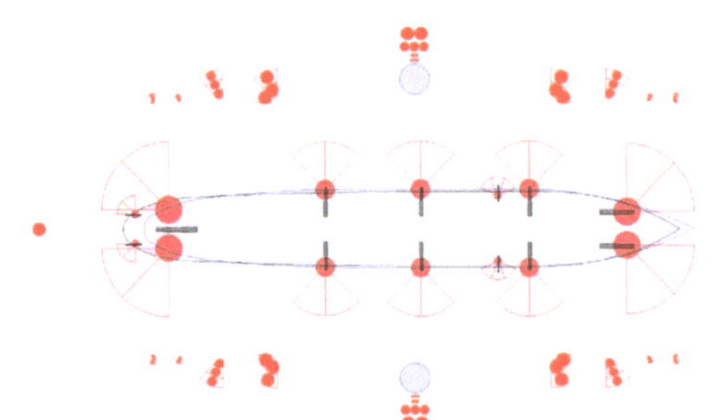

Heibetnuma (1881)
1881 01 01 - ***** - ***** | Constantinopol DYd, Constantinopol

1 | "Mehemet Selim" type (1877)
Wooden screw corvette, Turkey

Displacement: 1,280 tons at draft: 6.4 m | Length: 59.9 m pp | Beam: 10.01 m | Crew: 275 men

Machinery: 2 cylindrical boilers | 1 vertical compound E * 1 screw | 450 HP = 9 knots | Supplies: 250 tons of coal

Armament: 10-150/22, 4-37/23R, 4-25.4/40 Mg

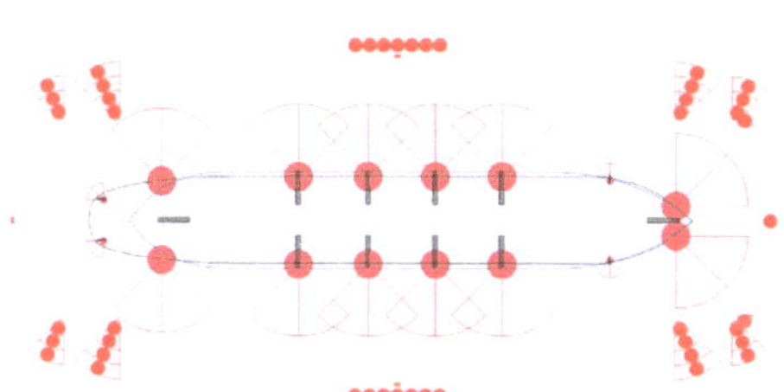

Mehemet Selim (1877)
1877 01 01 - 1879 08 14 - 1880 09 09 | Ismit DYd, Ismit

1 | "Ertogrul" type (1855)
Wooden screw frigate, Turkey

Displacement: 2,344 tons at draft: 7.19 m | Length: 76.2 m oa | Beam: 15.19 m | Crew: 400 men

Machinery: 2 rectangular boilers | 1 direct acting E * 1 screw | 2,200 HP = 10 knots | Supplies: 350 tons of coal

Rearmed (1888): 5-152.4/32, 8-150/30, 4-60/28, 4 24.5/40 Mg, 1-450 TT (^)

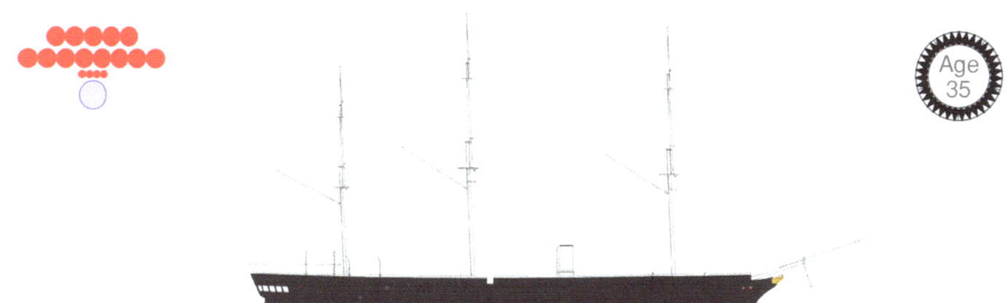

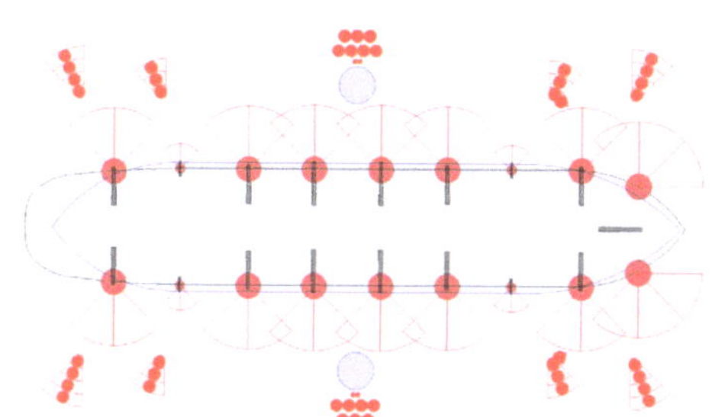

Ertogrul (1855)
1855 01 01 - 1863 01 01 - 1864 01 01 | Constantinopol DYd, Constantinopol

TORPEDO SHIPS

0 +1 | "Shahin-i Deria" type (1890)
Torpedo gunboat, Turkey

BUILDING

0Y:12M

Lufti Humayun

Displacement: 443 tons at draft: 2.44 m | Length: 60.96 m oa | Beam: 7.01 m | Crew: 75 men

Machinery: 4 loc. boilers | 2 * 1,750 HP vertical triple expansion EE * 2 screws | 3,500 HP = 22 knots | Supplies: 100 tons of coal

Armament: 1-105/40, 6-47/43, 4-356 TT (3*1 trainable, 1 >)

Age 0

NO PLAN AVAILABLE
Author's impression

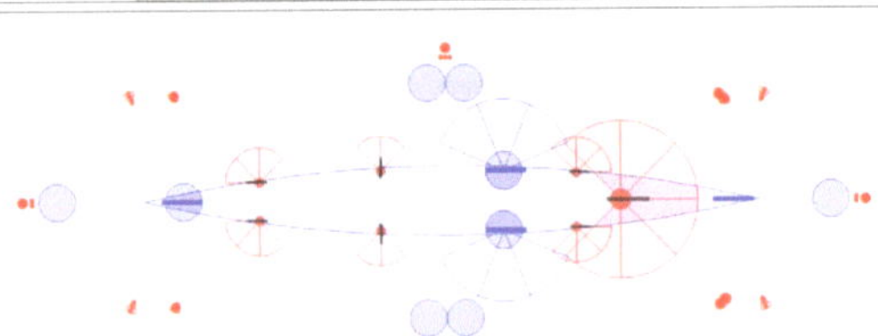

Sahin-I Deria (1890)
1890 01 01 - ***** - ***** | Constantinopol DYd

0 +2 | "Plenki Deria" type (1889)
Torpedo gunboats, Turkey

BUILDING IN GERMANY COLLECTING IN TURKEY

0Y:12M 0Y:12M

Plenki Deria Namet

Displacement: 755 tons at draft: 5.02 m, 900 tons full | Length: 70 m oa | Beam: 9.5 m | Crew: 80 men

Machinery: 4 loc. boilers | 2 * 2.350 HP vertical triple expansion EE * 2 screws | 4,700 HP = 18 knots | Supplies: 175 tons of coal

Armament: 2-105/40, 6-47/43, 3-356 TT (2*1 trainable, 1 >)

Age 0

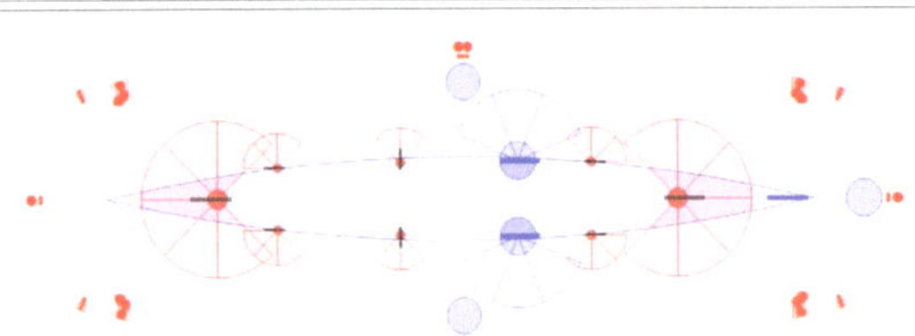

Namet (1889)
1889 01 01 - ***** - ***** | Germaniawerft, Kiel

Plenki Deria (1889)
1889 01 01 - ***** - ***** | Germaniawerft, Kiel

"Namet" is to be delivered in sections and completed in the Constantinopol DYd

0 +1 | "Ejder" type (1889)
Torpedo boat 1 class, Turkey

BUILDING IN GERMANY

0Y:12M

Ejder

Displacement: 138 tons at draft: 1.3 m | Length: 49 m oa | Beam: 5.6 m | Crew: 22 men

Machinery: 2 loc. boilers | 2 * 1,100 HP vertical triple expansion EE * 2 screws | 2,200 HP = 24 knots | Supplies: 50 tons of coal

Armament: 5-37/23R, 2-450 TT (1 trainable, 1 >)

NO PLAN AVAILABLE
Author's impression

Age 1

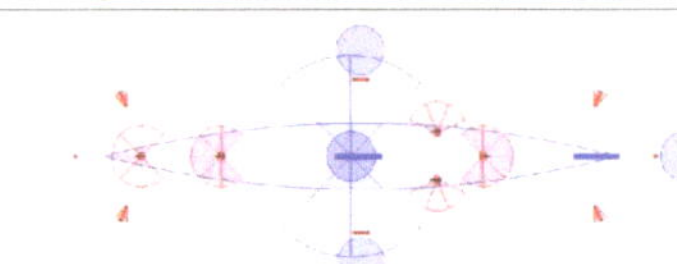

Ejder (1889)
1889 01 01 - 1890 01 01 - ***** | Germaniawerft, Kiel

2 +5 | "Nasir" type (1887)
Torpedo boats 1 class, Turkey

BUILDING IN GERMANY

Displacement: 87 tons at draft: 1.1 m | Length: 39 m oa | Beam: 4.8 m | Crew: 18 men

Machinery: 1 locomotive boiler | 1 vertical compound reciprocating E * 1 screw | 1,200 HP = 22 knots | Supplies: 16 tons of coal

Armament: 2-37/23, 2-450 TT (2 >>)

Age 3

Fatih (1887)
1887 01 01 - 1887 12 01 - 1889 12 01 | Germaniawerft, Kiel

Nasir (1887)
1887 01 01 - 1887 12 12 - 1889 12 01 | Germaniawerft, Kiel

Nussret (1888)
1888 01 01 - 1888 12 12 - ***** | Germaniawerft, Kiel

Pervins (1889)
1889 01 01 - 1889 12 12 - ***** | Germaniawerft, Kiel

Shehab (1888)
1888 01 01 - 1888 12 12 - ***** | Germaniawerft, Kiel

Seham (1889)
1889 01 01 - 1889 12 12 - ***** | Germaniawerft, Kiel

Tarik (1889)
1889 01 01 - 1889 09 01 - ***** | Germaniawerft, Kiel

2 | "Mahabet" type (1886)
Torpedo boats 1 class, Turkey

Displacement: 83 tons at draft: m | Length: 38.4 m oa | Beam: 4.57 m | Crew: 20 men

Machinery: 1 locomotive boiler | 1 vertical triple expansion E * 1 screw | 950 HP = 19 knots | Supplies: 22 tons of coal

Armament: 2-37/23R, 2-356 TT (2 >>)

Muhabet (1886)
1886 01 01 - 1887 01 01 - 1887 05 01 | Des Vignes, Chertsey

Satvet (1886)
1886 01 01 - 1887 01 01 - 1887 05 01 | Des Vignes, Chertsey

1 | "Timsah" type (1885)
Torpedo boat 1 class, Turkey

Displacement: 30 tons at draft: 1.4 m | Length: 28.6 m oa | Beam: 3.7 m | Crew: 17 men

Machinery: 1 locomotive boiler | 1 horizontal expansion E * 1 screw | 400 HP = 18 knots | Supplies: 4.8 tons of coal

Armament: 2-356 TT (2 >>)

Timsah (1885), ex-British yacht "Scirocco"
1885 01 01 - 1885 07 07 - 1885 12 12* | Des Vignes, Chertsey

1 | "Semsir-i Hucum" type (1885)
Torpedo boat 2 class, Turkey

Displacement: 14 tons at draft: 0.8 m | Length: 19 m oa | Beam: 2.3 m | Crew: 13 men

Machinery: 1 locomotive boiler | 1 horizontal compound direct acting E * 1 screw | 120 HP = knots | Supplies: 0.7 ton of coal

Armament: 2-356 TT (Djevetskiy)

Age 5

Semshur I-Hucum (1885)
1885 01 01 - 1885 08 01 - 1886 01 01 | F. Ch. Mediterr., la Seyne

3 +3 | "Burhan ed Din" type (1885)
Torpedo boats 1 class, Turkey

BUILDING

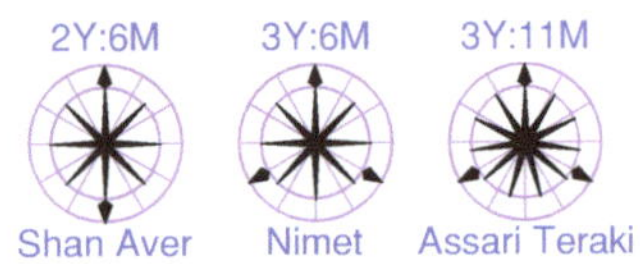

Displacement: 42 tons at draft: 1.1 m | Length: 33 m wl | Beam: 3.6 m | Crew: 11 men

Machinery: 1 locomotive boiler | 1 vertical compound reciprocating E * 1 screw | 550 HP = 20 knots | Supplies: 4.7 tons of coal

Armament: 1-35/40 Mg, 1-25,4/40 Mg, 2-356 TT (2 >>)

Age 5

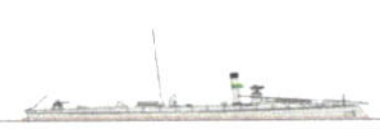

Assari Teraki (1886)
1886 02 01 - 1887 12 12 - ***** | Constantinopol DYd

Burhan ed Din (1885)
1885 01 01 - 1885 08 01 - 1885 10 01 | F. Ch. Mediterr., La Seyne

Mecidieh (1886)
1886 02 01 - 1887 01 01 - 1889 06 01 | Constantinopol DYd

Nimet (1886)
1886 05 01 - 1888 12 12 - ***** | Constantinopol DYd

Shan Aver (1887)
1887 05 01 - 1889 06 01 - ***** | Constantinopol DYd

Tevfik (1885)
1885 01 01 - 1885 08 01 - 1885 11 01 | F. Ch. Mediterr., La Seyne

5 | "Habicht"/"Giljom" type (1884)
Torpedo boats 1 class, Austria-Hungary

Displacement: 85 tons at draft: 1.1 m | Length: 39.9 m oa | Beam: 4.8 m | Crew: 18 men

Machinery: 1 locomotive boilers | 1 vertical triple expansion E * 1 screw | 980 HP = 21 knots | Supplies: 24 tons of coal

Armament: 2-37/23, 2-450 TT (2 >>)

Age 5

Giljom (1885)
1885 01 01 - 1886 01 01 - 1886 06 01 | Schichau, Elbing

Saiki (1885)
1885 01 01 - 1886 01 01 - 1886 06 01 | Schichau, Elbing

Seyf-i Bahri (1885)
1885 01 01 - 1886 01 01 - 1886 06 01 | Schichau, Elbing

Tir-i Zaffer (1885)
1885 01 01 - 1886 01 01 - 1886 06 01 | Schichau, Elbing

Vesile-I Nusret (1885)
1885 01 01 - 1886 01 01 - 1886 06 01 | Schichau, Elbing

Sister ships can be found in the Austria-Hungarian, German, Italian and Russian sections

2 | "Abdul Hamid" type (1886)
Submarines, Turkey

SUBMARINES

Displacement: 100 at draft: app. 3 m | 160 tons (submerged) | Length: 30.48 m oa | Beam: 3.7 m | Crew: 7 men

Machinery: 1 cylindrical boiler | 1 compound E * 1 screw | 250 HP = 6 (^) / 4 (v) knots | Supplies: 8 tons of coal

Diving depth: ?? m.

Armament: 2-25.4/40 Mg, 2-356 TT (2 >>)

Age 4

Abdul Hamid (1886)
1886 03 03 - 1886 09 06 - 1887 03 03 | Des Vignes, Chertsey

Abdul Medjid (1887)
1887 01 01 - 1887 03 03 - 1887 08 04 | Des Vignes, Chertsey

ARMOURED GUNBOATS

1 | "Hizber" type (1872)
River armoured battery, Turkey

Displacement: 404 tons at draft: 1.6 m | Length: 43.9 m pp | Beam: 9.4 m | Crew: 51 men
Machinery: 2 rectangular boilers | 1 horizontal reciprocating compound E * 1 screw | 400 HP = knots | Suppl.: 18 tons of coal
Armour: iron | Battery: 114 | WLBelt: 114
Armament: 2-120/25, 2-70/30, 2-35/40 Mg

Age 18

The author's impression.

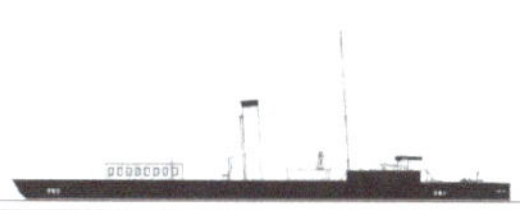

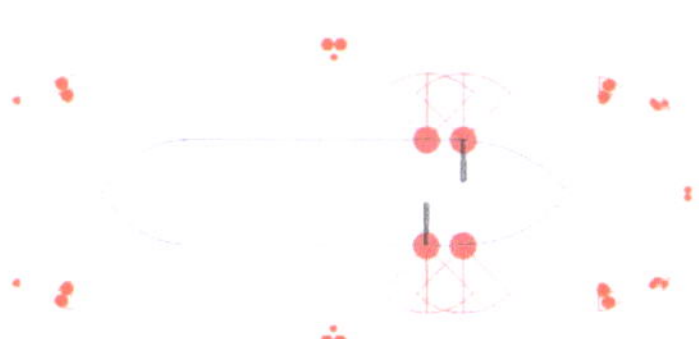

Hizber (1872)
1872 01 01 - 1873 09 01 - 1876 01 01 | Constantinopol DYd

2 | "Feth-ul Islam" type (1862)
River armoured batteries, Turkey

Displacement: 336 tons at draft: 1.6 m | Length: 31.4 m oa | Beam: 9.8 m | Crew: 50 men
Machinery: 2 rectangular boilers | 2 * 40 HP horizontal expansion EE * 2 screws | 80 HP = 8 knots | Supplies: 30 tons of coal
Armour: iron | Battery: 114 | WLBelt: 114
Armament: 2-177/25

Age 28

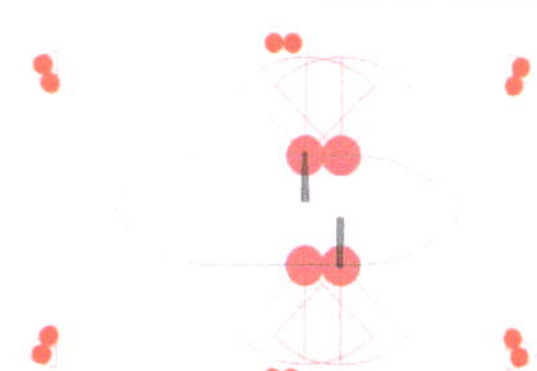

Fethul Islam (1862)
1862 01 01 - 1864 01 01 - 1866 01 01 | F.et Ch.de la M., La Seyne
Memduhiye (1862)
1862 01 01 - 1864 01 01 - 1866 01 01 | F.et Ch.de la M., La Seyne

Sister ship can be found in the Russian section

SLOOPS & GUNBOATS

2 | "Mansurah" type (1860)
Wooden screw sloops, Turkey

Displacement: 783 tons at draft: 4.4 m | Length: 55.1 m oa | Beam: 9.3 m | Crew: 130 men
Machinery: 2 rectangular boilers | 1 horizontal expansion E * 1 screw | 650 HP = 10 knots | Supplies: 120 tons of coal
Armament: 2-150/22, 2-120/25, 3-37/23, 2-25.4/40 Mg

Age 30

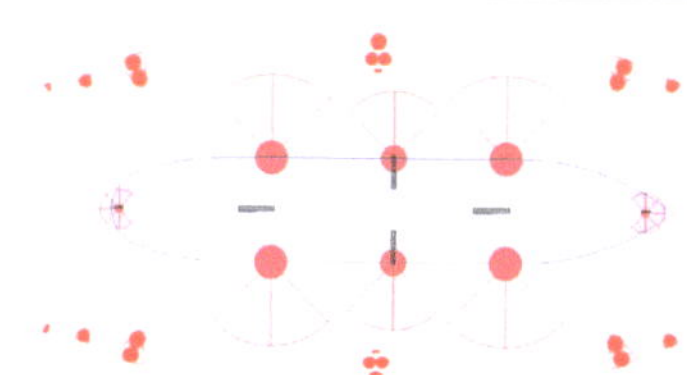

Mansurah (1860)
1860 01 01 - 1862 01 01 - 1865 01 01 | Constantinopol DYd
Muzaffer (1863)
1863 01 01 - 1864 01 01 - 1864 12 12 | Ismit DYd, Ismit

6 | "Beirut" type (1858)
Wooden screw sloops, Turkey

Displacement: 599 tons at draft: 3.91 m, 609 tons full | Length: 52.88 m oa | Beam: 8.1 m | Crew: 120 men

Machinery: 2 rectangular boilers | 1 horizontal expansion E * 1 screw | 450 HP = 10 knots | Supplies: 120 tons of coal

Rearmed (1879): 2-120/25, 2-76.2/26, 2-25.4/40 Mg

Age 32

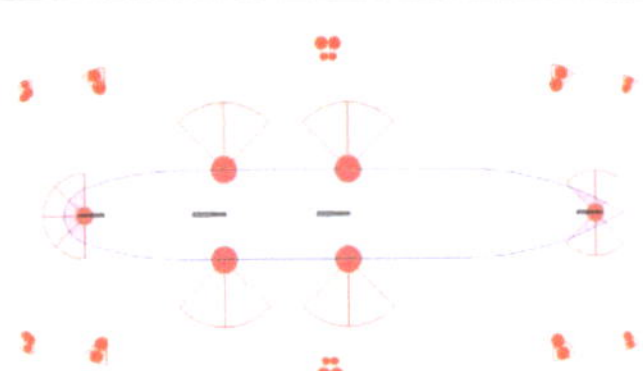

Beirut (1858)
1858 01 01 - 1859 01 01 - 1859 12 12 | Wigram, Blackwall

Iskenderieh (1860)
1860 01 01 - 1862 01 01 - 1863 01 01 | Constantinopol DYd

Meric (1862)
1862 01 01 - 1863 01 01 - 1864 01 01 | Gemlik Dyd, Gemlik

Sed-el-Bahr (1858)
1858 01 01 - 1859 01 01 - 1860 01 01 | White, Cowes

Utarit (1862)
1862 01 01 - 1864 01 01 - 1866 01 01 | Gemlik Dyd, Gemlik

Zuhaf (1860)
1860 01 01 - 1862 01 01 - 1865 01 01 | Constantinopol DYd

3 | "Sinop" type (1858)
Wooden screw sloops, Turkey

Displacement: 769 tons at draft: 4.32 m | Length: 52.43 m oa | Beam: 9.3 m | Crew: 153 men

Machinery: 2 rectangular boilers | 1 horizontal expansion E * 1 screw | 640 HP = 10 knots | Supplies: 120 tons of coal

Armament: 2-150/22, 2-120/25, 2-37/23, 2-25.4/40 Mg

Age 32

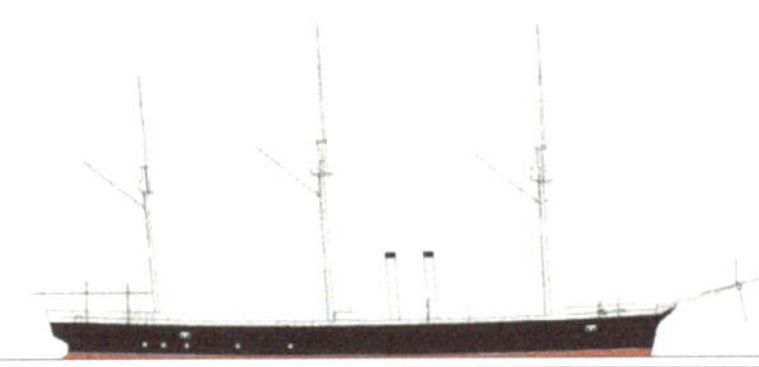

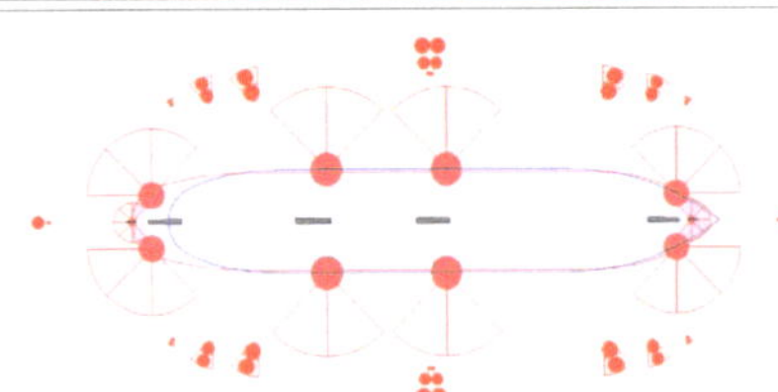

Bursa (1858)
1858 01 01 - 1859 01 01 - 1859 12 12 | Wigram, Blackwall

Edirneh (1858)
1858 01 01 - 1859 01 01 - 1859 12 12 | Wigram, Blackwall

Sinub (1858)
1858 01 01 - 1858 12 12 - 1859 11 26 | White, Cowes

Note: Sinub /// G (1888): 3-50/22, 3-37/23R

2 | "Rodos" type (1869)
Gunboats, Turkey

Displacement: 203 tons at draft: 1.8 m | Length: 35.7 m pp | Beam: 5.3 m | Crew: 35 men

Machinery: 1 rectangular boiler | 1 horizontal compound E * 1 screw | app. 100 HP = 12 knots | Supplies: 20 tons of coal

Rearmed (1885): 2-57/43, 2-25.4/40 Mg

Age 21

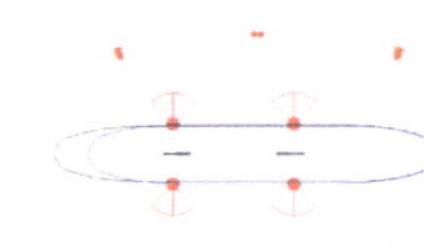

Istankoy (1871)
1871 01 01 - 1874 01 01 - 1874 12 12 | Constantinopol DYd

Rodos (1869)
1869 01 01 - 1870 01 01 - 1870 12 12 | Constantinopol DYd

2 | "Firat" type (1881)
Gunboats, Turkey

Displacement: 197 tons at draft: 1.9 m | Length: 37 m pp | Beam: 5.4 m | Crew: 55 men

Machinery: 1 locomotive boiler | 1 horizontal compound E * 1 screw | 280 HP = 10 knots | Supplies: 80 tons of coal

Armament: 4-76.2/28, 2-57/43

Firat (1881)
1881 01 01 - 1883 01 01 - 1885 01 01 | Constantinopol DYd

Sat (1881)
1881 01 01 - 1883 01 01 - 1885 01 01 | Constantinopol DYd

1 | "Saheddin" type (1867)
Gunboat, Turkey

Displacement: 163 tons at draft: 3.2 m | Length: 40.4 m pp | Beam: 6.7 m | Crew: 35 men

Machinery: 1 rectangular boiler | 1 horizontal compound E * 1 screw | app. 200 HP = 12 knots | Supplies: 25 tons of coal

Rearmed (1880): 4-76.2/26, 2-25,4/40 Mg

Shahiddin (1867)
1867 01 01 - 1868 01 01 - 1868 12 12 | Constantinopol DYd

2 | "Aynali Kavak" type (1867)
Gunboats, Turkey

Displacement: 195 tons at draft: 1.8 m | Length: 35.9 m pp | Beam: 5.3 m | Crew: 35 men

Machinery: 1 rectangular boiler | 1 horizontal compound E * 1 screw | app. 200 HP = 10 knots | Supplies: 20 tons of coal

Rearmed (1888): 1-57/43, 2-25.4/40 Mg

Aynali Kavak (1867)
1867 01 01 - 1869 01 01 - 1869 12 12 | Constantinopol DYd

Yali Kiosk (1867)
1867 01 01 - 1867 12 12 - 1869 01 01 | Constantinopol DYd

1 | "Sahir" type (1867)
Gunboat, Turkey

Displacement: 259 tons at draft: 3.2 m | Length: 40.4 m pp | Beam: 6.7 m | Crew: 35 men

Machinery: 1 rectangular boiler | 1 horizontal compound E * 1 screw | app. 100 HP = 8 knots | Supplies: 20 tons of coal

Rearmed (1880): 2-76.2/26, 2-25.4/40 Mg

Sahir (1867)
1867 01 01 - 1868 01 01 - 1869 01 01 | White, Cowes

2 | "Intibah" type (1865)
Wooden gunboats, Turkey

Displacement: 163 tons at draft: 3.2 m | Length: 40.4 m pp | Beam: 6.7 m | Crew: 35 men

Machinery: 1 rectangular boiler | 1 horizontal expansion E * 1 screw | app. 200 HP = 12 knots | Supplies: 20 tons of coal

Rearmed (1880): 4-76.2/26, 2-25.4/40 Mg

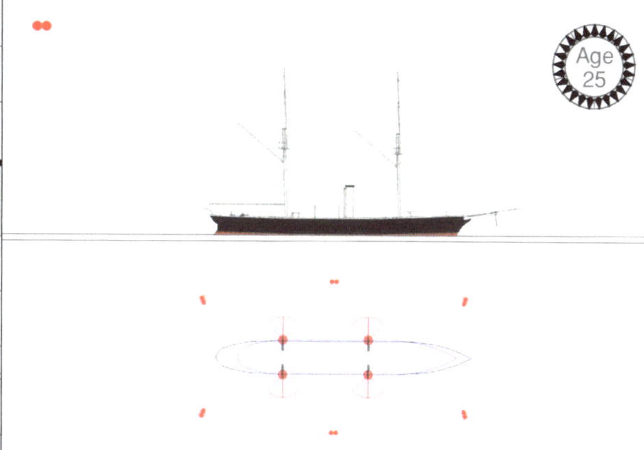

Mujde Resan (1866)
1866 01 01 - 1867 01 01 - 1868 01 01 | Constantinopol DYd

Ziver-i-Deria (1865)
1865 01 01 - 1866 01 01 - 1867 01 01 | Constantinopol DYd

1 | "Arkadi" type (1866)
Side wheel aviso, Turkey

Displacement: 755 tons at draft: 2.08 m | Length: 78.03 m oa | Beam: 8 m | Crew: 120 men

Machinery: 1 locomotive boiler | 1 horizontal compound E * side wheels | 1,050 HP = 10 knots | Supplies: 150 tons of coal

Armament: 2-76.2/26, 1-60/26

Arkadi (1866)
1866 01 01 - 1866 07 07 - 1866 12 12 | Potter W. & Son, London

2 | "Musul" type (1865)
Gunboats, Turkey

Displacement: 276 tons at draft: 2.8 m | Length: 39.9 m pp | Beam: 6.4 m | Crew: 35 men

Machinery: 1 rectangular boiler | 1 horizontal compound E * 1 screw | app. 200 HP = 9 knots | Supplies: 25 tons of coal

Rearmed (1881): 2-57/43, 2-37/23

Musul (1865)
1865 01 01 - 1865 09 09 - 1866 01 01 | Constantinopol DYd

Seyar (1865)
1865 01 01 - 1865 09 09 - 1866 01 01 | Constantinopol DYd

1 | "Sureyya" type (1865)
Side wheel aviso, Turkey

Displacement: 672 tons at draft: 2.13 m | Length: 56.08 m oa | Beam: 8.53 m | Crew: 90 men

Machinery: 4 rectangular boilers | 2 horizontal compound E * side wheels | 610 HP = 10 knots | Supplies: 120 tons of coal

Rearmed (1880): 2-57/43

Sureya (1865)
1865 01 01 - 1865 07 07 - 1865 12 12 | Samuda, Poplar

1 | "Ismail" type (1865)
Side wheel aviso, Turkey

Displacement: 1070 tons at draft: 3.6 m | Length: 76.2 m pp | Beam: 9.1 m | Crew: 135 men

Machinery: 2 rectangular boilers | 1 horizontal trunk E * side wheels | 300 HP = 10 knots | Supplies: 250 tons of coal

Armament: 1-120/23, 3-76.2/26

Ismail (1865)
1865 01 01 - 1865 06 06 - 1865 12 12 | ???, London

1 | "Kandiya" type (1865)
Side wheel aviso, Turkey

Displacement: 820 tons at draft: 2.01 m | Length: 73 m oa | Beam: 8 m | Crew: 90 men

Machinery: 2 rectangular boilers | 1 horizontal compound E * side wheels | 850 HP = 10.5 knots | Supplies: 120 tons of coal

Rearmed (1884): 3-57/43, 3-25.4/40 Mg

Kandiah (1865)
1865 01 01 - 1865 07 07 - 1865 12 12 | Britain

1 | "Fuad" type (1864)
Side wheel aviso, Turkey

Displacement: 1,075 tons at draft: 3.66 m | Length: 76.2 m oa | Beam: 9.14 m | Crew: 135 men

Machinery: 2 rectangular boilers | 1 horizontal compound E * side wheels | 1,200 HP = 12 knots | Supplies: 150 tons of coal

Rearmed (1880): 2-76.2/26, 2-57/43

Fuad (1864)
1864 01 01 - 1865 04 27 - 1865 12 12 | Millwall IW, Millwall

1 | "Ha Nieh" type (1863)
Side wheel aviso, Turkey

Displacement: 816 tons at draft: 1.93 m | Length: 70.1 m oa | Beam: 8.23 m | Crew: 105 men

Machinery: 2 rectangular boilers | 1 horizontal compound E * side wheels | 750 HP = 10 knots | Supplies: 120 tons of coal

Rearmed (1880): 3-47/43

Ha Nieh (1863)
1863 01 01 - 1863 07 07 - 1863 12 12 | Dudgeon, Poplar

1 | "Is ed Din" type (1864)
Side wheel gunboat, Turkey

Displacement: 1,058 tons at draft: 3.66 m | Length: 76.2 m oa | Beam: 9.14 m | Crew: 135 men

Machinery: 2 rect. boilers | 1 horizontal compound direct acting E * side wheels | 1,200 HP = 17 knots | Supplies: 150 tons of coal

Rearmed (1774): 1-120/23, 2-76.2/26

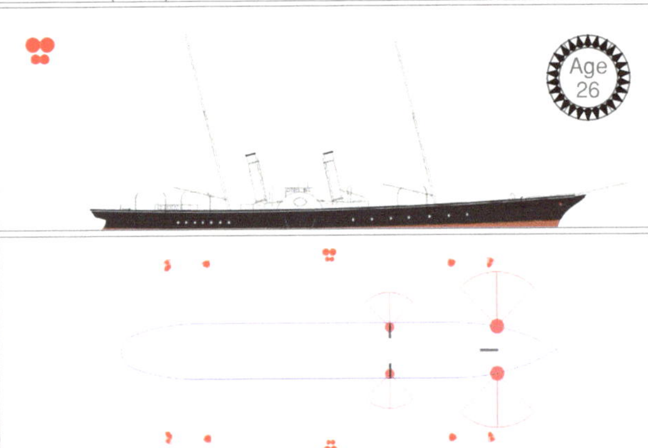

Is ed Din (1864)
1864 01 01 - 1865 01 01 - 1865 09 09 | Thames IW, Blackwall

1 | "Taliah" type (1863)
Side wheel aviso, Turkey

Displacement: 1,058 tons at draft: 3.66 m | Length: 76.2 m oa | Beam: 9.14 m | Crew: 130 men

Machinery: 2 rect. boilers | 1 horizontal compound direct acting E * side wheels | 1,250 HP = 17.7 knots | Supplies: 120 tons of coal

Rearmed (1880): 3-90/24, 2-37/23

Taliah (1863)
1863 01 01 - 1863 09 09 - 1864 03 01 | Samuda, Poplar

3 | "Akka" type (1859)
Gunboats, Turkey

Displacement: 196 tons at draft: 2.7 m | Length: 35.4 m oa | Beam: 6 m | Crew: 70 men

Machinery: 1 rectangular boilers | 1 single expansion E * 1 screw | 225 HP = 8 knots | Supplies: 80 tons of coal

Rearmed (1885): 3-76.2/26, 1-37/23

Akka (1859)
1859 01 01 - 1859 06 06 - 1859 09 24 | White, Cowes

Shevket Numa (1859)
1859 01 01 - 1859 09 09 - 1860 03 01 | Wigram, Blackwall

Necmifesan (1858)
1858 12 12 - 1859 12 12 - 1860 03 01 | Wigram, Blackwall

1 | "Sultaniye" type (1852)
Side wheel aviso, Turkey

Displacement: 2,909 tons at draft: 4.9 m | Length: 119.2 m pp | Beam: 12.2 m | Crew: 90 men

Machinery: 2 rectangular boilers | 1 horizontal trunk E * side wheels | 750 HP = 15 knots | Supplies: 300 tons of coal

Rearmed (1890): 2-120/43, 2-37/23

Sultaniye (1852)
1852 01 01 - 1852 12 23 - 1853 06 06 | Mare C.J., London

1 | "Serifiye" type (1873)
Side wheel yacht, Turkey

Displacement: 55 tons at draft: 2.7 m | Length: 28 m oa | Beam: 4.5 m | Crew: app. 30 men

Machinery: 1 locomotive boilers | 1 vertical compound E * side wheels | 50 HP = 12 knots | Supplies: app. 6 tons of coal

Armament: nil

Serifiye (1873)
1873 01 01 - 1873 07 07 - 1873 12 12 | Constanza

1 | "Taif" type (1869)
Side wheel armed yachts, Turkey

Displacement: 1,609 tons at draft: 5.1 m | Length: 71.6 m pp | Beam: 11 m | Crew: 250 men

Machinery: 2 rect. boilers | 1 horizontal compound direct acting E * side wheels | 450 HP = 12 knots | Suppl.: app. 180 tons of coal

Armament: 4-127/25 MLR

Asir (1871)
1871 01 01 - 1875 01 01 - 1876 01 01 | Constantinopol DYd

Taif (1869)
1869 01 01 - 1870 01 29 - 1872 01 01 | Constantinopol DYd

1 | "Ihsaniye" type (1869)
Side wheel armed yacht, Turkey

Displacement: 170 tons at draft: 2.3 m | Length: 31.7 m pp | Beam: 5.4 m | Crew: 137 men

Machinery: 1 rect. boiler | 1 horizontal compound direct acting E * side wheels | app. 100 HP = 8 knots | Supplies: 30 tons of coal

Rearmed (1884): 2-37/23

Ihsaniye (1869)
1869 01 01 - 1869 07 07 - 1870 01 01 | Maudsley, London

1 | "Medar-I Zaffer" type (1865)
Side wheel armed yacht, Turkey

Displacement: 1,343 tons at draft: 4.5 m | Length: 87.8 m pp | Beam: 10.8 m | Crew: 120 men

Machinery: 2 rectangular boilers | 1 single expansion E * side wheels | app. 500 HP = 13 knots | Supplies: 150 tons of coal

Rearmed (1880): 4-37/23

Eser-I Nussret (1865)
1865 01 01 - 1865 07 07 - 1865 12 12 | Quiggin J., London

Medar-I Zaffer (1865)
1865 01 01 - 1865 07 07 - 1865 12 12 | Quiggin J., London

1 | "Istanbul" type (1864)
Side wheel armed yacht, Turkey

Displacement: 909 tons at draft: 3.2 m | Length: 75 m oa | Beam: 8.8 m | Crew: 137 men

Machinery: 2 rectangular boilers | 1 horizontal compound E * side wheels | 2,400 HP = 16 knots | Supplies: 150 tons of coal

Armament: 2-120/25

Istanbul (1864)
1864 01 01 - 1865 01 01 - 1866 01 01 | Constantinopol DYd

1 | "Resmo" type (1862)
Side wheel armed yacht, Turkey

Displacement: 767 tons at draft: 2.1 m | Length: 69.8 m oa | Beam: 7.9 m | Crew: 80 men

Machinery: 1 rectangular boiler | 1 horizontal compound E * side wheels | 270 HP = 10 knots | Supplies: 150 tons of coal

Rearmed (1880): 3-70/22

Resmo (1862)
1862 01 01 - 1862 07 07 - 1862 12 12 | Britain

DUTCH FLEET

National flag

Fleet age 12

Ensign

Jack

Admiral

Months derivation -45

Warships built from 1858 01 01 to 1890 01 01

123 warships = 102,428 tons = 833 tons / average ship

Fleet age - 1878 11 21

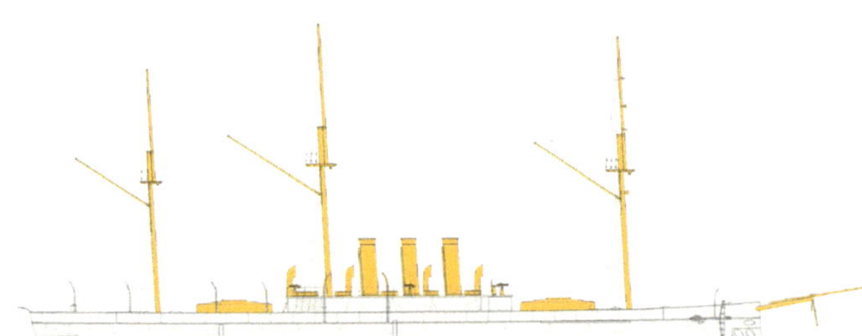

1889 fleet largest ship: Koning der Nederlanden (1871) coast defence battleship

Dutch naval construction

Amsterdam dockyard

11.6 tons / average ship / month

1890-Jan-01
8 ships building

t/ship/m

2,687 t.
tons / month = 104.4

Ship	t	t/m
Goentoer	85	3.5
Idjen	90	3.8
XXI	37	1.5
XXII	37	1.5
Sumatra	1,693	63.0
Lamongan	52	2.2
Makjan	52	2.2
Nobo	52	2.0

Amsterdam, Netherlands

32+ years of naval construction
61 ships | 79,196 tons

The plant's largest ship:
Koning der Nederlanden (1871) coast defence battleshp
5,315 tons

The plant's most effectively built ship:
Atjeh (1875) unprotected cruiser
107.3 tons/month

Specialisation in:
Unprotected cruisers | 8 ships | 27,400 tons
Screw sailing cruisers | 8 ships | 17,555 tons
Gunboats | 24 ships | 13,931 tons

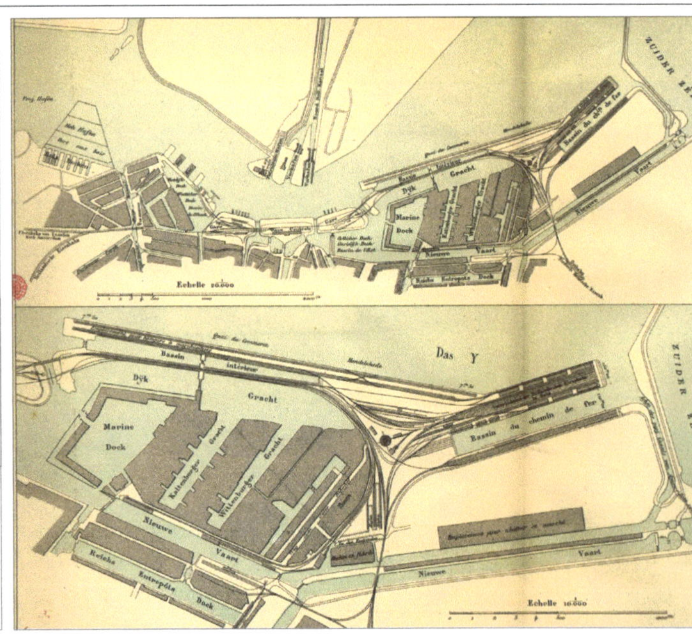

De Schelde

3.4 tons / average ship / month

1890-Jan-01
1 ship building

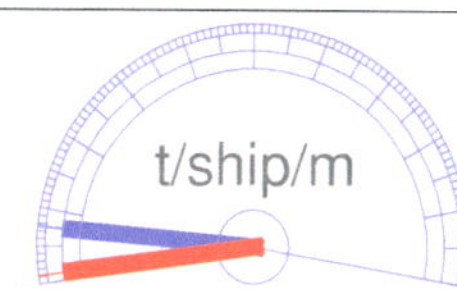

82 t.
tons / month = 3.4

t = 82
t/m =
3.4

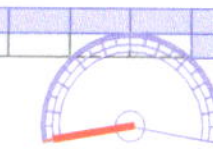

Cerberus

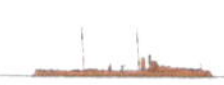

Flushing, Netherlands

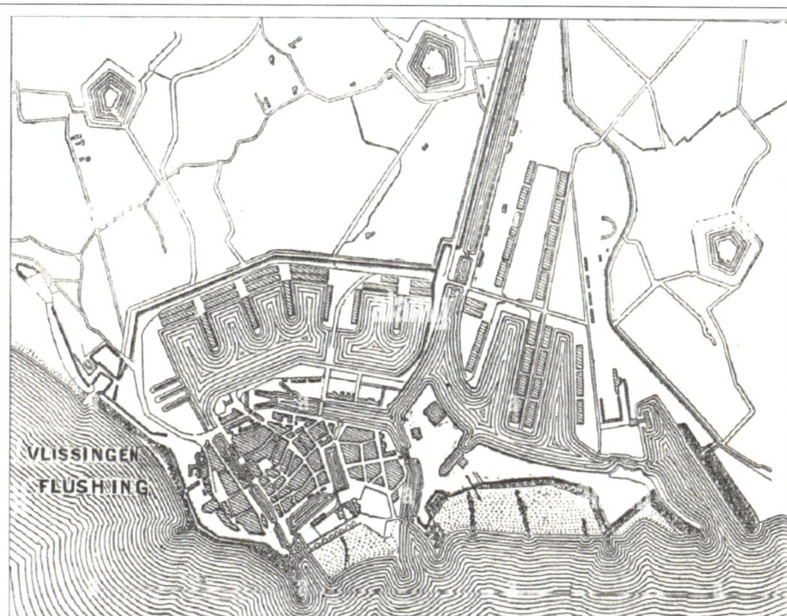

32+ years of naval construction
6 ships | 5,889 tons

The plant's largest ship:
Adolf Hertog van Nassau (1858) wooden screw frigate
3,690 tons

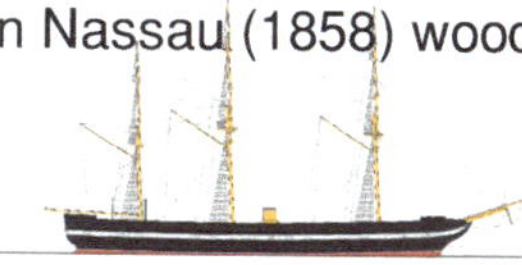

The plant's most effectively built ship:
Adolf Hertog van Nassau (1858) wooden screw frigate
75.6 tons/month

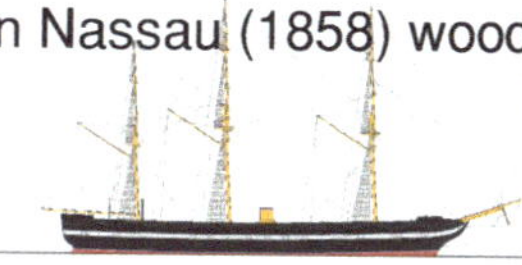

Specialisation in:
Screw sailing cruisers | 2 ships | 5,720 tons
Torpedo boats | 4 ships | 169 tons

Dutch fleet effective list

2 coast defence battleships

Koning der Nederlanden

Prins Hendrik der Nederlanden

16 monitors

Matador

Draak

Luipaard

Haai

Wesp

Hyena

Panter

Guinea

Cerberus

Buffel

Tijger

Bloedhond

Scorpioen

Heiligerlee

Krokodil

Stier

13 unprotected & wooden cruisers

Johan Willem Friso

Van Speyk

De Ruyter

Koningin Emma der Nederlanden

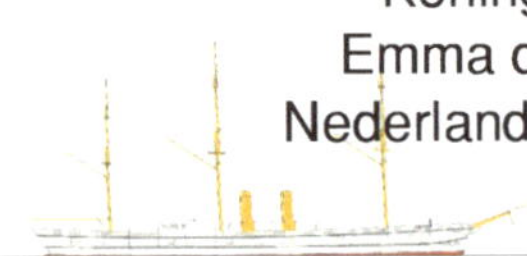

Tromp

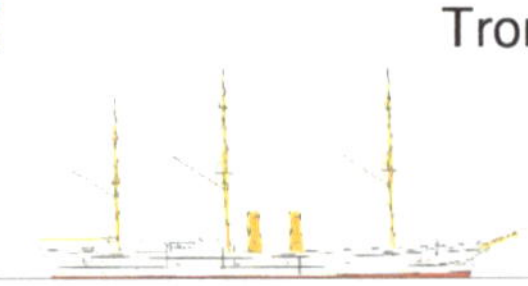

>>> unprotected & wooden cruisers (continued)

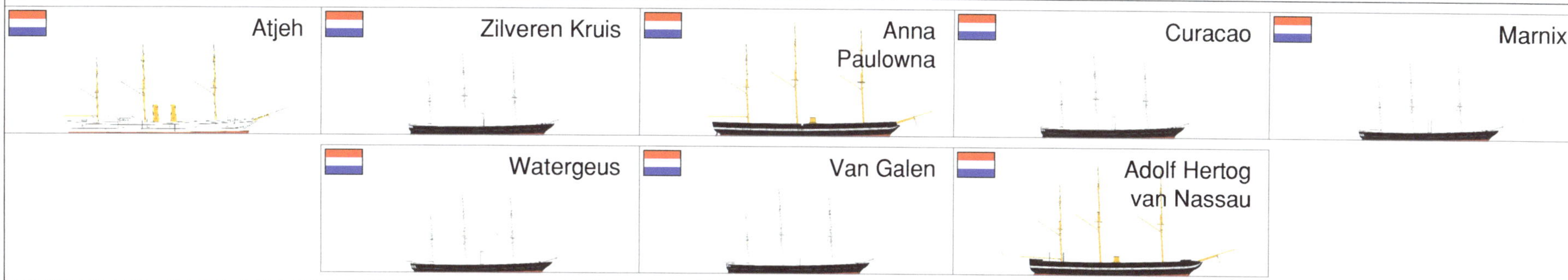

31 torpedo boats

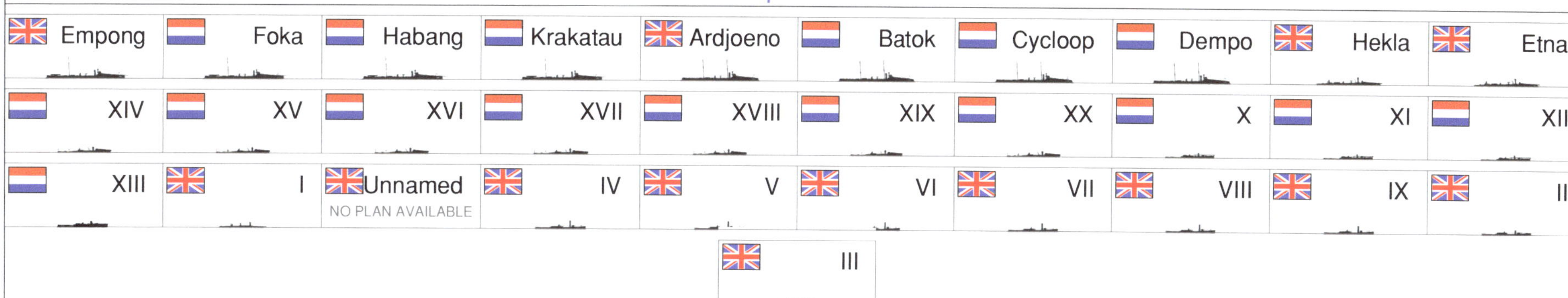

3 sloops

49 gunboats

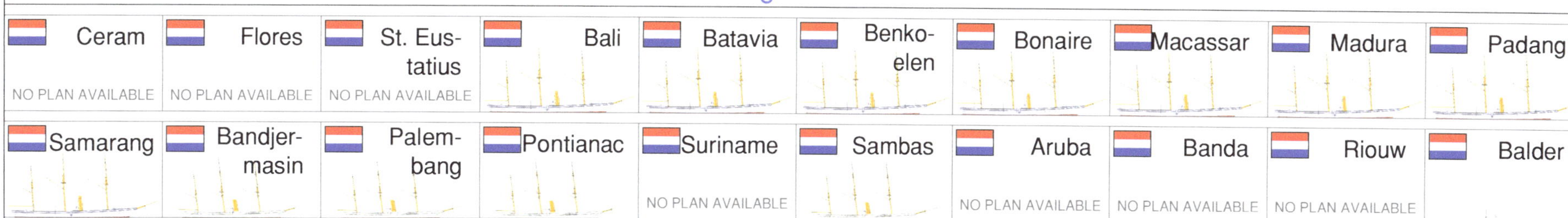

>>> gunboats (continued)

Braga	Bulgia	Dufa	Freyr	Hadda	Hefring	Heimdall	Njord	Thor	Tyr
Udur	Ulfr	Vali	Vidar	Wodan	Bever	Brak	Das	Dog	Fret
Geep	Gier	Havik	Lynx	Raaf	Sperwer	Vos	Hydra	Ever	

Dutch section ships names glossary

----- (TB2)

A
Adolf Hertog van Nassau (PS)
Alkmaar (CZ)
Anna Paulowna (CW)
Ardjoeno (TB1)
Aruba (GB)
Atjeh (CS)

B
Balder (GB)
Bali (GB)
Banda (GB)
Bandjermasin (GB)
Batavia (GB)
Batok (TB1)
Benkoelen (GB)
Bever (GB)
Bloedhond (MM)
Bonaire (GB)
Braga (GB)
Brak (GB)
Buffel (MM)
Bulgia (GB)

C
Ceram (GB)
Cerberus (MM)
Cerberus (TB1)
Curacao (CW)
Cycloop (TB1)

D
Das (GB)
De Ruyter (CS)
Dempo (TB1)
Dog (GB)
Draak (MM)
Dufa (GB)

E
Empong (TB1)
Etna (TB1)
Ever (hulk)

F
Flores (GB)
Foka (TB1)
Fret (GB)
Freyr (GB)

G
Geep (GB)
Gier (GB)
Goentoer (TB1)
Guinea (MM)

H
Haai (MM)
Habang (TB1)
Hadda (GB)
Havik (GB)
Hefring (GB)
Heiligerlee (MM)
Heimdall (GB)
Hekla (TB1)
Hydra (GB)
Hyena (MM)

I
I (TB1)
Idjen (TB1)
II (TB2)
III (TB2)
IV (TB2)
IX (TB2)

J
Java (CZ)
Johan Willem Friso (CS)

K
Koning der Nederlanden (CDB)
Koningin Emma der Nederlanden (CS)
Krakatau (TB1)
Krokodil (MM)

L
Lamongan (TB1)
Luipaard (MM)
Lynx (GB)

M
Macassar (GB)
Madura (GB)
Makjan (TB1)
Marnix (CZ)
Matador (MM)

N
Njord (GB)
Nobo (TB1)

P
Padang (GB)
Palembang (GB)
Panter (MM)
Pontianac (GB)
Prins Hendrik der Nederlanden (CDB)

R
Raaf (GB)
Riouw (GB)

S
Samarang (GB)
Sambas (GB)
Schorpioen (MM)
Sommelsdijk (CZ)
Sperwer (GB)
St. Eustatius (GB)
Stier (MM)
Sumatra (CP)
Suriname (GB)

T
Thor (GB)
Tijger (MM)
Tromp (CS)
Tyr (GB)

U
Udur (GB)
Ulfr (GB)

V
V (TB2)
Vali (GB)
Van Galen (CW)
Van Speyk (CS)
VI (TB2)
Vidar (GB)
VII (TB2)
VIII (TB2)
Vos (GB)

W
Watergeus (CZ)
Wesp (MM)
Wodan (GB)

X
X (TB1)
XI (TB1)
XII (TB1)
XIII (TB1)
XIV (TB1)
XIX (TB1)
XV (TB1)
XVI (TB1)
XVII (TB1)
XVIII (TB1)
XX (TB1)
XXI (TB1)
XXII (TB1)

Z
Zilveren Kruis (CW)

Dutch fleet naval guns

Calibre (mm)	Mark	Calibre (in)/pdr	Gun type	First Introduced	Weight (kg)	Gun lenght (in/cal)	Bore lenght (cal)	Charge weight (lbs)	Muzzle energy (ft/t)	Projectile weight (lbs)	Muzzle velocity (ft/s)	Wrought iron perforation (in)
280.0	Krupp	11.02"	BLR	1865	27 210	240.1 / 22.00	18.80	121.3	9 423	476.2	1 558	17.0
228.6	V	9"	MLR	1865	12 193	156.0 / 17.33	13.89	50.0	3 681	256.0	1 440	11.4
177.8	III	7"	MLR	1865	6 604	133.0 / 19.00	15.86	30.0	1 848	114.6	1 525	9.2
170.0	Krupp	6.8"	BLR	1864	5 510	167.3 / 25.00	21.90	27.6	2 226	112.4	1 558	10.5
165.0	France	6.46"	BLR		3 640	116.4 / 17.92	15.97			69.4		
160.0	III	64 cwt	MLR	1864	3 250	118.0 / 18.73	15.47	10.0	865	64.5	1 390	-
150.0	Krupp	5.9"	BLR	new	4 700	205.2 / 35.00	32.20	41.9	2 784	112.4	1 890	12.8
150.0	Krupp	5.9"	BLR	1865	3 940	151.6 / 26.00	23.00	20.9	1 447	69.4	1 558	9.1
120.0	Krupp	4.7"	BLR	new	2 310	165.4 / 35.00	-	19.8	1 290	57.3	1804	9.7
120.0	Krupp	4.7"	BLR	1859	1 380	115.2 / 25.00	-	9.9	680	36.1	1 493	7.0
120.0	Dutch '12	4.7"	BLR	1865	930	82.70 / 17.51	15.80	6.8	-	29.5	951	-
37.0	Hotchkiss R	1 pr	BLR	1873	209	33.5 / 23.00	29.13	0.2	-	1.1	1 450	-
37.0	Hotchkiss	1 pr	BLR	1884	72	33.5 / 23.00	33.15	0.2	-	1.1	1 450	-

WARSHIPS SECTION

Ship classes	Building	Active
Battleships		
Boardside ironclads		
Armoured cruisers		
Coast defence armoured ships		16
Protected cruisers	1	
Unprotected cruisers		6
Torpedo gunboats		
Torpedo boats 1 class	6	22
Torpedo boats 2 class	2	9
Submarines		
Wooden screw cruisers		7
Wooden screw sloops		5
Gunboats		19
Gunboats coastal		29

COAST DEFENCE BATTLESHIPS

1 | "Koning der Nederlanden" type (1871)
Coast defence battleship, Netherlands

Displacement: 5,315 tons at draft: 5.94 m | Length: 85.03 m oa | Beam: 15.14 m | Crew: 256 men

Machinery: 7 cylindrical boilers | 2 * 2,200 HP compound EE * 2 screws | 4,400 HP = 12 knots | Supplies: 650 tons of coal

Armour: iron | CT: 150; Turrets: 305-230; Barbettes: 230 | WLBelt: 150 200 150

Armament: 4-280/22, 4-120/16, 6-37/23R

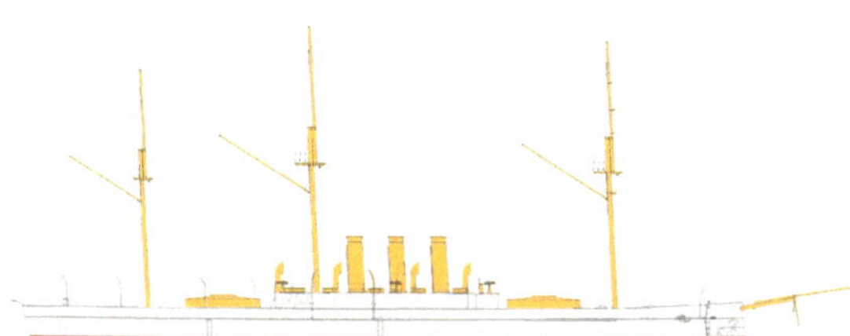

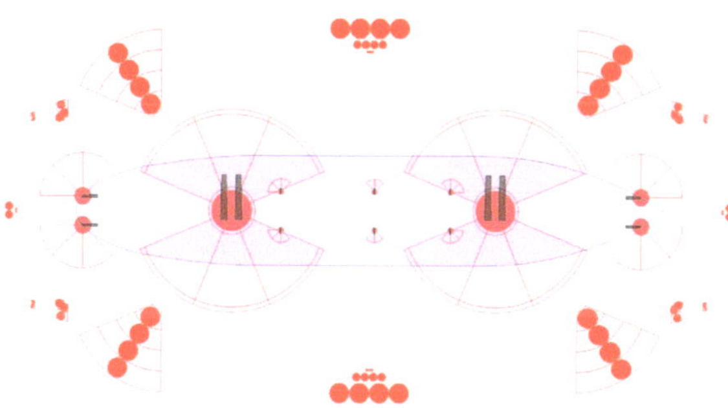

Koning der Nederlanden (1871)
1871 12 31 - 1874 10 20 - 1877 02 16 | Amsterdam DYd, Amsterdam

1 | "Prins Hendrik der Nederlanden" type (1865)
Coast defence battleship, Netherlands

Displacement: 3,320 tons at draft: 5.49 m | Length: 73.15 m oa | Beam: 13.41 m | Crew: 267 men

Machinery: 4 rectangular boilers | 1 horizontal return connecting rod E * 1 screw | 2,000 HP = 11.4 knots | Supplies: 467 tons of coal

Armour: iron | Turrets: 279-140; UPBelt: 0 114 0 | WLBelt: 114

Armament: 4-228.6/13.9 MLR, 4-120/16, 6-37/23R

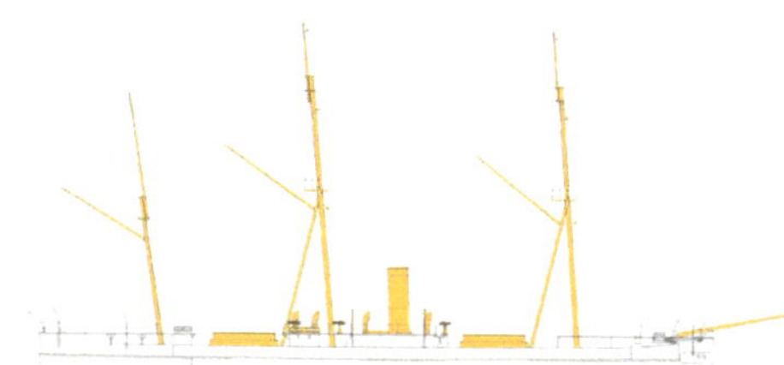

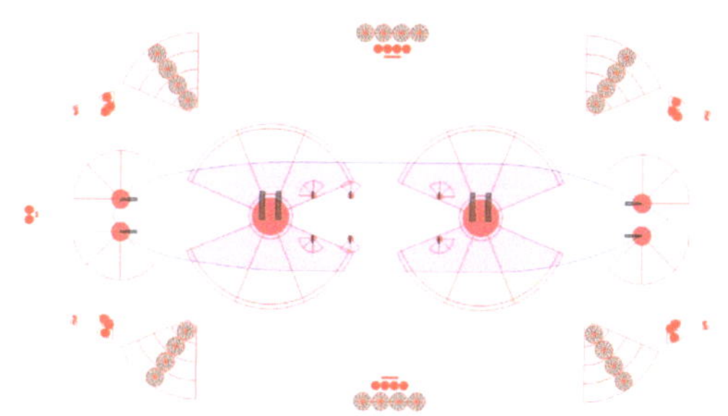

Prins Hendrik der Nederlanden (1865)
1865 01 01 - 1866 10 09 - 1867 12 12 | Laird, Birkenhead

2 | "Draak" type (1875)
Monitors, Netherlands

MONITORS

Displacement: 2,198 tons at draft: 3.63 m | Length: 65.1 m oa | Beam: 15.01 m | Crew: 131 men

Machinery: 4 cylindrical boilers | 2 * 400 HP horizontal compound EE * 2 screws | 807 HP = 8.4 knots | Supplies: 109 tons of coal

Armour: iron | CT: 140; Turret: 305-200 | WLBelt: 140(^25) 205(^25) 140(^25)

Armament: 2-280/22, 2-37/23

Age 15

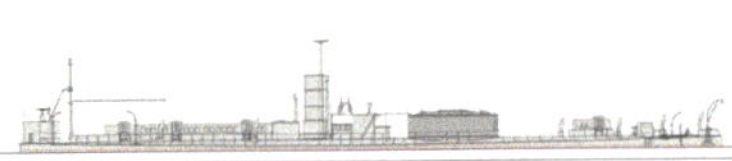

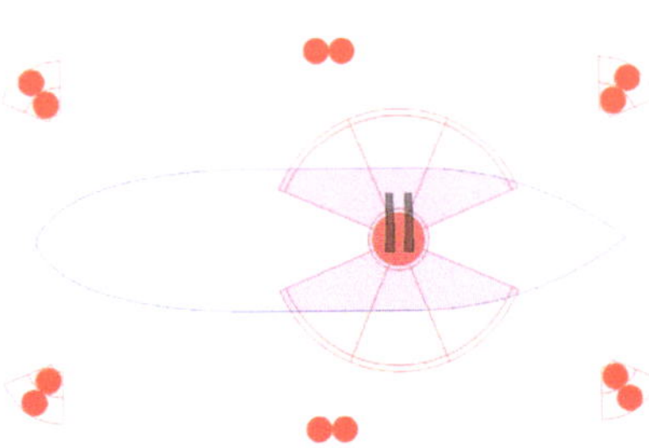

Draak (1875)
1875 11 15 - 1877 08 22 - 1878 06 06 | Amsterdam DYd

Matador (1876)
1876 01 01 - 1878 01 01 - 1879 01 01 | Fijenoord, Schiedam

5 | "Adder" type (1869)
Monitors, Netherlands

"Luipaard"

Displacement: 1,555 tons at draft: 2.97 m | Length: 58.6 m oa | Beam: 13.41 m | Crew: 117 men

Machinery: 4 rectangular boilers | 1 horizontal compound E * 1 screw | 560 HP = 7 knots | Supplies: 93 tons of coal

Armour: iron | CT: 140; Turrets: 285-203(^25) | WLBelt: 140(^25)

Armament: 2-228.6/13.9 MLR

Age 21

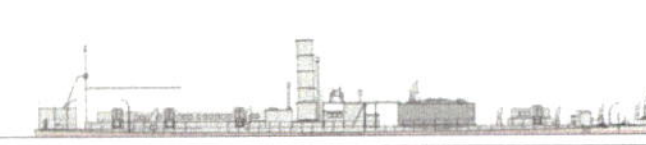

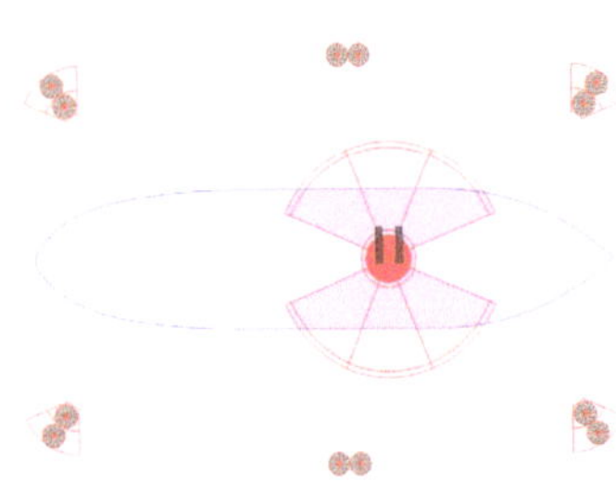

Haai (1870)
1870 05 01 - 1871 01 01 - 1873 01 01 | Fijenoord, Schiedam

Hyena (1869)
1869 11 21 - 1870 12 22 - 1872 01 01 | Amsterdam DYd

Luipaard (1874)
1874 01 01 - 1876 01 01 - 1877 01 01 | Fijenoord, Schiedam

5 | "Adder" type (1869)
Monitors, Netherlands (continued)

Panter (1869)
1869 11 21 - 1870 12 12 - 1872 01 01 | Amsterdam DYd

Wesp (1869)
1869 01 01 - 1871 01 01 - 1873 01 01 | Amsterdam DYd

Luipaard

Displacement: 1,555 tons at draft: 2.97 m | Length: 58.6 m oa | Beam: 13.41 m | Crew: 117 men

Machinery: 4 rectangular boilers | 1 horizontal compound E * 1 screw | 560 HP = 7 knots | Supplies: 93 tons of coal

Armour: iron | CT: 140; Turrets: 285-203(^25) | WLBelt: 140(^25)

Armament: 1-280/22

Age 21

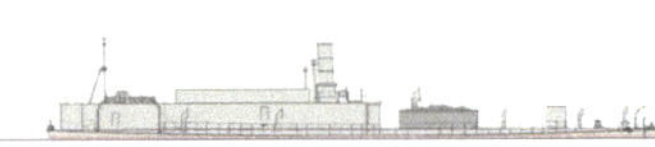

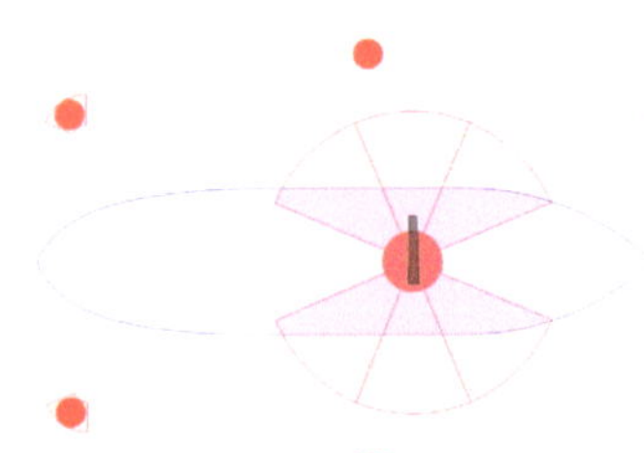

2 | "Cerberus" type (1867)
Monitors, Netherlands

Displacement: 1,559 tons at draft: 3 m | Length: 56.53 m oa | Beam: 13.36 m | Crew: 117 men

Machinery: 4 rect. boilers | 2 * 270 HP horizontal return connecting rod EE * 2 screws | 534 HP = 7 knots | Supplies: 98 tons of coal

Armour: iron | CT: 140; Turrets: 285-203 | WLBelt: 140(^25)

Armament: 2-228.6/13.9 MLR

Age 23

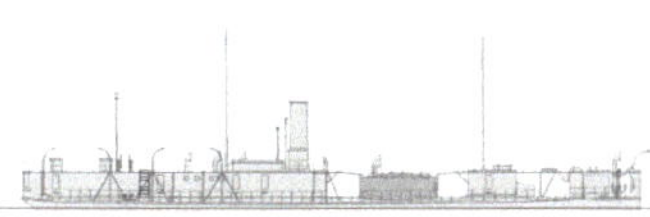

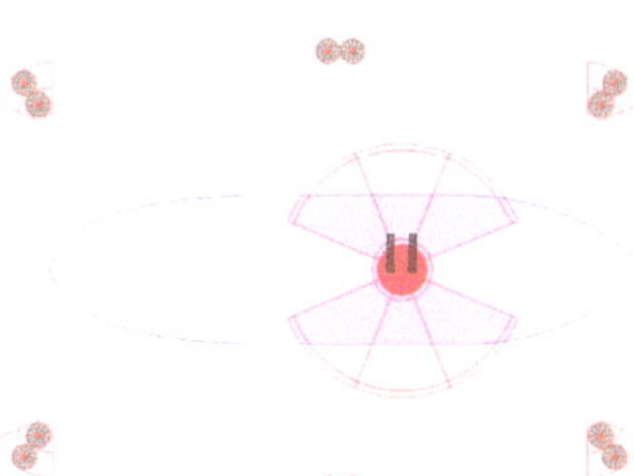

Bloedhond (1867)
1867 11 30 - 1869 01 14 - 1871 01 01 | Amsterdam DYd

Cerberus (1867)
1867 11 30 - 1869 01 14 - 1870 07 21 | Amsterdam DYd

3 | "Heiligerlee" type (1867)
Monitors, Netherlands

Displacement: 1,520 tons at draft: 3.05 m | Length: 57 m oa | Beam: 13.46 m | Crew: 117 men

Machinery: 4 rect. boilers | 2 * 315 HP horizontal return connecting rod EE * 2 screws | 630 HP = 8.5 knots | Supplies: 131 tons of coal

Armour: iron | CT: 140; Turrets: 285-203 | WLBelt: 140(^25)

Armament: 2-228.6/13.9 MLR

Age 23

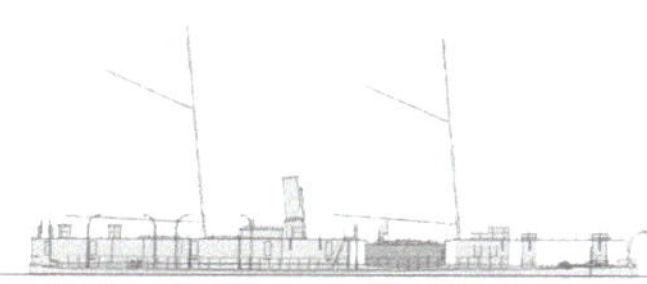

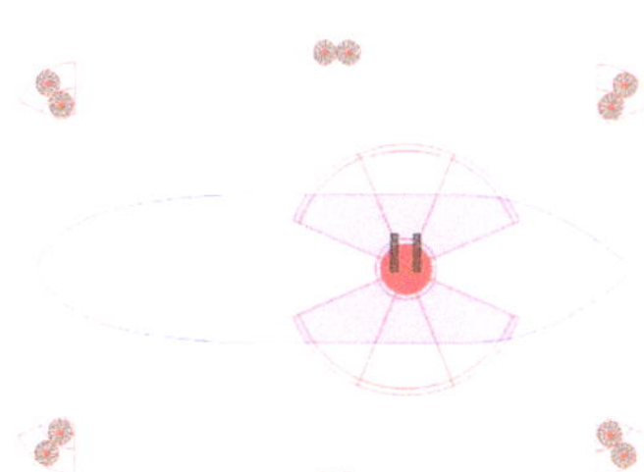

Heiligerlee (1867)
1867 08 01 - 1868 01 14 - 1868 05 26 | Laird, Birkenhead

Krokodil (1867)
1867 01 01 - 1868 02 13 - 1870 07 21 | Laird, Birkenhead

Tijger (1867)
1867 01 01 - 1868 02 22 - 1869 06 16 | Napier, Govan

1 | "Schorpioen" type (1867)
Monitor, Netherlands

Displacement: 2,140 tons at draft: 4.82 m | Length: 62.48 m oa | Beam: 11.89 m | Crew: 136 men

Machinery: 4 rect. boilers | 2 * 2,100 HP horizontal return connecting rod EE * 2 screws | 2,225 HP = 13 knots | Supplies: 246 tons of coal

Armour: iron | CT: 144; Turrets: 278-203 | WLBelt: 152(^25-29)

Armament: 2-228.6/13.9 MLR

Age 23

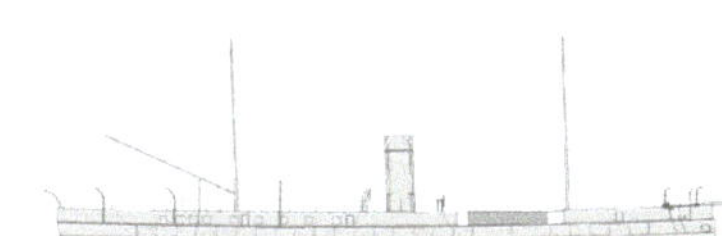

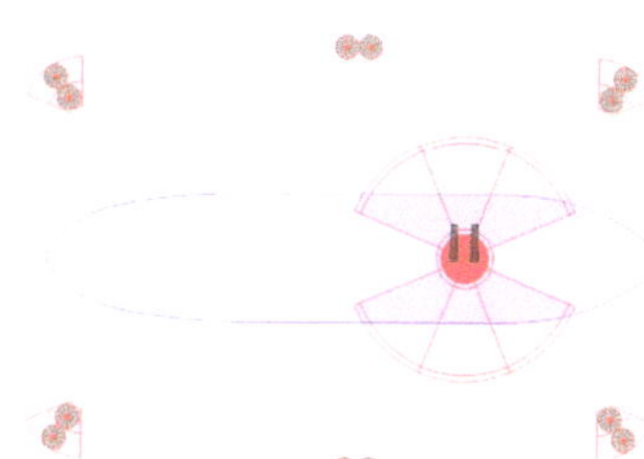

Schorpioen (1867)
1867 08 01 - 1868 01 18 - 1868 10 01 | F.et Ch. de la M., La Seyne

1 | "Stier" type (1867)
Monitor, Netherlands

Displacement: 2,078 tons at draft: 5.1 m | Length: 62.48 m oa | Beam: 11.89 m | Crew: 136 men

Machinery: 4 rect. boilers | 2 * 1,130 HP horizontal return connecting rod EE * 2 screws | 2,260 HP = 12.5 knots | Suppl.: 246 tons of coal

Armour: iron | Conning tower: 144; Turrets: 278-203 | WLBelt: 152(^25-29)

Armament: 2-228.6/13.9 MLR

Age 23

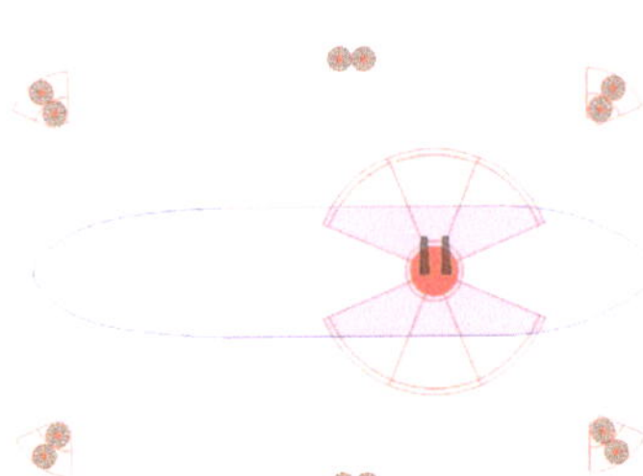

Stier (1867)
1867 06 07 - 1868 04 09 - 1869 01 01 | Laird, Birkenhead

1 | "Stier" type (1867)
Monitor, Netherlands

Displacement: 2,402 tons at draft: 5.1 m | Length: 59.69 m pp | Beam: 12.19 m | Crew: 159 men

Machinery: 4 rect. boilers | 2 * 1,000 HP horizontal expansion EE * 2 screws | 2,000 HP = 11.2 knots | Supplies: 150 tons of coal

Armour: iron | Turret: 203 | WLBelt: 152(^25)

Armament: 2-228.6/13.9 MLR, 4-120/16 MLR

Age 23

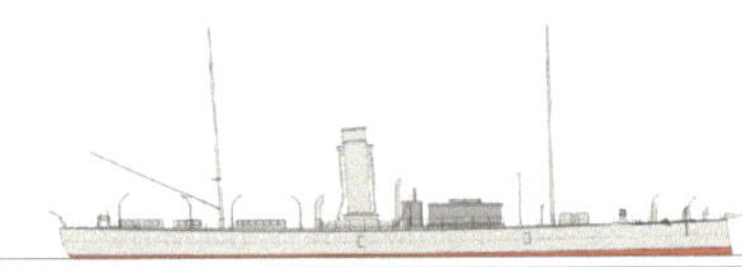

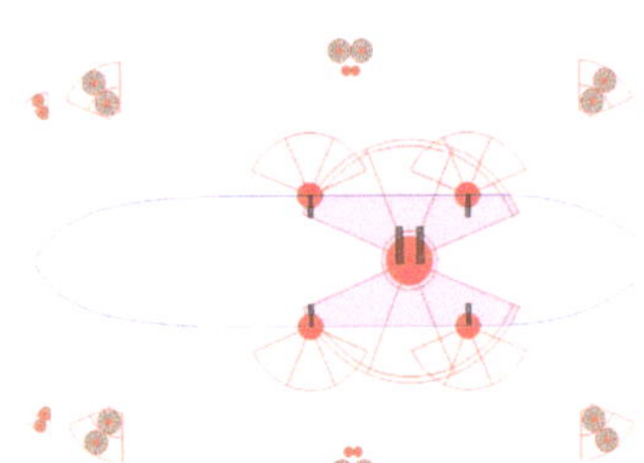

Buffel (1867)
1867 06 10 - 1868 03 10 - 1869 07 22 | Napier, Govan

Guinea (1868)
1868 01 01 - 1870 05 05 - 1873 10 16 | Amsterdam DYd

0 +1 | "Sumatra" type (1889)
Protected cruiser 2 class, Netherlands

PROTECTED CRUISER

BUILDING

0Y:10M

Sumatra

Displacement: 1,693 tons at draft: 4.67 m | Length: 69.97 m oa | Beam: 11.3 m | Crew: 181 men

Machinery: 4 cyl. boilers | 2 * 1,150 HP vertical triple expansion EE * 2 screws | 2,350 HP = 17 knots | Supplies: 207 / +69 tons of coal

Armour: steel | CT: 150(^25); Gun shield: 100 | WLDeck: 0(^38)

Arm.: 1-210/30, 1-150/35, 2-120/35, 2-37/23R, 4-37/23, 2-356 TT (^)

Age 1

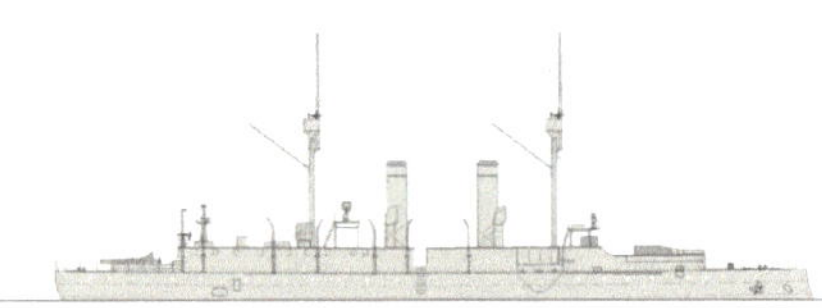

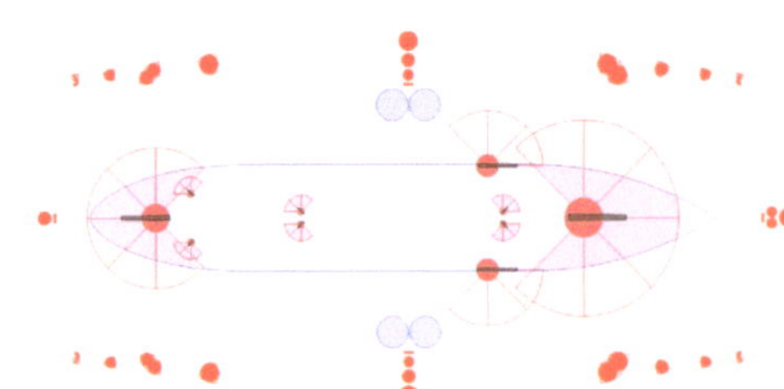

Sumatra (1889)
1889 03 05 - ***** - ***** | Amsterdam DYd

6 | "Atjeh" type (1875)
Unprotected cruisers 1 class, Netherlands

UNPROTECTED CRUISERS

Displacement: 3,425 tons at draft: 6.4 m | Length: 91.8 m oa | Beam: 12.5 m | Crew: 306 men
Machinery: 10 cylindrical boilers | 1 vertical reciprocating E * 1 screw | 2,700 HP = 12.5 knots | Supplies: 440 tons of coal
Armament: 4-170/25, 8-120/16

Age 15

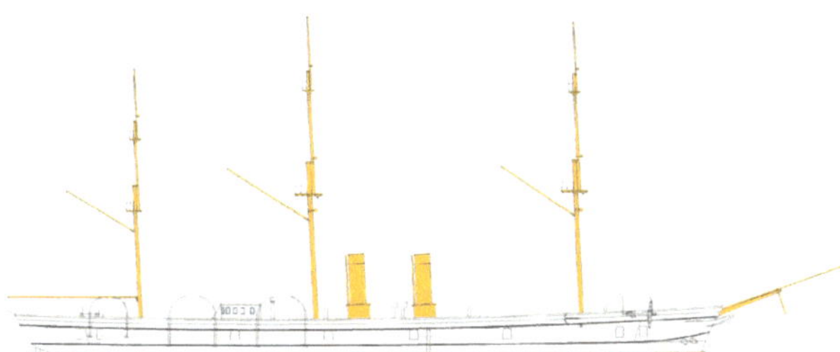

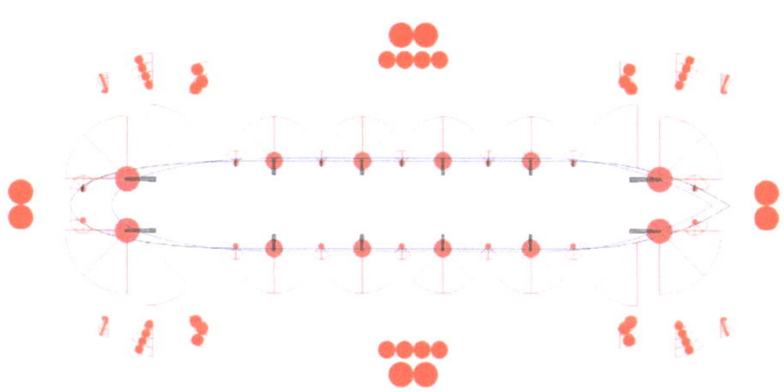

Atjeh (1875)
1875 03 03 - 1876 12 06 - 1877 11 01 | Amsterdam DYd, Amsterdam

De Ruyter (1876)
1876 11 11 - 1880 09 22 - 1885 08 01 | Amsterdam DYd, Amsterdam

Johan Willem Friso (1883)
1883 10 22 - 1886 06 10 - 1888 02 01 | Amsterdam DYd, Amsterdam

6 | "Atjeh" type (1875)
Unprotected cruisers 1 class, Netherlands

Koningin Emma der Nederlanden (1876)
1876 11 01 - 1879 01 20 - 1880 12 01 | Amsterdam DYd, Amsterdam

Tromp (1875)
1875 08 23 - 1877 12 12 - 1882 09 01 | Amsterdam DYd, Amsterdam

Van Speyk (1880)
1880 02 09 - 1882 06 07 - 1887 03 01 | Amsterdam DYd, Amsterdam

De Ruyter, Van Speyk, Johan Willem Friso

De Ruyter, Van Speyk, Johan Willem Friso

WOODEN CRUISERS

1 | "Anna Paulowna" type (1865)
Wooden screw frigate, Netherlands

NO PHOTO AVAILABLE

Displacement: 3,345 tons at draft: 6.2 m | Length: 91.85 m oa | Beam: 13.3 m | Crew: 325 men
Machinery: 4 rectangular boilers | 1 horizontal reciprocating E * 1 screw | 2,500 HP = app. 11 knots |
Supplies: app. 500 tons of coal
Armament: 10-165/14.5 MLR

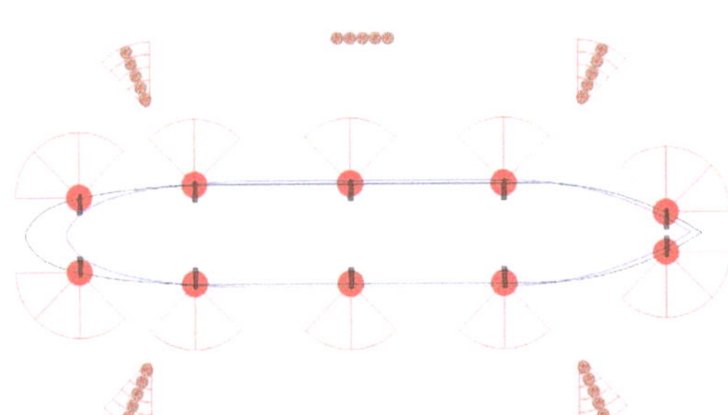

Anna Paulowna (1865)
1865 01 01 - 1867 01 01 - 1869 01 01 | Amsterdam DYd, Amsterdam

1 | "Adolf Hertog van Nassau" type (1858)
Wooden screw frigate, Netherlands

Displacement: 3,690 tons at draft: 6.8 m | Length: 72.86 m oa | Beam: 15.72 m | Crew: 500 men
Machinery: 4 rectangular boilers | 1 horizontal reciprocating E * 1 screw | 1,700 HP = app. 10 knots |
Supplies: 400 tons of coal
Armament: 8-165/14.5 MLR

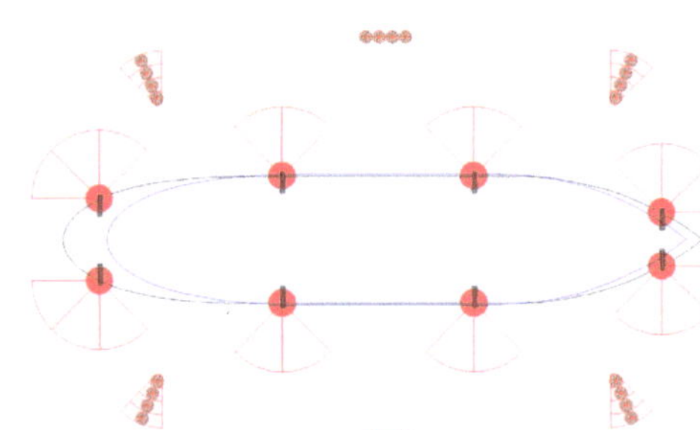

Adolf Hertog van Nassau (1858)
1858 12 04 - 1861 06 08 - 1863 01 01 | De Schelde, Flushing

1 | "Zilveren Kruis" type (1865)
Wooden screw corvette, Netherlands

Displacement: 2,180 tons at draft: 5.6 m | Length: 62.84 m oa | Beam: 12.75 m | Crew: 225 men

Machinery: 4 rectangular boilers | 1 horizontal reciprocating E * 1 screw | 830 HP = 8.5 knots | Supplies: 400 tons of coal

Armament: 4-177.8/15.8 ML, 8-165/15.9 ML

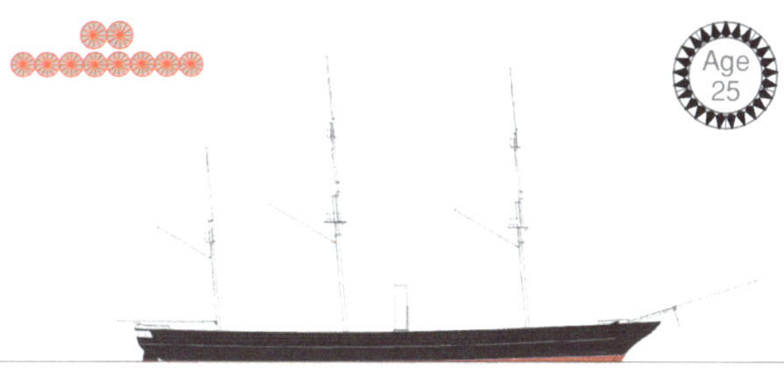

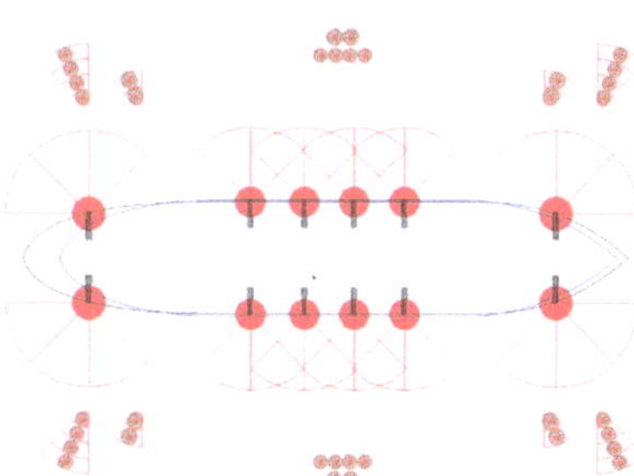

Zilveren Kruis (1865)
1865 06 18 - 1869 06 18 - 1870 08 21 | Amsterdam DYd

2 | "Watergeus" type (1863)
Wooden screw corvettes, Netherlands

Displacement: 1,405 tons at draft: 4.9 m | Length: 58 m oa | Beam: 9.74 m | Crew: 130 men

Machinery: 4 rectangular boilers | 1 horizontal expansion E * 1 screw | 720 HP = 10 knots | Supplies: 84 tons of coal

Armament: 6-160/16 MLR

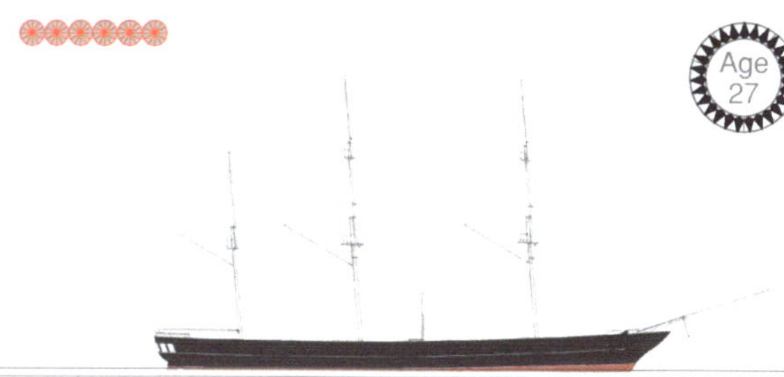

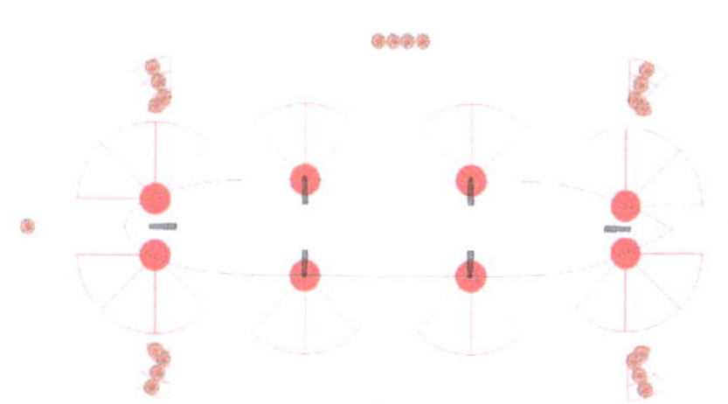

Watergeus (1863)
1863 01 01 - 1864 01 01 - 1865 01 01 | Amsterdam DYd

Marnix (1866)
1866 01 01 - 1867 01 01 - 1868 01 01 | Amsterdam DYd

2 | "Djambi" type (1858)
Wooden screw corvettes, Netherlands

Displacement: 2,030 tons at draft: 5.5 m | Length: 62.84 m oa | Beam: 12.25 m | Crew: 225 men

Machinery: 4 rectangular boilers | 1 horizontal reciprocating E * 1 screw | 830 HP = 8.5 knots | Supplies: 340 tons of coal

Armament: 8-165/14.5 MLR

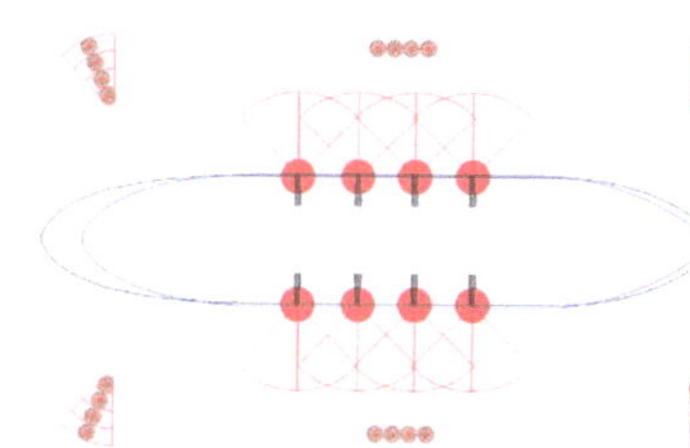

Curacao (1860)
1860 07 14 - 1863 12 10 - 1867 05 16 | De Schelde, Flushing

Van Galen (1868)
1868 02 19 - 1872 07 09 - 1876 04 01 | Amsterdam DYd

TORPEDO SHIPS

0 +3 | "Lamongan" type (1890)
Torpedo boats 1 class, Netherlands

BUILDING

Displacement: 51.6 tons at draft: 0.9 m | Length: 32.86 m oa | Beam: 4.04 m | Crew: 15 men

Machinery: 1 locomotive boiler | 1 vertical double expansion E * 1 screw | 800 HP = 20.4 knots | Supplies: 8 tons of coal

Armament: 2-37/23, 2-356 TT (>>)

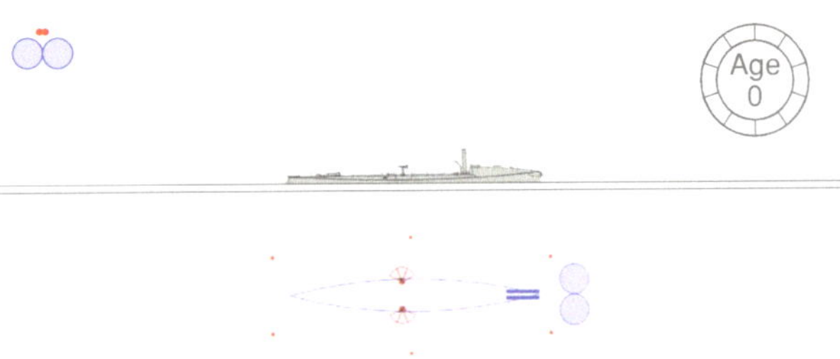

Lamongan (1890)
1890 01 01 - ***** - ***** | Amsterdam DYd

Makjan (1890)
1890 01 01 - ***** - ***** | Amsterdam DYd

Nobo (1890)
1890 01 01 - ***** - ***** | Amsterdam DYd

0 +2 | "XXI" type (1889)
Torpedo boats 2 class, Netherlands

BUILDING

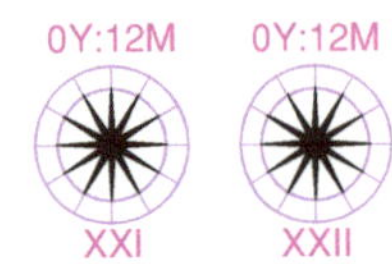

Displacement: 37 tons at draft: 0.76 m | Length: 25.6 m oa | Beam: 3.17 m | Crew: 12 men

Machinery: 1 locomotive boiler | 1 vertical compound E * 1 screw | 427 HP = 17.5 knots | Supplies: 5 tons of coal

Armament: 1-37/23, 2-356 TT (1*2 trainable), 1 spar torpedo

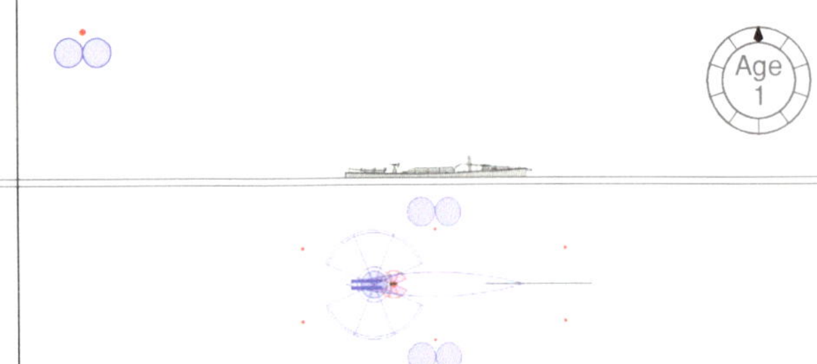

XXI (1889)
1889 01 01 - ***** - ***** | Amsterdam DYd

XXII (1889)
1889 01 01 - ***** - ***** | Amsterdam DYd

4 +2 | "Empong" type (1887)
Torpedo boats 1 class, Netherlands

BUILDING

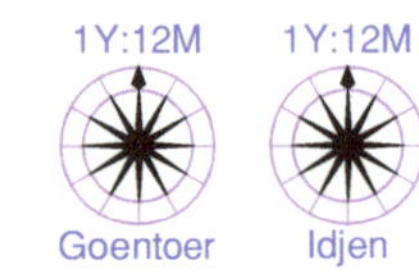

Displacement: 85 tons at draft: 1.68 m | Length: 39.22 m oa | Beam: 3.96 m | Crew: 16 men

Machinery: 1 locomotive boiler | 1 vertical triple expansion E * 1 screw | 1,000 HP = 22 knots | Supplies: 15 tons of coal

Armament: 2-37/23, 2-457,2 TT (1 trainable, 1 >)

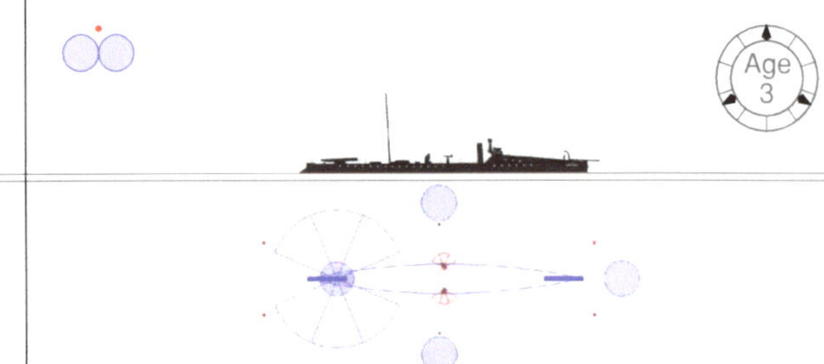

Empong (1887)
1887 01 01 - 1888 01 01 - 1889 01 01 | Yarrow, Poplar

Foka (1887)
1887 01 01 - 1888 01 01 - 1889 01 01 | Amsterdam DYd

Goentoer (1888)
1888 01 01 - 1888 11 07 - ***** | Amsterdam DYd

Habang (1887)
1887 01 01 - 1888 01 01 - 1889 01 01 | Amsterdam DYd

Idjen (1888)
1888 01 01 - 1889 07 31 - ***** | Amsterdam DYd

Krakatau (1887)
1887 01 01 - 1888 01 01 - 1889 01 01 | Amsterdam DYd

4 +2 | "Empong" type (1887)
Torpedo boats 1 class, Netherlands (continued)

4 +1 | "Ardjoeno" type (1886)
Torpedo boats 1 class, Netherlands (continued)

4 +1 | "Ardjoeno" type (1886)
Torpedo boats 1 class, Netherlands

BUILDING

1Y:12M

Idjen

Displacement: 83 tons at draft: m | Length: 38.38 m oa | Beam: 3.88 m | Crew: 16 men

Machinery: 1 locomotive boiler | 1 vertical compound E * 1 screw | 800 HP = 21 knots | Supplies: 15 tons of coal

Armament: 2-37/23, 2-457,2 TT (1 trainable, 1 >)

Age 4

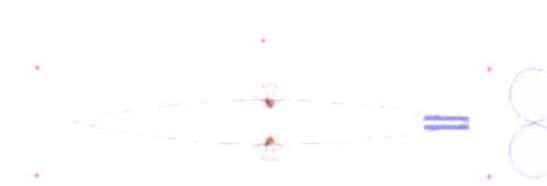

Ardjoeno (1886)
1886 01 01 - 1886 06 06 - 1887 01 01 | Yarrow, Poplar

Batok (1886)
1886 01 01 - 1887 01 01 - 1888 01 01 | Amsterdam DYd

Cerberus (1888)
1888 01 01 - 1888 08 29 - ***** | De Schelde, Flushing

Cycloop (1886)
1886 01 01 - 1887 01 01 - 1888 01 01 | Amsterdam DYd

Dempo (1886)
1886 01 01 - 1887 01 01 - 1888 01 01 | Fijenoord, Schiedam

Sister ship can be found in the Chilean section

2 | "Batum"/"Etna" type (1879)
Torpedo boats 1 class, Russia

Displacement: 45 tons at draft: 0.95 m | Length: 30.48 m oa | Beam: 3.81 m | Crew: 13 men

Machinery: 1 locomotive boiler | 1 horozontal compound E * 1 screw | 500 HP = 21.5 knots | Supplies: 7 tons of coal

Armament: 2-37/23, 2-356 TT (2 >>)

Age 11

Etna (1882)
1882 01 01 - 1882 06 06 - 1883 01 01 | Yarrow, Poplar

Hekla (1882)
1882 01 01 - 1882 06 06 - 1883 01 01 | Yarrow, Poplar

Sister ships can be found in the Argentinean, Austria-Hungarian, Brazilean, British, Chilean, Greek, Italian, Japanese and Russian sections

7 | "XIV" type (1878)
Spar torpedo boats 1 class, Netherlands

Displacement: 29 tons at draft: 0.8 m | Length: 24.6 m oa | Beam: 3.15 m | Crew: 10 men

Machinery: 1 locomotive boiler | 1 horizontal direct acting E * 1 screw | 250 HP = 16 knots | Supplies: app. 4 tons of coal

Rearmed: 1-356 TT (1 >)

Age 12

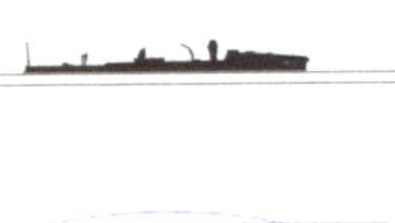

XIV (1878)
1878 01 01 - 1879 01 01 - 1880 01 01 | Fijenoord, Schiedam

XV (1880)
1880 01 01 - 1881 01 01 - 1882 01 01 | Fijenoord, Schiedam

XVI (1880)
1880 01 01 - 1881 01 01 - 1882 01 01 | Fijenoord, Schiedam

XVII (1880)
1880 01 01 - 1881 01 01 - 1882 01 01 | Fijenoord, Schiedam

XVIII (1880)
1880 01 01 - 1881 01 01 - 1882 01 01 | De Schelde, Flushing

XIX (1880)
1880 01 01 - 1881 01 01 - 1882 01 01 | De Schelde, Flushing

XX (1880)
1880 01 01 - 1881 01 01 - 1882 01 01 | De Schelde, Flushing

4 | "X" type (1878)
Spar torpedo boats, Netherlands

NO PHOTO AVAILABLE

Displacement: 29 tons at draft: 0.8 m | Length: 24.16 m oa | Beam: 3.15 m | Crew: 10 men

Machinery: 1 locomotive boiler | 1 horizontal direct acting E * 1 screw | 250 HP = 15.8 knots | Supplies: app. 4 tons of coal

Armament: 1 spar torpedo

Age 12

X (1878)
1878 01 01 - 1879 01 01 - 1880 01 01 | Fijenoord, Schiedam

XI (1878)
1878 01 01 - 1879 01 01 - 1880 01 01 | Rijkswerf, Willemsoord

XII (1878)
1878 01 01 - 1879 01 01 - 1880 01 01 | De Schelde, Vlissingen

XIII (1878)
1878 01 01 - 1879 01 01 - 1880 01 01 | Fijenoord, Schiedam

1 | "III" type (1878)
Spar torpedo boat, Netherlands

NO PHOTO AVAILABLE

Displacement: 26 tons at draft: 0.72 m | Length: 22.98 m oa | Beam: 3.16 m | Crew: 10 men

Machinery: 1 locomotive boiler | 1 horizontal direct acting E * 1 screw | 220 HP = 12.2 knots | Supplies: app. 3 tons of coal

Armament: 1 spar torpedo

Age 12

III (1878)
1878 01 01 - 1878 06 06 - 1879 01 01 | Yarrow, Poplar

1 | "White" type (1883)
Spar torpedo boat 2 class, Netherlands

Displacement: 10 tons at draft: 1.19 m | Length: 13.86 m oa | Beam: 2.96 m | Crew: 5 men

Machinery: 1 locomotive boiler | 1 horizontal direct acting E * 1 screw | 60 HP = 12 knots | Supplies: app. 1 ton of coal

Armament: 1 spar torpedo

unnamed (1883)
1883 01 01 - 1883 06 06 - 1884 01 01 | White, Cowes

6 | "II" type (1876)
Spar torpedo boats, Netherlands

Displacement: 26 tons at draft: 1.6 m | Length: 24.16 m oa | Beam: 3.14 m | Crew: 10 men

Machinery: 1 locomotive boiler | 1 horizontal direct acting E * 1 screw | 250 HP = 18 knots | Supplies: app. 4 tons of coal

Armament: 2 spar torpedoes

Age 12

IV (1878)
1878 01 01 - 1879 01 01 - 1879 06 06 | Thornycroft, Woolston

V (1878)
1878 01 01 - 1879 01 01 - 1879 06 06 | Thornycroft, Woolston

VI (1878)
1878 01 01 - 1879 01 01 - 1879 06 06 | Thornycroft, Woolston

VII (1878)
1878 01 01 - 1879 01 01 - 1879 06 06 | Thornycroft, Woolston

VIII (1878)
1878 01 01 - 1879 01 01 - 1879 06 06 | Thornycroft, Woolston

IX (1878)
1878 01 01 - 1879 01 01 - 1879 06 06 | Thornycroft, Woolston

1 | "II" type (1876)
Spar torpedo boat, Netherlands

Displacement: 26 tons at draft: 0.75 m | Length: 23.3 m oa | Beam: 3.04 m | Crew: 10 men

Machinery: 1 locomotive boiler | 1 horizontal direct acting E * 1 screw | 240 HP = 14.5 knots | Supplies: app. 3 tons of coal

Armament: 1 spar torpedo

Age 14

II (1876)
1876 01 01 - 1877 01 01 - 1877 06 06 | Thornycroft, Woolston

Sister ship can be found in the Italian section

1 | "I" type (1875)
Spar torpedo boat, Netherlands

Displacement: 25 tons at draft: 0.8 m | Length: 21.8 m oa | Beam: 2.74 m | Crew: 9 men

Machinery: 1 locomotive boiler | 1 compound E * 1 screw | 220 HP = 13 knots | Supplies: 6 tons of coal

Armament: 1 spar torpedo

Age 15

I (1875)
1875 01 01 - 1875 06 06 - 1876 01 01 | Yarrow, Poplar

1 | "Java" type (1884)
Composite screw sloop, Netherlands

Displacement: 1,280 tons at draft: 3.74 m | Length: 62.56 m oa | Beam: 9.53 m | Crew: 104 men

Machinery: 4 rectangular boilers | 1 horizontal expansion E * 1 screw | 1,017 HP = 12.6 knots | Supplies: 200 tons of coal

Armament: 1-150/22, 3-120/16, 2-37/23R

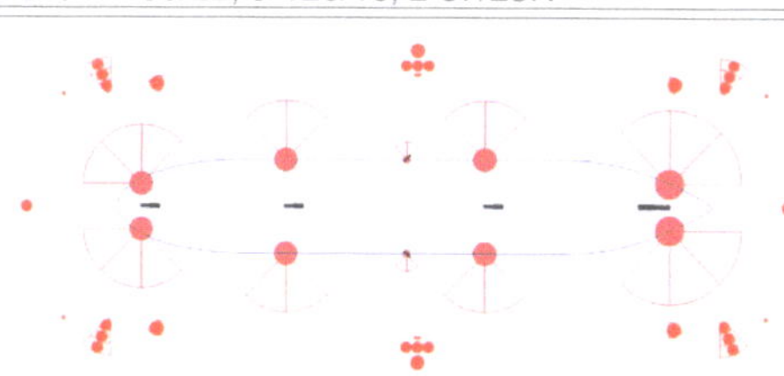

Java (1884)
1884 01 01 - 1885 01 01 - 1886 01 01 | Fijenoord, Schiedam

SLOOPS & GUNBOATS

1 | "Sommelsdijk" type (1881)
Composite screw sloop, Netherlands

Displacement: 997 tons at draft: 3.58 m | Length: 54.5 m oa | Beam: 9.51 m | Crew: 96 men

Machinery: 4 rectangular boilers | 1 horizontal expansion E * 1 screw | 759 HP = 11.4 knots | Supplies: 180 tons of coal

Armament: 1-150/22, 3-120/18, 2-37/23, 2-37/23R

Age 9

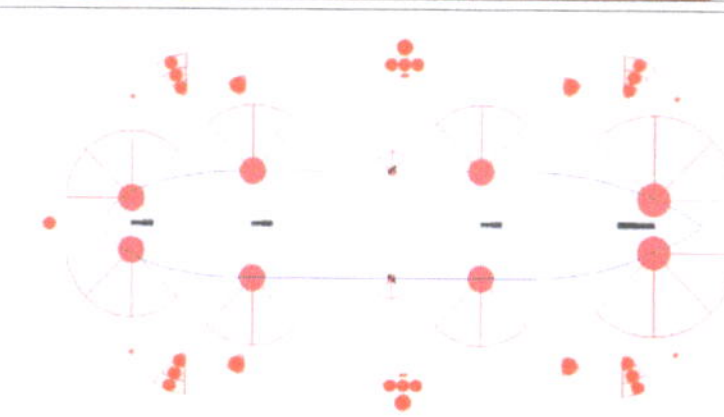

Sommelsdijk (1881)
1881 01 01 - 1882 01 01 - 1883 01 01 | Amsterdam DYd

1 | "Alkmaar" type (1873)
Composite screw sloop, Netherlands

Displacement: 1,051 tons at draft: 4.8 m | Length: 54.46 m oa | Beam: 9.2 m | Crew: 106 men

Machinery: 4 rectangular boilers | 1 horizontal expansion E * 1 screw | 686 HP = 10.3 knots | Supplies: 160 tons of coal

Armament: 1-150/22, 6-120/18

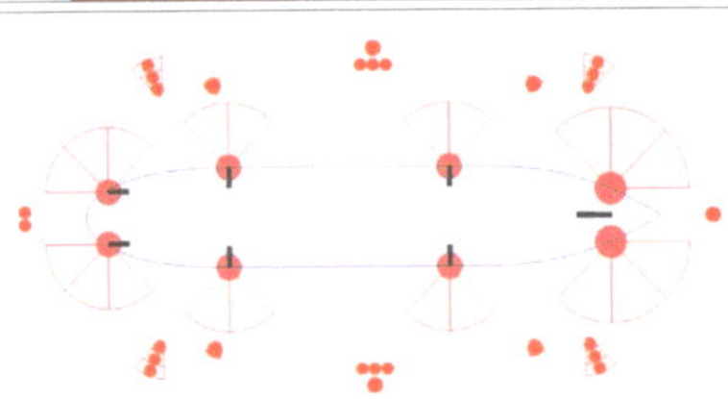

Alkmaar (1873)
1873 01 01 - 1874 01 01 - 1875 01 01 | Amsterdam DYd

2 | "Ceram" type (1886)
Composite gunboats, Netherlands

Displacement: 541 tons at draft: 3.13 m | Length: 52.83 m oa | Beam: 7.8 m | Crew: 83 men

Machinery: 2 cylindrical boilers | 1 horizontal compound E * 1 screw | 800 HP = 12.8 knots | Supplies: 90 tons of coal

Armament: 3-120/25, 2-37/23

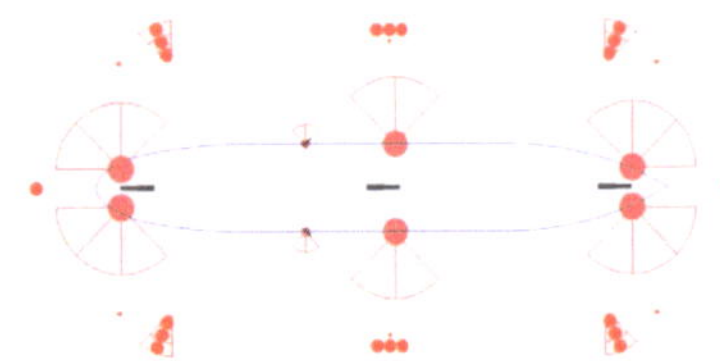

Ceram (1886)
1886 01 01 - 1887 01 01 - 1888 01 01 | Fijenoord, Schiedam

Flores (1886)
1886 01 01 - 1887 01 01 - 1888 01 01 | Amsterdam DYd

2 | "Suriname" type (1876)
Composite gunboats, Netherlands

Displacement: 870 tons at draft: 3.61 m | Length: 54 m oa | Beam: 9 m | Crew: 90 men

Machinery: 2 rectangular boilers | 1 horizontal expansion E * 1 screw | 450 HP = 9.5 knots | Supplies: 130 tons of coal

Armament: 1-150/22, 3-120/18, 2-37/23

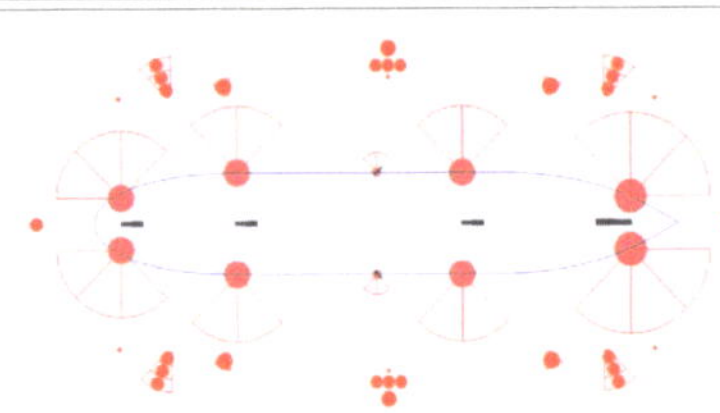

St. Eustatius (1877)
1877 01 01 - 1878 01 01 - 1879 01 01 | Fijenoord, Schiedam

Suriname (1876)
1876 01 01 - 1877 01 01 - 1878 01 01 | Amsterdam DYd

8 | "Batavia" type (1875)
Composite gunboats, Netherlands

Displacement: 837 tons at draft: 3.69 m | Length: 53.6 m oa | Beam: 9 m | Crew: 84 men

Machinery: 2 rectangular boilers | 1 horizontal expansion E * 1 screw | 400 HP = 8.4 knots | Supplies: 130 tons of coal

Armament: 1-150/22, 3-120/18, 2-37/23

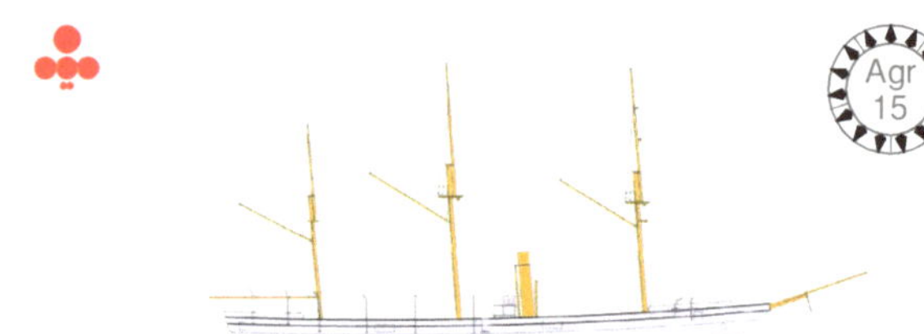

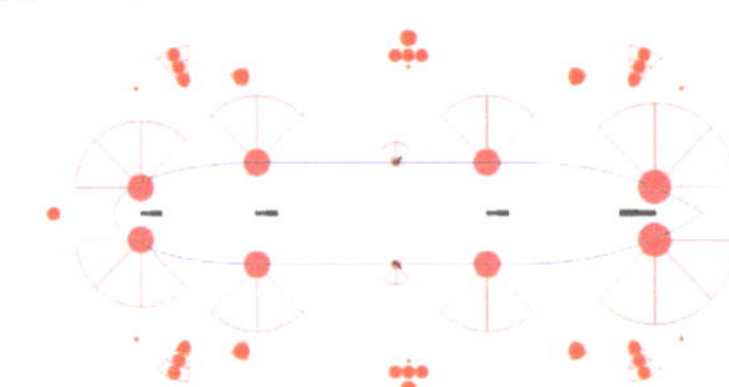

Bali (1877)
1877 01 01 - 1878 01 01 - 1879 01 01 | Fijenoord, Schiedam

Batavia (1875)
1875 01 01 - 1876 01 01 - 1877 01 01 | Amsterdam DYd

Benkoelen (1878)
1878 01 01 - 1879 01 01 - 1880 01 01 | Fijenoord, Schiedam

Bonaire (1876)
1876 01 01 - 1877 01 01 - 1878 01 01 | Fijenoord, Schiedam

Macassar (1876)
1876 01 01 - 1877 01 01 - 1878 01 01 | Amsterdam DYd

Madura (1879)
1879 01 01 - 1880 01 01 - 1881 01 01 | Amsterdam DYd

Padang (1877)
1877 01 01 - 1878 01 01 - 1879 01 01 | Fijenoord, Schiedam

8 | "Batavia" type (1875)
Composite gunboats, Netherlands (continued)

Samarang (1875)
1875 01 01 - 1876 01 01 - 1877 01 01 | Fijenoord, Schiedam

4 | "Pontianac" type (1873)
Composite gunboats, Netherlands

Displacement: 718 tons at draft: 4.5 m | Length: 49.86 m oa | Beam: 9 m | Crew: 106 men

Machinery: 4 rectangular boilers | 1 horizontal expansion E * 1 screw | 370 HP = 9 knots | Supplies: app. 80 tons of coal

Armament: 1-160/18 MLR, 2-120/18

Age 15

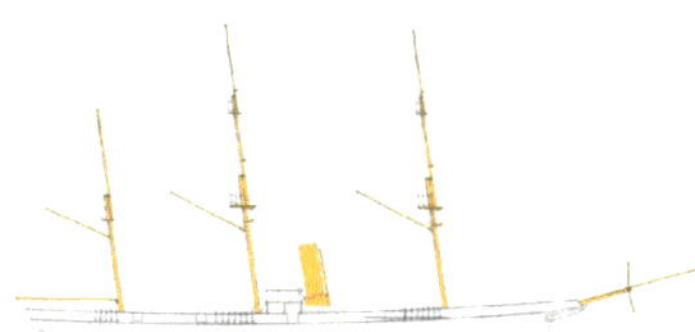

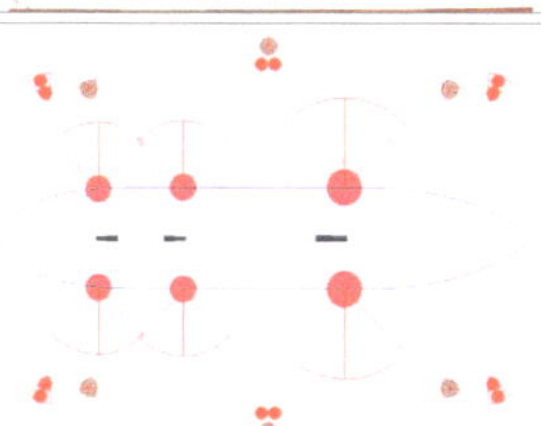

Bandjermasin (1873)
1873 07 29 - 1874 03 07 - 1874 06 16 | Amsterdam DYd

Palembang (1873)
1873 08 05 - 1874 03 03 - 1874 06 16 | Fijenoord, Schiedam

Pontianac (1873)
1873 07 29 - 1873 12 20 - 1874 05 25 | Amsterdam DYd

Sambas (1873)
1873 07 12 - 1874 01 24 - 1874 05 10 | Fijenoord, Schiedam

1 | "Aruba" type (1872)
Composite gunboat, Netherlands

Displacement: 786 tons at draft: 4.5 m | Length: 50 m oa | Beam: 9 m | Crew: 90 men

Machinery: 2 cylindrical boilers | 1 horizontal expansion E * 1 screw | 413 HP = 9.15 knots | Supplies: 100 tons of coal

Armament: 1-150/22, 2-120/18

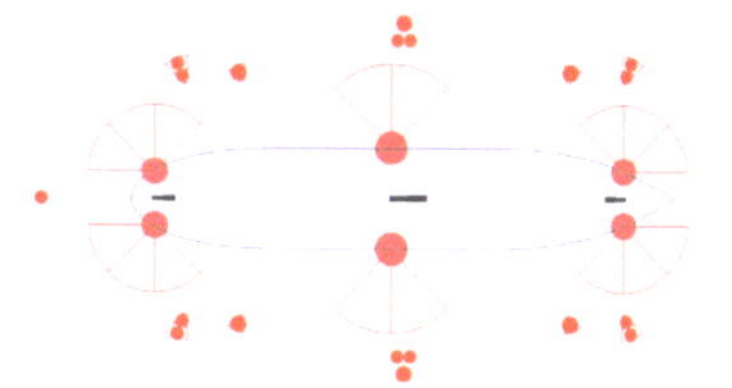

Aruba (1872)
1872 01 01 - 1873 12 30 - 1876 01 16 | Amsterdam DYd

2 | "Banda" type (1871)
Composite gunboats, Netherlands

Displacement: 689 tons at draft: 4.4 m | Length: 48.97 m oa | Beam: 9 m | Crew: 90 men

Machinery: 4 rectangular boilers | 1 horizontal expansion E * 1 screw | 200 HP = 7 knots | Supplies: app. 80 tons of coal

Armament: 1-160/16 MLR, 2-120/18

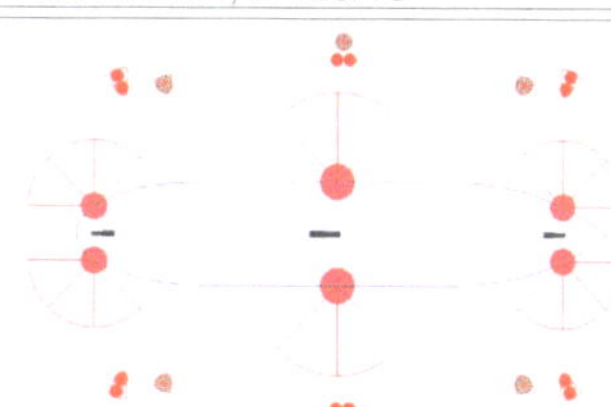

Banda (1871)
1871 01 01 - 1872 01 01 - 1873 01 01 | Fijenoord, Schiedam

Riouw (1871)
1871 01 01 - 1872 01 01 - 1873 01 01 | Fijenoord, Schiedam

16 | "Wodan" type (1873)
Gunboats, Netherlands

Displacement: 264 tons at draft: 2.3 m, 280 tons full | Length: 27.8 m oa | Beam: 8.2 m | Crew: 34 men

Machinery: 2 cylindrical boilers | 2 * 50 HP vertical compound EE * 2 screws | 101 HP = 7 knots | Supplies: 35 tons of coal

Armament: 1-283/20, 2-37/23, 1-37/23R

Age 15

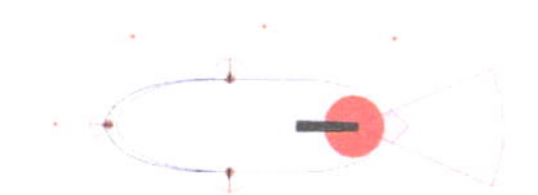

Balder (1877)
1877 01 01 - 1878 01 01 - 1879 01 01 | Christie, Delfshaven

Braga (1878)
1878 01 01 - 1879 01 01 - 1880 01 01 | Christie, Delfshaven

Bulgia (1876)
1876 01 01 - 1877 01 01 - 1878 01 01 | Amsterdam DYd

Dufa (1877)
1877 01 01 - 1878 01 01 - 1879 01 01 | Amsterdam DYd

Freyr (1876)
1876 01 01 - 1877 01 01 - 1878 01 01 | Christie, Delfshaven

Hadda (1878)
1878 01 01 - 1879 01 01 - 1880 01 01 | Amsterdam DYd

Hefring (1876)
1876 01 01 - 1877 01 01 - 1878 01 01 | Fijenoord, Schiedam

16 | "Wodan" type (1873)
Gunboats, Netherlands (continued)

Heimdall (1877)
1877 01 01 - 1878 01 01 - 1879 01 01 | Christie, Delfshaven

Njord (1878)
1878 01 01 - 1879 01 01 - 1880 01 01 | Christie, Delfshaven

Thor (1877)
1877 01 01 - 1878 01 01 - 1879 01 01 | Christie, Delfshaven

Tyr (1878)
1878 01 01 - 1879 01 01 - 1880 01 01 | Christie, Delfshaven

Udur (1876)
1876 01 01 - 1877 01 01 - 1878 01 01 | Amsterdam DYd

Ulfr (1877)
1877 01 01 - 1878 01 01 - 1879 01 01 | Amsterdam DYd

Vali (1877)
1877 01 01 - 1878 01 01 - 1879 01 01 | Fijenoord, Schiedam

Vidar (1878)
1878 01 01 - 1879 01 01 - 1880 01 01 | Fijenoord, Schiedam

Wodan (1876)
1876 01 01 - 1877 01 01 - 1878 01 01 | Christie, Delfshaven

12 | "Dog" type (1874)
Gunboats, Netherlands (continued)

Havik (1875)
1875 01 01 - 1876 01 01 - 1877 01 01 | Amsterdam DYd

Lynx (1875)
1875 01 01 - 1876 01 01 - 1877 01 01 | Fijenoord, Schiedam

Raaf (1873)
1873 01 01 - 1874 01 01 - 1875 01 01 | Amsterdam DYd

Sperwer (1875)
1875 01 01 - 1876 01 01 - 1877 01 01 | Christie, Delfshaven

Vos (1873)
1873 01 01 - 1874 01 01 - 1875 01 01 | Fijenoord, Schiedam

12 | "Das" type (1873)
Gunboats, Netherlands

Displacement: 218 tons at draft: 2.3 m | Length: 25.88 m oa | Beam: 8.2 m | Crew: 34 men

Machinery: 2 cylindrical boilers | 2 * 60 HP vertical compound EE * 2 screws | 124 HP = 7.5 knots | Supplies: 35 tons of coal

Armament: 1-228.6/13.9 MLR

Age 16

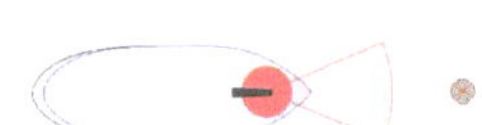

Bever (1874)
1874 01 01 - 1875 01 01 - 1876 01 01 | Christie, Delfshaven

Brak (1874)
1874 01 01 - 1875 01 01 - 1876 01 01 | Fijenoord, Schiedam

Das (1873)
1873 01 01 - 1874 01 01 - 1875 01 01 | Amsterdam DYd

Dog (1874)
1874 01 01 - 1875 01 01 - 1876 01 01 | Amsterdam DYd

Fret (1875)
1875 01 01 - 1876 01 01 - 1877 01 01 | Amsterdam DYd

Geep (1873)
1873 01 01 - 1874 01 01 - 1875 01 01 | Amsterdam DYd

Gier (1874)
1874 01 01 - 1875 01 01 - 1876 01 01 | Amsterdam DYd

2 | "Ever" type (1871)
Gunboats, Netherlands

Displacement: 170 tons at draft: 2.2 m | Length: 25.9 m oa | Beam: 7.62 m | Crew: 22 men

Machinery: 2 cylindrical boilers | 2 * 55 HP vertical compound EE * 2 screws | 110 HP = 8 knots | Supplies: 35 tons of coal

Armament: 1-228.6/13.9 MLR

Age 19

Ever (1871)
1871 11 03 - 1873 01 01 - 1873 09 22 | Armstrong, Elswick

Hydra (1871)
1871 11 03 - 1873 01 01 - 1873 09 30 | Armstrong, Elswick

JAPANESE FLEET

National flag

Fleet age 8

Ensign

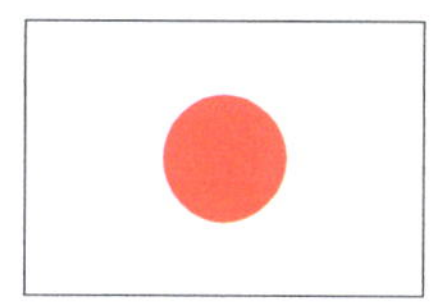

Jack

Admiral

Months derivation +4

Warships built from 1858 01 01 to 1890 01 01

35 warships = 46,724 tons = 1,335 tons / average ship

Fleet age - 1882 12 26

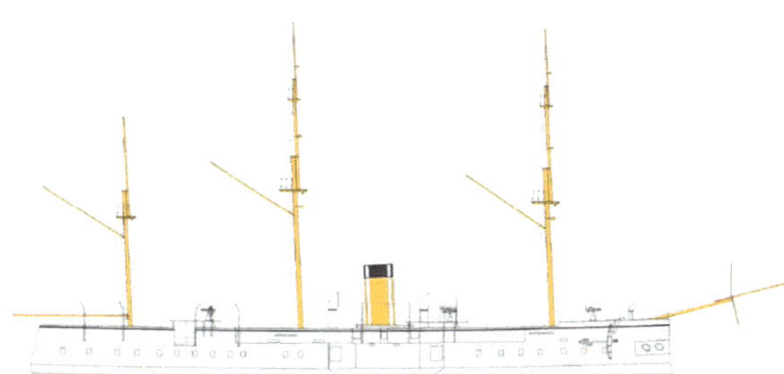

1889 fleet largest ship: Fuso (1875) battery ironclad

Japanese naval construction

Yokosuka dockyard

44.6 tons / average ship / month

1890-Jan-01
2 ships building

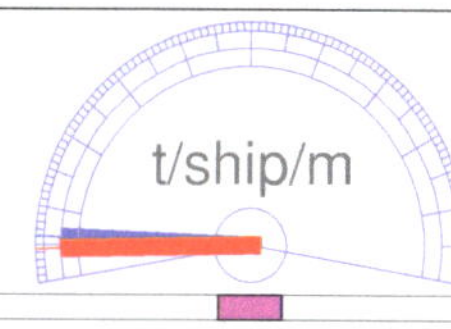

5,801 t.
tons / month = 89.1

t = 1,584
t/m =
27.8

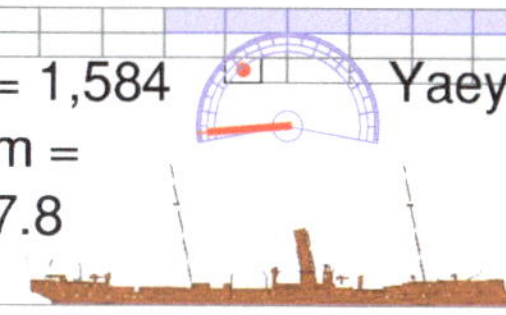

t = 4,217
t/m =
61.3

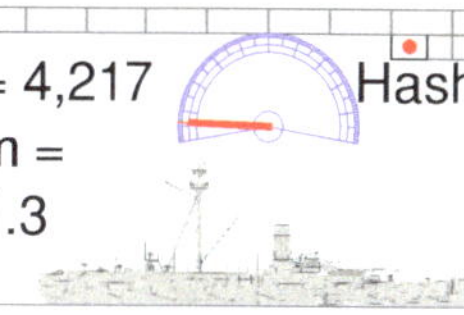

Yokosuka, Japan

Yokosuka map dates 1942. I can't use the Google map instead — the beach line, city quarters, everythig is absolutely different.

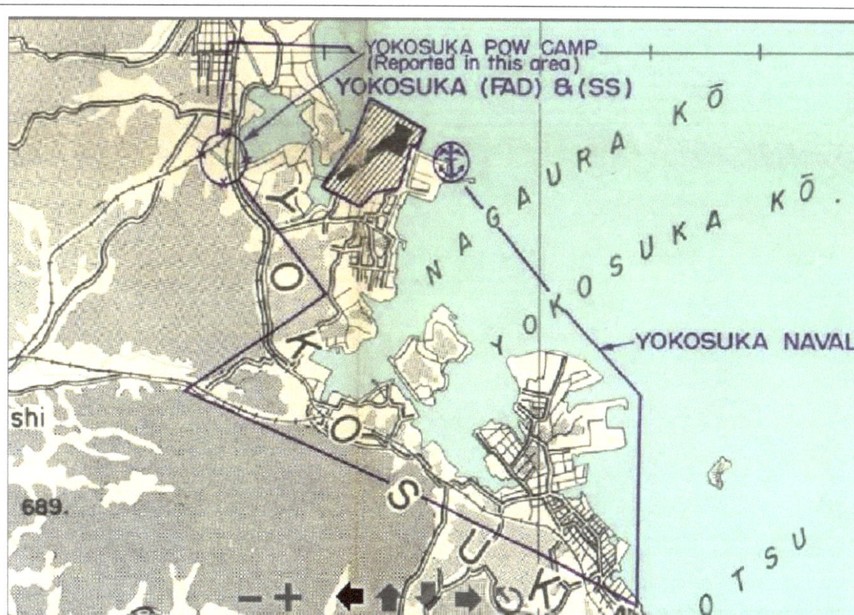

17 years of naval construction
12 ships | 17,919 tons

The plant's largest ship:
Hashidate (1888) protected cruiser
4,217 tons

The plant's most effectively built ship:
Hashidate (1888) protected cruiser
61.3 tons/month

Specialisation in:
Screw sailing cruisers | 4 ships | 5,827 tons
Protected cruiser | 1 ship | 4,217 tons
Unprotected cruisers | 2 ships | 3,334 tons

Onohama

16.5 tons / average ship / month

1890-Jan-01

2 ships building

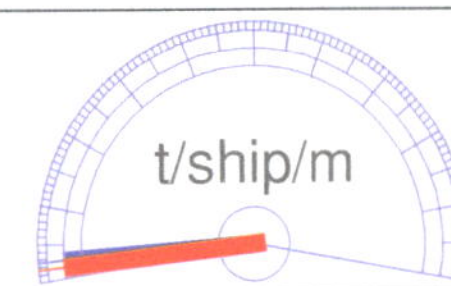

1,242 t.

tons / month = 32.9

t = 612 t/m = 12.5 Akagi	t = 630 t/m = 20.4 Oshima

Kobe, Japan

7 years of naval construction

4 ships | 3,330 tons

The plant's largest ship:

Yamato (1883) screw corvette

1,476 tons

The plant's most effectively built ship:

Yamato (1883) screw corvette

26.4 tons/month

Specialisation in:

Gunboats | 3 ships | 1,854 tons

Screw sailing cruiser | 1 ship | 1,476 tons

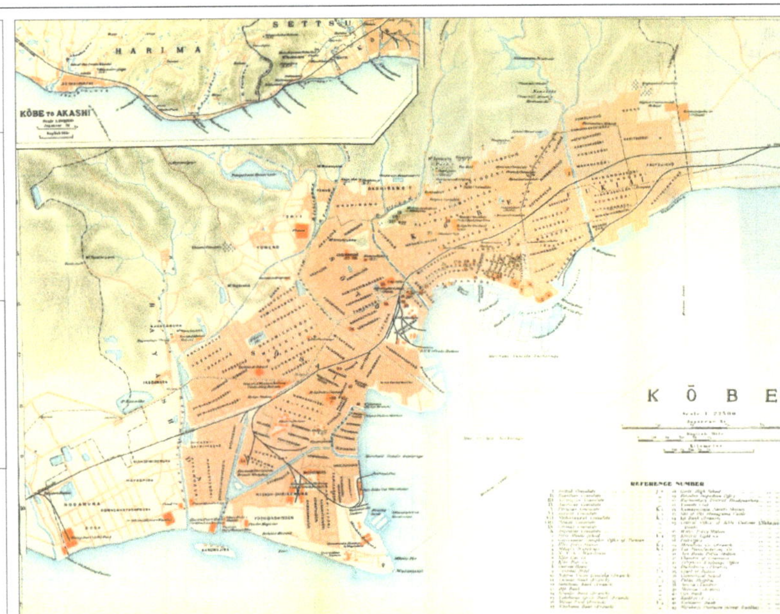

Japanese fleet effective list

4 boardside ironclads

Fuso

Kongo

Hiei
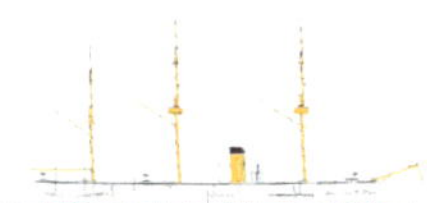

Ryujo
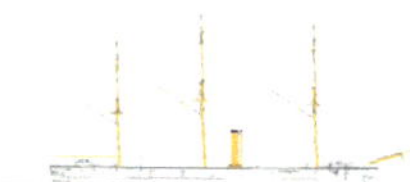

2 protected cruisers

Takachiho
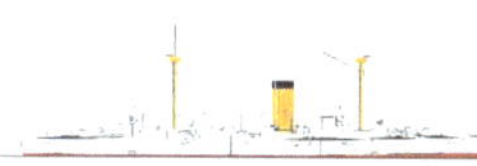

Naniwa
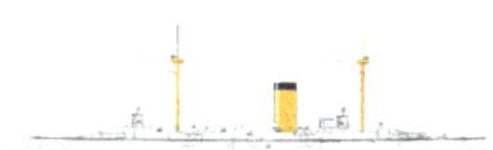

10 unprotected & wooden cruisers

Takao

Musashi
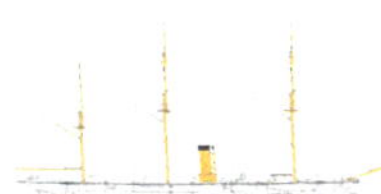

Yamato
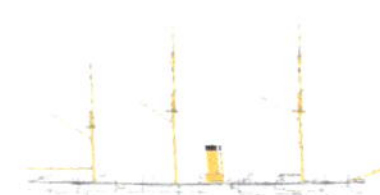

Katsuragi
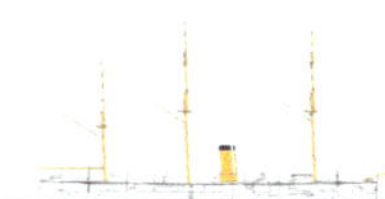

Tsukushi

Kaimon

Tenryu

Nisshin

Asama

Tsukuba

5 torpedo boats

Kotaka

1

2

3

4

5 gunboats

Atago

Chokai

Maya

Banjo

Hosho

2 paddle avisos

Kasuga	Jingei
NO PLAN AVAILABLE	

Fleet auxilaries

Fuji	Adzuma
Tender 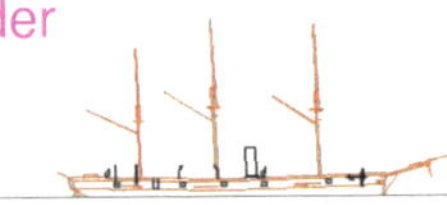	Accomodation ship 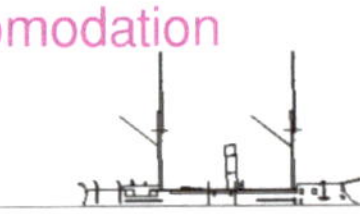

Japanese section ship names glossary

1 (TB1)
2 (TB1)
3 (TB1)
4 (TB1)

A
Adzuma (accomS)
Akagi (GB)
Amagi (CZ)
Asama (CW)
Atago (GB)

B
Banjo (GB)

C
Chokai (GB)

F
Fujijama (tendr)
Fuso (BO)

H
Hashidate (CP)
Hiei (BO)
Hosho (GB)

J
Jingei (CWP)

K
Kaimon (CW)
Kasuga (CWP)
Katsuragi (CW)
Kongo (BO)
Kotaka (TB1)

M
Maya (GB)
Moshin (hulk)
Musashi (CW)
Naniwa (CP)
Nisshin (CW)

O
Oshima (GB)

R
Ryujo (BO)

T
Takachiho (CP)
Takao (CS)
Tenryu (CW)
Tsukuba (CW)
Tsukushi (CS)

Y
Yaeyama (CS)
Yamato (CW)

Japanese fleet naval guns

Calibre (mm)	Mark	Calibre (in)/pdr	Gun type	First Introduced	Weight (kg)	Gun lenght (in/cal)	Bore lenght (cal)	Charge weight (lbs)	Muzzle energy (ft/t)	Projectile weight (lbs)	Muzzle velocity (ft/s)	Wrought iron perforation (in)
320.0	France	12.6"	BLR		47 300	464.4 / 36.86		463.0	29 850	879.6	2 034	-
280.0	Krupp	11.02"	BLR	1865	27 210	240.1 / 21.78	18.80	121.3	9 423	476.2	1 558	17.0
260.0	Krupp	10.24"	BLR		27 600	393.6 / 38.45	32.00	191.8	12 770	451.9	2 018	20.1
254.0	Armstrong	10"	BLR		25 000		26.00	180.0		450.0	1 873	
210.0	Krupp	8.24"	BLR		8 680	165.4 / 20.00	17.20	30.9	3 306	172.0	1 519	11.6
177.8	III	7"	MLR	1865	6 604	133.0 / 19.00	15.86	30.0	1 848	114.6	1 525	9.2
177.8	-	7"	BLR	1864	4 166	120.0 / 17.14	14.21	11.0	909	91.3	1 100	-
170.0	Krupp	6.8"	BLR	1864	5 510	167.3 / 25.00	21.90	27.6	2 226	112.4	1 558	10.5
150.0	Krupp	5.9"	BLR		4 700	205.2 / 34.71	32.20	41.9	2 784	112.4	1 890	12.8
150.0	Germany	5.9"	BLR	1878	4 040	176.0 / 29.80	27.20	33.1	111.5	112.4	1 624	11.0
150.0	Krupp	5.9"	BLR	1865	3 940	151.6 / 25.66	23.00	20.9	1 447	69.4	1 558	9.1
140.0		5.45"	BLR	1870	2 600	123.6 / 22.42	21.00	11.2		61.7	1 529	
127.0	II, III - IV, V	5"	BLR	1870	1 931 - 2 032	129.2 / 25.83	25.00	15.5	1 124	50.0	1 770	8.7
120.0	II, L	4.7"	BLR	1884	2 083	194.1 / 41.09	40.00	12.0	486	45.0	2150	8.5
120.0	Krupp	4.7"	BLR	-	2 310	165.4 / 35.00	-	19.8	1 290	57.3	1804	9.7
120.0	Krupp	4.7"	BLR	1859	1 380	115.2 / 25.00		9.9	680	36.1	1 493	7.0
120.0	II, L	4.7"	MLR	1865	1 778	120.0 / 25.53	22.00	6.8	312	38.8	1 425	-
76.2	I	12 pr	BLR	1865	356	92.4 / 30.78	28.00	4.0	249	12.5	1 530	-
57.0	I & II	6 pr	BLR	1876	406	96.5 / 43.00	40.00	1.2	345	6.0	1 818	3.7
47.0	I & II	3 pr	BLR	1876	254	79.6 / 43.00	40.00	1.8	321	3.3	1 873	3.1
37 R	Hotchkiss	1 pr	BLR	1873	209	33.5 / 23.00	19.13	0.2	-	1.1	1 450	-
37.0	Hotchkiss	1 pr	BLR	1884	72	33.5 / 23.00	19.15	0.2	-	1.1	1 450	-

WARSHIPS SECTION

Ship classes	Building	Active
Battleships		
Boardside ironclads		4
Armoured cruisers		
Coast defence armoured ships		
Protected cruisers	1	2
Unprotected cruisers	1	3
Torpedo gunboats		
Torpedo boats 1 class		5
Torpedo boats 2 class		
Submarines		
Wooden screw cruisers		8
Wooden screw sloops		1
Gunboats	2	5
Gunboats coastal		

BATTERY IRONCLADS

1 | "Fuso" type (1875)
Battery ironclad, Japan

Displacement: 3,717 tons at draft: 5.5 m | Length: 68.5 m wl | Beam: 14.6 m | Crew: 250 men

Machinery: 8 cylindrical boilers | 2 * 2,000 HP horizontal compound EE * 2 screws | 3,932 HP = 13 knots | Supplies: 250 / +110 tons of coal

Armour: iron | Battery: 229-152(^25) |
WLBelt: 114(^25) 229 114(^25)

Armament: 4-210/22, 2-170/25, 6-47/23, 1-25/40

Age 15

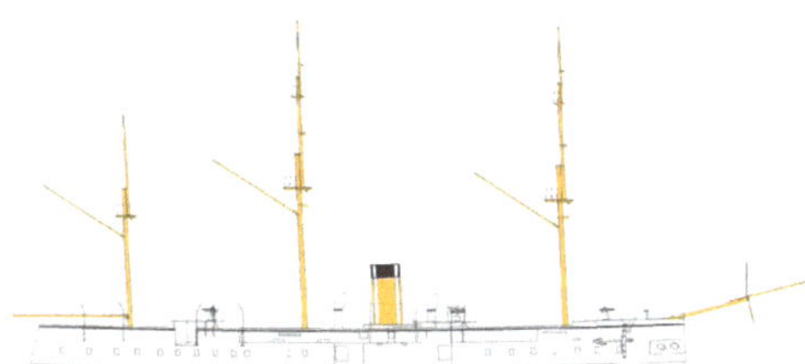

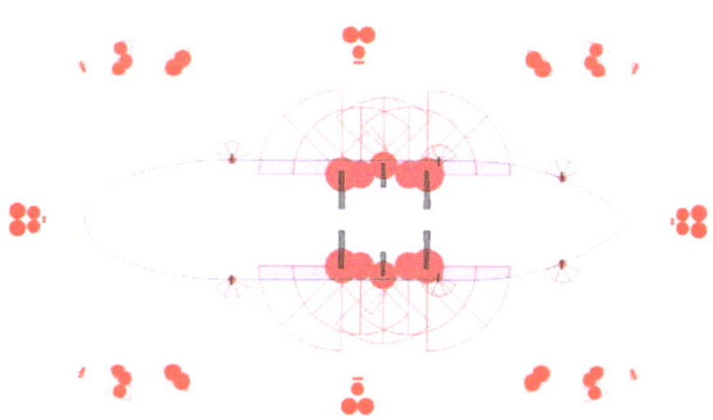

Fuso (1875)
1875 09 01 - 1877 04 14 - 1878 01 01 | Samuda, Poplar

2 | "Kongo" type (1875)
Battery ironclads, Japan

Displacement: 2,200 tons at draft: 5.3 m | Length: 70 m wl | Beam: 12.4 m | Crew: 308 men

Machinery: 6 cylindrical boilers | 1 horizontal compound E * 1 screw | 2,500 HP = 14 knots | Supplies: 280 tons of coal

Armour: iron |
WLBelt: 76(^13) 114(^13) 76(^13)

Armament: 3-170/17, 6-150/25, 6-37/23

Age 15

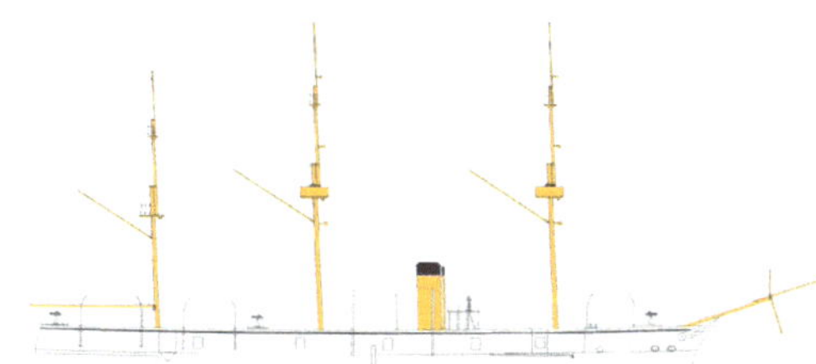

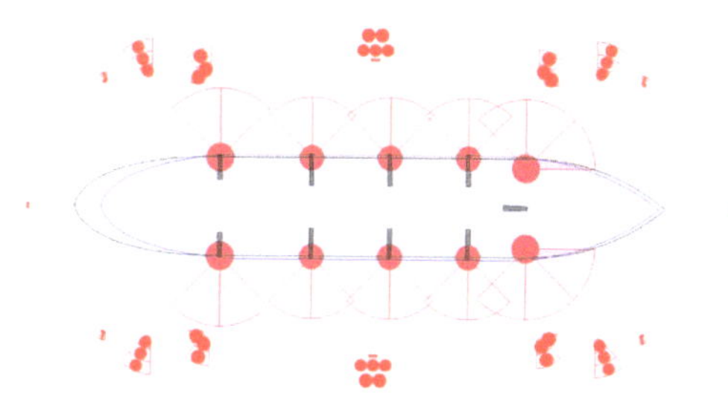

Hiei (1875)
1875 09 01 - 1877 06 12 - 1878 03 01 | Milford Hav., Pembroke

Kongo (1875)
1875 09 01 - 1877 04 01 - 1878 01 01 | Earle, Hull

1 | "Ryujo" type (1864)
Battery ironclad, Japan

Displacement: 1,429 tons at draft: 5.3 m | Length: 65 m pp | Beam: 10.5 m | Crew: 275 men
Machinery: 4 rectangular boilers | 1 compound E * 1 screw | 800 HP = 9 knots | Supplies: app. 135 tons of coal
Armour: iron | Battery: 101 |
WLBelt: 114(^25)
Armament: 2-160/14.8, 10-140/22

Age 26

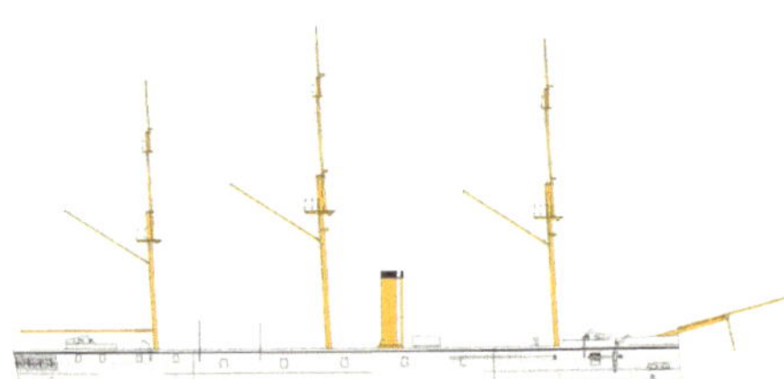

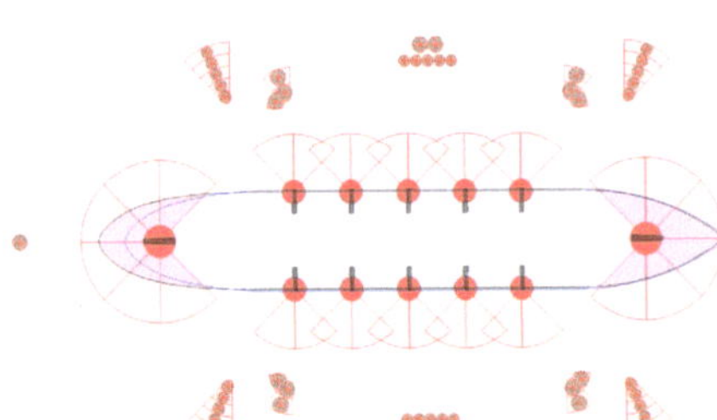

Ryujo (1864)
1864 01 01 - 1864 12 12 - 1869 04 27 | Hall Russell, Aberdeen

0 +1 | "Chiyoda" type (1888)
Belted cruiser, Japan

BUILDING IN BRITAIN

ARMOURED CRUISER

Displacement: 2,400 tons at draft: 4.27 m | Length: 94.49 m wl | Beam: 12.98 m | Crew: 350 men
Machinery: 6 locomotive boilers | 2 * 2,800 HP vertical triple expansion EE * 2 screws | 5,600 HP = 19 knots | Supplies: 240 tons of coal
Armour: steel | CT: 114(^25) |
WLBelt: 0(^38-25) 114(^38-25) 0(^38-25)
Armament: 10-120/40, 14-47/43, 3-25,4/40, 3-356 TT (^)

Age 2

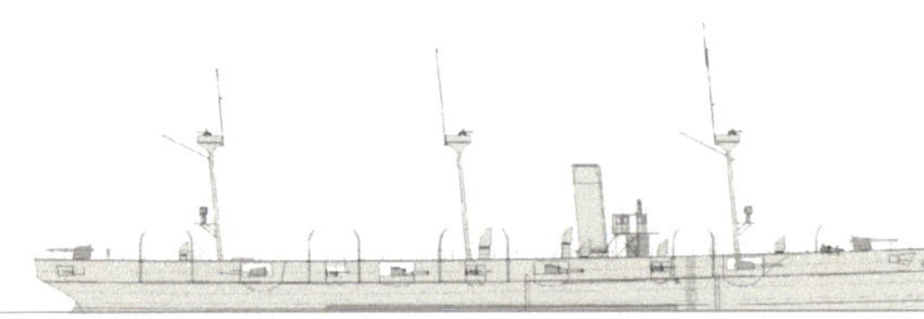

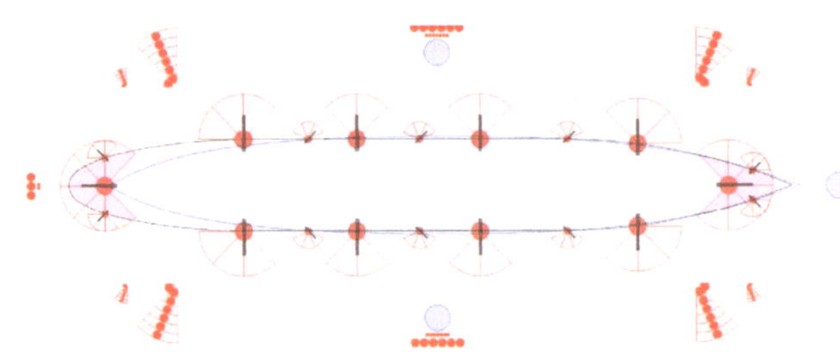

Chiyoda (1888)
1888 11 01 - ***** - ***** | Thompson, Clydebank

PROTECTED CRUISERS

0 +1 | "Matsushima" type (1888)
Protected cruiser 2 class, Japan

BUILDING IN FRANCE

Displacement: 4,217 tons at draft: 6.05 m | Length: 91.81 m wl | Beam: 15.59 m | Crew: 360 men

Machinery: 8 Belleville boilers | 2 * 2,700 HP horizontal expansion EE * 2 screws | 5,400 HP = 16.5 knots | Supplies: 405 / +275 tons of coal

Armour: steel | CT: 152(^102); Barbettes: 305/100; Hatches: 125; Hoists: 100 | WLDeck: 0(^37)

Armament: 1-320/38, 12-120/40, 5-57/43, 12-37/43, 4-350 TT (^)

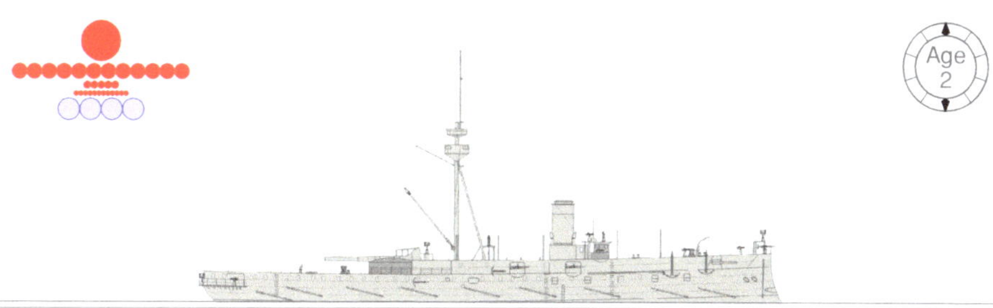

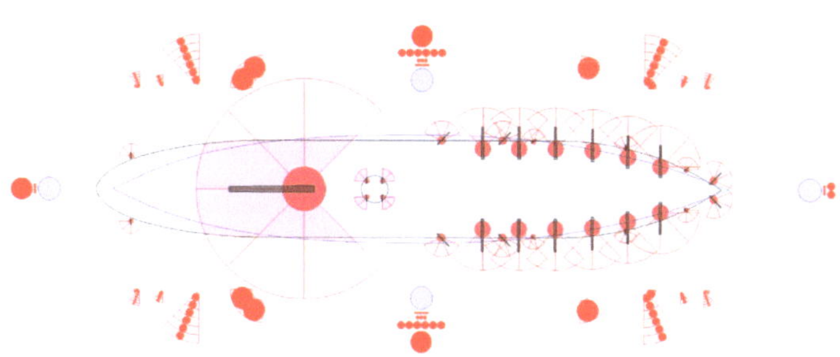

Matsushima (1888)
1888 02 01 - ***** - ***** | Forges et Chantiers de la Méditerranée, La Seyne

0 +2 | "Itsukushima" type (1888)
Protected cruisers 2 class, Japan

1 BUILDING IN FRANCE

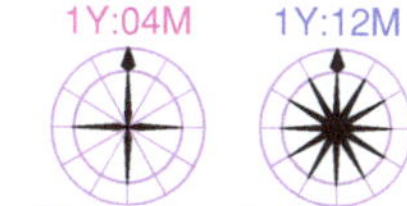

Displacement: 4,217 tons at draft: 6.05 m | Length: 91.81 m wl | Beam: 15.59 m | Crew: 360 men

Machinery: 8 Belleville boilers | 2 * 2,700 HP horizontal expansion EE * 2 screws | 5,400 HP = 16.5 knots | Supplies: 405 / +275 tons of coal

Armour: steel | CT: 152(^102); Barbettes: 305/100; Hatches: 125; Hoists: 100 | WLDeck: 0(^37)

Armament: 1-320/38, 11-120/40, 6-57/43, 11-47/43, 4-350 TT (^)

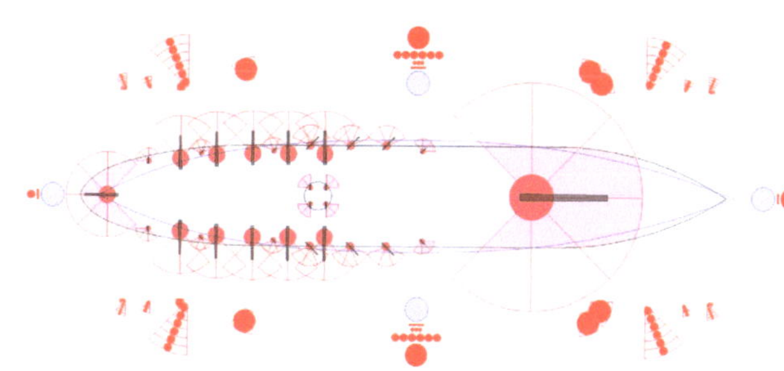

Hashidate (1888)
1888 09 01 - ***** - ***** | Yokosuka DYd, Yokosuka

Itsukushima (1888)
1888 01 01 - 1889 07 11 - ***** | Forges et Chantiers de la Méditerranée, La Seyne

UNPROTECTED CRUISERS

2 | "Naniwa" type (1884)
Protected cruisers 2 class, Japan

Displacement: 3,650 tons at draft: 6.1 m | Length: 91.44 m pp | Beam: 14 m | Crew: 325 men

Machinery: 6 cylindrical boilers | 2 * 3,750 horizontal expansion EE * 2 screws | 7,500 HP = 18 knots | Supplies: 350 / +450 tons of coal

Armour: steel | CT: 51; Gun shields: 37 | WLDeck: 0(/76^51)

Armament: 2-260/35, 6-150/35, 2-57/43, 10-25.4/40 Mg, 4-356 TT (^)

Age 6

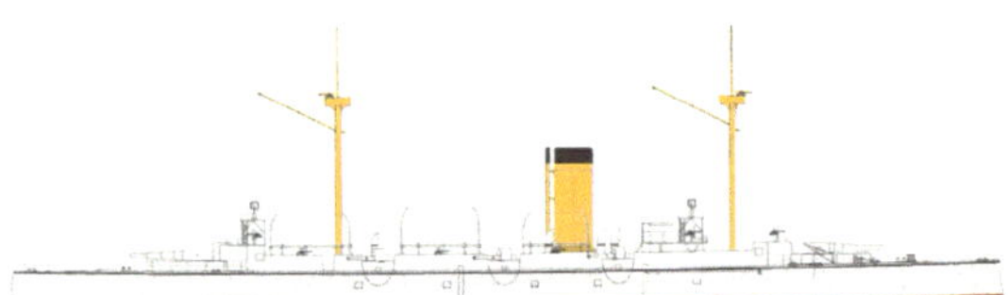

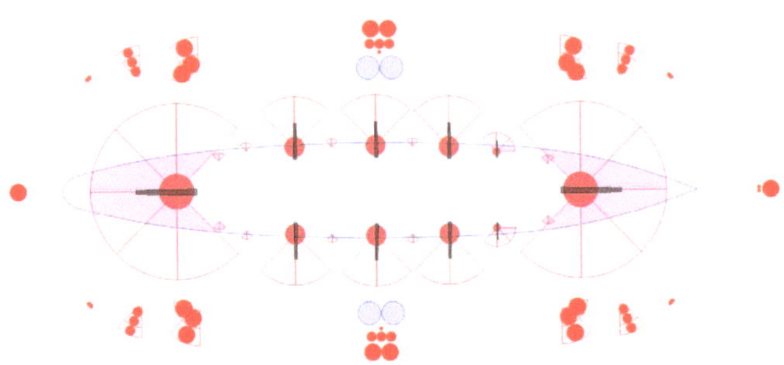

Naniwa (1884)
1884 03 27 - 1885 03 18 - 1885 12 01 | Armstrong, Elswick

Takachiho (1884)
1884 04 10 - 1885 05 16 - 1886 03 26 | Armstrong, Elswick

Sister ship can be found in the United Sates section

0 +1 | "Yaeyama" type (1887)
Unprotected cruiser 2 class, Japan

BUILDING

Displacement: 1,584 tons at draft: 4 m | Length: 96.9 m pp | Beam: 10.5 m | Crew: 217 men

Machinery: 6 cylindrical boilers | 2 * 2,700 HP horizontal expansion EE * 2 screws | 5,400 HP = 21 knots | Supplies: 350 tons of coal

Armour: nil

Armament: 3-120/35, 8-47/43, 2-457,2 TT (^)

Age 17

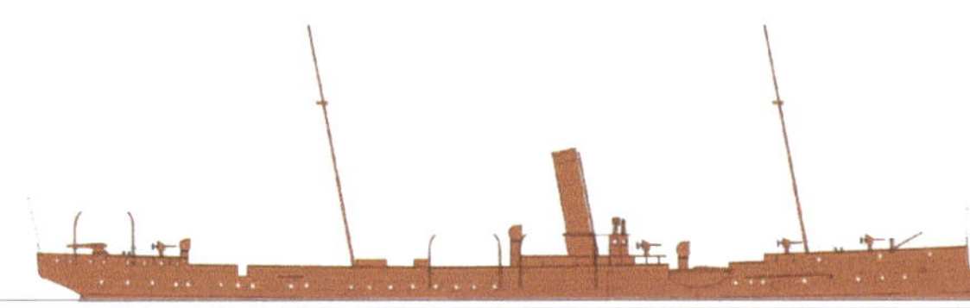

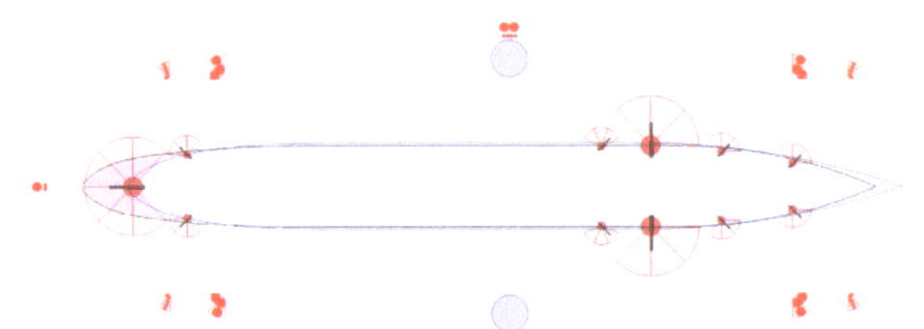

Yaeyama (1887)
1887 06 01 - 1889 05 01 - ***** | Yokosuka DYd, Yokosuka

1 | "Takao" type (1886)
Unprotected cruiser 2 class, Japan

Displacement: 1,750 tons at draft: 4 m, 1,927 tons full | Length: 70.99 m wl | Beam: 10.5 m | Crew: 222 men

Machinery: 2 cylindrical boilers | 2 * 1,660 HP horizontal compound EE * 2 screws | 2,330 HP = 15 knots | Supplies: 300 tons of coal

Armament: 4-150/35, 1-120/35, 3-57/43, 2-25.4/40 Mg, 2-356 TT (^)

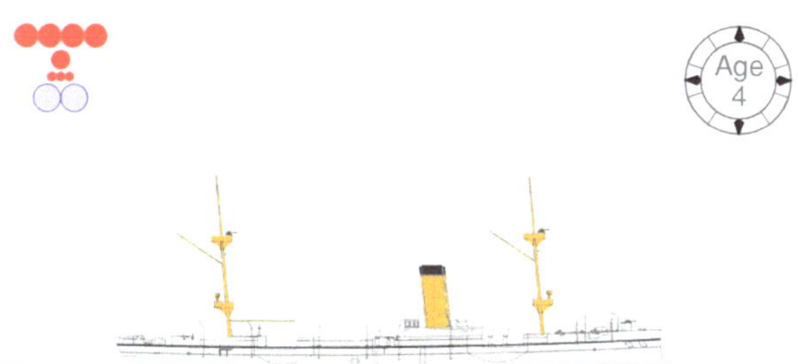

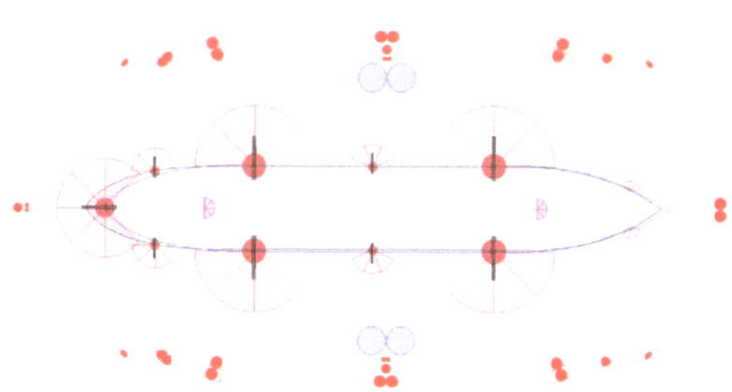

Takao (1886)
1886 10 01 - 1888 10 15 - 1889 11 16 | Yokosuka DYd

1 | "Arturo Prat"/"Tsukushi" type (1879)
Unprotected cruiser 2 class, Japan

Displacement: 1,350 tons at draft: 4.4 m, 1,542 tons full | Length: 64 m pp | Beam: 9.7 m | Crew: 186 men

Machinery: 4 cyl. boilers | 2 *1,440 HP horizontal return connecting rod EE * 2 screws | 2,887 HP = 16.5 knots | Suppl.: 300 tons coal

Armour: steel | CT: 12 | WLDeck: 0(^11)

Armament: 2-254/26, 4-120/32, 2-76/19, 4-25.4/40 Mg

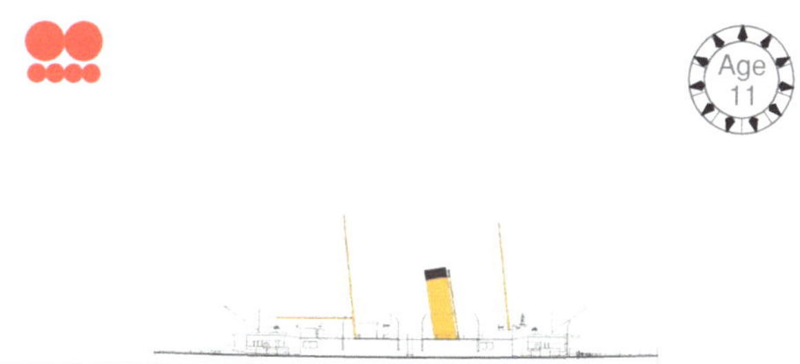

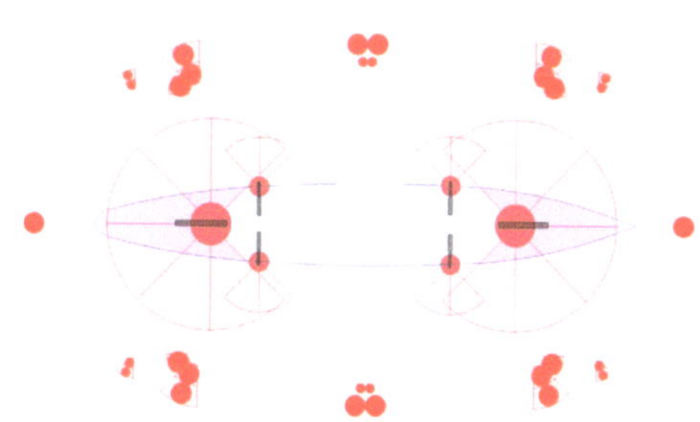

Tsukushi (1879)
1879 10 02 - 1880 08 11 - 1883 06 01 | Armstrong, Elswick

Sister ships can be found in the Chinese section

3 | "Katsuragi" type (1882)
Composite screw corvettes, Japan

WOODEN CRUISERS

Displacement: 1,476 tons at draft: 4.6 m | Length: 67.78 m wl | Beam: 10.7 m | Crew: 231 men

Machinery: 6 cylindrical boilers | 2 * 800 HP horizontal compound direct acting EE * 2 screws | 1;622 HP = 13 knots | Supplies: 145 tons of coal

Armament: 2-170/25, 5-120/35, 1-75/25, 4-25.4/40 Mg, 2-356 TT (^)

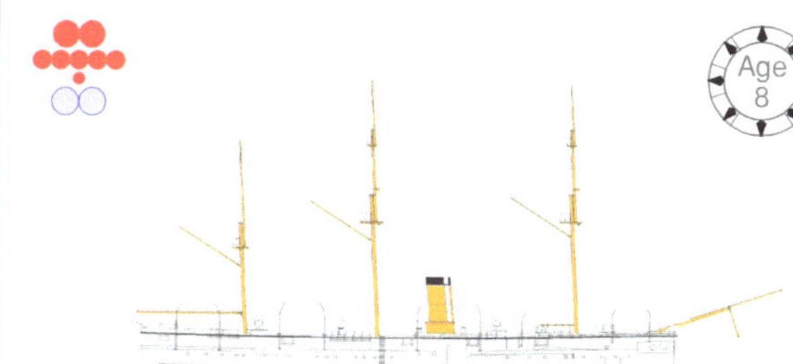

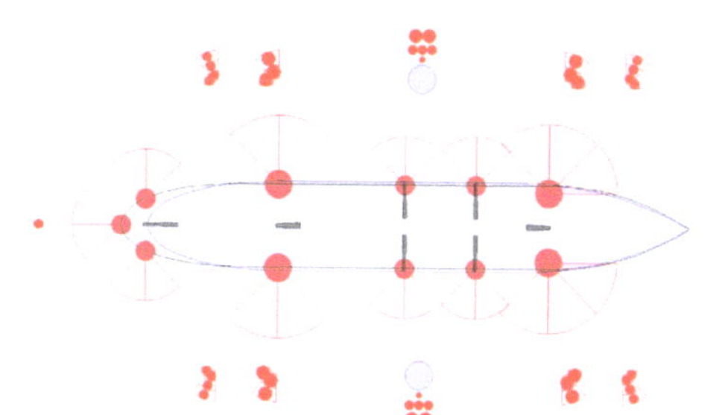

Katsuragi (1882)
1882 12 01 - 1885 03 31 - 1887 10 01 | Yokosuka DYd

Musashi (1884)
1884 10 01 - 1886 03 30 - 1888 02 01 | Yokosuka DYd

Yamato (1883)
1883 02 01 - 1885 04 01 - 1887 10 01 | Onohama, Kobe

2 | "Kaimon" type (1877)
Wooden screw corvettes, Japan

Displacement: 1,350 tons at draft: 4.98 m, 1,429 tons full | Length: 64.26 m pp | Beam: 9.91 m | Crew: 210 men
Machinery: 4 cylindrical boilers | 1 horizontal compound direct acting E * 1 screw | 1,267 HP = 12 knots | Supplies: 197 tons of coal
Armament: 1-170/25, 2-120/25, 1-75/30, 4-25.4/40 Mg

Age 13

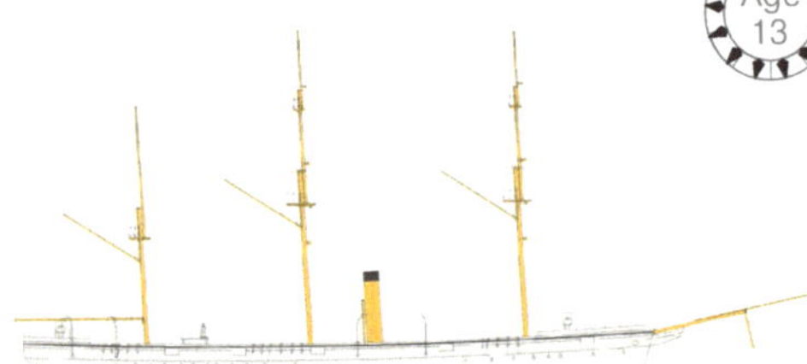

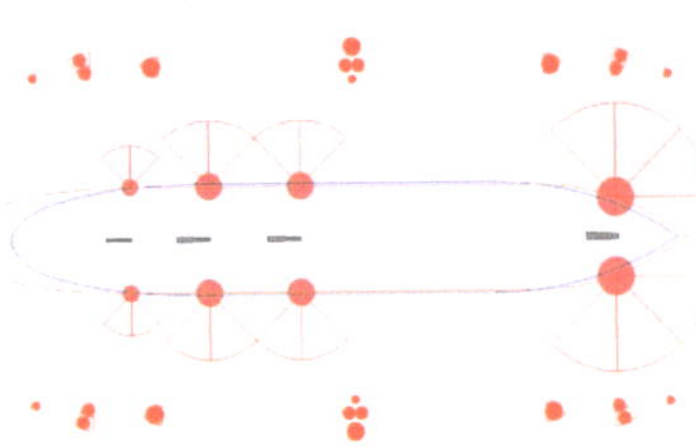

Kaimon (1877)
1877 08 01 - 1882 09 01 - 1884 08 13 | Yokosuka DYd

Tenryu (1878)
1878 01 01 - 1883 09 01 - 1885 03 01 | Yokosuka DYd

1 | "Asama" type (1868)
Wooden screw corvette, Japan

NO PHOTO AVAILABLE

Displacement: 1,423 tons at draft: 4.26 m | Length: 69.72 m pp | Beam: 8.76 m | Crew: 150 men
Machinery: 4 cylindrical boilers | 1 horizontal compound E * 1 screw | 300 HP = 11 knots | Supplies: 197 tons of coal
Armament: 8-170/25, 4-114.5/25

Age 22

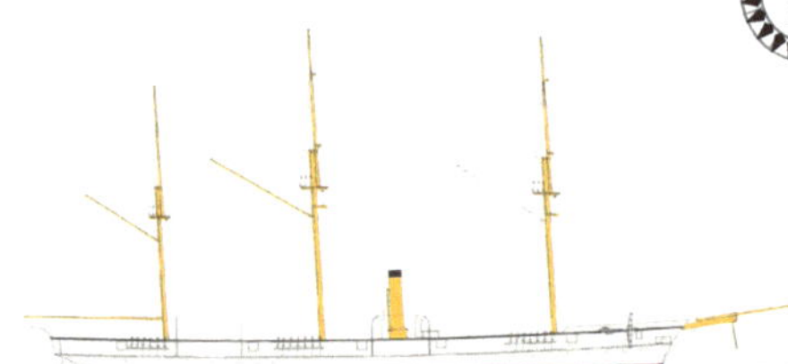

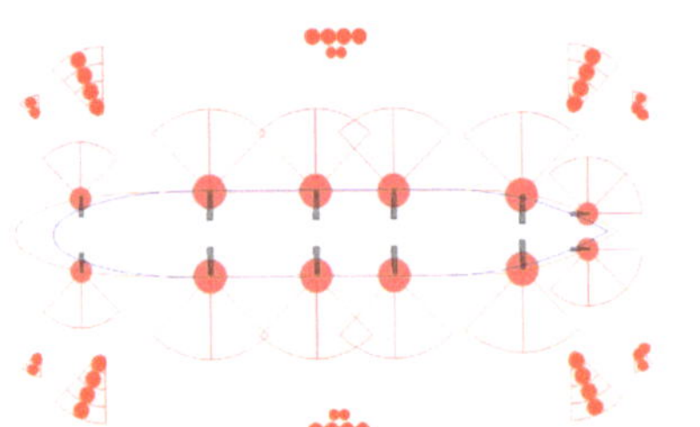

Asama (1868)
1868 01 01 - 1869 01 01 - 1874 07 01 | Arman, Bordeaux

1 | "Nisshin" type (1868)
Wooden screw corvette, Japan

Displacement: 1,490 tons at draft: 4.2 m | Length: 61.9 m pp | Beam: 8.76 m | Crew: 145 men
Machinery: 2 rectangular boilers | 1 horizontal compound reciproc. E * 1 screw | 710 HP = 11 knots | Supplies: 180 tons of coal
Armament: 1-170/22, 6-165/22

Age 22

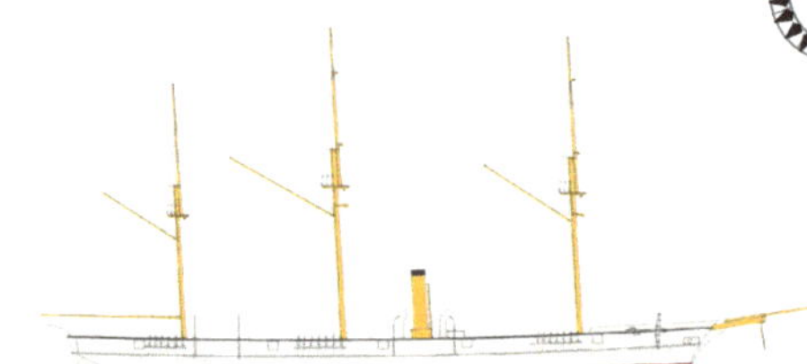

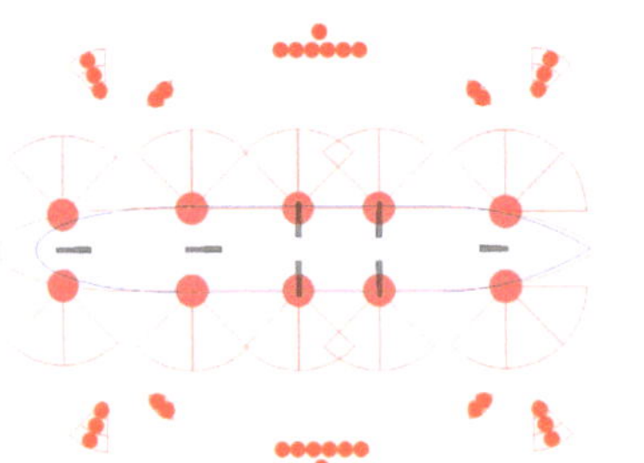

Nisshin (1868)
1868 01 01 - 1869 01 01 - 1870 05 01 | Gips & Son, Dordrecht

1 | "Amagi" type (1875)
Wooden screw corvette, Japan

Displacement: 911 tons at draft: 4.36 m | Length: 65.3 m wl | Beam: 10.89 m | Crew: 159 men

Machinery: 4 rectangular boilers | 1 horizontal compound E * 1 screw | 720 HP = 11 knots | Supplies: 150 tons of coal

Armament: 1-170/22, 4-120/25, 3-75/21, 3-25.4/40 Mg

Age 15

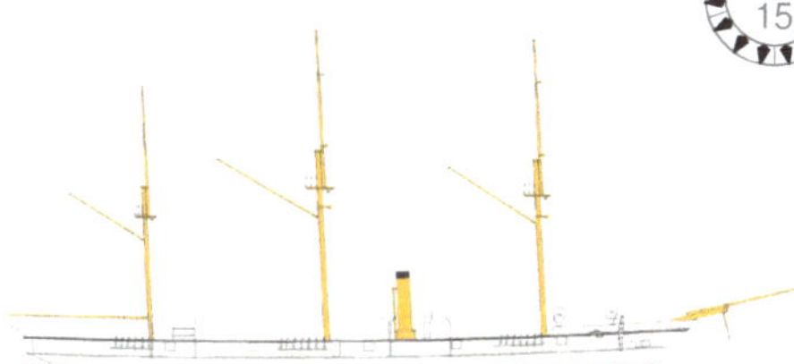

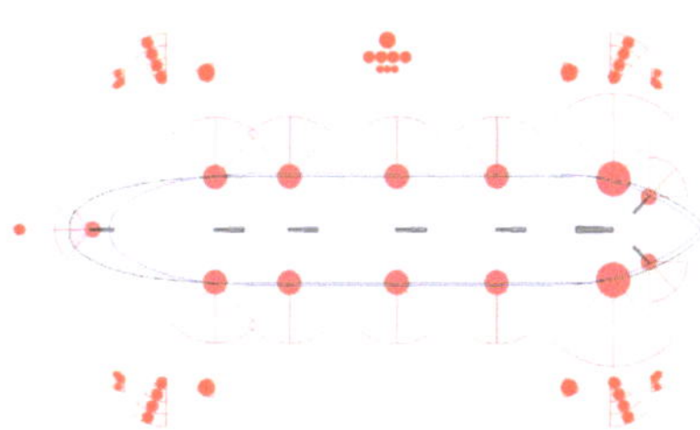

Amagi (1875)
1875 09 01 - 1877 03 01 - 1878 04 01 | Yokosuka DYd

1 | "Jingei" type (1873)
Wooden paddle corvette, Japan

NO PHOTO AVAILABLE

Displacement: 1,465 tons at draft: 4.39 m | Length: 75.9 m wl | Beam: 9.5 m | Crew: 170 men

Machinery: 4 rectangular boilers | 1 recyprocating direct acting E * side wheels | 1,450 HP = 14 knots | Suppl.: app. 160 tons of coal

Armament: 2-120/25

Age 17

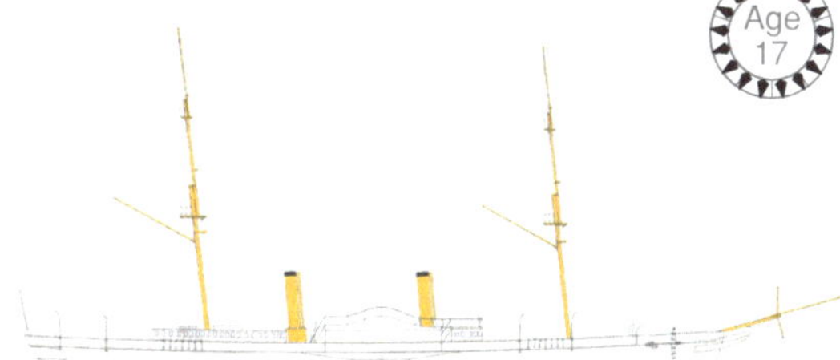

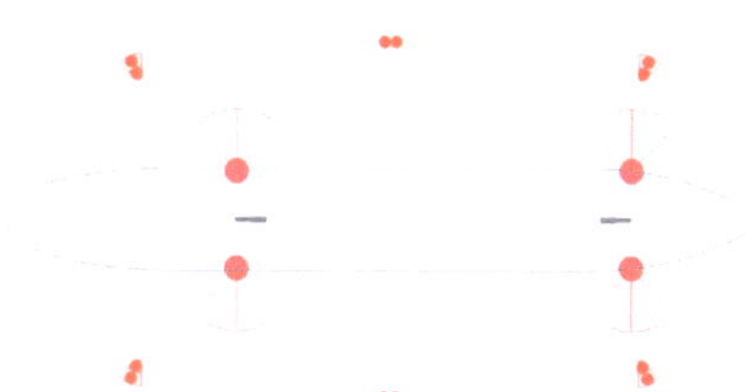

Jingei (1873)
1873 09 01 - 1875 01 01 - 1881 01 01 | Yokosuka DYd

1 | "Spartan"/"Tsukuba" type (1852)
Wooden screw corvette, Japan

Displacement: 1,978 tons at draft: 5.48 m | Length: 60.3 m wl | Beam: 10.6 m | Crew: 301 men

Machinery: 4 rectangular boilers | 1 horizontal direct acting E * 1 screw | 526 HP = 10 knots | Supplies: app. 200 tons of coal

Armament: 6-127/19

Age 48

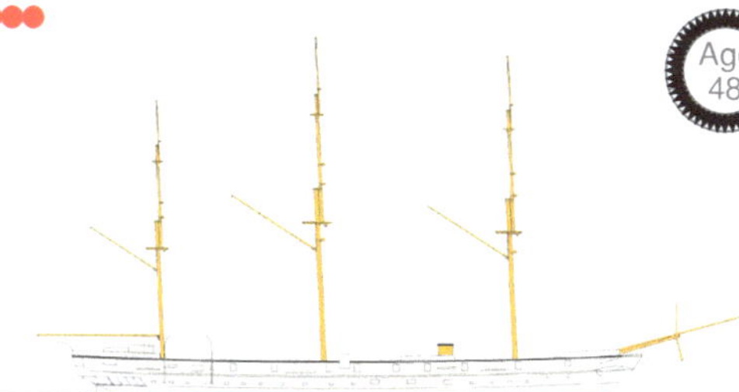

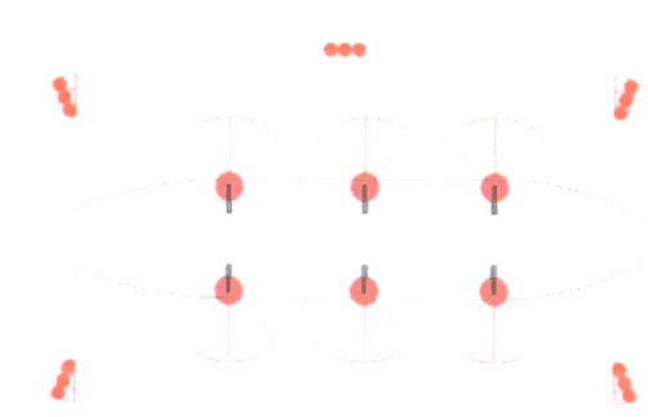

Tsukuba (1852)
1852 01 01 - 1853 04 09 - 1854 01 01 | Moulmein, Burma

TORPEDO SHIPS

0 +3 | "75"/"5" modified type (1886)
Torpedo boats, France

BUILDING IN FRANCE

0Y:0M 0Y:0M 0Y:0M

5 6 7

Displacement: 54 tons at draft: 0.89 m | Length: 35 m oa | Beam: 3.35 m | Crew: 16 men

Machinery: 1 locomotive boiler | 1 horizontal E * 1 screw | 525 HP = 19 knots | Supplies: 3 tons of coal

Armament: 1-37/23, 2-356 TT (1*1 trainable, 1 >)

Age 0

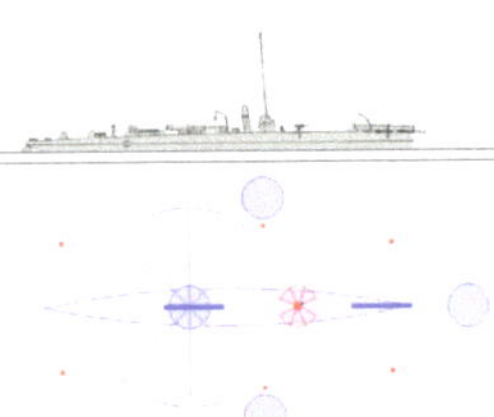

5 (1890)
1890 01 01 - ***** - ***** | Schneider, Chalon

6 (1890)
1890 01 01 - ***** - ***** | Schneider, Chalon

7 (1890)
1890 01 01 - ***** - ***** | Schneider, Chalon

Sister ships can be found in the Romanian and French sections

1 | "Kotaka" type (1886)
Torpedo boat, Japan

Displacement: 203 tons at draft: 1.7 m | Length: 50.29 m pp | Beam: 5.79 m | Crew: 20 men

Machinery: 2 locomotive boilers | 2 * 600 HP compound EE * 2 screws | 1,217 HP = 19 knots | Supplies: 30 tons of coal

Armament: 4-35.6/40 Mg, 6-356 TT (2*2 trainable, 2 >>)

Age 4

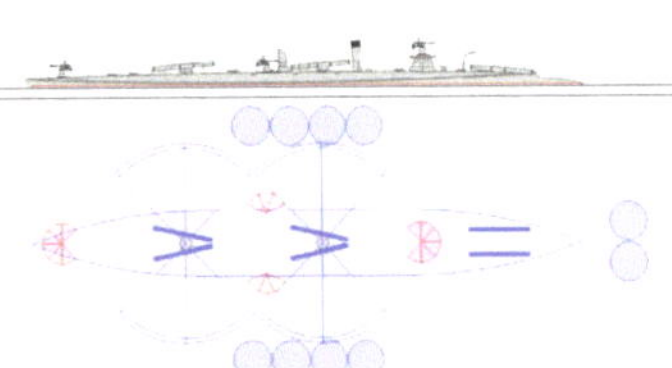

Kotaka (1886)
1886 01 01 - 1887 01 01 - 1888 10 01 | Yarrow, Poplar

4 | "Batum"/"1" type (1879)
Torpedo boats, Russia

Displacement: 45 tons at draft: 0.94 m | Length: 30.48 m oa | Beam: 3.68 m | Crew: 13 men

Machinery: 1 locomotive boiler | 1 horizontal compound E * 1 screw | 430 HP = 20 knots | Supplies: 8.3 tons of coal

Rearmed: 1-25.4/40 MG, 1-356 TT (1 trainable)

Age 11

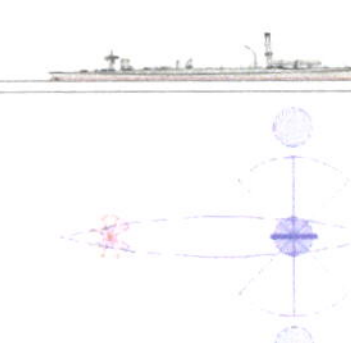

1 (1879)
1879 01 01 - 1879 06 06 - 1880 01 01 | Yarrow, Poplar

2 (1879)
1879 01 01 - 1879 06 06 - 1880 01 01 | Yarrow, Poplar

3 (1879)
1879 01 01 - 1879 06 06 - 1880 01 01 | Yarrow, Poplar

4 (1879)
1879 01 01 - 1879 06 06 - 1880 01 01 | Yarrow, Poplar

Sister ships can be found in the Argentinean, Austria-Hungarian, Brazilean, British, Chilean, Dutch, Greek, Italian and Russian sections

SLOOPS & GUNBOATS

0 +1 | "Oshima" type (1889)
Gunboat, Japan

BUILDING

0Y:5M

Oshima

Displacement: 630 tons at draft: 2.75 m | Length: 53.5 m pp | Beam: 8 m | Crew: 130 men

Machinery: 2 cyl. boilers | 2 * 600 HP vertical triple expansion EE * 2 screws | 1,200 HP = 16 knots | Supplies: 140 tons of coal

Armament: 4-120/43, 4*47/43

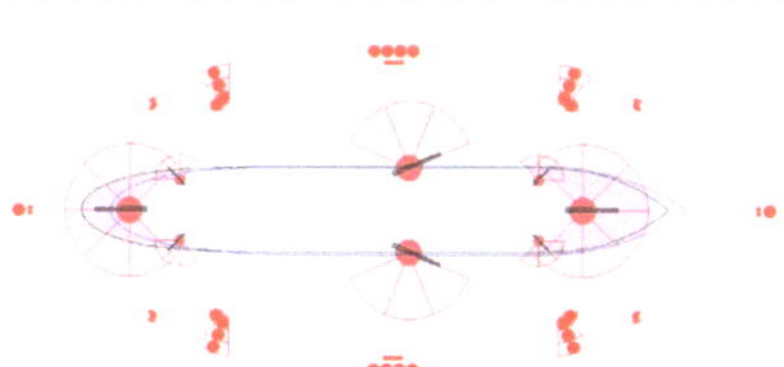

Oshima (1889)
1889 08 01 - ***** - ***** | Onohama, Kobe

3 +1 | "Maya" type (1885)
Gunboats, Japan

BUILDING

3Y:7M

Akagi

Displacement: 612 tons at draft: 2.95 m | Length: 47 m pp | Beam: 8.2 m | Crew: 104 men

Machinery: 2 cyl. boilers | 2 * 480 HP horizontal compound direct acting EE * 2 screws | 960 HP = 12 knots | Suppl.: 60 tons of coal

Armament: 1-210/22, 1-120/30

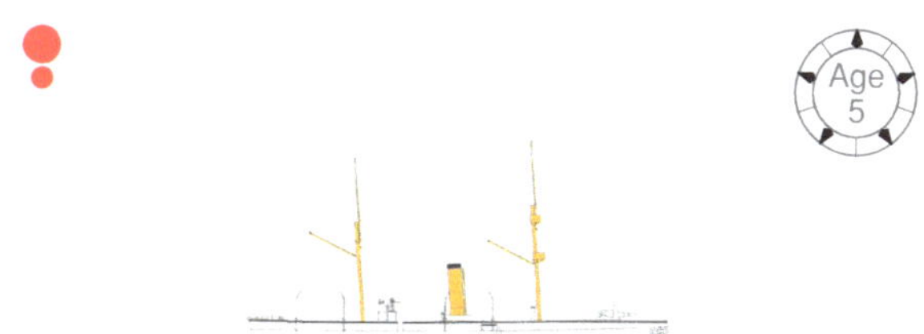

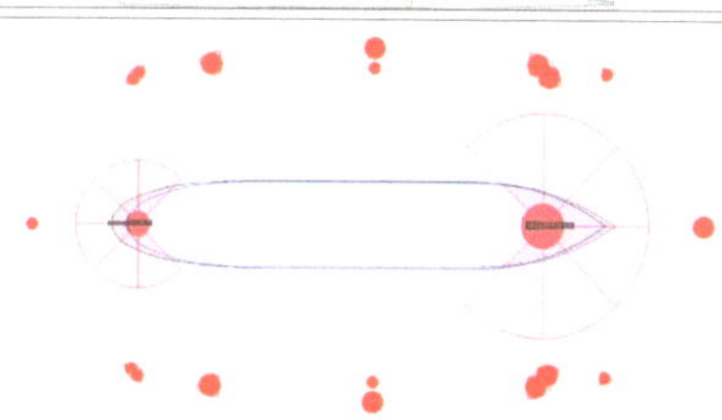

Akagi (1886)
1886 06 01 - 1888 08 01 - ***** | Onohama, Kobe

Atago (1886)
1886 07 01 - 1887 06 01 - 1889 03 01 | Yokosuka DYd,

Chokai (1885)
1885 12 01 - 1887 09 20 - 1888 10 01 | Ishiwakajima, Tokyo

Maya (1885)
1885 05 01 - 1886 08 18 - 1887 12 01 | Onohama, Kobe

3 +1 | "Maya" type (1885)
Gunboats, Japan (continued)

Atago

1 | "Hosho" type (1867)
Wooden gunboat, Japan

Displacement: 316 tons at draft: 2.05 m | Length: 43.9 m pp | Beam: 6.7 m | Crew: 65 men

Machinery: 2 rectangular boilers | 1 horizontal direct acting E * 1 screw | 240 HP = 11 knots | Supplies: app. 81 tons of coal

Armament: 1-177.8/14.8 MLR, 1-140/25

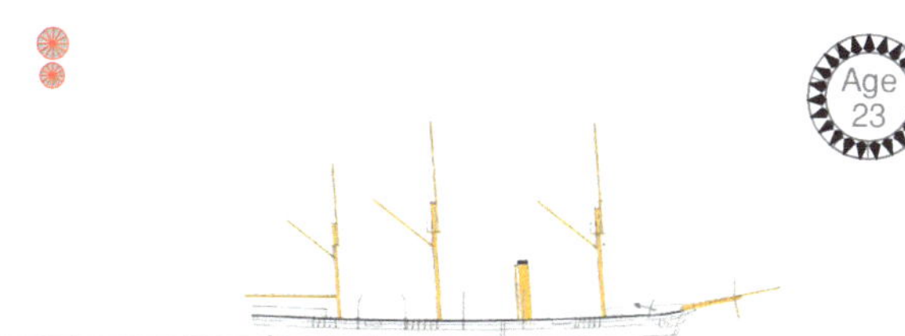

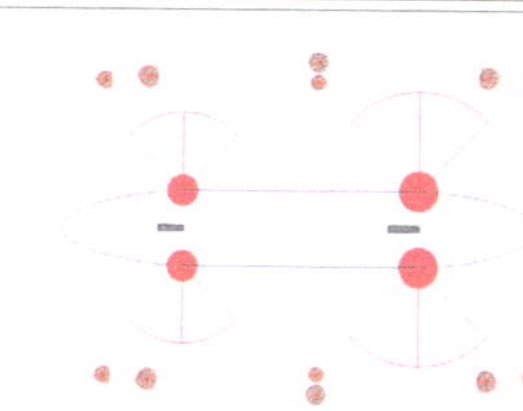

Hosho (1867)
1867 01 01 - 1868 01 01 - 1869 01 01 | Hall Russell, Aberdeen

1 | "Banjo" type (1877)
Wooden gunboat, Japan

Displacement: 656 tons at draft: 3.9 m | Length: 46.9 m oa | Beam: 7.88 m | Crew: 112 men

Machinery: 4 rectangular boilers | 1 horizontal compound E * 1 screw | 659 HP = 10 knots | Supplies: 60 / +60 tons of coal

Armament: 1-150/25, 1-120/25, 2-75/30, 3-25.4/40 Mg

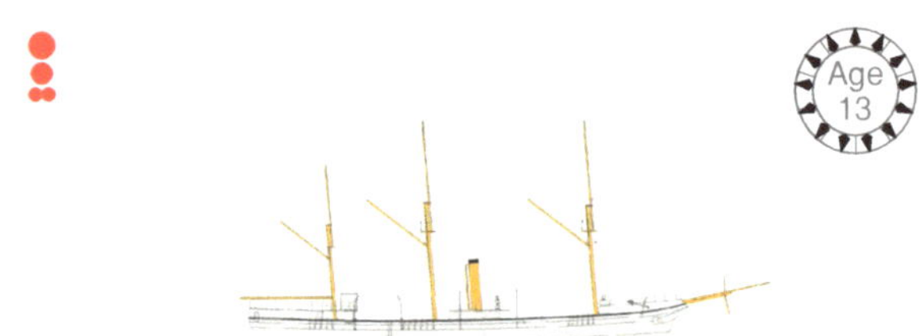

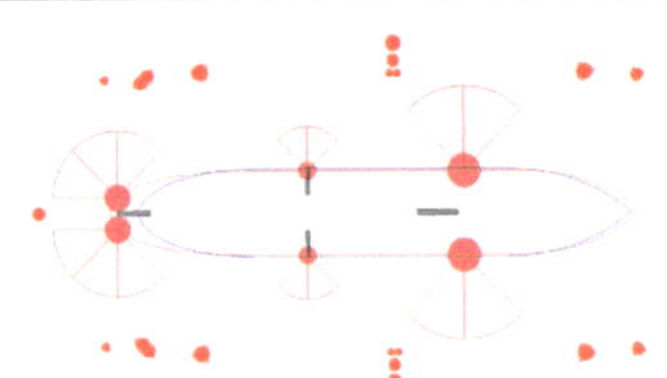

Banjo (1877)
1877 02 01 - 1878 07 01 - 1880 08 01 | Yokosuka DYd

DANISH FLEET

National flag

Fleet age 10

Ensign

Jack

Admiral

Months derivation -29

Warships built from 1858 01 01 to 1890 01 01

49 warships = 40,604 tons = 829 tons / average ship

Fleet age - 1880 04 07

1889 fleet largest ship: Helgoland (1876) coast defence battleship

Danish naval construction

Copenhagen dockyard

32.1 tons / average ship / month

1890-Jan-01
4 ships building

t/ship/m

4,553 t.
tons / month = 128.6

Ship	t	t/m
Valkyrien	2,972	78.0
Hjaelperen	200	4.2
Hekla	1,300	41.9
Springeren	81	4.5

Copenhagen, Denmark

32+ years of naval construction
24 ships | 36,083 tons

The plant's largest ship:
Helgoland (1876) coast defence battleship
5,332 tons

The plant's most effectively built ship:
Helgoland (1876) coast defence battleship
137.0 tons/month

Specialisation in:
Coast defence battleships | 2 ships | 8,724 tons
Monitors | 3 ships | 7,531 tons
Protected cruisers | 3 ships | 6,734 tons

Danish fleet effective list

2 boardside ironclads

Danmark

Dannebrog
Training ship

6 coast defence armoured ships

Iver Hvitfeldt

Helgoland

Odin

Gorm

Lindormen

Rolf Krake

1 protected cruiser

Tordenskjold

1 unprotected & 4 wooden cruisers

Fyen

St. Thomas

Ingolf

Dagmar

Jylland

22 torpedo boats

Havhesten

Narhvalen

Soloven

Storen

Hvalrossen

Delfinen

Svaerdfisken

Soulven

Hajen

12

13

10

11

8

9

6

7

4

5

P 2

>>> torpedo boats (continued)

P 3

1

11 gunboats

Gronsund
NO PLAN AVAILABLE

Guldorg-sund
NO PLAN AVAILABLE

Little Belt

Oresund

Store Belt

Falster

Moen

Diana
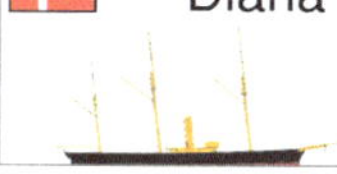

Fylla
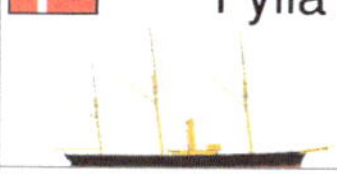

Absalon

Esbern Snare
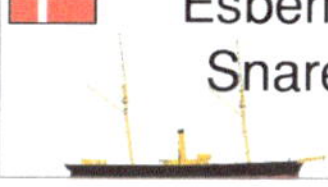

Danish section ship names glossary

1 (TB2)
10 (TB2)
11 (TB2)
12 (TB2)
13 (TB2)
4 (TB2)
5 (TB2)
6 (TB2)
7 (TB2)
8 (TB2)
9 (TB2)

A
Absalon (GB)

D
Dagmar (CZ)
Danmark (BO),
Dannebrog (accomS)
Delfinen (TB1)
Diana (GB)

E
Esbern Snare (GB)

F
Falster (GB)
Fyen (CS)
Fylla (GB)

G
Gorm (MM)
Gronsund (GB)
Guldorgsund (GB)

H
Hajen (TB1)
Havhesten (TB1)
Hekla (CP)
Helgoland (CDB)
Hjaelperen (ML)
Hvalrossen (TB1)

I
Ingolf (CZ)
Iver Hvitfeldt (CDB)

J
Jylland (CW)

L
Lindormen (MM)
Little Belt (GB)

M
Moen (GB)

N
Narhvalen (TB1)

O
Odin (MM)
Oresund (GB)

P
P 2 (TB2)
P 3 (TB2)

R
Rolf Krake (MM)

S
Soloven (TB1)
Soulven (TB1)
Springeren (TB1)
St. Thomas (CW)
Store Belt (GB)
Storen (TB1)
Svaerdfisken (TB1)

T
Tordenskjold (CP)

V
Valkyrien (CP)

Danish fleet naval guns

Calibre (mm)	Mark	Calibre (in)/pdr	Gun type	First Introduced	Weight (kg)	Gun lenght (in/cal)	Bore lenght (cal)	Charge weight (lbs)	Muzzle energy (ft/t)	Projectile weight (lbs)	Muzzle velocity (ft/s)	Wrought iron perforation (in)
355.0	Krupp	13.98"	BLR		51 300	349.2 / 25.00	21.80	330.7	24 910	1 157.4	1 762	24.8
305.0	Krupp	12.01"	BLR		35 400	264.0 / 22.00	18.90	180.2	14 110	725.3	1 675	20.0
260.0	Krupp	10.24"	BLR		27 600	393.6 / 38.45	32.00	191.8	12 770	451.9	2 018	20.1
260.0	Krupp	10.24"	BLR		21 600	225.2 / 22.00	19.00	112.4	8 428	451.9	1 640	16.7
254.0	II	10"	MLR	1868	18 289	180.0 / 18.00	14.55	70.0	5 408	410.0	1 379	13.1
210.0	Krupp	8.24"	BLR		13 300	288.5 / 35.00	35.00	105.8	6 745	451.9	1 640	16.7
203.2	III	8"	MLR	1862	9 144	144.0 / 18.00	14.75	35.0	2 398	179.0	1 390	9.7
152.4	VI	6"	BLR	1873	5 080	173.5 / 28.92	26.00	45.0	2 424	100.0	1 870	11.9
152,4		6"	BLR		2 460	114.0 / 19.00	16.70	60.6	-	55.1	1 076	-
150.0	Krupp	5.9"	BLR		4 700	205,2 / 35,00	32.20	41.9	2 784	112.4	1 890	12.8
150.0	Krupp	5.9"	BLR	1865	3 940	151.6 / 25.67	23.00	20.9	1 447	69.4	1 558	9.1
150.0	Krupp	5.9"	BLR	1865	3 500	128.4 / 21.72	19.10	21.8	1 418	69.4	1 542	8.8
140.0		5.5"	BLR		2 690	123.4 / 22.39	21.30			50.7	1 900	-
120.0	Krupp	4.7"	BLR		2 130	141.6 / 29.97	27.30	17.4		57.3	1 720	-
120.0	Krupp	4.7"	BLR	1859	1 380	115.2 / 25.00	-	9.9	680	36.1	1 493	7.0
86.0	Krupp	3 pr	BLR		490	82.8 / 29.97	21.30	15.2	-	15.2	1 457	-
57.0	I & II	6 pr	BLR	1876	406	97.6 / 43.50	40.00	1.2	345	6.0	1 818	3.7
37.0	Hotchkiss R	1 pr	BLR	1873	209	33.5 / 23.00	29.13	0.2	-	1.1	1 450	-

WARSHIPS SECTION

Ship classes	Building (exports including)	Active
Battleships		
Boardside ironclads		1
Armoured cruisers		
Coast defence armoured ships		6
Protected cruisers	2	1
Unprotected cruisers		1
Torpedo gunboats		
Torpedo boats 1 class	1	9
Torpedo boats 2 class		13
Submarines		
Wooden screw cruisers		2
Wooden screw sloops		2
Gunboats		
Gunboats coastal		11
Minelayers	1	

1 | "Iver Hvitfeldt" type (1884)
Coast defence battleship, Denmark

Displacement: 3,392 tons at draft: 5.59 m | Length: 74 m pp | Beam: 15.11 m | Crew: 277 men

Machinery: 8 cylindrical boilers | 2 * 2,550 HP vertical compound EE * 2 screws | 5,100 HP = 15.25 knots | Supplies: 290 tons of coal

Armour: steel | CT: 115; Barbettes: 215 |
WLBelt: 0(^53) 175(^53) 295(^53) 175(^53) 0(^53)

Armament: 2-260/35, 4-120/30, 2-57/43, 12-37/23R, 2-381 TT (^), 2-356 TT (^)

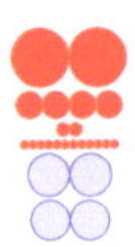

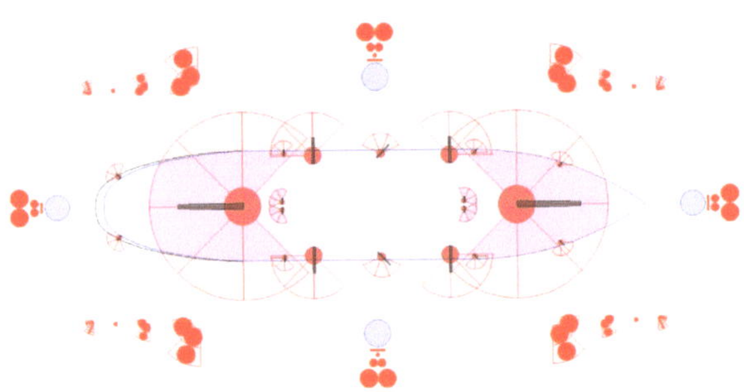

Iver Hvitfeldt (1884)
1884 04 09 - 1886 04 14 - 1887 06 01 | Copenhagen DYd, Copenhagen

1 | "Helgoland" type (1876)
Coast defence battleship, Denmark

Displacement: 5,332 tons at draft: 5.89 m | Length: 79.12 m oa | Beam: 18.03 m | Crew: 331 men

Machinery: 8 cylindrical boilers | 2 * 2,000 HP horizontal return connecting rod EE * 2 screws | 4,000 HP = 13.75 knots | Supplies: 224 tons of coal

Armour: iron | CT: 305; Barbette: 254(^25); Battery: 254 |
WLBelt: 203(^25) 305(^25) 203(^25)

Armament: 1-305/22, 4-260/22, 5-120/25, 10-37/23R, 2-381 TT (v), 3-356 TT (^)

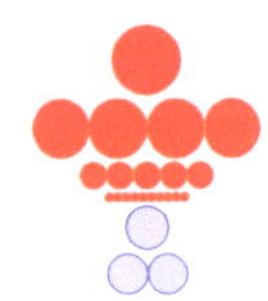

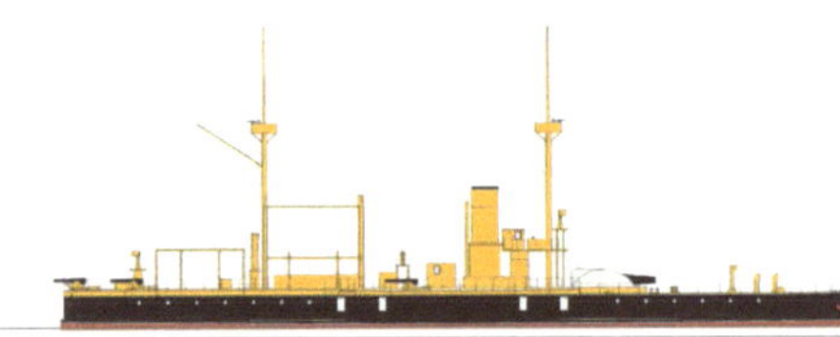

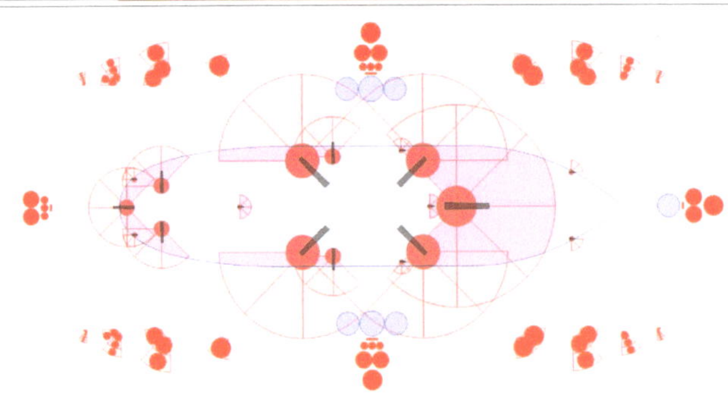

Helgoland (1876)
1876 05 20 - 1878 05 09 - 1879 08 20 | Copenhagen DYd, Copenhagen

MONITORS

1 | "Odin" type (1871)
Coast defence battery, Denmark

Displacement: 3,170 tons at draft: 5 m | Length: 73.4 m pp | Beam: 14.78 m | Crew: 206 men

Machinery: 4 cyl. boilers | 2 * 1,150 HP horizontal return connecting rod EE * 2 screws | 2,300 HP = 12 knots | Supplies: 177 tons of coal

Armour: iron | CT: 142; Battery: 203 | WLBelt: 101(^13), 203(^13) 101(^13)

Armament: 4-254/14.5 MLR, 4-87/24, 4-37/23R, 2-37/23

Age 19

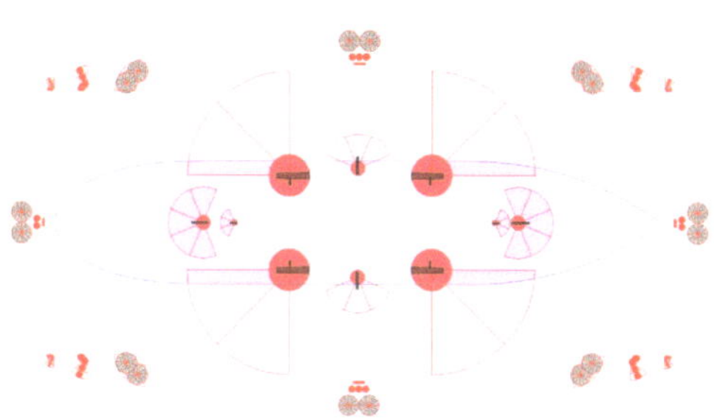

Odin (1871)
1871 04 13 - 1872 12 12 - 1874 09 07 | Copenhagen DYd

1 | "Gorm" type (1867)
Monitor, Denmark

Displacement: 2,313 tons at draft: 4.37 m | Length: 71.11 m pp | Beam: 12.19 m | Crew: 160 men

Machinery: 4 rect. boilers | 2 * 800 HP horizontal return connecting rod EE * 2 screws | 1,600 HP = 12.5 knots | Suppl.: 113 tons of coal

Armour: iron | Turret: 203 | WLBelt: 178

Armament: 2-254/14.5 MLR, 3-47/43, 4-37/23

Age 23

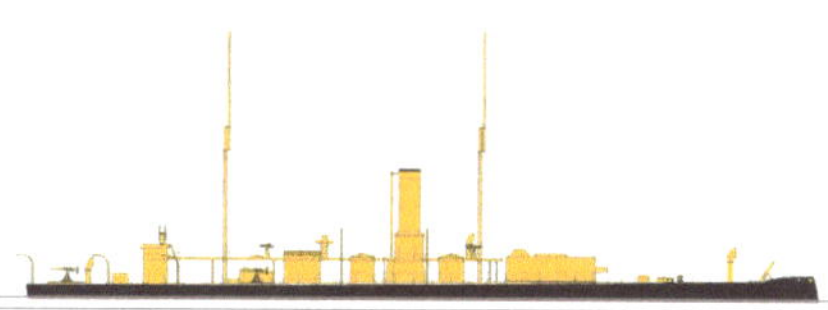

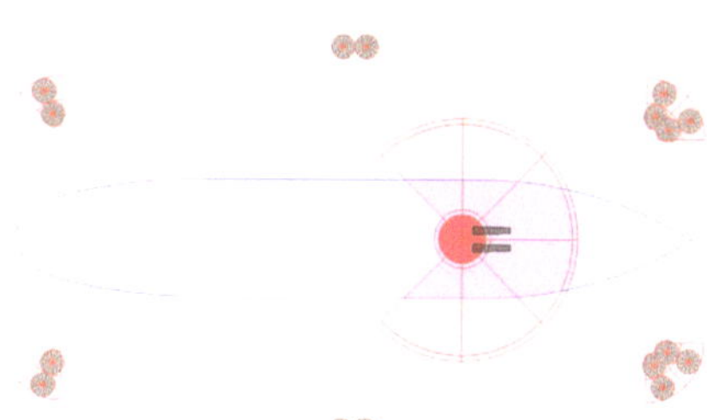

Gorm (1867)
1867 11 18 - 1870 05 17 - 1871 06 23 | Copenhagen DYd

1 | "Lindormen" type (1866)
Monitor, Denmark

Displacement: 2,048 tons at draft: 4.44 m | Length: 66.42 m oa | Beam: 11.99 m | Crew: 150 men

Machinery: 4 rect. boilers | 2 * 750 HP horizontal return connecting rod EE * 2 screws | 1,500 HP = 12 knots | Supplies: 125 tons of coal

Armour: iron | Turret: 140 | WLBelt: 127

Armament: 2-228.6/13.9, 4-37/23R

Age 24

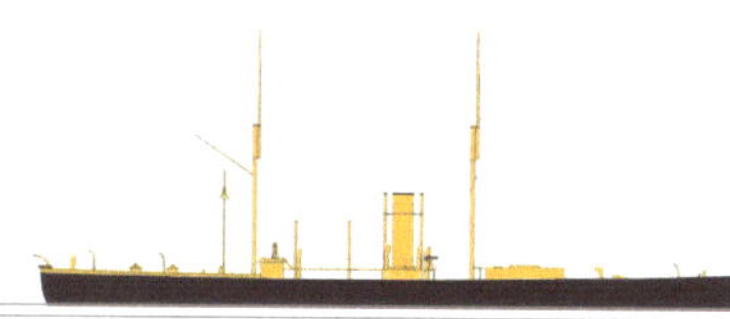

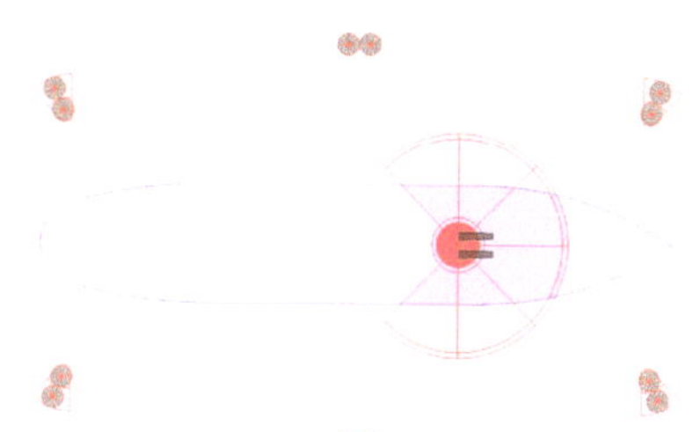

Lindormen (1866)
1866 07 20 - 1868 08 08 - 1869 08 15 | Copenhagen DYd

BATTERY IRONCLADS

1 | "Rolf Krake" type (1862)
Monitor, Denmark

Displacement: 1,320 tons at draft: 3.2 m | Length: 56 m pp | Beam: 11.63 m | Crew: 150 men

Machinery: 2 rectangular boilers | 1 single expansion E * 1 screw | 750 HP = 9.2 knots | Supplies: 135 tons of coal

Armour: iron | Turrets 110 | WLBelt: 110(^13)

Armament: 2-203.2/14.8 MLR, 2-100/24, 4-37/23R

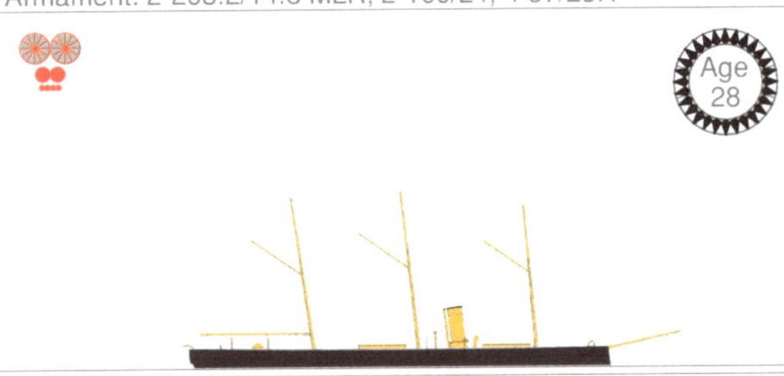

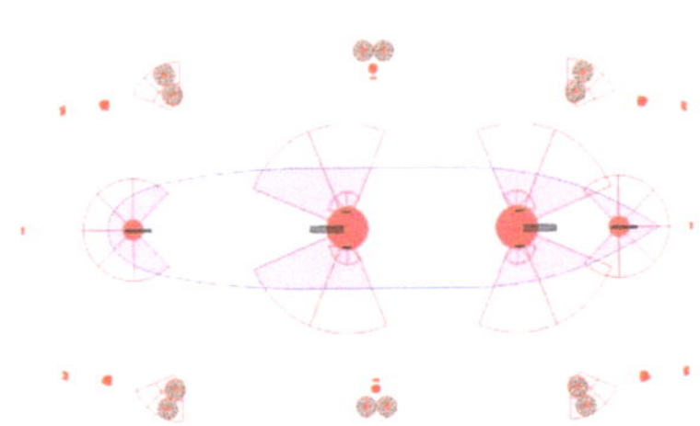

Rolf Krake (1862)
1862 01 01 - 1863 05 06 - 1863 07 01 | Napier, Govan

1 | "Danmark" type (1862)
Battery ironclad, Denmark

Displacement: 4,670 tons at draft: 5.94 m | Length: 82.29 m pp | Beam: 15.24 m | Crew: 530 men

Machinery: 4 rectangular boilers | 1 single expansion E * 1 screw | 1,000 HP = 8.5 knots | Supplies: 625 tons of coal

Armour: iron | Battery: 0 =89= 114 =89= 0 | WLBelt: 89 114 89

Rearmed: 12-203.2/14.8 MLR, 12-140/14.8 MLR

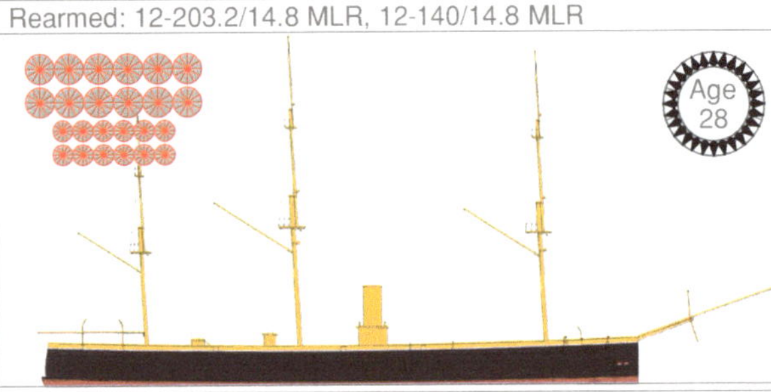

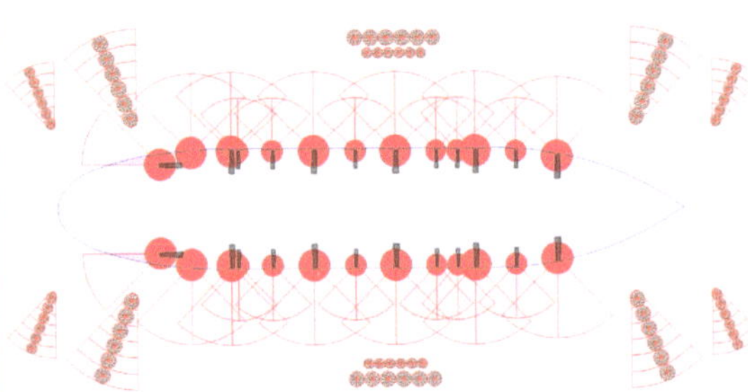

Danmark (1862)
1862 09 09 - 1864 02 23 - 1864 10 01 | Thompson, Clydebank

1 | "Dannebrog" type (1848/1862)
Battery ironclad, Denmark

Displacement: 3,057 tons at draft: 6.76 m | Length: 56.5 m pp | Beam: 16.3 m | Crew: 350 men

Machinery: 4 rectangular boilers | 1 single expansion E * 1 screw | 1,150 HP = 9 knots | Supplies: app. 400 tons of coal

Armour: iron | Battery: 114 | WLBelt: 114

Armament: disarmed

Nil

Disarmed

Dannebrog (1848/1862)
1848 04 25 - 1850 09 25 - 1853 06 17 | Copenhagen DYd

PROTECTED CRUISERS

0 +1 | "Hekla" type (1889)
Protected cruiser 2 class, Denmark

BUILDING

0Y:8M

Valkyrien

Displacement: 1,300 tons at draft: 3.45 m | Length: 70.58 m pp | Beam: 10.36 m | Crew: 156 men

Machinery: 8 Thornycroft boilers | 2 * 1,500 HP triple expansion EE * 2 screws | 3,000 HP = 17 knots | Supplies: 113 tons of coal

Armour: steel | CT: 25(^25) | WLDeck: 0(/45^25)

Armament: 2-150/35, 4-57/43, 6-37/23R, 4-381 TT (^)

Age 1

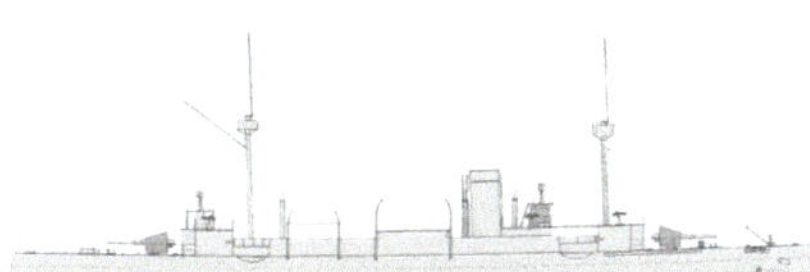

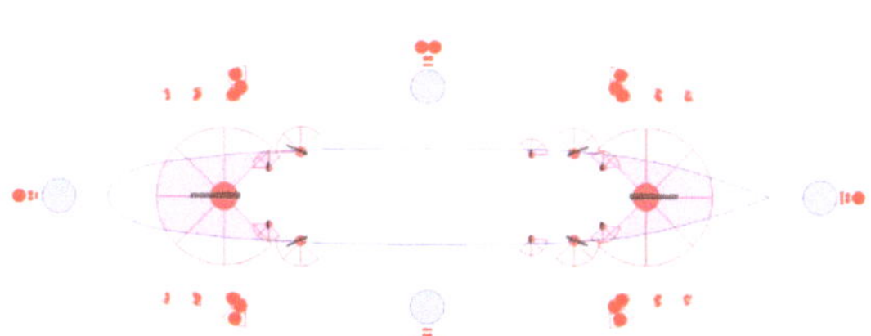

Hekla (1889)
1889 05 09 - ***** - ***** | Copenhagen DYd

0 +1 | "Valkyrien" type (1886)
Protected cruiser 2 class, Denmark

BUILDING

3Y:3M

Valkyrien

Displacement: 2,972 tons at draft: 5.54 m | Length: 81.28 m pp | Beam: 13.23 m | Crew: 282 men

Machinery: 4 cyl. boilers | 2 * 2,600 HP vertical triple expansion EE * 2 screws | 5,200 HP = 17 knots | Supplies: 488 tons of coal

Armour: steel | CT: 50 | WLDeck: 0(^65)

Armament: 2-210/35, 6-150/35, 4-57/43, 8-37/23R, 5-381 TT (^)

Age 1

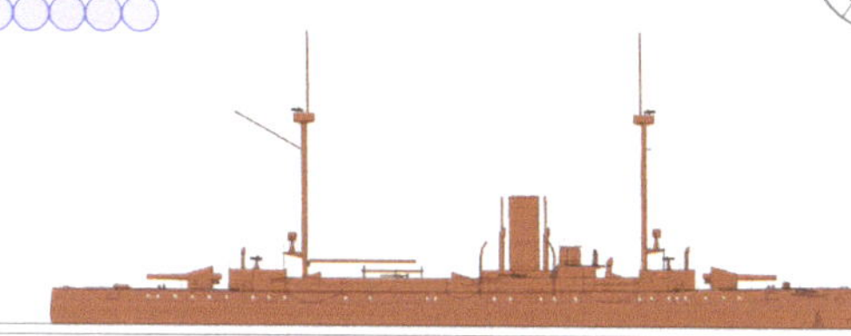

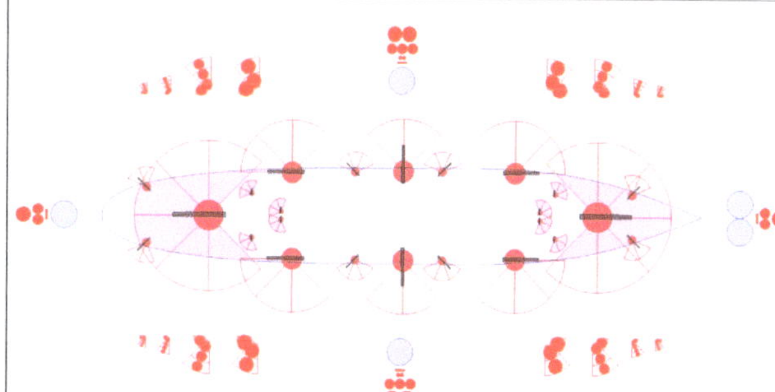

Valkyrien (1886)
1886 10 27 - 1888 09 08 - ***** | Copenhagen DYd

1 | "Tordenskjold" type (1879)
Protected cruiser 2 class, Denmark

Displacement: 2,462 tons at draft: 4.8 m | Length: 67.75 m oa | Beam: 13.23 m | Crew: 206 men

Machinery: 8 cyl. boilers | 2 *1,300 HP vertical compound EE * 2 screws | 2,600 HP = 12.75 knots | Supplies: 170 tons of coal

Armour: compound | CT: 203; Barbettes: 203 | WLDeck: 0 / 97(^38)

Armament: 1-356/26, 4-120/24, 8-37/23, 1-381 TT (v), 3-356 TT (^)

Age 1

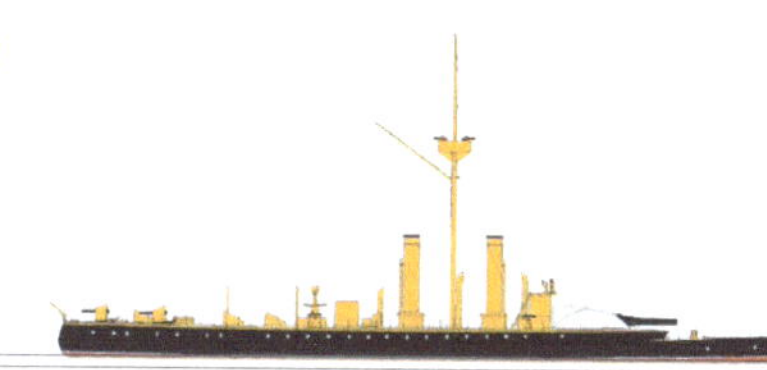

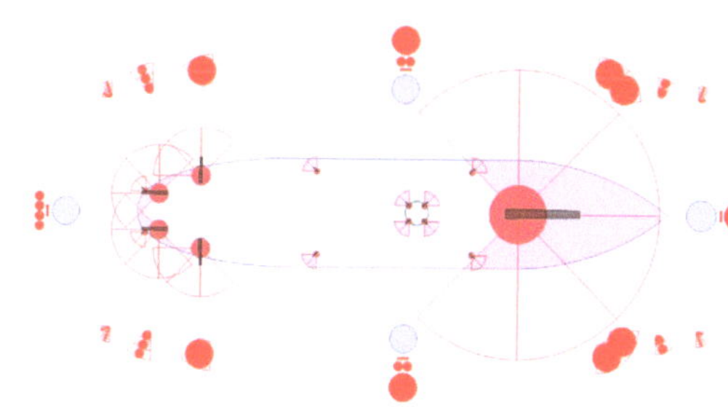

Tordenskjold (1879)
1879 06 05 - 1880 09 30 - 1882 09 29 | Copenhagen DYd

UNPROTECTED CRUISER

1 | "Fyen" type (1881)
Unprotected cruiser 1 class, Denmark

Displacement: 2,663 tons at draft: 6.1 m | Length: 69 m pp | Beam: 13.82 m | Crew: 300 men

Machinery: 4 cylindrical boilers | 1 vertical double expansion E * 1 screw | 2,600 HP = 12.5 knots | Supplies: 260 tons of coal

Armour: nil

Armament: 4-150/35, 14-150/22, 8-37/23R, 2-356 TT (^)

Age 1

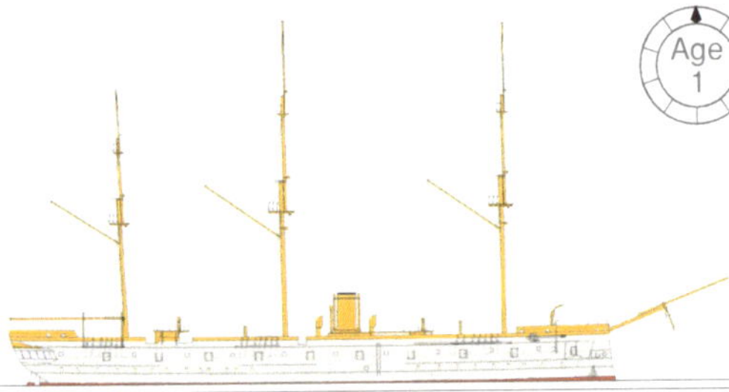

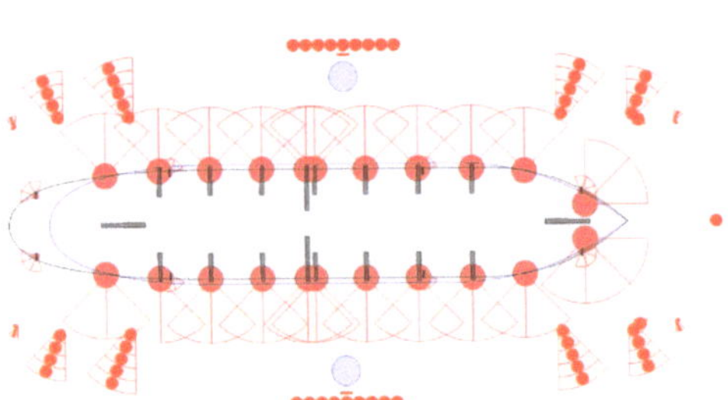

Fyen (1881)
1881 03 14 - 1882 09 27 - 1884 09 09 | Copenhagen DYd

WOODEN CRUISERS

1 | "Jylland" type (1857)
Wooden screw frigate, Denmark

Displacement: 2,420 tons at draft: 5.89 m | Length: 60.96 m oa | Beam: 12.98 m | Crew: 406 men

Machinery: 4 rectangular boilers | 1 horizontal single expansion E * 1 screw | 1,350 HP = 11 knots | Supplies: app. 300 tons of coal

Armament: 2-203.2/14.8 MLR, 24-152.4/21 MLR

Age 33

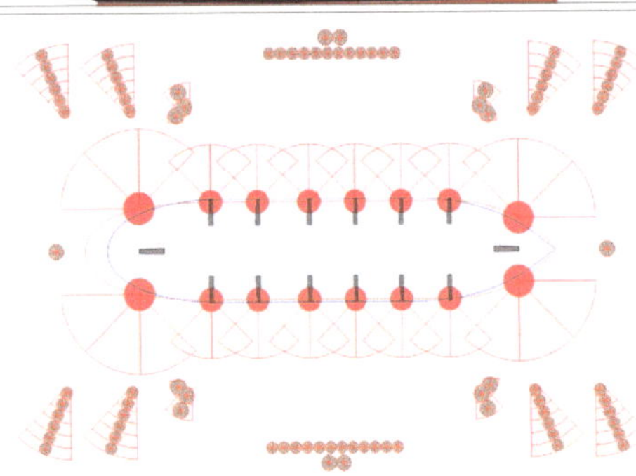

Jylland (1857)
1857 11 06 - 1860 11 20 - 1862 05 01 | Copenhagen DYd

1 | "St. Thomas" type (1869)
Wooden screw corvette, Denmark

Displacement: 1,550 tons at draft: 5.18 m | Length: 70.1 m oa | Beam: 9.98 m | Crew: 182 men

Machinery: 4 cylindrical boilers | 1 vertical double expansion E * 1 screw | 1,800 HP = 13 knots | Supplies: 190 tons of coal

Armament: 1-203.2/16 MLR, 4-152.4/16 MLR, 1-75/17 MLR

Age 21

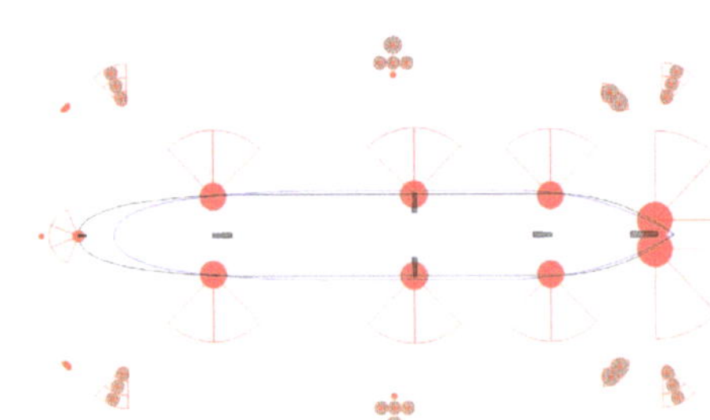

St. Thomas (1869)
1869 01 01 - 1871 07 26 - 1872 12 12 | Copenhagen DYd

1 | "Ingolf" type (1875)
Wooden screw corvette, Denmark

Displacement: 996 tons at draft: 4.16 m, 1,012 tons full | Length: 63.9 m oa | Beam: 8.53 m | Crew: 156 men

Machinery: 4 cylindrical boilers | 1 compound E * 1 screw | 630 HP = 10.2 knots | Supplies: app. 140 tons of coal

Armament: 3-150/25

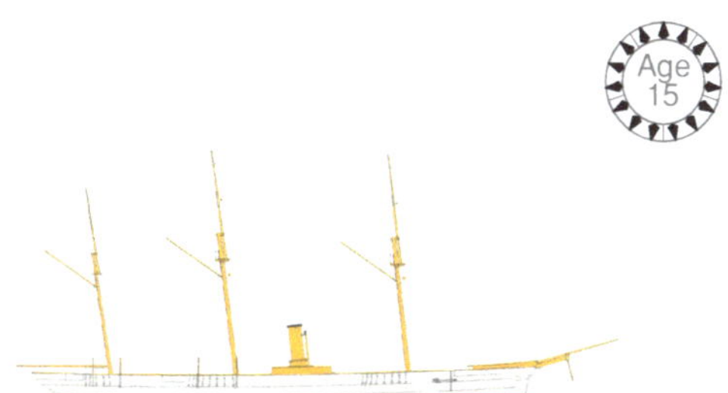

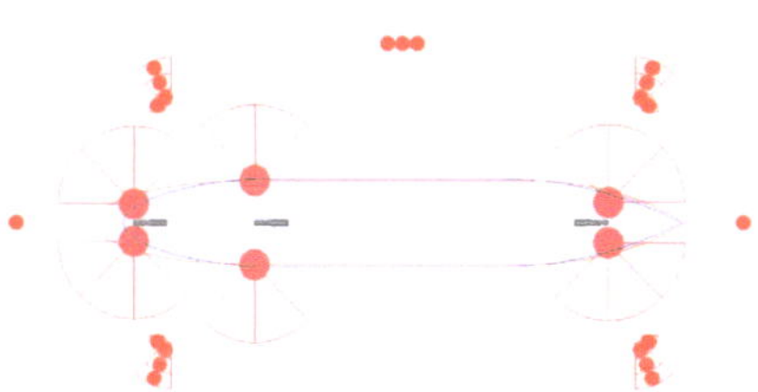

Ingolf (1875)
1875 01 01 - 1876 09 01 - 1878 01 01 | Copenhagen DYd

0 +1 | "Springeren" type (1890)
Torpedo boat 1 class, Denmark

TORPEDO SHIPS

BUILDING

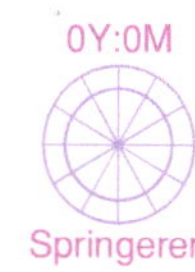

Displacement: 81 tons at draft: 2.11 m | Length: 36.27 m oa | Beam: 3.96 m | Crew: 15 men

Machinery: 1 locomotive boiler | 1 vertical compound E * 1 screw | 800 HP = 17.5 knots | Supplies: 14 tons of coal

Armament: 2-37/23R, 2-457,2 TT (2 >>)

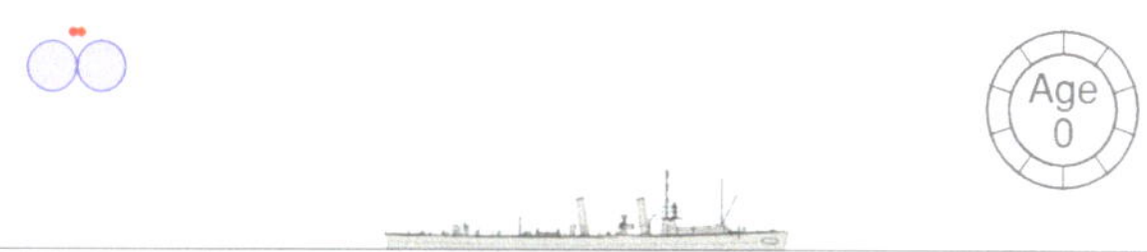

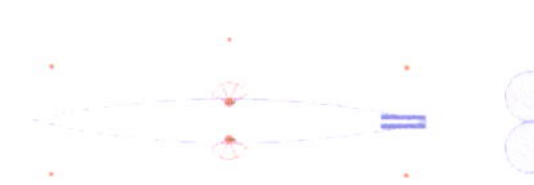

Springeren (1890)
1890 01 01 - ***** - ***** | Copenhagen DYd

2 | "Havhesten" type (1887)
Torpedo boats 1 class, Denmark

Displacement: 94 tons at draft: 2.17 m | Length: 42.03 m oa | Beam: 4.2 m | Crew: 20 men

Machinery: 1 locomotive boiler | 1 vertical compound E * 1 screw | 1,200 HP = 18.5 knots | Supplies: 15 tons of coal

Armament: 2-37/23R, 4-381 TT (1*2 trainable, 2 >>)

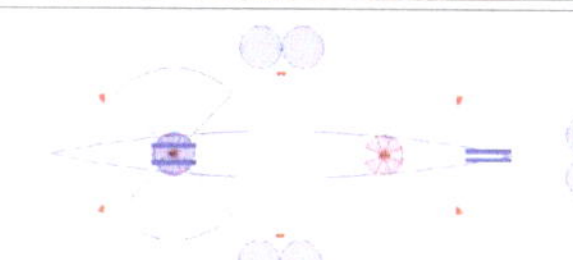

Havhesten (1887)
1887 01 01 - 1888 01 01 - 1889 01 01 | Thornycroft, Woolston

Narhvalen (1887)
1887 01 01 - 1888 01 01 - 1889 01 01 | Thornycroft, Woolston

2 | "Storen" type (1886)
Torpedo boats 1 class, Denmark

Displacement: 89 tons at draft: 0.61 m | Length: 39.92 m oa | Beam: 4.51 m | Crew: 14 men

Machinery: 1 locomotive boiler | 1 vertical compound E * 1 screw | 1,200 HP = 18.5 knots | Supplies: 14 tons of coal

Armament: 2-37/23R, 4-381 TT (1*2 trainable, 2 >>)

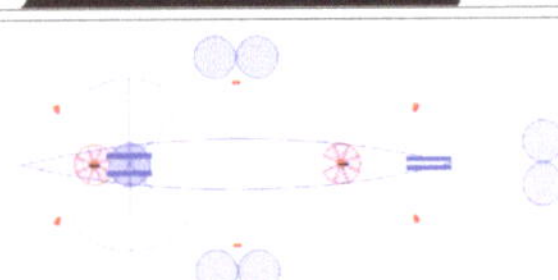

Soloven (1886)
1886 01 01 - 1887 01 01 - 1888 01 01 | Thornycroft, Woolston

Storen (1886)
1886 01 01 - 1887 01 01 - 1888 01 01 | Thornycroft, Woolston

1 | "Hvalrossen" type (1883)
Torpedo boat 1 class, Denmark

Displacement: 64 tons at draft: 0.61 m | Length: 34.74 m oa | Beam: 3.81 m | Crew: 14 men

Machinery: 1 locomotive boiler | 1 vertical compound E * 1 screw | 660 HP = 17 knots | Supplies: 10 tons of coal

Armament: 1-37/23R, 2-381 TT (2 >>)

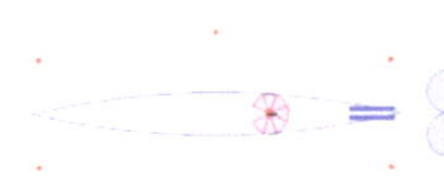

Hvalrossen (1883)
1883 01 01 - 1884 01 01 - 1885 01 01 | Thornycroft, Woolston

1 | "Delfinen" type (1882)
Torpedo boat 1 class, Denmark

Displacement: 66 tons at draft: 0.76 m | Length: 34.44 m oa | Beam: 3.81 m | Crew: 15 men

Machinery: 1 locomotive boiler | 1 vertical compound E * 1 screw | 700 HP = 17 knots | Supplies: 12 tons of coal

Armament: 1-37/23R, 2-381 TT (2 >>)

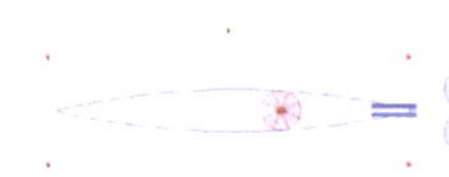

Delfinen (1882)
1882 01 01 - 1883 01 01 - 1883 06 06 | Thornycroft, Woolston

1 | "Svaerdfisken" type (1880)
Torpedo boat 1 class, Denmark

Displacement: 49 tons at draft: 1.83 m | Length: 33.53 m oa | Beam: 3.66 m | Crew: 14 men

Machinery: 1 locomotive boiler | 1 vertical compound E * 1 screw | 600 HP = 17 knots | Supplies: app. 10 tons of coal

Armament: 1-37/23R, 2-381 TT (2 >>)

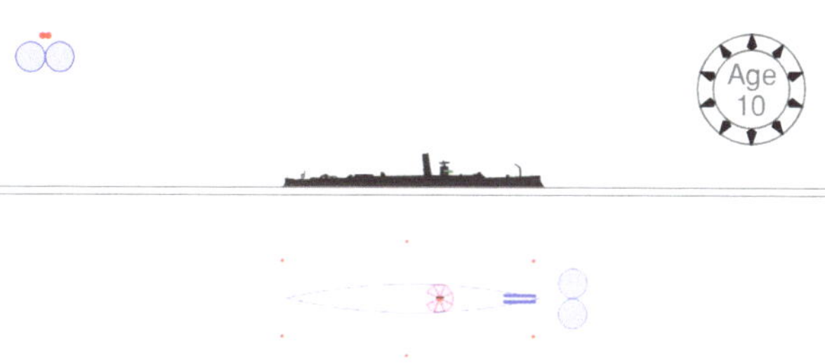

Svaerdfisken (1880)
1880 01 01 - 1881 01 01 - 1881 06 06 | Thornycroft, Woolston

1 | "Soulven" type (1879)
Torpedo boat 1 class, Denmark

NO PHOTO AVAILABLE

Displacement: 37 tons at draft: 0.94 m | Length: 27.54 m oa | Beam: 3.32 m | Crew: 12 men

Machinery: 1 locomotive boiler | 1 compound reciprocating E * 1 screw | 450 HP = 15 knots | Supplies: 5 tons of coal

Armament: 1-37/23R, 2-356 TT (2 >>)

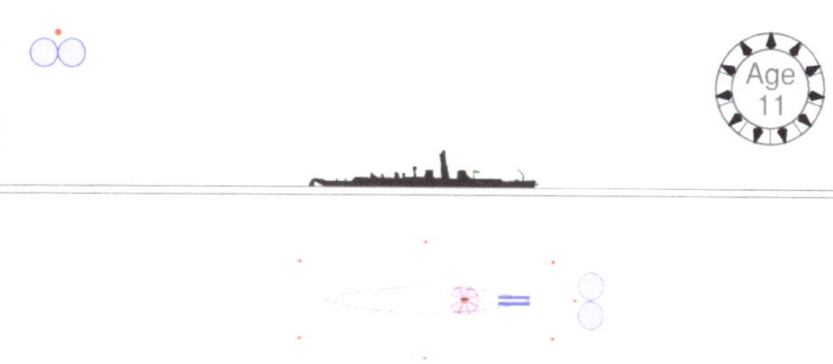

Soulven (1879)
1879 01 01 - 1880 01 01 - 1881 01 01 | F. Ch. Mediterr., La Seyne

Sister ship can be found in the Greek section

1 | "Haien" type (1878)
Torpedo boat, Denmark

Displacement: 32 tons at draft: 0.76 m | Length: 27.8 m oa | Beam: 3.03 m | Crew: 12 men

Machinery: 1 locomotive boiler | 1 vertical compound E * 1 screw | 350 HP = 15.5 knots | Supplies: app. 6 tons of coal

Armament: 1-37/23R, 1-356 TT (1 >)

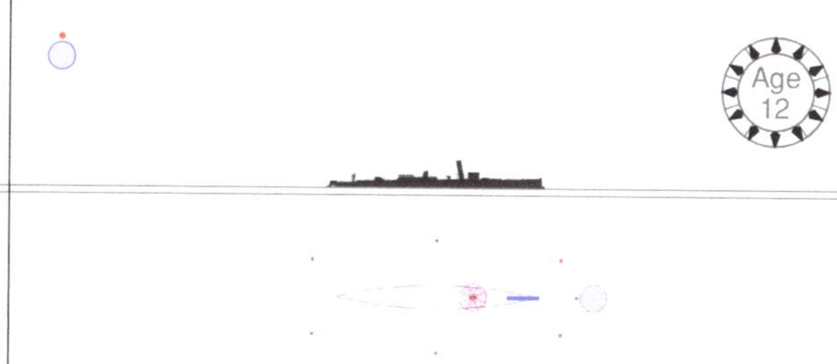

Haien (1878)
1878 01 01 - 1879 01 01 - 1879 08 19 | Thornycroft, Woolston

2 | "12" type (1888)
Torpedo boats 2 class, Denmark

Displacement: 24 tons at draft: 0.9 m | Length: 24.48 m oa | Beam: 2.74 m | Crew: 8 men

Machinery: 1 locomotive boiler | 1 vertical double expansion E * 1 screw | 350 HP = 15 knots | Supplies: 3 tons of coal

Armament: 1-37/23R, 2-356 TT (Djevetskiy)

Age 2

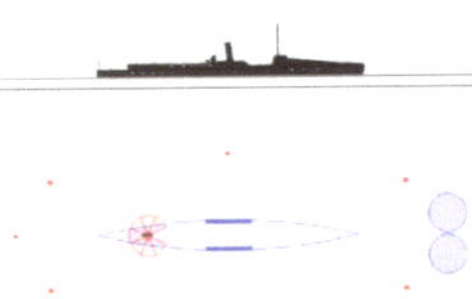

12 (1888)
1888 01 01 - 1888 06 06 - 1889 01 01 | Thornycroft, Woolston

13 (1888)
1888 01 01 - 1888 06 06 - 1889 01 01 | Thornycroft, Woolston

2 | "10" type (1887)
Torpedo boats 2 class, Denmark

NO PHOTO AVAILABLE

Displacement: 16 tons at draft: 0.9 m | Length: 21.44 m oa | Beam: 2.44 m | Crew: 6 men

Machinery: 1 locomotive boiler | 1 vertical double expansion E * 1 screw | 180 HP = 13.5 knots | Supplies: 1 ton of coal

Armament: 1-37/23R, 2-356 TT (2 >>)

des 3

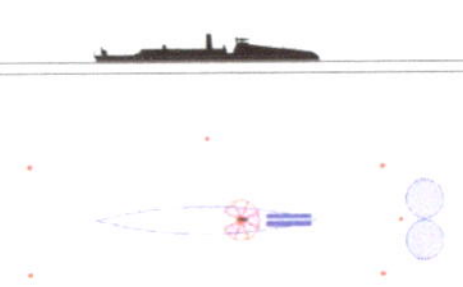

10 (1887)
1887 01 01 - 1888 01 01 - 1888 06 06 | Thornycroft, Woolston

11 (1887)
1887 01 01 - 1888 01 01 - 1888 06 06 | Thornycroft, Woolston

Sister ships can be found in the British sections

2 | "8" type (1885)
Torpedo boats 2 class, Denmark

Displacement: 17 tons at draft: 0.8 m | Length: 21.88 m oa | Beam: 2.47 m | Crew: 6 men

Machinery: 1 locomotive boiler | 1 vertical double expansion E * 1 screw | 170 HP = 15.7 knots | Supplies: 1 ton of coal

Armament: 1-37/23R, 2-356 TT (2 >>)

des 5

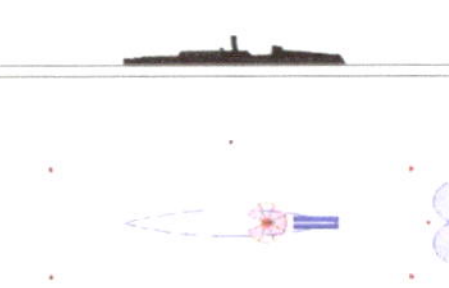

8 (1885)
1885 01 01 - 1886 01 01 - 1886 06 06 | Thornycroft, Woolston

9 (1885)
1885 01 01 - 1886 01 01 - 1886 06 06 | Thornycroft, Woolston

2 | "6" type (1883)
Torpedo boats 2 class, Denmark

Displacement: 16 tons at draft: 0.8 m | Length: 20.12 m oa | Beam: 2.44 m | Crew: 6 men

Machinery: 1 locomotive boiler | 1 vertical double expansion E * 1 screw | 150 HP = 13 knots | Supplies: app. 1 ton of coal

Armament: 1-37/23R, 2-356 TT (>>)

des 7

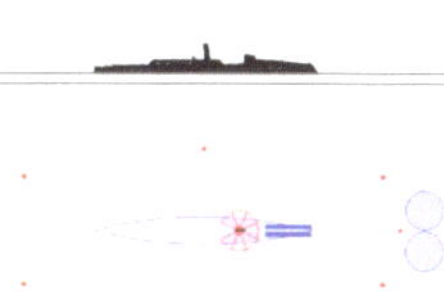

6 (1883)
1883 01 01 - 1884 01 01 - 1885 01 01 | Thornycroft, Woolston

7 (1883)
1883 01 01 - 1884 01 01 - 1885 01 01 | Thornycroft, Woolston

2 | "Euterpe"/"4" type (1882)
Torpedo boats 2 class, Italy

Euterpe

Displacement: 13 tons at draft: 0.76 m | Length: 19.2 m oa | Beam: 2.25 m | Crew: 6 men

Machinery: 1 locomotive boiler | 1 vertical double expansion E * 1 screw | 150 HP = 13 knots | Supplies: 1 ton of coal

Armament: 1-37/23R, 2-356 TT (>>)

Age 8

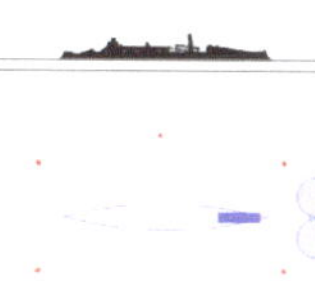

4 (1882)
1882 01 01 - 1882 06 06 - 1883 01 01 | Thornycroft, Woolston

5 (1882)
1882 01 01 - 1882 06 06 - 1883 01 01 | Thornycroft, Woolston

Sister ships can be found in the British and Italian sections

2 | "P 2" type (1877)
Spar torpedo boats 2 class, Denmark

NO PHOTO AVAILABLE

Displacement: 14 tons at draft: 0.8 m | Length: 14.6 m oa | Beam: 2.8 m | Crew: 7 men

Machinery: 1 locomotive boiler | 1 vertical double expansion E * 1 screw | 100 HP = 14 knots | Supplies: 1 ton of coal

Armament: 1 Spar torpedo

Age 13

NO PLAN AVAILABLE

P 2 (1877)
1877 01 01 - 1877 06 06 - 1878 01 01 | White, Cowes

P 3 (1877)
1877 01 01 - 1877 06 06 - 1878 01 01 | White, Cowes

1 | "Rap"/"1" type (1874)
Spar torpedo boat, Norway

Displacement: 15 tons at draft: 0.9 m | Length: 17.67 m oa | Beam: 2.28 m | Crew: 7 men
Machinery: 1 locomotive boiler | 1 vertical double expansion E * 1 screw | 100 HP = 15.13 knots | Supplies: 1.5 tons of coal
Armament: 1 Spar torpedo

Age 16

1 (1874)
1874 01 01 - 1874 06 06 - 1875 01 01 | Thornycroft, Woolston

Sister ships can be found in the Norwegian and Swedish sections

SLOOPS & GUNBOATS

0 +1 | "Hjaelperen" type (1888)
Minelayer, Denmark

BUILDING

Displacement: 283 tons at draft: 2,28 m | Length: 37.03 m oa | Beam: 6.77 m | Crew: 25 men
Machinery: 4 cyl. boilers | 2 * 165 HP vertical double expansion * 2 screws | 330 HP = 10 knots | Supplies: app. 30 tons of coal
Armament: 3-37/23R, 33 mines

Age 2

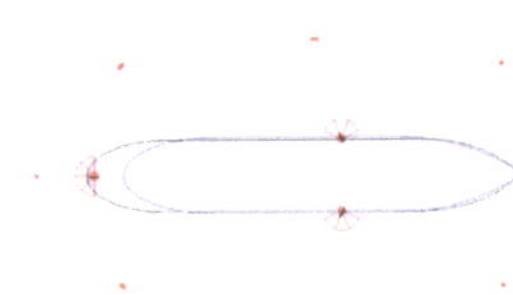

Hjaelperen (1888)
1888 01 01 - ***** - ***** | Copenhagen DYd

1 | "Dagmar" type (1859)
Wooden sloop, Denmark

Displacement: 1,175 tons at draft: 5 m | Length: 54.4 m oa | Beam: 10.3 m | Crew: 190 men
Machinery: 4 rectangular boilers | 1 horizontal direct acting E * 1 screw | 800 HP = 9 knots | Supplies: app. 190 tons of coal
Armament: 14-160/16 MLR

Age 31

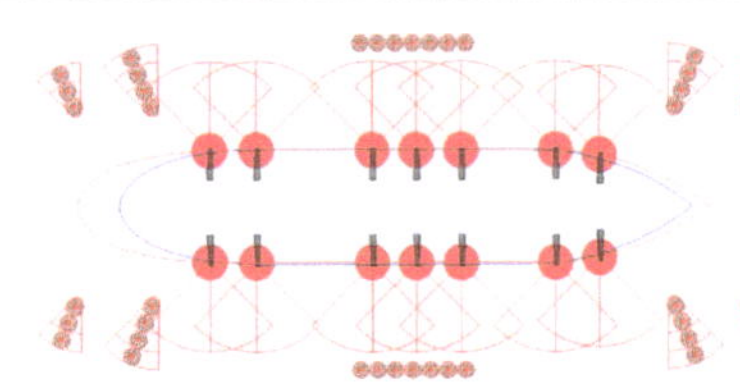

Dagmar (1859)
1859 01 01 - 1861 01 01 - 1863 01 01 | Copenhagen DYd

2 | "Gronsund" type (1882)
Gunboats, Denmark

Displacement: 243 tons at draft: 2.48 m | Length: 36.27 m oa | Beam: 6.12 m | Crew: 49 men

Machinery: 2 cylindrical boilers | 1 vertical compound E * 1 screw | 300 HP = 10 knots | Supplies: 60 tons of coal

Armament: 2-120/22, 2-37/23R

Age 8

NO PLAN AVAILABLE

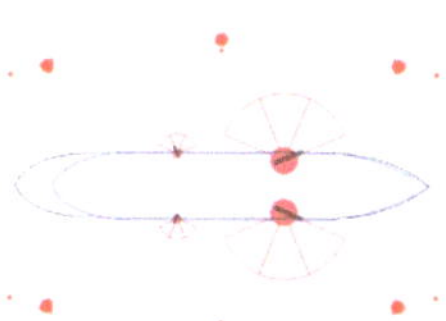

Gronsund (1882)
1882 01 01 - 1883 01 01 - 1884 01 01 | Copenhagen DYd

Guldorgsund (1883)
1883 01 01 - 1884 01 01 - 1885 01 01 | Copenhagen DYd

1 | "Moen" type (1874)
Gunboat, Denmark

Displacement: 410 tons at draft: m | Length: 34.3 m oa | Beam: 8.84 m | Crew: 50 men

Machinery: 2 cylindrical boilers | 2 * 250 HP vertical compound EE * 2 screws | 500 HP = 9 knots | Supplies: 20 tons of coal

Armament: 1-254/18, 2-84/13, 2-37/23R

Age 16

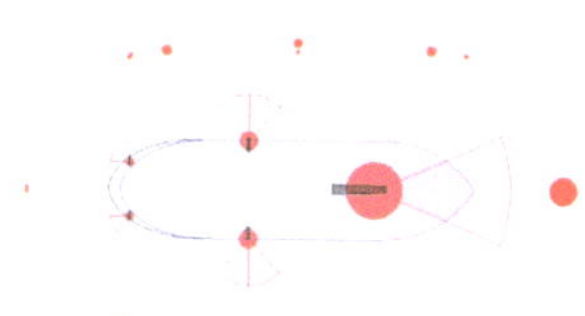

Moen (1874)
1874 01 01 - 1875 01 01 - 1876 01 01 | Copenhagen DYd

3 | "Oresund" type (1873)
Gunboats, Denmark

Displacement: 240 tons at draft: 2.14 m | Length: 26.1 m oa | Beam: 8 m | Crew: 36 men

Machinery: 2 cyl. boilers | 2 * 100 HP horizontal triple expansion EE * 2 screws | 200 HP = 7.5 knots | Supplies: 19 tons of coal

Armament: 1-254/18 MLR, 2-84/13, 2-37/23R

Age 17

Little Belt (1874)
1874 01 01 - 1875 01 01 - 1876 01 01 | Copenhagen DYd

Oresund (1873)
1873 01 01 - 1874 01 01 - 1875 01 01 | Copenhagen DYd

Store Belt (1873)
1873 06 06 - 1874 06 06 - 1875 06 06 | Copenhagen DYd

1 | "Falster" type (1871)
Gunboat, Denmark

Displacement: 356 tons at draft: 2.64 m | Length: 34.3 m oa | Beam: 8.84 m | Crew: 50 men

Machinery: 2 rect. boilers | 2 * 250 HP vertical double expansion EE * 2 screws | 500 HP = 10.5 knots | Supplies: 20 tons of coal

Armament: 1-254/18 MLR, 2-84/13 MLR, 2-37/23R

Age 19

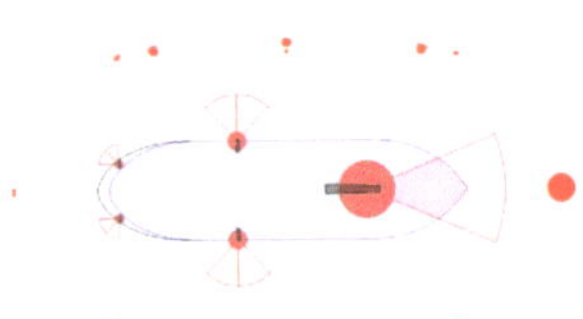

Falster (1871)
1871 01 01 - 1872 01 01 - 1873 01 01 | Copenhagen DYd

2 | "Fylla" type (1861)
Wooden gunboats, Denmark

Displacement: 550 tons at draft: 3.1 m | Length: 47 m oa | Beam: 7.9 m | Crew: 58 men

Machinery: 4 rectangular boilers | 1 horizontal triple expansion E * 1 screw | 500 HP = 8.7 knots | Supplies: 60 tons of coal

Armament: 3-152,4/14 MLR

Age 29

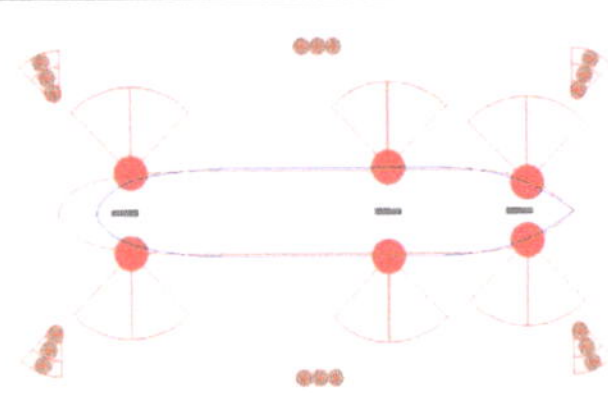

Diana (1862)
1862 01 01 - 1863 11 11 - 1865 01 01 | Copenhagen DYd

Fylla (1861)
1861 01 01 - 1862 07 01 - 1863 01 01 | Copenhagen DYd

2 | "Absalon" type (1861)
Wooden gunboats, Denmark

Displacement: 516 tons at draft: 3.26 m | Length: 46 m oa | Beam: 7.92 m | Crew: 75 men

Machinery: 4 rectangular boilers | 1 horizontal triple expansion E * 1 screw | 500 HP = 11 knots | Supplies: 65 tons of coal

Armament: 3-152.4/14 MLR

Age 29

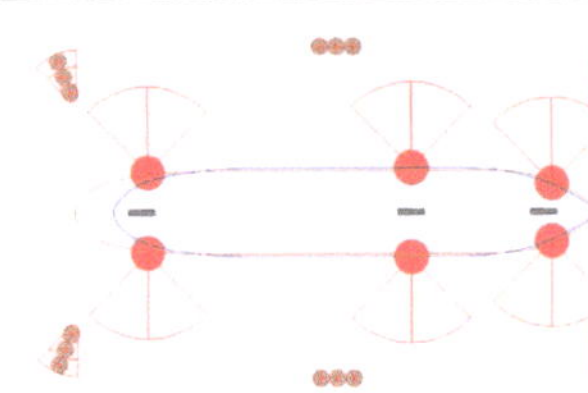

Absalon (1861)
1861 01 01 - 1862 01 01 - 1862 06 25 | Thames IW, Blackwall

Esbern Snare (1861)
1861 01 01 - 1862 01 01 - 1863 01 01 | Thames IW, Blackwall

SWEDISH FLEET

National flag

Ensign

Jack

Admiral

Warships built from 1858 01 01 to 1890 01 01

50 warships = 29,625 tons = 593 tons / average ship

Fleet age - 1880 02 15

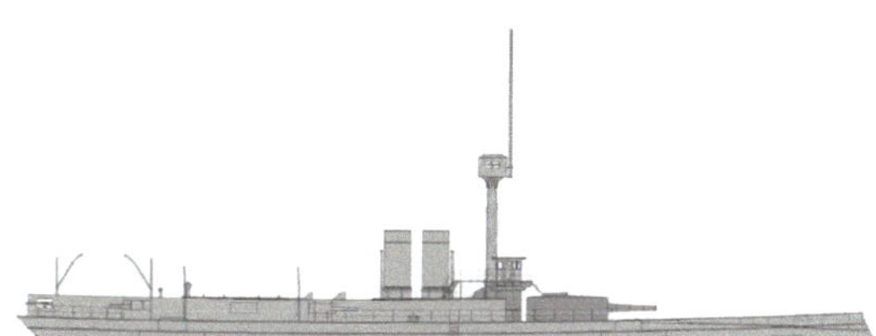

1889 fleet largest ship: Svea (1884) coast defence battleship

Swedish naval construction

Karlskrona dockyard

7.2 tons / average ship / month

1890-Jan-01
1 ship building

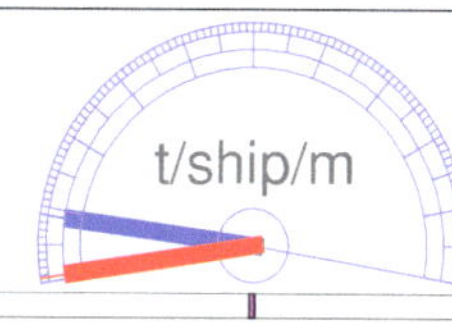

260 t.
tons / month = 7.2

t = 260
t/m =
7.2

Svensksund

NO PLAN AVAILABLE

Karlskrona, Sweden

32+ years of naval construction
11 ships | 8,105 tons

The plant's largest ship:
Vanadis (1861) wooden screw frigate
2,140 tons

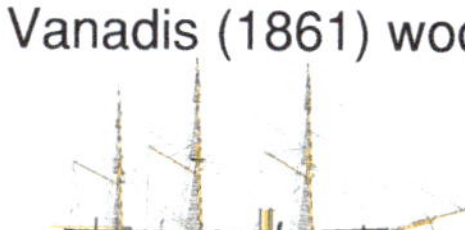

The plant's most effectively built ship:
Vanadis (1861) wooden screw frigate
89.4 tons/month

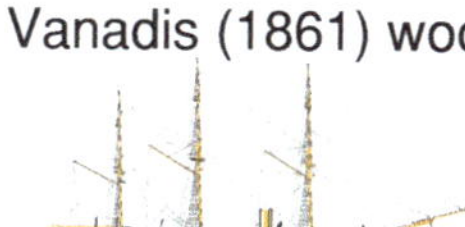

Specialisation in:
Screw sailing cruisers | 2 ships | 3,990 tons
Gunboats | 5 ships | 2,436 tons
Auxillaries | 1 ship | 1,590 tons

Lindholmen

90.1 tons / average ship / month

1890-Jan-01

1 ship building

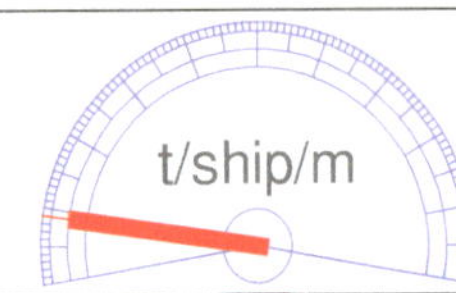

3,328 t.

tons / month = 90.1

t = 3,238

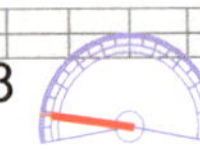

Gota

t/m =

90.1

Goteborg, Sweden

17 years of naval construction

3 ships | 6,781 tons

The plant's largest ship:

Gota (1888)

3,238 tons

The plant's most effectively built ship:

Gota (1888)

90.1 tons/month

Specialisation in:

Coast defence battleships | 2 ships | 6,289 tons

Gunboat | 1 ship | 492 tons

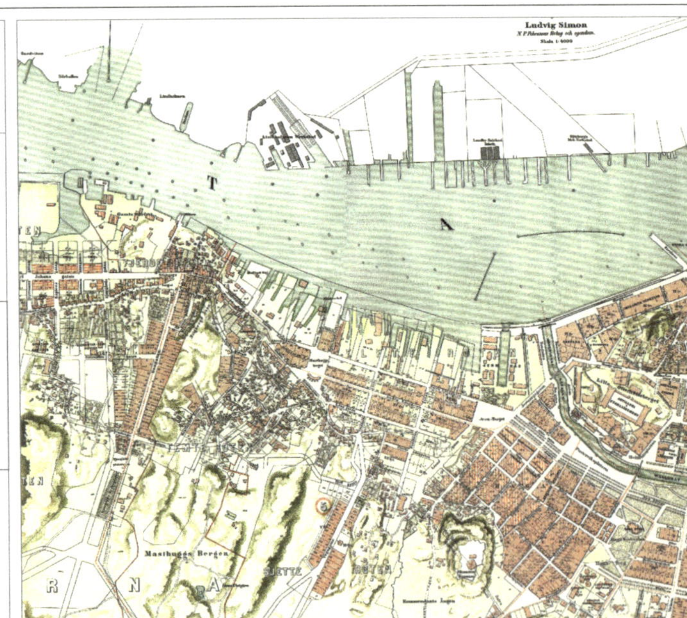

Swedish fleet effective list

5 coast defence armoured ships

Svea
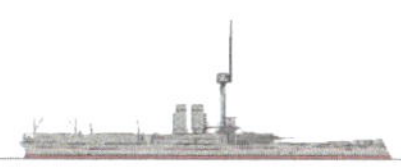
Loke

Tirfing

Thordon

John Ericsson

4 wooden cruisers

Freja
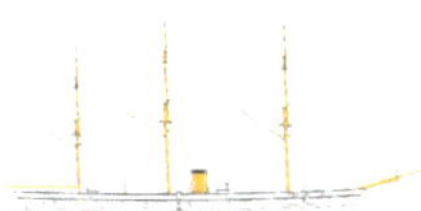
Saga
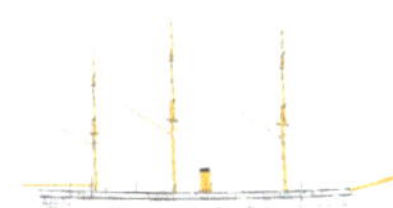
Balder

Vanadis

10 armoured gunboats

Berserk, Bjorn, Folke, Gerda, Hildur, Solve, Ulf, Fenris, Skold, Garmer

20 torpedo boats

Agda, Agne, Munin, Blink, Freke, Gere, Hugin, Blixt, Bygve, Bylgia

Galdr, Narf, Norve, 141, 143, 145, 147, 149, 151, Glimt

9 gunboats

Edda

Rota

Skaggald

Skagul

Skuld

Urd

Verdande

Blenda
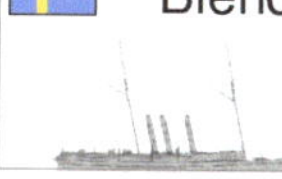
Disa

Swedish section ship names glossary

141 (TB2)
143 (TB2)
145 (TB2)
147 (TB2)
149 (TB2)
151 (TB2)

A
Agda (TB1)
Agne (TB1)

B
Balder (CW)
Berserk (GBA)
Bjorn (GBA)
Blenda (GB)
Blink (TB1)
Blixt (TB1)
Bygve (TB1)
Bylgia (TB1)

D
Disa (GB)

E
Edda (GB)

F
Fenris (MR)
Folke (GBA)
Freja (trailS)
Freke (TB1)

G
Galdr (TB1)
Garmer (GBA)
Gerda (GBA)
Gere (TB1)
Glimt (TB2)
Gota (CDB)

H
Hildur (GBA)
Hugin (TB1)

J
John Ericsson (MM)

L
Loke (MM)

M
Munin (TB1)

N
Narf (TB1)
Norve (TB1)

R
Rota (GB)

S
Saga (trailS)
Skaggald (GB)
Skagul (GB)
Skold (MR)
Skuld (GB)
Solve (GBA)
Svea (CDB)
Svensksund (GB)

T
Thordon (MM)
Tirfing (MM)

U
Ulf (GBA)
Urd (GB)

V
Vanadis (CW)
Verdande (GB)

Swedish fleet naval guns

Calibre (mm)	Mark	Calibre (in)/pdr	Gun type	First Introduced	Weight (kg)	Gun lenght (in/cal)	Bore lenght (cal)	Charge weight (lbs)	Muzzle energy (ft/t)	Projectile weight (lbs)	Muzzle velocity (ft/s)	Wrought iron perforation (in)
274.3	M81	10.8"	BLR		27 100	277.2 / 25.67	23.90	145.5	10 550	396.8	1 788	18.4
274.3	M76	10.8"	BLR		23 600	211.8 / 19.61	17.80	90.4	6 272	396.8	1 378	13.8
254.0	M85	10"	BLR		29 800	340.0 / 34.00	32.90	242.5	13 750	401.2	2 100	21.9
240.0	M76	9.45"	BLR		16 400	194.9 / 20.63	18.90	242.5	4 102	273.4	1 365	11.9
240.0	-	9.45"	BLR		14 400	179.5 / 19.00	17.10	59.5	3 789	224.9	1 312	11.4
167.0	-	6.58"	BLR		5 200	135.2 / 20.56	18.70	22.2	1 384	97.7	1 365	8.3
152.4	M89	6"	BLR		5 200	203.8 / 34.00	32.00	54.0	2 964	100.0	2 067	13.1
120.0	M81	4.72"	BLR		1 900	123.5 / 26.14	24.00	16.0	-	48.5	1 640	-
75.0	Krupp	2.95"	BLR		1 060	147.6 / 50.00	50.00	3.1	627	11.5	2 805	11.1
57.0	I & II	6 pr	BLR	1876	406	96.5 / 43.00	40.00	1.2	345	6.0	1 818	3.7
37.0	Hotchkiss	1 pr	BLR	1873	209	33.5 / 23.0	19.13	0.2	-	1.1	1 450	-
37.0	Hotchkiss	1 pr	BLR	1884	72	33.5 / 23.0	19.15	0.2	-	1.1	1 450	-

WARSHIPS SECTION

Ship classes	Building	Active
Battleships		
Boardside ironclads		
Armoured cruisers		
Coast defence armoured ships	1	5
Protected cruisers		
Unprotected cruisers		
Torpedo gunboats		
Torpedo boats 1 class		13
Torpedo boats 2 class		7
Submarines		
Wooden screw cruisers		4
Wooden screw sloops		
Gunboats		7
Gunboats coastal	1	11
Minelayers		

1 | +1 | "Svea" type (1884)
Coast defence battleships, Sweden

COAST DEFENCE BATTLESHIPS

BUILDING

1Y:12M

Gota

Displacement: 3,238 tons at draft: 5.18 m | Length: 78.7 m wl | Beam: 14.78 m | Crew: 252 men

Machinery: 6 cylindrical boilers | 2 * 2,350 HP vertical compound EE * 2 screws | 4,700 HP = 16 knots | Supplies: 300 tons of coal

Armour: compound | CT: 263 | Turret: 297-246 | Barbette: 297-147 | WLBelt: 0(v51) 147(^51) 297(^51) 147(^51) 0(v51)

Armament: 2-254/32, 4-152.4/34, 11-57/43, 6-25.4/40 Mg, 3-356 TT (v)

Age 2

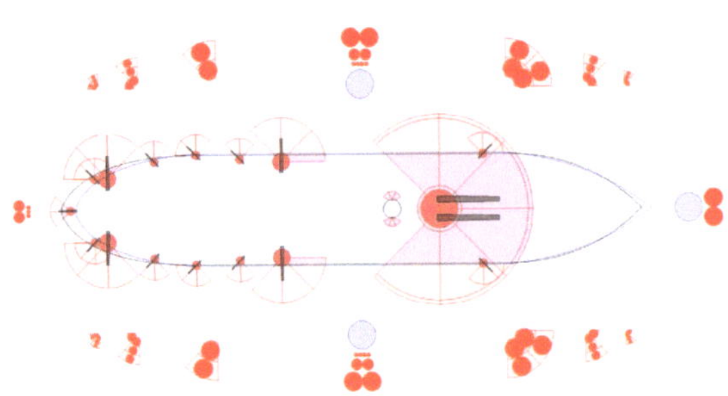

Gota (1888)

1888 01 01 - 1889 09 30 - ***** | Lindholmen, Goteborg

1 | +1 | "Svea" type (1884)
Coast defence battleships, Sweden (continued)

Displacement: 3,051 tons at draft: 5.18 m; 3,273 tons full | Length: 75.7 m wl | Beam: 14.78 m | Crew: 252 men

Machinery: 6 cylindrical boilers | 2 * 1,800 HP vertical compound EE * 2 screws | 3,640 HP = 14.7 knots | Supplies: 288 tons of coal

Armour: compound | CT: 263 | Turret: 297-246 | Barbette: 297-147 | WLBelt: 0(v51) 147(^51) 297(^51) 147(^51) 0(v51)

Armament: 2-254/32, 6-152.4/28, 6-57/43, 4-25.4/40 Mg, 1-356 TT (v)

Age 6

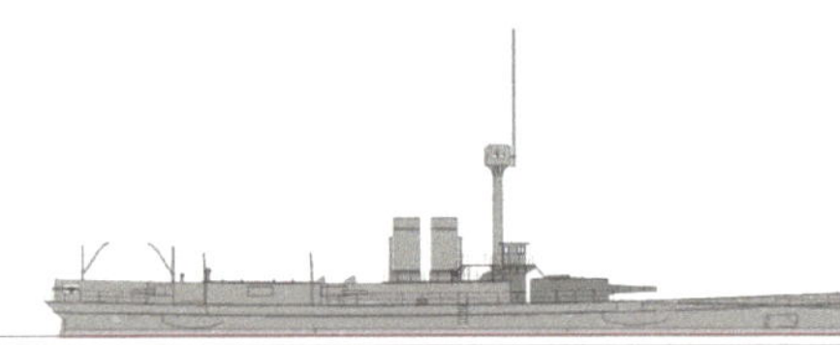

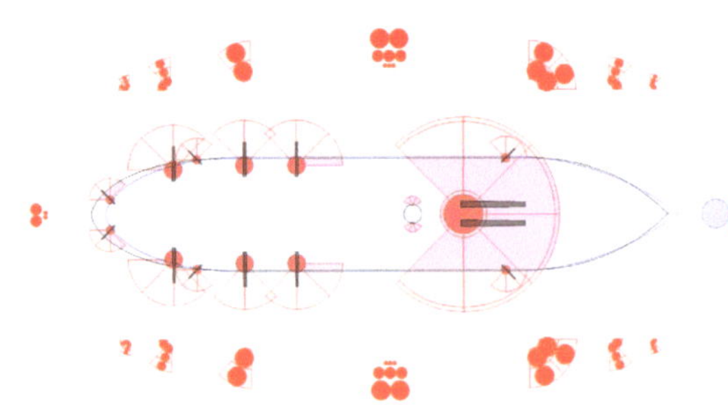

Svea (1884)

1884 03 01 - 1886 12 12 - 1887 09 20 | Lindholmen, Goteborg

1 | "Loke" type (1867)
Monitor, Sweden

MONITORS

Displacement: 1,574 tons at draft: 3.61 m | Length: 62.4 m wl | Beam: 13.72 m | Crew: 80 men

Machinery: 4 cylindrical boilers | 1 vibrating lever E * 1 screw | 430 HP = 8.3 knots | Supplies: 134 tons of coal

Armour: iron | CT: 229, Turret: 365 | WLBelt: 175(^25)

Armament: 2-240/17

Age 16

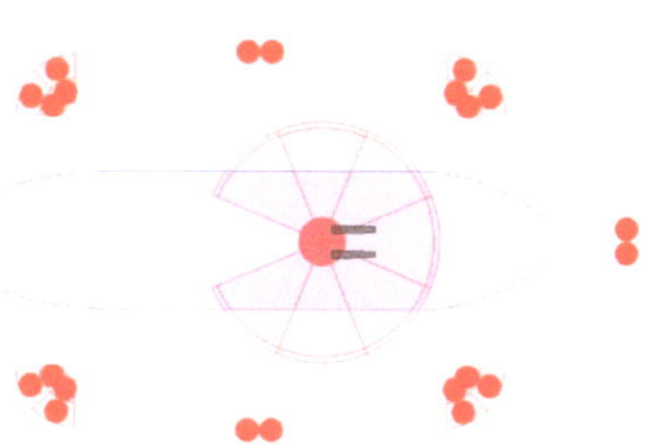

Loke (1867)
1867 03 01 - 1869 09 04 - 1871 11 01 | Motala Verks., Norrkoping

3 | "John Ericsson" type (1864)
Monitors, Sweden

Displacement: 1,476 tons at draft: 3.66 m | Length: 60.9 m wl | Beam: 13.8 m | Crew: 80 men

Machinery: 2 rectangular boilers | 1 vibrating lever E * 1 screw | 380 HP = 7 knots | Supplies: 120 tons of coal

Armour: iron | CT: 152, Turret: 261 | WLBelt: 114(^25)

Rearmed (1880): 2-240/17

Age 26

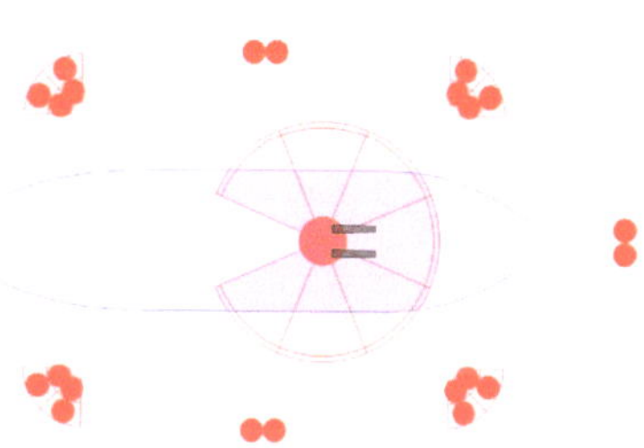

John Ericsson (1864)
1864 07 01 - 1865 03 17 - 1865 11 01 | Motala Verks., Norrkoping

Thordon (1865)
1865 04 01 - 1865 12 01 - 1866 08 01 | Motala Verks., Norrkoping

Tirfing (1865)
1865 02 01 - 1866 10 06 - 1868 02 01 | Motala Verks., Norrkoping

1 | "Vanadis" type (1861)
Wooden screw frigate, Sweden

WOODEN CRUISERS

Displacement: 2,140 tons at draft: 6.1 m | Length: 64 m wl | Beam: 12.8 m | Crew: 316 men

Machinery: 2 rectangular boilers | 1 single expantion E * 1 screw | 1,400 HP = 11.5 knots | Supplies: 250 tons of coal

***

Armament: 8-167.6/15 MLR

Age 20

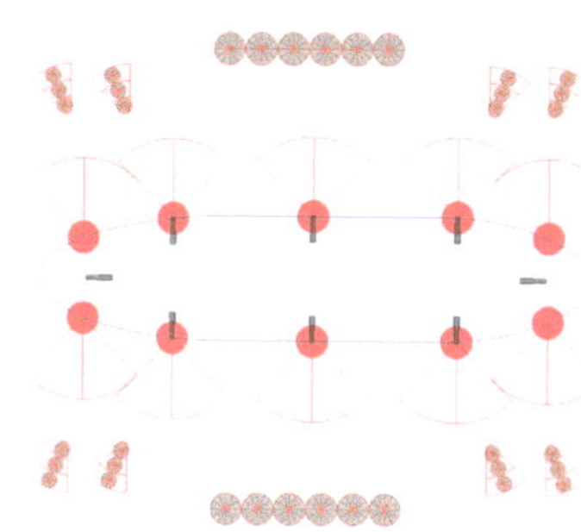

Vanadis (1861)
1861 01 01 - 1862 01 01 - 1863 01 01 | Karlskrona DYd

1 | "Freja" type (1884)
Wooden screw corvette, Sweden

Displacement: 2,000 tons at draft: 5.72 m | Length: 66 m pp | Beam: 12 m | Crew: 284 men
Machinery: 4 cylindrical boilers | 1 vertical triple expansion E * 1 screw | 1,750 HP = 12 knots | Supplies: 180 tons of coal
Armament: 4-152.4/25, 8-122/22, 4-65/16

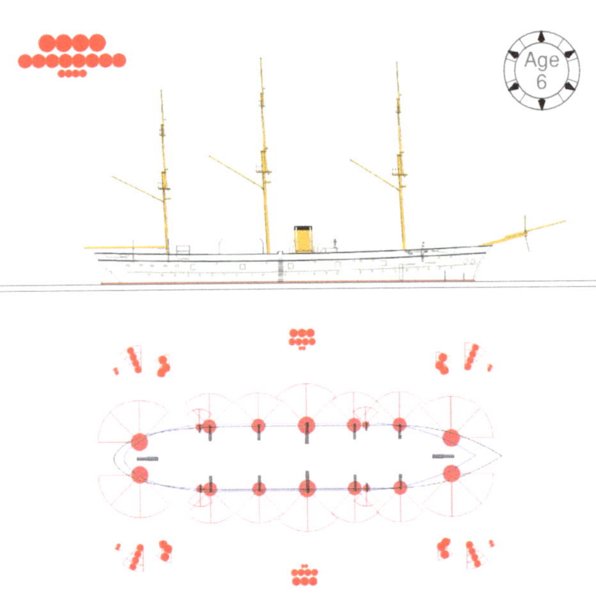

Freja (1884)
1884 01 01 - 1885 07 25 - 1886 06 05 | Kockums, Malmo

1 | "Saga" type (1874)
Wooden screw corvette, Sweden

Displacement: 1,590 tons at draft: 5.59 m | Length: 61 m wl | Beam: 10.2 m | Crew: 188 men
Machinery: 2 cylindrical boilers | 1 vertical triple expansion E * 1 screw | 900 HP = 11 knots | Supplies: 170 tons of coal
Armament: 1-167.6/18 MLR, 6-122/22

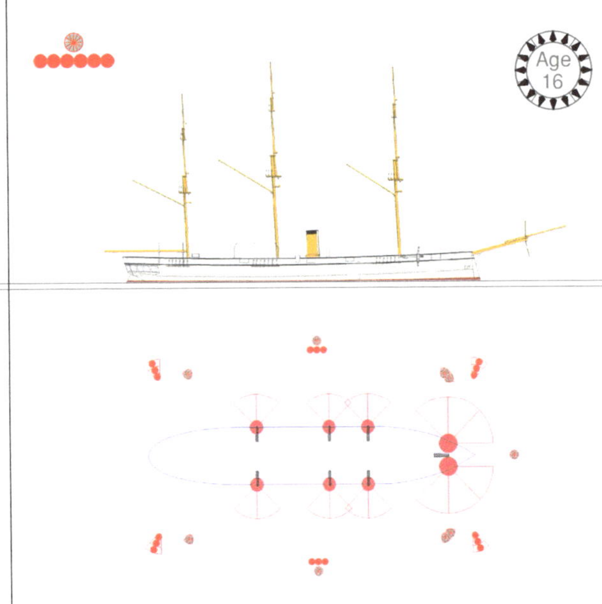

Saga (1874)
1874 08 07 - 1877 11 12 - 1878 06 20 | Karlskrona DYd

1 | "Balder" type (1869)
Wooden screw corvette, Sweden

Displacement: 1,850 tons at draft: 5.59 m | Length: 61.8 m wl | Beam: 11.1 m | Crew: 218 men
Machinery: 2 cylindrical boilers | 1 horizontal return connecting rod E * 1 screw | 1,380 HP = 12 knots | Supplies: 270 tons of coal
Armament: 6-167.6/18 MLR

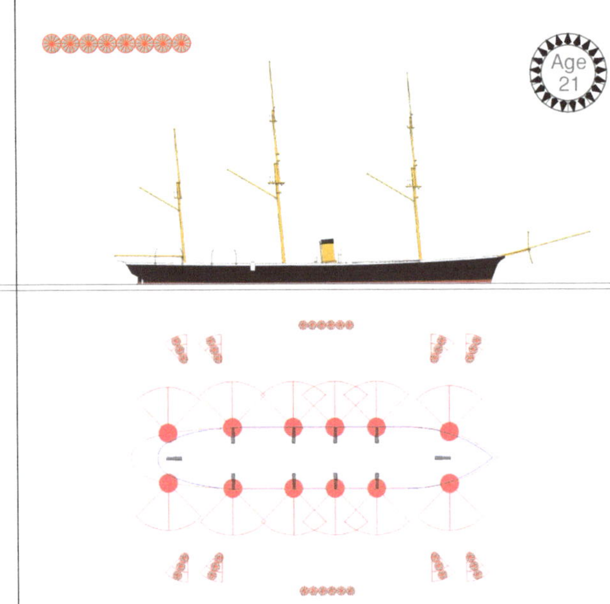

Balder (1869)
1869 01 01 - 1870 01 01 - 1871 01 01 | Karlskrona DYd

TORPEDO SHIPS

3 | "Munin" type (1885)
Torpedo boats 1 class, Sweden

Displacement: 66 tons at draft: 0.61 m | Length: 34.44 m oa | Beam: 3.81 m | Crew: 16 men

Machinery: 1 locomotive boilers | 1 compound E * 1 screw | 620 HP = 18.8 knots | Supplies: 12 tons of coal

Armament: 1-25.4/40, 2-381 TT (2 >>)

Age 5

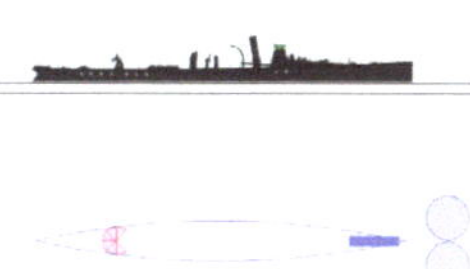

Munin (1885)
1885 01 01 - 1887 01 01 - 1888 01 01 | Bergsund, Stockholm

Freke (1885)
1885 01 01 - 1886 01 01 - 1887 01 01 | Bergsund, Stockholm

Gere (1886)
1886 01 01 - 1887 01 01 - 1888 01 01 | Bergsund, Stockholm

2 | "Agne" type (1886)
Torpedo boats 1 class, Sweden

Displacement: 40 tons at draft: 1,85 m | Length: 31.5 m oa, 30.7 m pp | Beam: 3.55 m | Crew: 12 men

Machinery: 1 cylindrical boiler | 1 vertical double expantion E * 1 screw | 460 HP = 19 (19.5) knots | Supplies: 8 tons of coal

Armament: 1-25.4/40 Mg, 2-381 TT (1 trainable, 1 >)

Age 4

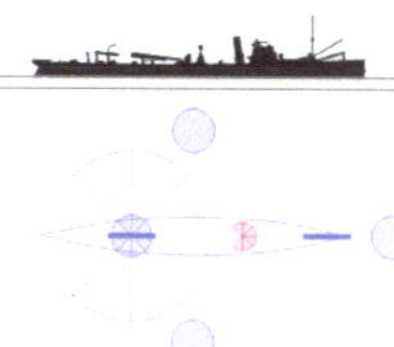

Agda (1886)
1886 01 01 - 1887 01 01 - 1888 01 01 | Karlskrona DYd

Agne (1886)
1886 01 01 - 1887 01 01 - 1888 01 01 | Stockholm DYd

2 | "Bygve" type (1886)
Torpedo boats 1 class, Sweden

Displacement: 40 tons at draft: 1.85 m; 45 tons full | Length: 31.5 m oa; 30.7 m pp | Beam: 3.55 m | Crew: 12 men

Machinery: 1 cylindrical boiler | 1 vertical double expantion E * 1 screw | 440 HP = 19 (19.5) knots | Supplies: 8 tons of coal

Armament: 1-25.4/40 Mg, 2-381 TT (2 >>)

Age 4

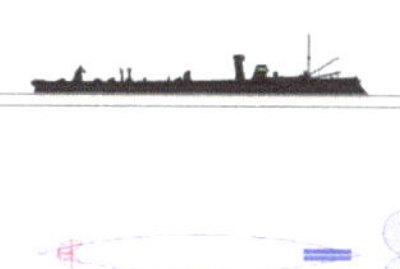

Bygve (1886)
1886 01 01 - 1887 01 01 - 1888 01 01 | Stockholm DYd

Bylgia (1886)
1886 01 01 - 1887 01 01 - 1888 01 01 | Stockholm DYd

1 | "Blink" type (1884)
Torpedo boat 1 class, Sweden

NO PHOTO AVAILABLE

Displacement: 34 tons at draft: 0.86 m | Length: 28.64 m oa | Beam: 3.66 m | Crew: 10 men

Machinery: 1 cylindrical boiler | 1 vertical double expantion E * 1 screw | 390 HP = 16 knots | Supplies: 6.5 tons of coal

Armament: 1-25.4/40 Mg, 1-381 TT (1 >)

Age 6

Blink (1884)
1884 01 01 - 1885 01 01 - 1886 01 01 | Stockholm DYd

1 | "Suhum"/"Hugin" type (1882)
Torpedo boat 1 class, Russia

Displacement: 65 tons at draft: 0.72 m | Length: 34.44 m oa | 34.49 m wl; 34.04 m pp | Beam: 3.81 m | Crew: 16 men

Machinery: 1 cylindrical boiler | 1 compound expantion E * 1 screw | 620 (650) HP = 18.5 (19) knots | Suppl.: 10 tons of coal

Armament: 1-25.4/40 Mg, 2-381 TT (2 >>)

Age 8

Hugin (1883)
1883 01 01 - 1884 01 01 - 1885 01 01 | Thornycroft, Woolston

Sister ships can be found in the British and Russian sections

4 | "Adlebaran"/"Blixt" type (1881)
Torpedo boats 1 class, Italy

Displacement: 40 tons at draft: 1.85 m | Length: 30.48 m oa | Beam: 3.55 m | Crew: 12 men

Machinery: 1 locomotive boiler | 1 vertical double expansion E * 1 screw | 420 HP = 19 knots | Supplies: app. 8 tons of coal

Armament: 2-25.4/40, 2-356 TT (2 >>)

Age 8

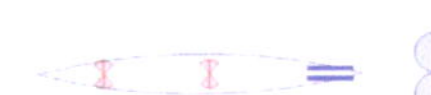

Blixt (1882)
1882 01 01 - 1882 06 06 - 1883 01 01 | Thornycroft, Woolston

Galdr (1885)
1885 01 01 - 1886 01 01 - 1887 01 01 | Stockholm DYd

Narf (1885)
1885 01 01 - 1886 01 01 - 1887 01 01 | Stockholm DYd

Norve (1885)
1885 01 01 - 1886 01 01 - 1887 01 01 | Stockholm DYd

Sister ships can be found in the Italian section

6 | "2" type (1877)
Spar torpedo boats, Sweden

NO PHOTO AVAILABLE

Displacement: 24.3 tons at draft: 1.3 m | Length: 17.07 m oa | Beam: 3.35 m | Crew: 10 men

Machinery: 1 locomotive boiler | 1 compound expansion E * 1 screw | 80 HP = 9.5 knots | Supplies: app. 1 ton of coal

Armament: 2 spar torpedoes

Age 13

141 (1877)
1877 01 01 - 1878 01 01 - 1879 01 01 | Karlskrona DYd

143 (1877)
1877 01 01 - 1878 01 01 - 1879 01 01 | Karlskrona DYd

145 (1877)
1877 01 01 - 1878 01 01 - 1879 01 01 | Sweden

147 (1877)
1877 01 01 - 1878 01 01 - 1879 01 01 | Sweden

149 (1877)
1877 01 01 - 1878 01 01 - 1879 01 01 | Sweden

151 (1877)
1877 01 01 - 1878 01 01 - 1879 01 01 | Sweden

SLOOPS & GUNBOATS

1 | "Rap"/"Glimt" type (1874)
Spar torpedo boat, Norway

Rap

Displacement: 15.3 tons at draft: 0.9 m | Length: 17.67 m oa | Beam: 2.28 m | Crew: 7 men

Machinery: 1 locomotive boilers | 1 vertical double expansion E * 1 screw | 100 HP = 18 knots | Supplies: 1.5 tons of coal

Armament: 2 spar torpedoes

Age 16

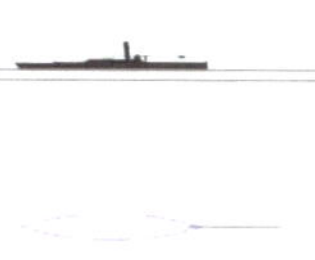

Glimt (1874)
1874 01 01 - 1874 06 06 - 1875 01 01 | Thornycroft, Woolston

Sister ships can be found in the Danish and Norwegian sections

7 | "Hildur" type (1870)
Armoured gunboats, Sweden

Gerda

Displacement: 453 tons at draft: 2.59 m | Length: 39.7 m wl | Beam: 8 m | Crew: 42 men

Machinery: 2 rectangular boilers | 2 * 65 HP horizontal compound EE * 2 screws | 133 HP = 8 knots | Supplies: 22 tons of coal

Armour: iron | Turret: 418-75 | WLBelt: 76(^38) 85(^38) 76(^38)

Armament: 1-240/18

Age 20

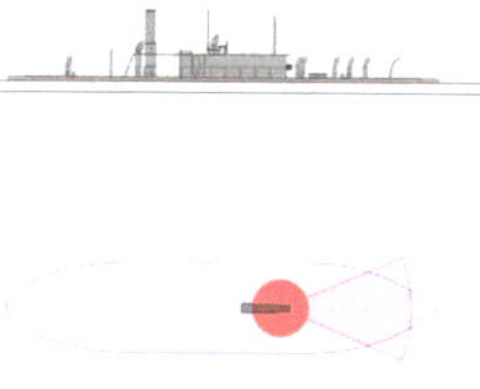

Berserk (1873)
1873 01 01 - 1874 01 01 - 1875 01 01 | Motala Verks., Norrkoping

Bjorn (1872)
1872 01 01 - 1873 06 23 - 1874 07 16 | Motala Verks., Norrkoping

Folke (1873)
1873 01 01 - 1874 01 01 - 1875 01 01 | Motala Verks., Norrkoping

Gerda (1870)
1870 09 09 - 1871 09 01 - 1872 06 16 | Motala Verks., Norrkoping

Hildur (1870)
1870 01 01 - 1871 07 14 - 1872 05 23 | Motala Verks., Norrkoping

Solve (1874)
1874 01 01 - 1875 01 01 - 1875 05 10 | Motala Verks., Norrkoping

Ulf (1872)
1872 01 01 - 1872 11 11 - 1873 06 22 | Motala Verks., Norrkoping

2 | "Skold" type (1867)
Armoured gunboats, Sweden

Displacement: 250 tons at draft: 2.2 m | Length: 32.7 m oa | Beam: 6.9 m | Crew: 50 men

Machinery: 1 rectangular boiler | 1 horizontal compound E * 1 screw | 25 HP = 3.75 knots / 24 menP | Supplies: 5 tons of coal

Armour: iron | Fixed turret: 229 | WLBelt: 25

Armament: 1-240/18

Age 23

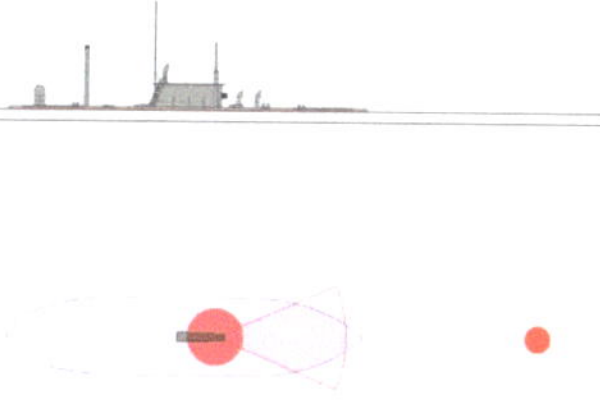

Fenris (1871)
1871 06 18 - 1872 07 01 - 1873 06 06 | Bergsund, Stockholm

Skold (1867)
1867 05 19 - 1868 11 11 - 1871 02 27 | Bergsund, Stockholm

Folke

Note gunhouse situated back.

1 | "Garmer" type (1865)
Armoured gunboat, Sweden

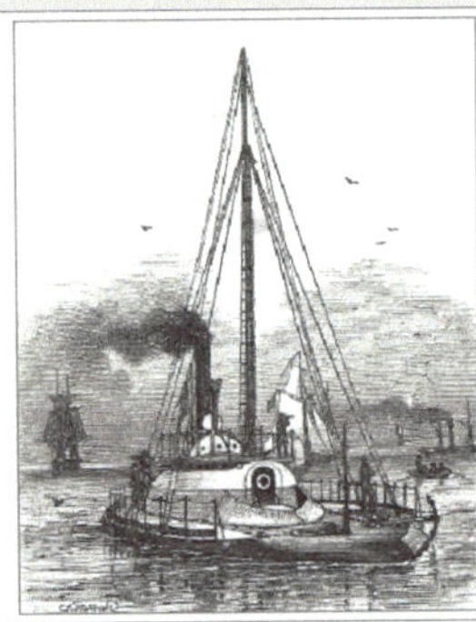

Displacement: 271 tons at draft: 2.21 m | Length: 28.5 m oa | Beam: 6.98 m | Crew: 20 men

Machinery: 1 rectangular boiler | 1 horizontal compound E * 1 screw | 90 HP = 5.5 knots | Supplies: app. 20 tons of coal

Armour: iron | Fixed turret: 149 | WLBelt: 39(^20)

Armament: 1-240/18; 2-12.7 Palmkranz MG

Garmer (1865)
1865 07 01 - 1867 05 19 - 1868 03 17 | Motala Verks., Norrkoping

0 +1 | "Svensksund" type (1890)
Gunboat-minelayer, Sweden

BUILDING

Displacement: 260 tons at draft: 3 m | Length: 38.5 m oa | Beam: 7.9 m | Crew: 50 men

Machinery: 2 cylindrical boilers | 1 vertical compound E * 1 screw | 440 HP = 12.5 knots | Supplies: app. 15 tons of coal

Armament: 2-57/43, app. 30 mines

Age
0

NO PLAN AVAILABLE

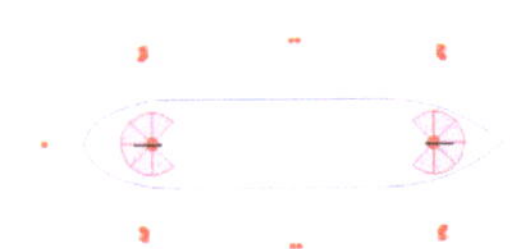

Svensksund (1890)
1890 01 01 - 1891 09 30 - 1893 01 01 | Karlskrona DYd

1 | "Edda" type (1884)
Gunboat, Sweden

Displacement: 613 tons at draft: 3.1 m | Length: 56 m oa | Beam: 8.26 m | Crew: 76 men

Machinery: 4 cylindrical boilers | 2 * 480 HP vertical compound EE * 2 screws | 960 HP = 13.6 knots | Supplies: 80 tons of coal

Armament: 1-274/23, 1-152.4/25, 2-25.4/40 Mg

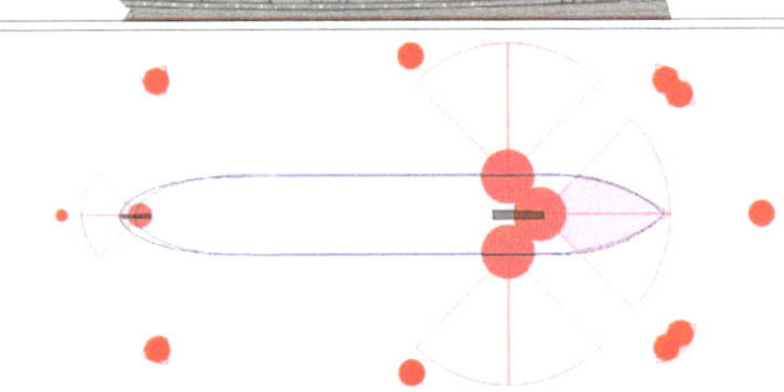

Edda (1884)
1884 01 01 - 1885 01 01 - 1886 01 01 | Karlskrona DYd

6 | "Urd" type (1876)
Gunboats, Sweden

Displacement: 527 tons at draft: 2.82 m | Length: 52.3 m oa | Beam: 8.05 m | Crew: 72 men

Machinery: 4 cylindrical boilers | 2 * 390 HP horizontal compound EE * 2 screws | 780 HP = 13 knots | Supplies: 80 tons of coal

Armament: 1-274/18, 1-122/23, 2-25.4/40 Mg

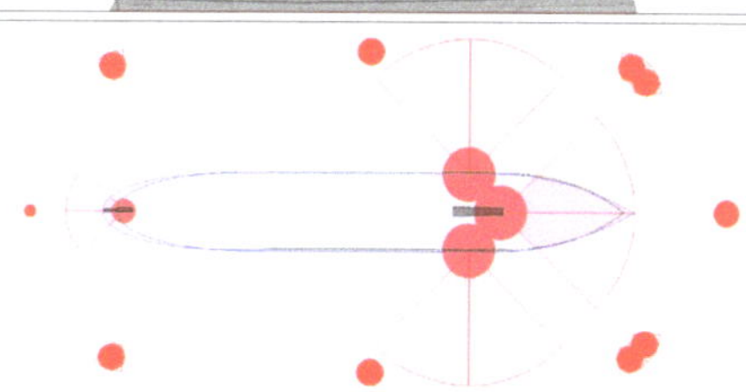

Rota (1877)
1877 01 01 - 1878 01 01 - 1879 01 01 | Stockholm DYd

Skaggald (1876)
1876 01 01 - 1877 01 01 - 1878 01 01 | Stockholm DYd

Skagul (1878)
1878 01 01 - 1879 01 01 - 1880 01 01 | Stockholm DYd

Skuld (1878)
1878 01 01 - 1879 01 01 - 1880 01 01 | Karlskrona DYd

Urd (1876)
1876 01 01 - 1877 01 01 - 1878 01 01 | Kockums, Malmo

Verdande (1877)
1877 01 01 - 1878 01 01 - 1879 01 01 | Karlskrona DYd

2 | "Blenda" type (1873)
Gunboats, Sweden

Displacement: 492 tons at draft: 3.35 m | Length: 51 m oa | Beam: 8.05 m | Crew: 71 men

Machinery: 4 cylindrical boilers | 2 * 290 HP vertical compound EE * 2 screws | 590 HP = 11.5 knots | Supplies: 81 tons of coal

Armament: 1-274/18, 1-122/23, 2-25.4/40 Mg

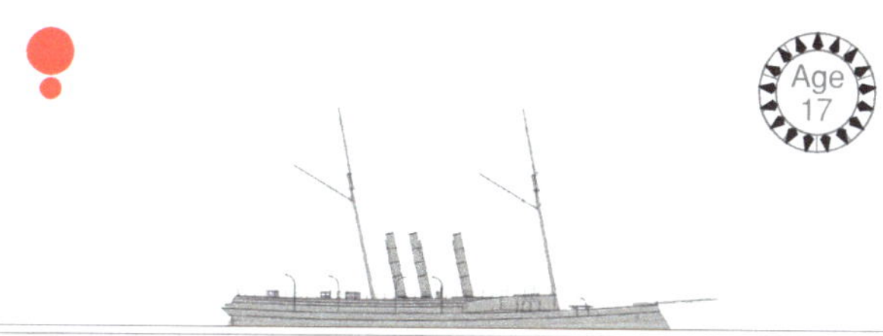

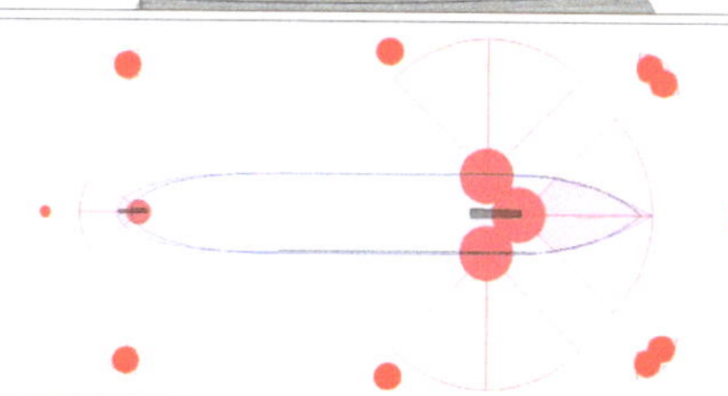

Blenda (1873)
1873 12 12 - 1874 10 16 - 1875 12 12 | Lindholmen, Goteborg

Disa (1875)
1875 01 01 - 1877 04 21 - 1878 01 01 | Karlskrona DYd

NORWEGIAN FLEET

National flag

Fleet age 12

Ensign Jack Admiral

Months derivation -47

Warships built from 1858 01 01 to 1890 01 01

27 warships = 16,939 tons = 627 tons / average ship

Fleet age - 1878 09 25

1889 fleet largest ship: Kong Sverre (1858) wooden screw frigate

Norwegian naval construction

Horten dockyard

25.6 tons / average ship / month

1890-Jan-01
2 ships building

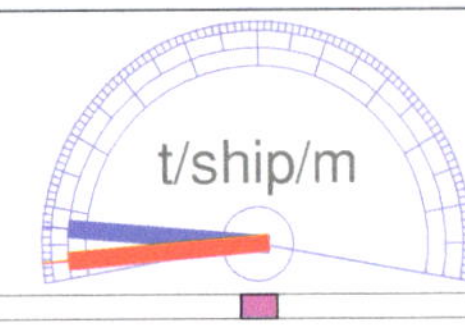

1,226 t.
tons / month = 51.2

t = 45
t/m =
1.9

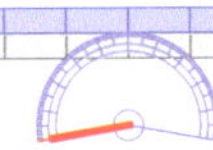

Otter

NO PLAN AVAILABLE

t = 1,181
t/m =
49.3

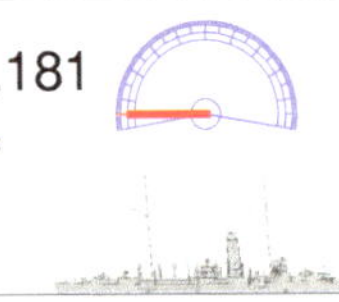

Viking

Horten, Norway

32+ years of naval construction
24 ships | 15,418 tons

The plant's largest ship:
Kong Sverre (1858) wooden screw frigate
3,485 tons

The plant's most effectively built ship:
Kong Sverre (1858) wooden screw frigate
72.8 tons/month

Specialisation in:
Gunboats | 12 ships | 5,563 tons
Monitors | 3 ships | 4,890 tons
Screw sailing cruiser | 1 ship | 3,485 tons

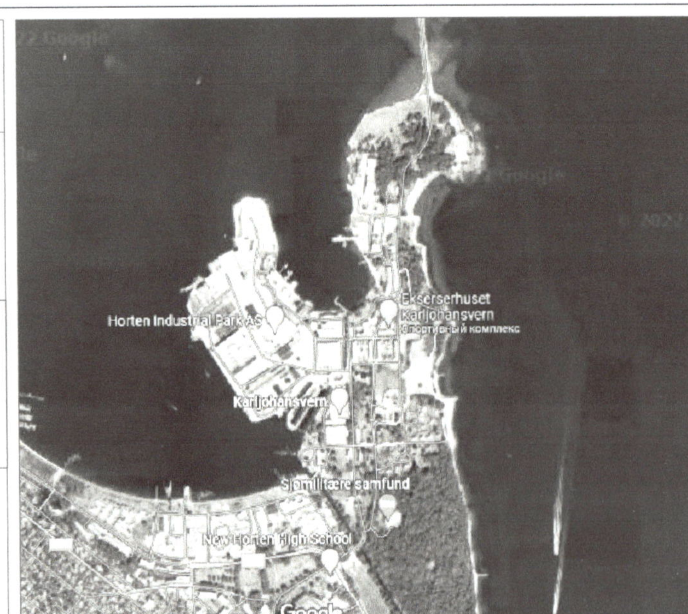

Norwegian fleet effective list

5 coast defence armoured ships

Thor

Trudvang

Mjolner

Skorpionen

1 wooden cruiser

Kong Sverre

8 torpedo boats

Orm
NO PLAN AVAILABLE

Snar
NO PLAN AVAILABLE

Pil
NO PLAN AVAILABLE

Rask
NO PLAN AVAILABLE

Springer
NO PLAN AVAILABLE

Od
NO PLAN AVAILABLE

Ulven

Rap

2 sloops & 10 gunboats

Ellida

Nordst-jerna
NO PLAN AVAILABLE

Gor

Tyr

Vidar
NO PLAN AVAILABLE

Brage
NO PLAN AVAILABLE

Nor
NO PLAN AVAILABLE

Sleipner
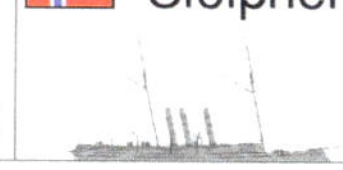

Uller

Vale

Glommen
NO PLAN AVAILABLE

Lougen
NO PLAN AVAILABLE

Norwegian section ship names glossary

B
- Brage (GB)

E
- Ellida (CZ)

G
- Glommen (GB)
- Gor (GB)

K
- Kong Sverre (CW)

L
- Lougen (GB)

M
- Mjolner (MM)

N
- Nor (GB)
- Nordstjerna (CZ)

O
- Od (TB1)
- Orm (TB1)
- Otter (TB1)

P
- Pil (TB1)

R
- Rap (TB2)
- Rask (TB1)

S
- Scorpionen (MM)
- Sleipner (GB)
- Snar (TB1)
- Springer (TB1)

T
- Thor (MM)
- Trudvang (MM)
- Tyr (GB)

U
- Uller (GB)
- Ulven (TB1)

V
- Vale (GB)
- Vidar (GB)
- Viking (CP)

Norwegian fleet naval guns

Calibre (mm)	Mark	Calibre (in)/pdr	Gun type	First Introduced	Weight (kg)	Gun lenght (in/cal)	Bore lenght (cal)	Charge weight (lbs)	Muzzle energy (ft/t)	Projectile weight (lbs)	Muzzle velocity (ft/s)	Wrought iron perforation (in)
267.0	Armstr., [1] 3	10.51"	MLR		21 700	202.4 / 19.26	16.70	77.2	7 463	316.4	1 549	15.5
267.0	Armstr., [1] 2	10.51"	MLR		19 700	175.8 / 16.72	13.80	77.2	5 692	316.4	1 444	13.4
260.0	Krupp	10.24"	BLR		21 700	225.2 / 22.00	19.00	81.6	7 966	381.4	1 575	16.2
202.0	Armstrong	7.94"	MLR		7 400	129.8 / 16.32	13.20	29.8	1 696	153.9	1 247	8.3
177.8	III	7"	MLR	1865	6 604	133.0 / 19.00	15.86	30.0	1 848	114.6	1 525	9.2
167.0	Palliser	6.58"	BLR	1865	4 900	139.0 / 21.14	17.00	16.5	1 345	82.7	1 329	8.2
155.0	Palliser	6.11"	BLR	1865	3 400	123.6 / 20.26	16.80	7.7	-	59.1	1 116	-
150.0	Krupp	5.87"	BLR	-	6 640	234.6 / 40.00	36.94	22.7	4 461	112.4	2 392	14.8
150.0	Krupp	5.87"	BLR	-	4 700	205.2 / 34.74	32.20	41.9	2 784	112.4	1 890	12.8
150.0	Krupp	5.87"	BLR	1865	3 940	151.6 / 25.67	23.00	20.9	1 447	69.4	1 558	9.1
70.0			BLR				25.00		-			
57.0	I & II	6 pr	BLR	1876	406	96.5 / 43.00	40.00	1.2	345	6.0	1 818	3.7
37.0	Hotchkiss R	1 pr	BLR	1873	209	33.5 / 23.0	29.13	0.2	-	1.1	1 450	-
37.0	Hotchkiss	1 pr	BLR	1884	72	33.5 / 23.0	33.15	0.2	-	1.1	1 450	-

WARSHIPS SECTION

Ship classes	Building (exports including)	Active
Battleships		
Boardside ironclads		
Armoured cruisers		
Coast defence armoured ships		4
Protected cruisers	1	
Unprotected cruisers		
Torpedo gunboats		
Torpedo boats 1 class	1	7
Torpedo boats 2 class		1
Submarines		
Wooden screw cruisers		1
Wooden screw sloops		1
Gunboats		
Gunboats coastal		10
Minelayers		

3 | "Thor" type (1870)
Monitor, Norway

MONITORS

NO PHOTO AVAILABLE

Displacement: 1,490 tons at draft: 3.5 m | Length: 63.33 m oa | Beam: 13.87 m | Crew: 85 men

Machinery: 2 rectangular boilers | 1 vibrating lever E * 1 screw | 450 HP = 8 knots | Supplies: 140 tons of coal

Armour: iron | CT: 152; Turret: 261 | WLBelt: 114(^25)

Armament: 2-267/13 MLR

Age 25

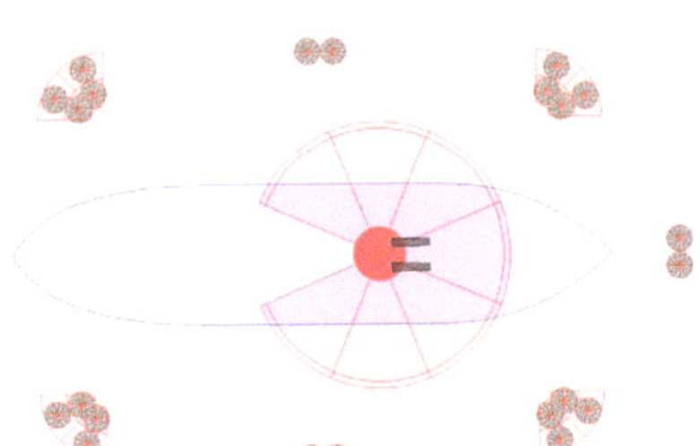

Mjolner (1866)
1866 01 01 - 1868 05 02 - 1870 01 01 | Motala Verks., Norrkoping

Thor (1870)
1870 01 01 - 1872 06 05 - 1874 01 01 | Horten DYd

Trudvang (1868)
1868 06 06 - 1869 05 03 - 1872 01 01 | Horten DYd

1 | "Scorpionen" type (1865)
Monitor, Norway

Displacement: 1,975 tons at draft: 3.81 m | Length: 62.33 m oa | Beam: 14.48 m | Crew: 95 men

Machinery: 4 rectangular boilers | 1 vibrating lever E * 1 screw | 600 HP = 8 knots | Supplies: 200 tons of coal

Armour: iron | CT: 229; Turret: 365 | WLBelt: 175(^25)

Armament: 2-267/13 MLR

Age 20

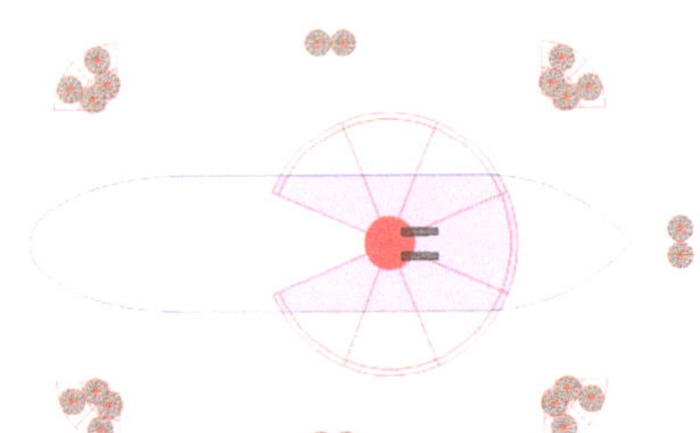

Scorpionen (1865)
1865 01 01 - 1866 10 30 - 1868 01 01 | Horten DYd

0 +1 | "Viking" type (1890)
Protected cruiser 3 class, Norway

BUILDING

PROTECTED CRUISER

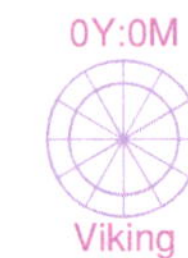

Displacement: 1,181 tons at draft: 3.66 m | Length: 63.48 m oa | Beam: 9.3 m | Crew: 125 men

Machinery: 4 cyl.l boilers | 2 * 1,000 HP vertical triple expansion EE * 2 screws | 2,000 HP = 15 knots | Supplies: 140 tons of coal

Armour: steel | CT: 35(^20); Casemates: 25 | WLDeck: 0(^35)

Armament: 2-150/40, 4-57/43, 6-37/23R, 2-356 TT (^)

Age 0

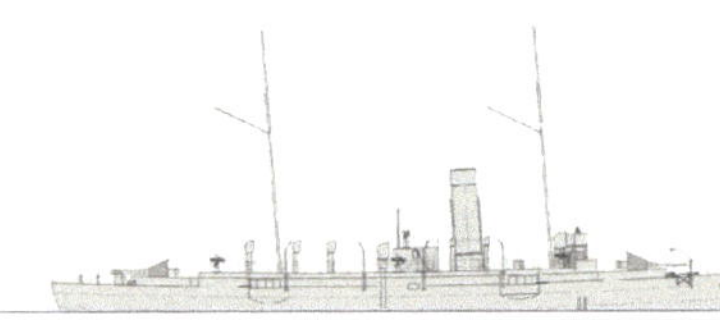

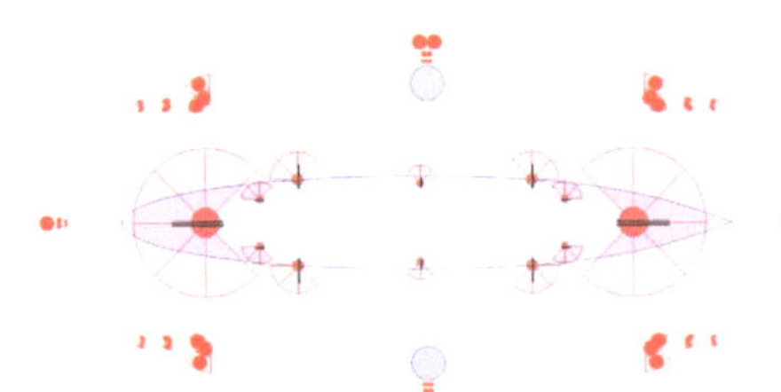

Viking (1890)
1890 01 01 - 1891 04 02 - 1892 01 01 | Horten DYd

WOODEN CRUISER

1 | "Kong Sverre" type (1858)
Wooden screw frigate, Norway

Displacement: 3,485 tons at draft: 6.81 m | Length: 66.5 m pp | Beam: 15.49 m | Crew: 550 men

Machinery: 2 rectangular boilers | 1 horizontal single expansion * 1 screw | 1,800 HP = 11.5 knots | Supplies: app. 500 tons of coal

Armament: 2-203.2/14.8 MLR, 4-170/19 MLR, 4-155/21 MLR

Age 32

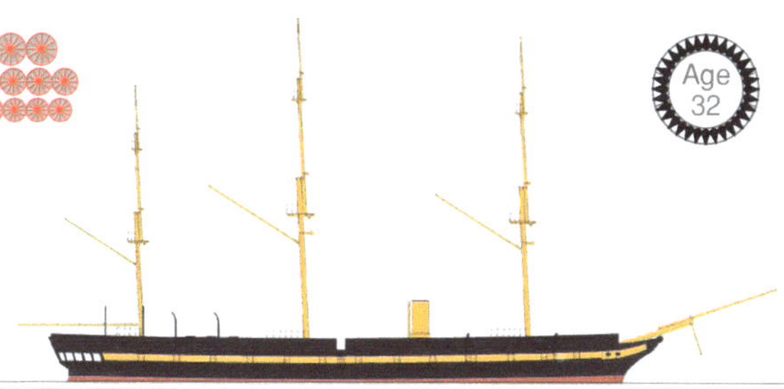

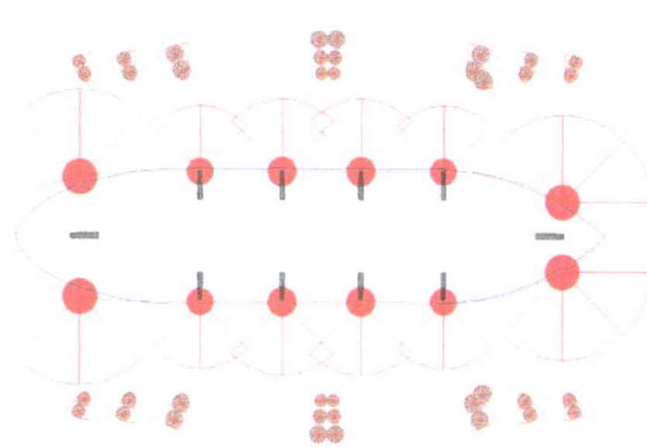

Kong Sverre (1858)
1858 01 01 - 1860 10 23 - 1862 01 01 | Horten DYd

TORPEDO SHIPS

1 +2 | "Snar" type (1886)
Torpedo boats, Norway

BUILDING

1Y:12M

Otter

Displacement: 45 tons at draft: 1.71 m | Length: 33 m oa | Beam: 3.72 m | Crew: 10 men

Machinery: 1 locomotive boiler | 1 compound expansion E * 1 screw | 500 HP = 18.5 knots | Supplies: 3 tons of coal

Armament: 1-37/23R, 2-356 TT (2 >>)

Orm (1887)
1887 06 06 - 1888 04 28 - 1889 06 06 | Horten DYd

Otter (1888)
1888 01 01 - 1888 12 13 - ***** | Horten DYd

Snar (1886)
1886 06 06 - 1887 07 25 - 1888 06 06 | Horten DYd

2 | "Rask" type (1884)
Torpedo boats, Norway

Displacement: 40 tons at draft: 0.76 m | Length: 30 m oa | Beam: 3.84 m | Crew: 13 men

Machinery: 1 locomotive boiler | 1 compound expansion E * 1 screw | 450 HP = 18 knots | Supplies: 3 tons of coal

Armament: 1-37/23, 2-356 TT (2 >>)

Pil (1885)
1885 01 01 - 1886 01 16 - 1887 01 01 | Horten DYd

Rask (1884)
1884 09 09 - 1885 10 31 - 1886 09 09 | Horten DYd

1 | "Springer" type (1882)
Torpedo boat, Norway

Displacement: 40 tons at draft: 0.76 m | Length: 30 m oa | Beam: 3.84 m | Crew: 13 men

Machinery: 1 locomotive boiler | 1 compound expansion E * 1 screw | 450 HP = 18 knots | Supplies: 3 tons of coal

Armament: 1-37/23R, 2-356 TT (2 >)

Springer (1882)
1882 09 09 - 1883 09 03 - 1884 09 09 | Horten DYd

1 | "Od" type (1881)
Torpedo boat, Norway

Displacement: 41 tons at draft: 0.76 m | Length: 28.98 m oa | Beam: 3.41 m | Crew: 13 men

Machinery: 1 locomotive boiler | 1 compound expansion E * 1 screw | 420 HP = 18 knots | Supplies: 3 tons of coal

Armament: 1-37/23, 1-356 TT (1 >)

Od (1881)
1881 01 01 - 1882 03 14 - 1883 06 06 | Horten DYd

1 | "Ulven" type (1877)
Spar torpedo boat, Norway

Displacement: 16 tons at draft: 0.76 m | Length: 17.07 m oa | Beam: 3 m | Crew: 7 men

Machinery: 1 locomotive boiler | 1 compound expansion E * 1 screw | 100 HP = 10 knots | Supplies: 1.5 tons of coal

Armament: 1 spar torpedo

Age 13

Ulven (1877)
1877 01 01 - 1878 01 01 - 1879 01 01 | Stockholm DYd

1 | "Rap" type (1873)
Spar torpedo boat, Norway

Displacement: 15 tons at draft: 0.9 m | Length: 17.34 m oa | Beam: 2.28 m | Crew: 7 men

Machinery: 1 locomotive boilers | 1 vertical double expansion E * 1 screw | 100 HP = 14 knots | Supplies: 1.5 tons of coal

Armament: 1 spar torpedo

Age 17

Rap (1873)
1873 01 01 - 1873 06 06 - 1874 01 01 | Thornycroft, Woolston

Sister ships can be found in the Danish and Swedish sections

SLOOPS & GUNBOATS

1 | "Ellida" type (1879)
Composite screw sloop, Norway

Displacement: 1,028 tons at draft: 4.4 m | Length: 58 m oa | Beam: 9.9 m | Crew: 150 men

Machinery: 2 cylindrical boilers | 2 * 450 HP vertical compound EE * 2 screws | 900 HP = 10.5 knots | Supplies: 97 tons of coal

Armament: 5-150/22, 1-120/32, 2-37/23R, 1-356 TT (1 >)

Age 11

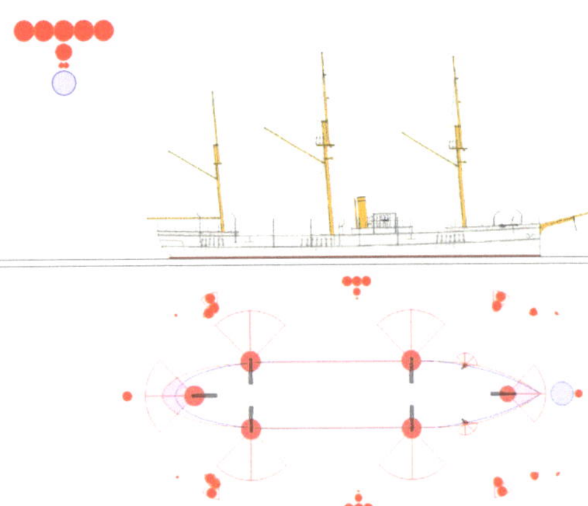

Elida (1879)
1879 01 01 - 1880 08 25 - 1882 01 01 | Horten DYd

1 | "Nordstjerna" type (1860)
Wooden screw sloop, Norway

Displacement: 1,610 tons at draft: 5.43 m | Length: 55.48 m pp | Beam: 11.48 m | Crew: 220 men

Machinery: 2 rectangular boilers | 1 horizontal triple expansion E * 1 screw | 800 HP = 10 knots | Supplies: 97 tons of coal

Armament: 6-155/18.5 MLR

Nordstjerna (1860)
1860 01 01 - 1862 01 01 - 1864 01 01 | Horten DYd

2 | "Gor" type (1882)
Gunboats, Norway

Displacement: 273 tons at draft: 2.2 m | Length: 31.27 m oa | Beam: 8.5 m | Crew: 44 men

Machinery: 2 cylindrical boilers | 2 * 210 HP vertical compound EE * 2 screws | 420 HP = 10.5 knots | Supplies: 22 tons of coal

Armament: 1-260/22, 1-57/43, 2-37/27R, 1-356 TT (1 >)

Age 8

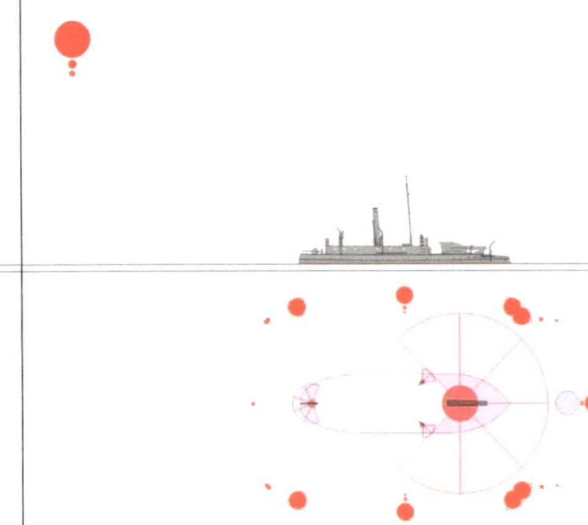

Gor (1882)
1882 06 06 - 1884 05 07 - 1886 01 01 | Horten DYd

Tyr (1885)
1885 01 01 - 1887 03 16 - 1889 01 01 | Horten DYd

1 | "Vidar" type (1879)
Wooden gunboat, Norway

Displacement: 262 tons at draft: 2.2 m | Length: 28.88 m oa | Beam: 7.9 m | Crew: 41 men

Machinery: 2 cylindrical boilers | 2 * 165 HP vertical compound EE * 2 screws | 330 HP = 9.5 knots | Supplies: 22 tons of coal

Armament: 1-270/14 MLR, 1-37/23, 2-37/27R

Vidar (1879)
1879 01 01 - 1882 01 03 - 1883 01 01 | Horten DYd

2 | "Brage" type (1877)
Wooden gunboats, Norway

Displacement: 264 tons at draft: 2.2 m | Length: 28.88 m oa | Beam: 7.9 m | Crew: 38 men

Machinery: 2 cylindrical boilers | 2 * 160 HP vertical compound EE * 2 screws | 220 HP = 8.5 knots | Supplies: 22 tons of coal

Armament: 1-270/14 MLR, 1-37/23, 2-37/23R

Brage (1877)
1877 01 01 - 1878 11 01 - 1880 01 01 | Horten DYd

Nor (1877)
1877 01 01 - 1878 10 23 - 1880 01 01 | Horten DYd

2 | "Glommen" type (1861)
Wooden gunboats, Norway

Displacement: 280 tons at draft: 2.8 m | Length: 41.1 m oa | Beam: 7.4 m | Crew: 50 men

Machinery: 1 rectangular boilers | 1 horizontal triple expansion E * 1 screw | 180 HP = 10 knots | Supplies: 22 tons of coal

Armament: 1-170/14.5 MLR, 1-155/18.5 MLR, 4-93.5/14 MLR

Glommen (1861)
1861 01 01 - 1863 01 01 - 1865 01 01 | Horten DYd

Lougen (1861)
1861 01 01 - 1863 01 01 - 1865 01 01 | Horten DYd

1 | "Blenda"/"Sleipner" type (1873)
Gunboat, Sweden

Displacement: 571 tons at draft: 3.35 m | Length: 53.26 m oa | Beam: 8.05 m | Crew: 71 men

Machinery: 4 cylindrical boilers | 2 * 315 HP vertical compound EE * 2 screws | 630 HP = 10 knots | Supplies: 81 tons of coal

Armament: 1-260/22, 1-150/25, 2-25.4/40 Mg

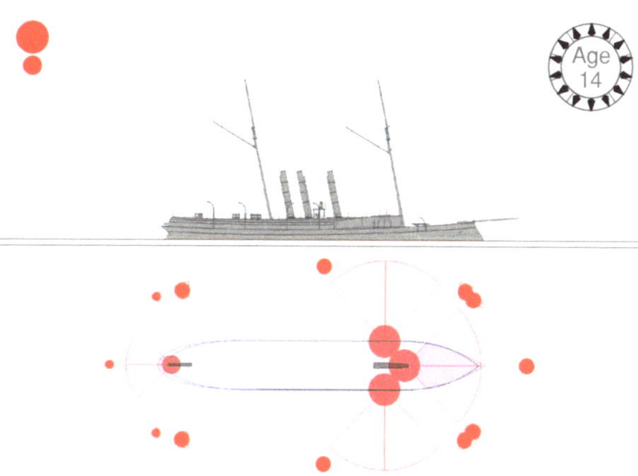

Sleipner (1877)
1877 01 01 - 1878 11 01 - 1880 01 01 | Horten DYd

2 | "Vale" type (1872)
Wooden gunboats, Norway

Displacement: 229 tons at draft: 2.2 m | Length: 27.28 m oa | Beam: m | Crew: 41 men

Machinery: 2 cylindrical boilers | 2 * 100 HP vertical compound EE * 2 screws | 200 HP = 8 knots | Supplies: 22 tons of coal

Armament: 1-270/14 MLR, 1-37/23, 2-37/27R

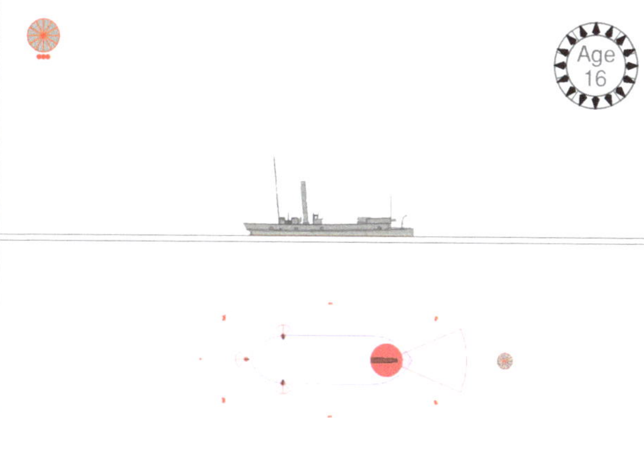

Uller (1874)
1874 06 06 - 1876 07 21 - 1878 01 01 | Horten DYd

Vale (1872)
1872 01 01 - 1874 04 14 - 1876 01 01 | Horten DYd

CHILEAN FLEET

National flag

Ensign

Jack

Admiral

Warships built from 1858 01 01 to 1890 01 01

20 warships = 16,536 tons = 827 tons / average ship

Fleet age - 1881 06 11

1889 fleet largest ship: Almirante Cochrane (1873) battery ironclad

Chilean fleet effective list

2 boardside ironclads / 1 coast defence armoured ship

- Almirante Cochrane
- Valparaiso
- Huascar

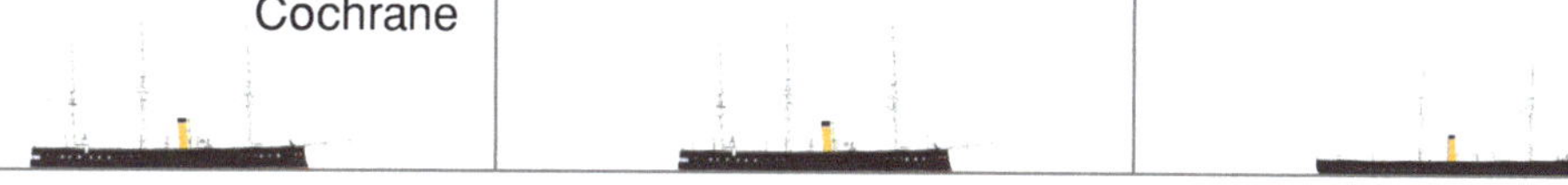

1 protected cruiser

- Esmeralda

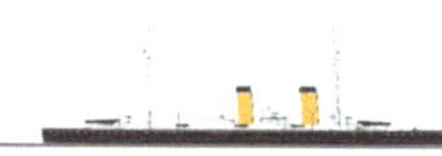

2 wooden cruisers

- Chacabuco
- Abtao (Training ship)

12 torpedo boats

- Sergente Aldea
- Glaura
- Guale
- Jane-quero
- Ricamilla
- Teguelda
- ??? — NO PLAN AVAILABLE
- Colocolo — NO PLAN AVAILABLE
- Tucapel — NO PLAN AVAILABLE
- Fresia
- Lanca
- Quidore

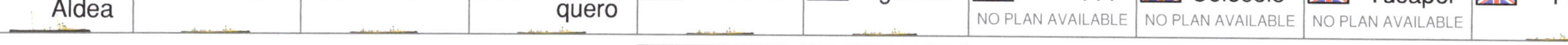

2 sloops

- Pilcomayo
- Magallanes

Chilean section ship names glossary

----- (TB2)

A
Abtao (trailS)
Almirante Cochrane (BO)

C
Chacabuco (CW)
Colocolo (TB2)

E
Esmeralda (CP)

F
Fresia (TB2)

G
Glaura (TB1)
Guale (TB1)

H
Huascar (MM)

J
Janequero (TB1)

L
Lanca (TB2)

M
Magallanes (GB)

P
Pilcomayo (GB)

Q
Quidore (TB2)

R
Ricamilla (TB1)

S
Sergente Aldea (TB1)

T
Teguelda (TB1)
Tucapel (TB2)

V
Valparaiso (BO)

Chilean fleet naval guns

Calibre (mm)	Mark	Calibre (in)/pdr	Gun type	First Introduced	Weight (kg)	Gun lenght (in/cal)	Bore lenght (cal)	Charge weight (lbs)	Muzzle energy (ft/t)	Projectile weight (lbs)	Muzzle velocity (ft/s)	Wrought iron perforation (in)
254.0	II, III, IV	10"	BLR	1887	29 465	342.4 / 34.24	32.00	252.0	14 430	500.0	1 914	22.3
240.0	Krupp	9.45"	BLR	1887	21 700	330.7 / 35.00			11 910	474.0	1 903	21.0
203.2	III	8"	MLR	1862	9 144	144.0 / 18.00	14.75	35.0	2 398	179.0	1 390	9.7
152.4	VI	6"	BLR	1873	5 080	173.5 / 28.92	26.00	45.0	2 424	100.0	1 870	11.9
150.0	Germany	5.9"	BLR	1864	3 440	128.8 / 21.80	19.10	14.3	1 418	65.0	1 555	8.0
120.0	40 pr / 35 cwt	4.7"	BLR		1 778	120.0 / 25.53	22.00	6.8	312	38.8	1 425	
80.0	Krupp	3.96"	BLR		230	61.8 / 20.00	17.40	0.9		8.3	1 053	
57.0	I & II	6 pr	BLR	1876	406	97.6 / 43.49	40.00	1.2	345	6.0	1 818	3.7
37.0	Hotchkiss R	1 pr	BLR	1873	209	33.5 / 23.20	20.00	0.2	-	1.1	1 450	-
37.0	Hotchkiss	1 pr	BLR	1884	72	33.5 / 23.00	20.00	0.2	-	1.1	1 450	-

WARSHIPS SECTION

Ship classes	Building	Active
Battleships		
Boardside ironclads		2
Armoured cruisers		
Coast defence armoured ships		1
Protected cruisers		1
Unprotected cruisers		
Torpedo gunboats		
Torpedo boats 1 class		6
Torpedo boats 2 class		6
Submarines		
Wooden screw cruisers		2
Wooden screw sloops		2
Gunboats		
Gunboats coastal		

0 +1 | "Capitan Prat" type (1889)
Battleship, Chile

BUILDING IN FRANCE

BATTLESHIP

0Y:12M

Capitan Prat

Displacement: 6,901 tons at draft: 6.96 m | Length: 99.97 m pp | Beam: 18.49 m | Crew: 480 men

Machinery: 5 cylindrical boilers | 2 horizontal triple expansion * 2 screws | 12,000 HP = 18.3 knots | Supplies: 400 tons of coal

Armour: steel | Conning tower: 270; Turrets: 275/100; Barbettes: 275/50; UPBelt: 0 =80= 100 =80= 0 | WLBelt: 150>100(^50) 300>200 125>100(^50)

Armament: 4-240/35, 8-120/45, 6-57/43, 4-47/43, 10-37/23, 4-450 TT (^)

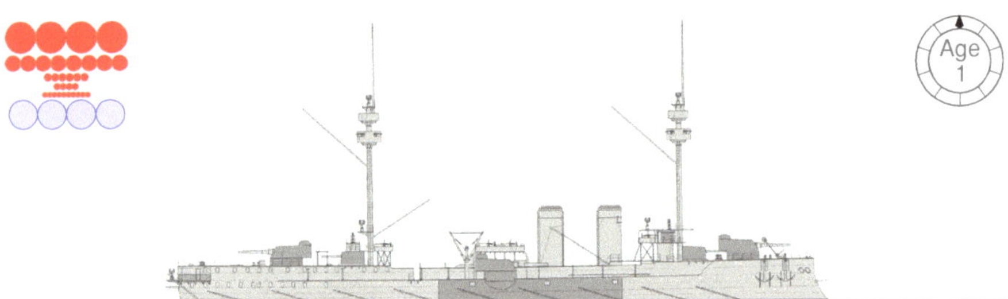

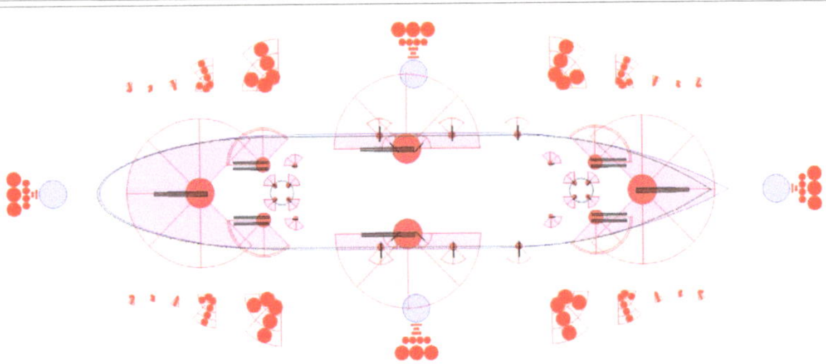

Capitan Prat (1889)

1889 01 01 - ***** - ***** | Forges et Chantiers de la Méditerranée, La Seyne

1 | "Huascar" type (1864)
Monitor, Peru

MONITOR

Displacement: 2,030 tons at draft: 5.5 m | Length: 57.91 m oa | Beam: 10.67 m | Crew: 170 men

Machinery: 4 rectangular boilers | 1 single expansion * 1 screw | 1,650 HP = 12.3 knots | Supplies: 200 tons of coal

Armour: iron | Conning tower: 76; Turret: 203-140 | WLBelt: 51 114 51

Armament: 2-254/14.5 MLR, 2-148/14.8 MLR

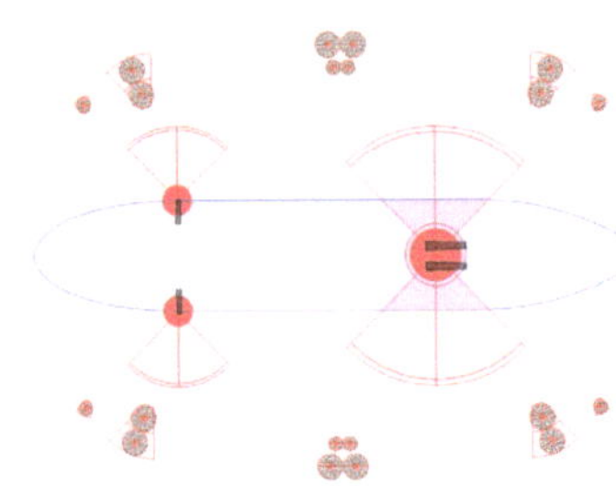

Huascar (1864)

1864 01 01 - 1865 10 06 - 1866 11 08 | Laird, Birkenhead

2 | "Almirante Cochrane" type (1873)
Battery ironclads, Chile

BATTERY IRONCLADS

Displacement: 3,370 tons at draft: 6.65 m | Length: 64.01 m pp | Beam: 13.94 m | Crew: 300 men

Machinery: 6 cylindrical boilers | 2 horizontal trunk * 2 screws | 2,920 HP = 12.75 knots | Supplies: 500 tons of coal

Armour: iron | Conning tower: 76; Battery: 203-152(^13) | WLBelt: 114(^13) 152<229>152(^13) 114(^13)

Armament: 6-228.6/13.9 MLR, 1-120/22, 2-63.5/19, 1-25.4/40 Mg

Age 17

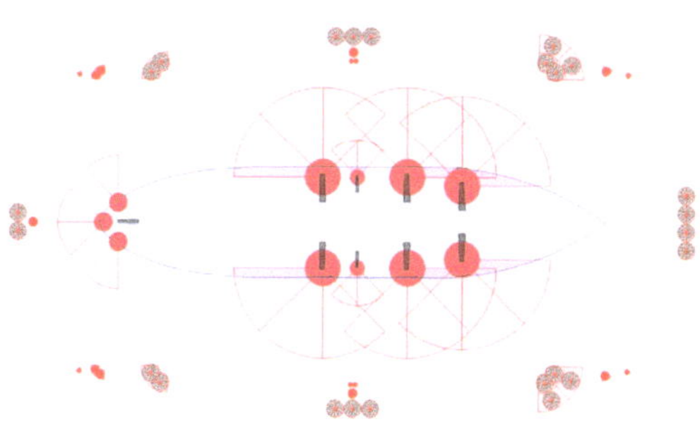

Almirante Cochrane (1873)
1873 01 01 - 1874 01 25 - 1874 12 12 | Earle, Hull

Valparaiso (1873)
1873 01 01 - 1875 01 01 - 1875 12 12 | Earle, Hull

0 +2 | "Presidente Errazuriz" type (1889)
Protected cruisers 2 class, Chile

PROTECTED CRUISERS

BUILDING IN FRANCE

Displacement: 2,047 tons at draft: 4.39 m | Length: 81.79 m pp | Beam: 7.9 m | Crew: 170 men

Machinery: 4 cylindrical boilers | 2 horizontal triple expansion * 2 screws | 5,400 HP = 18.35 knots | Supplies: 200 tons of coal

Armour: steel | Conning tower: 50 | WLDeck: 0(35^25) 0(60^35) 0(35^25)

Armament: 4-150/35, 2-120/35, 4-57/43, 4-37/23, 3-356 TT (^)

Age 1

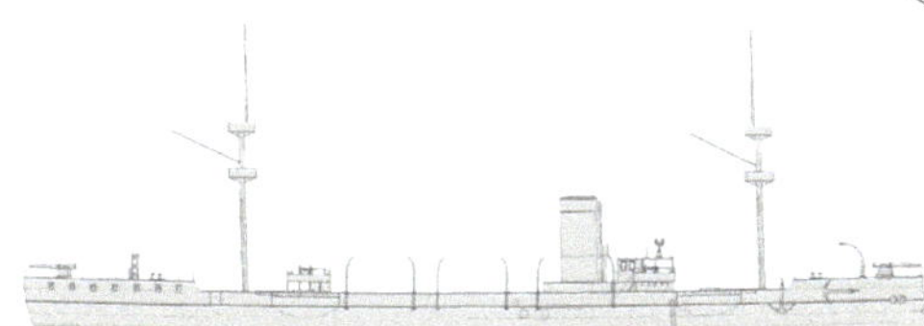

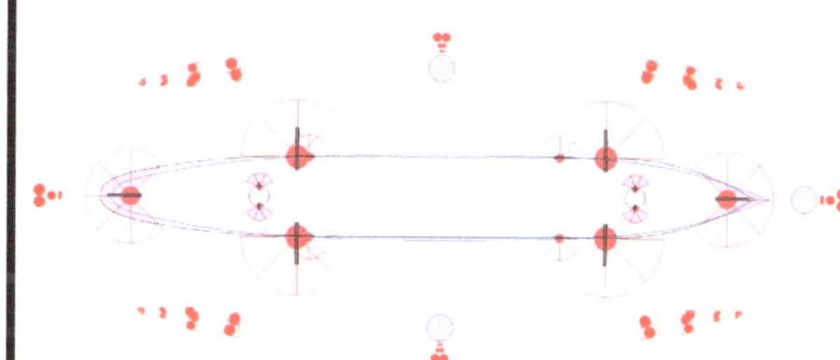

Presidente Errazuriz (1889)
1889 01 01 - ***** - ***** | F. et Ch. de la Méditerranée, La Seyne

Presidente Pinto (1889)
1889 01 01 - ***** - ***** | F. et Ch. de la Méditerranée, La Seyne

1 | "Esmeralda" type (1881)
Protected cruiser 2 class, Chile

Displacement: 2,950 tons at draft: 5.64 m | Length: 82.3 m pp | Beam: 12.8 m | Crew: 296 men

Machinery: 4 cylindrical boilers*2 | 2 horizontal compound * 2 screws | 6,803 HP = 18 knots | Supplies: 400 / +200 tons of coal

Armour: steel | Conning tower: 51; Hosts: 25 | WLDeck: 0/25(^13)

Armament: 2-254/32, 6-152.4/26, 2-57/43, 6-37/23

Age 9

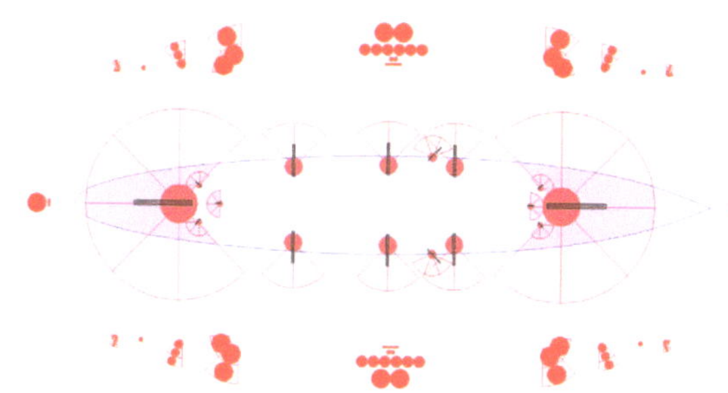

Esmeralda (1881)
1881 01 01 - 1883 06 06 - 1884 07 15 | Armstrong, Elswick

WOODEN CRUISERS

1 | "Chacabuco" type (1865)
Wooden screw corvette, Chile

Displacement: 1,101 tons at draft: 5.3 m | Length: 66.82 m pp | Beam: 10.18 m | Crew: 200 men

Machinery: 4 rectangular boilers | 1 horizontal expansion * 1 screw | 1,200 HP = 10 knots | Supplies: 347 tons of coal

Armament: 2-215/15, 3-210/15, 1-127/15

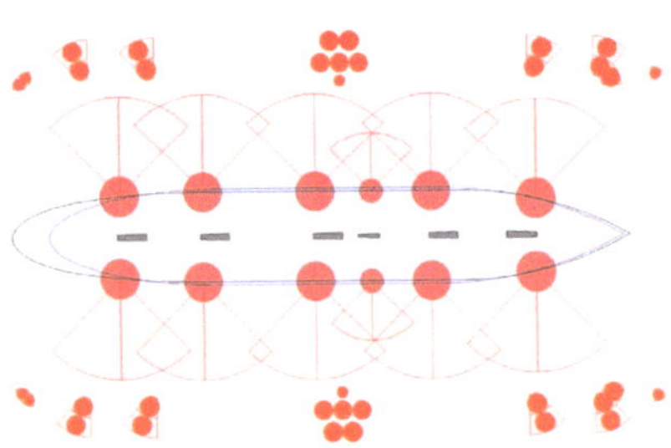

Chacabuco (1865)
1865 01 01 - 1866 01 01 - 1868 01 01 | Ravenhill, Low Walker

1 | "Abtao" type (1864)
Composite screw corvette, Chile

Displacement: 1,600 tons at draft: 5.13 m | Length: 69.29 m oa | Beam: 8.99 m | Crew: 130 men

Machinery: 4 rectangular boilers | 1 single expansion * 1 screw | 800 HP = 10 knots | Supplies: 300 tons of coal

Armament: 1-150/17, 4-120/18

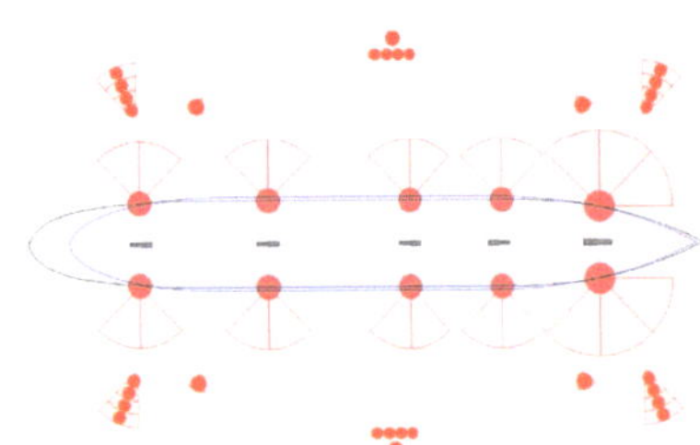

Abtao (1864)
1864 01 01 - 1865 01 01 - 1866 01 01 | Denny, Glasgow

TORPEDO SHIPS

0 +2 | "Almirante Lynch" type (1889)
Torpedo gunboats, Chile

BUILDING IN BRITAIN

Displacement: 713 tons at draft: 2.53 m | Length: 70.1 m pp | Beam: 8.38 m | Crew: 87 men

Machinery: 4 locomotive boilers | 2 triple expansion * 2 screws | 4,532 HP = 20.3 knots | Supplies: 100 /+150 tons of coal

Armour: steel | WLDeck: 0(^25)

Armament: 3-76,2/40, 4-47/43, 5-356 TT (4*1 trn, 1 >)

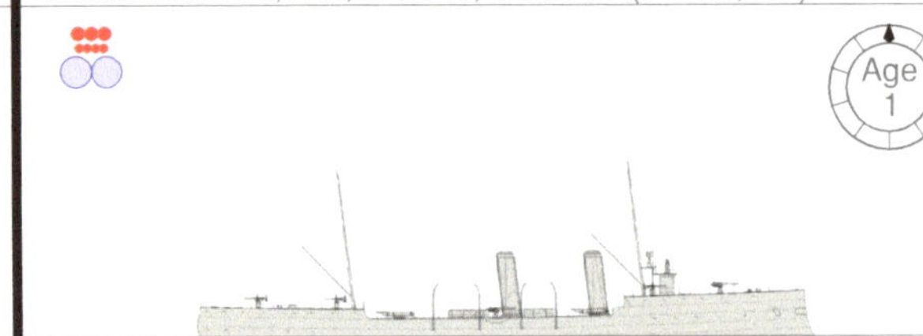

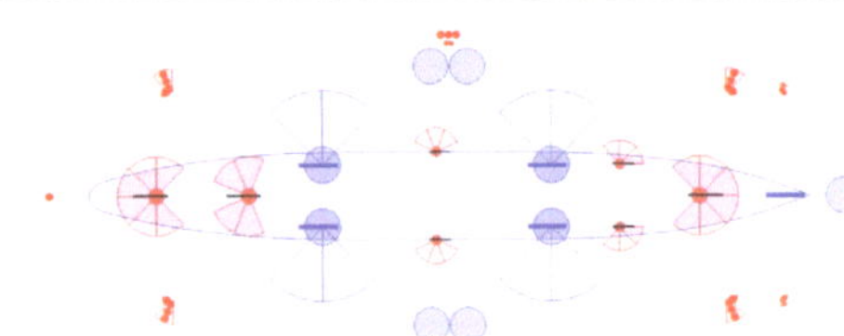

Almirante Condell (1889)
1889 01 01 - 1890 01 01 - 1890 12 12 | Laird, Birkenhead

Almirante Lynch (1889)
1889 01 01 - 1890 01 01 - 1890 12 12 | Laird, Birkenhead

0 +3 | "Ariete"/"7" type (1886)
Torpedo boats 1 class, Spain

BUILDING IN BRITAIN

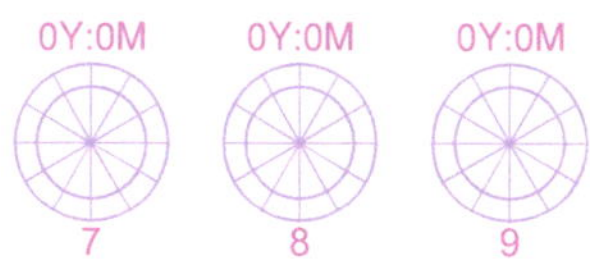

Displacement: 97 tons at draft: 1.07 m | Length: 44.96 m oa | Beam: 4.34 m | Crew: 27 men

Machinery: 2 Thornycroft boilers | 2 compound * 2 screws | 1,800 HP = 25 knots | Supplies: 25 tons of coal

Armament: 2-47/43, 3-381 TT (2*1 trn, 1 >)

Age 4

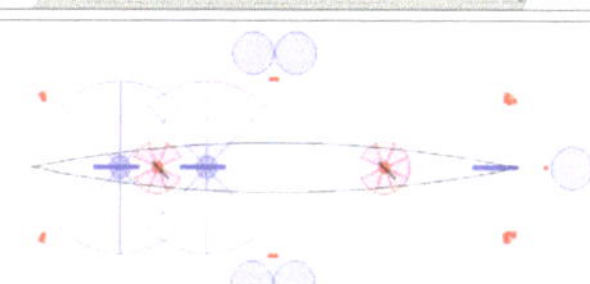

7 (1890)
1890 01 01 - ***** - ***** | Thornycroft, Woolston

8 (1890)
1890 01 01 - ***** - ***** | Thornycroft, Woolston

9 (1890)
1890 01 01 - ***** - ***** | Thornycroft, Woolston

Sister ships can be found in the Argentinean, Brazilean, French and Spanish sections

1 | "Ardjoeno"/"Sergente Aldea" type (1886)
Torpedo boat 1 class, Netherlands

Displacement: 75 tons at draft: 2.13 m | Length: 38.1 m oa | Beam: 3.96 m | Crew: 16 men

Machinery: 1 locomotive boiler | 1 vertical compound * 1 screw | 700 HP = 20 knots | Supplies: 12 / +8 tons of coal

Armament: 2-25.4/40 Mg, 2-356 TT (>>)

Age 4

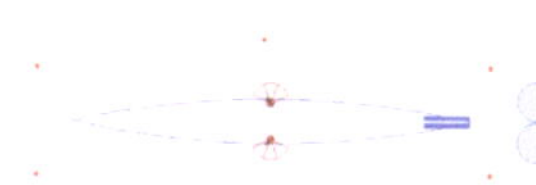

Sergente Aldea (1886)
1886 01 01 - 1886 06 06 - 1887 01 01 | Yarrow, Poplar

Sister ships can be found in the Dutch section

1 | White type (1887)
Torpedo boat 3 class, Chile

Displacement: 5 tons at draft: 0.45 m | Length: 15.24 m oa | Beam: 2.74 m | Crew: 5 men

Machinery: 1 locomotive boilers | 1 horizontal direct acting * 1 screw | 80 HP = 15 knots | Supplies: app. 1 ton of coal

Armament: 1 Spar torpedo

unnamed (1887)
1887 01 01 - 1887 06 06 - 1888 01 01 | White, Cowes

5 | "Батум"/"Glaura" type (1879)
Torpedo boats 1 class, Russia

Displacement: 40 tons at draft: 0.95 m | Length: 30.48 m oa | Beam: 3.81 m | Crew: 15 men

Machinery: 1 locomotive boiler | 1 horizontal compound * 1 screw | 400 HP = 18.5 knots | Supplies: 9 tons of coal

Armament: 2-25.4/40 Mg, 2-356 TT (>>), 1 Spar torpedo

Age 9

Glaura (1881)
1881 01 01 - 1881 06 06 - 1882 01 01 | Yarrow, Poplar

Guale (1881)
1881 01 01 - 1881 06 06 - 1882 01 01 | Yarrow, Poplar

Janequero (1881)
1881 01 01 - 1881 06 06 - 1882 01 01 | Yarrow, Poplar

Ricamilla (1879)
1879 01 01 - 1879 06 06 - 1880 01 01 | Yarrow, Poplar

Teguelda (1879)
1879 01 01 - 1879 06 06 - 1880 01 01 | Yarrow, Poplar

Sister ships can be found in the Argentinean, Austria-Hungarian, Brazilean, British, Dutch, Greek, Japanese, and Russian sections

SLOOPS

3 | "III"/"Fresia" type (1879)
Spar torpedo boats 2 class, Austria-Hungary

Displacement: 25 tons at draft: 0.95 m | Length: 26.6 m oa | Beam: 3.3 m | Crew: 15 men
Machinery: 1 locomotive boiler | 1 compound * 1 screw | 400 HP = 19 knots | Supplies: app. 4 tons of coal
Armament: 1 Spar torpedo

Fresia (1879)
1879 01 01 - 1879 06 06 - 1880 01 01 | Yarrow, Scottstown

Lanca (1879)
1879 01 01 - 1879 06 06 - 1880 01 01 | Yarrow, Scottstown

Quidore (1879)
1879 01 01 - 1879 06 06 - 1880 01 01 | Yarrow, Scottstown

2 | "Colocolo" type (1879)
Spar torpedo boats 3 class, Chile

Displacement: 5 tons at draft: 0.45 m | Length: 14.63 m oa | Beam: 2.44 m | Crew: 5 men
Machinery: 1 locomotive boiler | 1 horizontal direct acting * 1 screw | 40 HP = 12 knots | Supplies: app. 1 ton of coal
Armament: 1 Spar torpedo

Colocolo (1879)
1879 01 01 - 1880 01 01 - 1881 01 01 | Yarrow, Scottstown

Tucapel (1879)
1879 01 01 - 1880 01 01 - 1881 01 01 | Yarrow, Scottstown

1 | "Pilcomayo" type (1873)
Wooden screw sloop, Peru

Displacement: 800 tons at draft: 3.51 m | Length: 55.4 m oa | Beam: 7.47 m | Crew: 130 men
Machinery: 2 cylindrical boilers | 1 horizontal compound * 1 screw | 1,080 HP = 11 knots | Supplies: app. 50 tons of coal
Armament: 2-150/22, 2-101.6/22

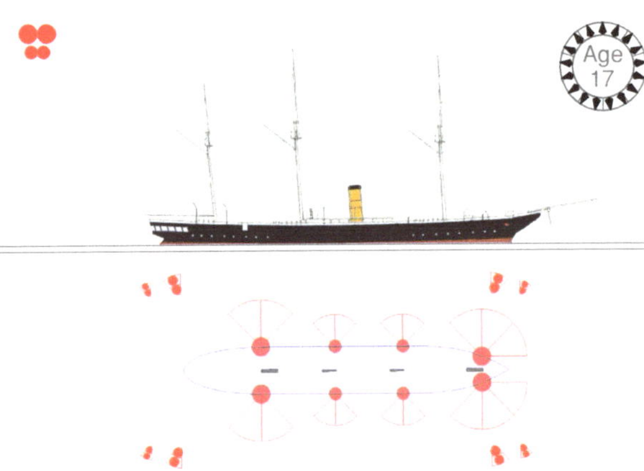

Pilcomayo (1873)
1873 01 01 - 1874 01 01 - 1875 01 01 | Wigram, Blackwall

1 | "Magallanes" type (1871)
Wooden screw sloop, Chile

Displacement: 950 tons at draft: 3.51 m | Length: 61 m oa | Beam: 9.09 m | Crew: 143 men
Machinery: 4 cylindrical boilers | 2 horizontal compound * 2 screws | 1,040 HP = 11 knots | Supplies: 300 tons of coal
Armament: 1-177.8/14.2 MLR, 1-160/16.5 MLR, 2-95.3/16

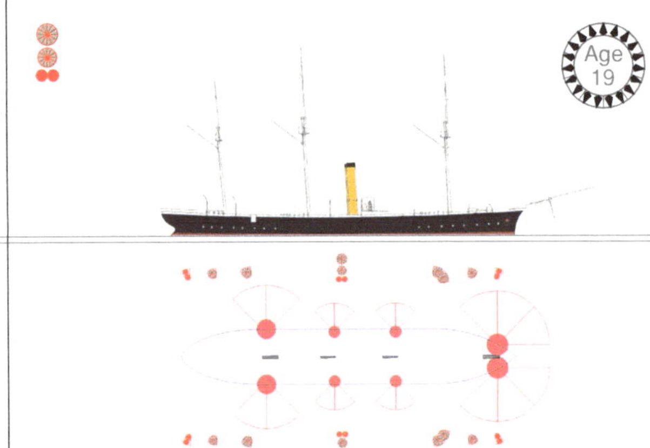

Magallanes (1871)
1871 01 01 - 1872 01 01 - 1874 01 01 | Green, Blackwall

GREEK FLEET

National flag

Fleet age 9

Ensign

Jack

Admiral

Months derivation -10

Warships built from 1858 01 01 to 1890 01 01

34 warships = 10,907 tons = 321 tons / average ship

Fleet age - 1881 11 04

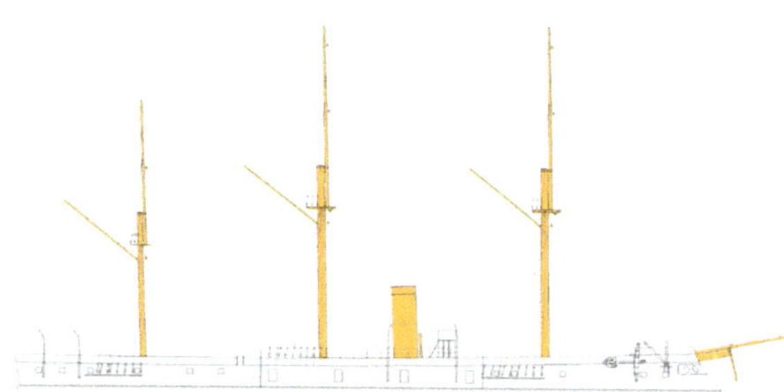

1890 fleet largest ship: Basilissa Olga (1868) battery ironclad

Greek fleet effective list

2 boardside ironclads

Basilissa Olga

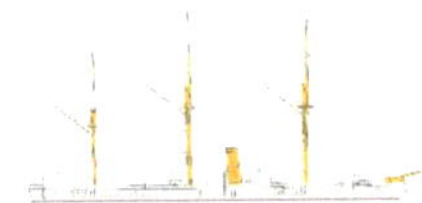

Basileos Georgious

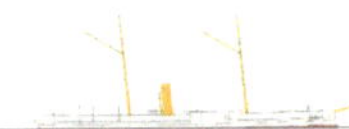

1 unprotected / 1 wooden cruisers

Nauarchos Miaoulis

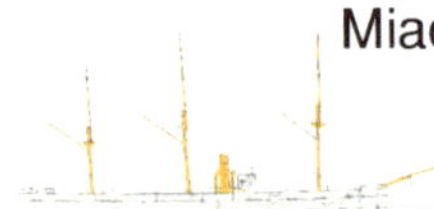

Hellas

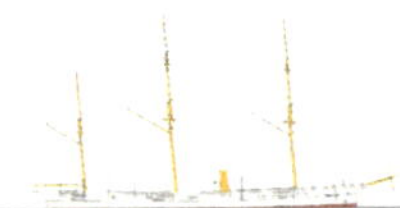

20 torpedo boats

V 11 | V 12 | V 13 | Chios | Kas | Kypros | Mytelene | Rodos | Samos | Kalithea

Perinika | Perse-phone | Terpsi-chore | Terpsi-thea | Alfa (NO PLAN AVAILABLE) | Beta (NO PLAN AVAILABLE) | Delta (NO PLAN AVAILABLE) | Gamma (NO PLAN AVAILABLE) | Ionia | Sphinx

1 submarine

Norden-feldt 1

14 gunboats

Aidon

Kilchi

Kissa

Achelaos

Alpheos

Eurotas

Peneos

Akteon

Ambrakia

Paralos — NO PLAN AVAILABLE

Salamnia — NO PLAN AVAILABLE

Afroessa — NO PLAN AVAILABLE

Plixavra — NO PLAN AVAILABLE

Syros — NO PLAN AVAILABLE

Greek section ship names glossary

A
Achelaos (GB)
Afroessa (GB)
Aidon (GB)
Akteon (GB)
Alfa (TB2)
Alpheos (GB)
Ambrakia (GB)

B
Basileos Georgious (BZA)
Basilissa Olga (BZA)
Beta (TB2)

C
Chios (TB1)

D
Delta (TB2)

E
Eurotas (GB)

G
Gamma (TB2)

H
Hellas (CZ)

I
Ionia (TB2)

K
Kalithea (TB1)
Kas (TB1)
Kilchi (GB)
Kissa (GB)
Kypros (TB1)

M
Mytelene (TB1)

N
Nauarchos Miaoulis (CS)
Nordenfeldt I (SS)

P
Paralos (GB)
Peneos (GB)
Perinika (TB1)
Persephone (TB1)
Plixavra (GB)

R
Rodos (TB1)

S
Salamnia (GB)
Samos (TB1)
Sphinx (TB2)
Syros (GB)

T
Terpsichore (TB1)
Terpsithea (TB1)

V
V 11 (TB1)
V 12 (TB1)
V 13 (TB1)

Greek fleet naval guns

Calibre (mm)	Mark	Calibre (in)/pdr	Gun type	First Introduced	Weight (kg)	Gun lenght (in/cal)	Bore lenght (cal)	Charge weight (lbs)	Muzzle energy (ft/t)	Projectile weight (lbs)	Muzzle velocity (ft/s)	Wrought iron perforation (in)
275.0		10.8"	BLR	1870	22 800	212.4 / 19.62	18.00	126.8	8 515	396.8	1 608	16.2
170.0	Krupp	6.8"	BLR	1864	5 510	167.3 /	21.90	27.6	2 226	112.4	1 558	10.5
170.0						/ 20.00						
150.0	Krupp	5.9"	BLR		4 700	205,2 / 34,71	32.20	41.9	2 784	112.4	1 890	12.8
150.0	Krupp	5.9"	BLR	1878	4 040	176.0 / 29.80	27.20	33.1		112.4	1 624	11.0
150.0	Krupp	5.9"	BLR	1865	3 940	151.6 /	23.00	20.9	1 447	69.4	1 558	9.1
100.0	1881	3.94"	BLR		1 180	103.2 /	26.00	9.9	-	9.9	1 673	-
95.3	I	20 pr	BLR	-	610	107.5 / 28.65	27.00	6.0	380	19.5	1 677	-
37.0	Hotchkiss R	1 pr	BLR	1873	209	33.5 / 23.00	19.13	0.2	-	1.1	1 450	-
37.0	Hotchkiss	1 pr	BLR	1884	72	33.5 / 23.00	19.15	0.2	-	1.1	1 450	-

WARSHIPS SECTION

Ship classes	Building (exports including)	Active
Battleships		
Boardside ironclads		2
Armoured cruisers		
Coast defence armoured ships		
Protected cruisers		
Unprotected cruisers		1
Torpedo gunboats		
Torpedo boats 1 class		14
Torpedo boats 2 class		6
Submarines		1
Wooden screw cruisers		1
Wooden screw sloops		
Gunboats		
Gunboats coastal		14
Minelayers		

0 +3 | "Spetsai" type (1888)
Battleships, Greece

BUILDING IN FRANCE

BATTLESHIPS

0Y:12M Psara | 1Y:12M Hydra | 1Y:12M Spetsai

Displacement: 4,808 tons at draft: 5.49 m | Length: 102 m pp | Beam: 15.8 m | Crew: 440 men

Machinery: 4 cylindrical boilers | 2 * 3,350 HP vertical triple expansion EE * 2 screws | 6,700 HP = 17 knots | Supplies: 500 / +190 tons of coal

Armour: steel | CT: 100; Barbettes: 355-300 |
WLBelt: 100(^57) 300>76(^57) 100(^57)

Armament: 2-275/34, 1-275/28, 5-150/35, 4-90/22, 6-37/23R, 4-37/23, 3-356 TT (^)

Age 2

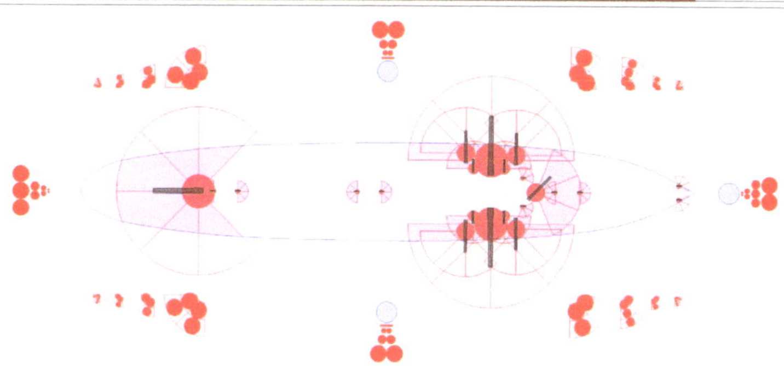

Hydra (1888)
1888 01 01 - 1889 05 15 - ***** | Loire, Saint-Nazaire

Psara (1889)
1889 01 01 - ***** - ***** | Forges et Chantiers de la Méditerranée, La Seyne

Spetsai (1888)
1888 01 01 - 1889 10 26 - ***** | Forges et Chantiers de la Méditerranée, La Seyne

1 | "Basilissa Olga" type (1868)
Battery ironclad, Greece

BATTERY IRONCLADS

Displacement: 2,030 tons at draft: 5.79 m | Length: 76 m pp | Beam: 11.89 m | Crew: 258 men

Machinery: 4 rectangular boilers | 1 horizontal compound E * 1 screw | 1,950 HP = 10 knots | Supplies: 240 tons of coal

Armour: iron | Battery: 121 |
WLBelt: 152(^25)

Armament: 2-228.6/13.9 MLR, 8-165/14

Age 22

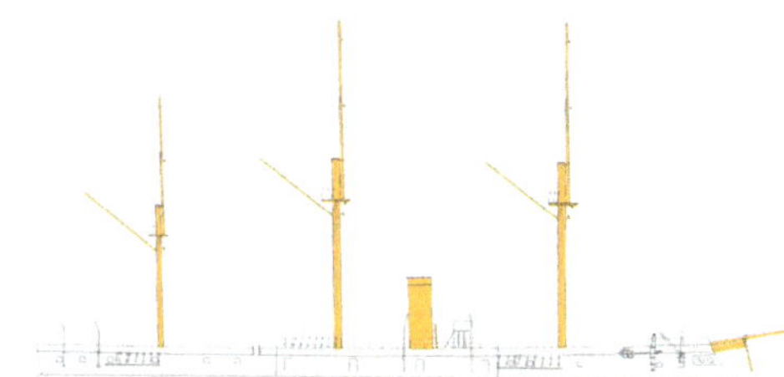

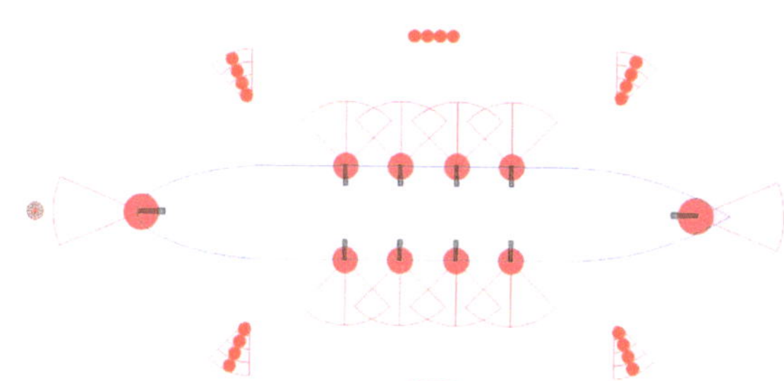

Basilissa Olga (1868)
1868 01 01 - 1869 01 26 - 1870 01 01 | STT, Trieste

1 | "Basileos Georgious" type (1866)
Battery ironclad, Greece

Displacement: 1774 tons at draft: 4.88 m | Length: 60.96 m pp | Beam: 10.06 m | Crew: 152 men

Machinery: 4 rect. boilers | 2 * 1,050 HP horizontal compound EE * 2 screws | 2,100 HP = 12.2 knots | Supplies: 210 tons of coal

Armour: iron | Battery: 152 | WLBelt: 114(^25) 178(^25) 114(^25)

Armament: 2-228.6/13.9 MLR, 2-95.3/13

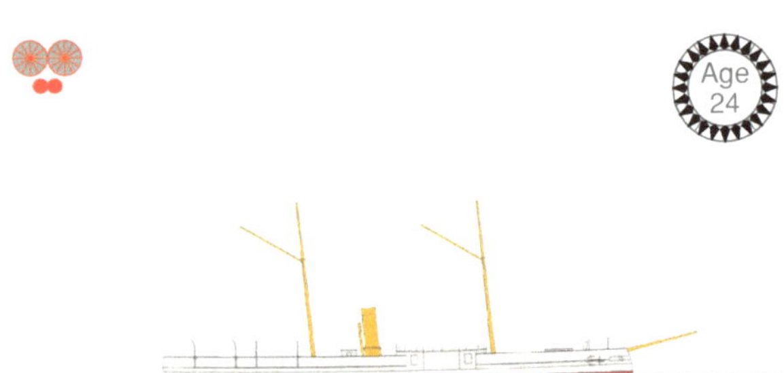

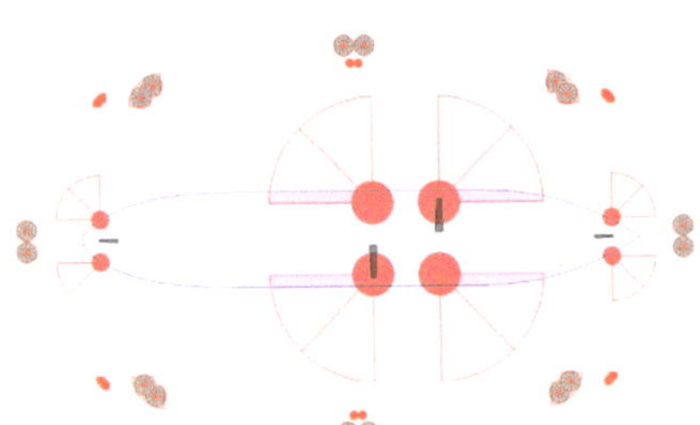

Basileos Georgious (1866)
1866 01 01 - 1867 12 28 - 1868 12 12 | Thames IW, Blackwall

1 | "Nauarchos Miaoulis" type (1877)
Unprotected cruiser, Greece

UNPROTECTED CRUISER

Displacement: 1,820 tons at draft: 4.4 m | Length: 75 m pp | Beam: 11 m | Crew: 180 men

Machinery: 6 cylindrical boilers | 1 horizontal compound E * 1 screw | 2,440 HP = 13.5 knots | Supplies: 220 tons of coal

Armament: 3-165/25, 1-165-/20, 6-37/23R, 3-350 TT (^)

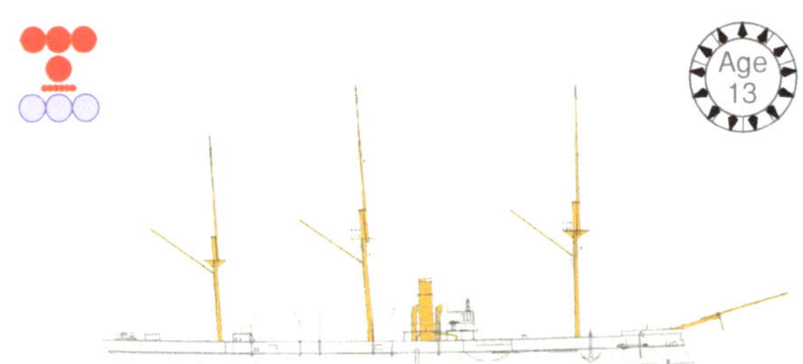

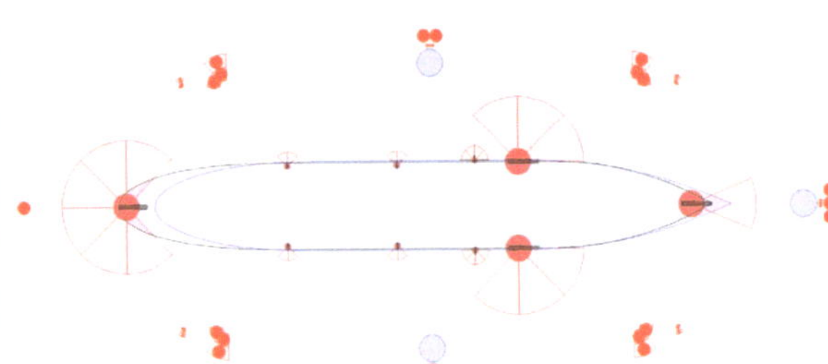

Nauarchos Miaoulis (1877)
1877 01 01 - 1879 01 01 - 1881 01 01 | F.et Ch.de la M., La Seyne

1 | "Hellas" type (1860)
Wooden screw corvette, Greece

WOODEN CRUISER

NO PHOTO AVAILABLE

Displacement: 1,627 tons at draft: 4.27 m | Length: 61 m oa | Beam: 11 m | Crew: 150 men

Machinery: 2 rectangular boilers | 1 horizontal single expansion E * 1 screw | 700 HP = 8 knots | Supplies: app. 150 tons of coal

Armament: 12-150/16

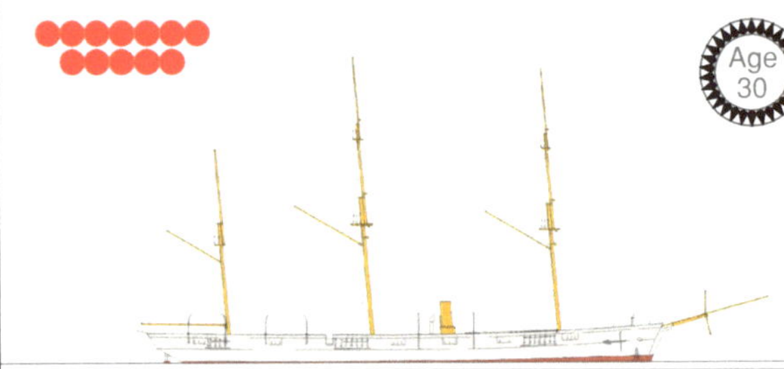

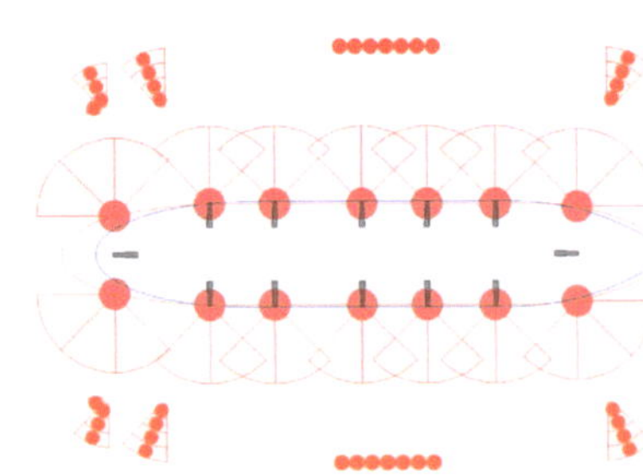

Hellas (1860)
1860 01 01 - 1861 01 01 - 1862 01 01 | Pitcher, Northfleet

TORPEDO SHIPS

3 +3 | "V 11" type (1887)
Torpedo boats, Greece

BUILDING IN GERMANY

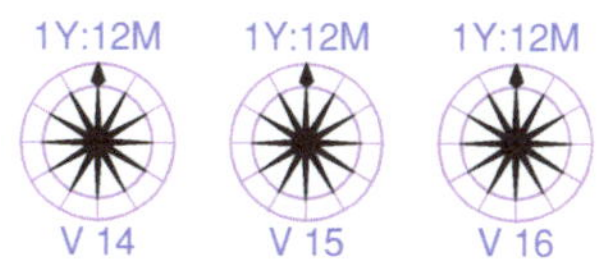

Displacement: 85 tons at draft: 1.65 m | Length: 39 m oa | Beam: 4.66 m | Crew: 20 men

Machinery: 1 locomotive boiler | 1 compound reciprocating E * 1 screw | 1,000 HP = 18 knots | Supplies: 20 tons of coal

Armament: 1-37/23R, 3-356 TT (1 trainable, 2 >>)

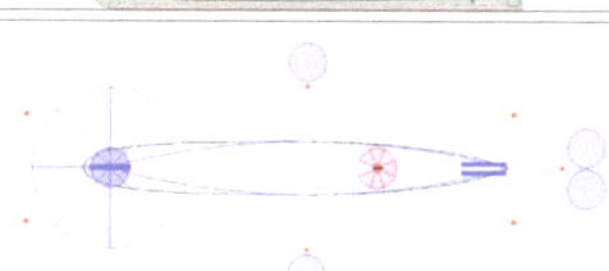

V 11 (1887)
1887 01 01 - 1888 01 01 - 1889 01 01 | Vulcan, Hamburg

V 12 (1887)
1887 01 01 - 1888 01 01 - 1889 01 01 | Vulcan, Hamburg

V 13 (1887)
1887 01 01 - 1888 01 01 - 1889 01 01 | Vulcan, Hamburg

V 14 (1888)
1888 07 07 - 1889 01 01 - ***** | Vulcan, Hamburg

V 15 (1888)
1888 07 07 - 1889 01 01 - ***** | Vulcan, Hamburg

V 16 (1888)
1888 07 07 - 1889 01 01 - ***** | Vulcan, Hamburg

3 +3 | "V 11" type (1887)
Torpedo boats, Greece (continued)

V 11

V type torpedo boats in the base

6 | "Batum"/"Chios" type (1879)
Torpedo boats, Russia

Displacement: 48.4 tons at draft: 0.8 m | Length: 30.48 m oa | Beam: 3.8 m | Crew: 13 men

Machinery: 1 locomotive boiler | 1 horizontal compound E * 1 screw | 450 HP = 19.5 knots | Supplies: 8.3 tons of coal

Armament: 1-37/23R, 2-356 TT (2 >>), 1 spar torpedo

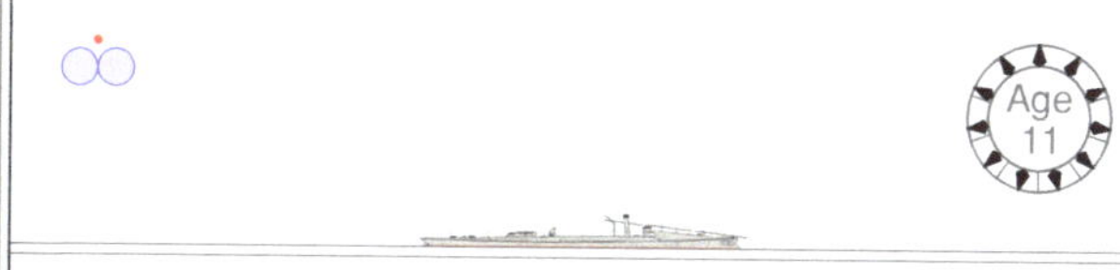

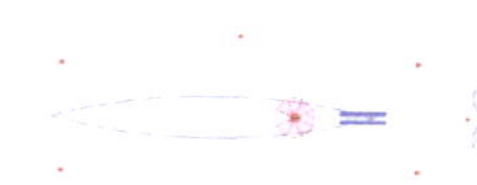

Chios (1881)
1881 01 01 - 1881 06 06 - 1882 01 01 | Yarrow, Poplar

Kas (1881)
1881 01 01 - 1881 06 06 - 1882 01 01 | Yarrow, Poplar

Kypros (1881)
1881 01 01 - 1881 06 06 - 1882 01 01 | Yarrow, Poplar

Mytelene (1881)
1881 01 01 - 1881 06 06 - 1882 01 01 | Yarrow, Poplar

Rodos (1881)
1881 01 01 - 1881 06 06 - 1882 01 01 | Yarrow, Poplar

Samos (1881)
1881 01 01 - 1881 06 06 - 1882 01 01 | Yarrow, Poplar

Sister ships can be found in the Argentinean, Austria-Hungarian, Brazilean, British, Chilean, Dutch, Japanese, and Russian sections

5 | "Soulven"/"Kalithea" type (1880)
Torpedo boats, Greece

NO PHOTO AVAILABLE

Displacement: 35 tons at draft: 0.94 m | Length: 27.54 m oa | Beam: 3.35 m | Crew: 16 men

Machinery: 1 locomotive boiler | 1 compound reciprocating E * 1 screw | 500 HP = 16 knots | Supplies: 5 tons of coal

Armament: 1-37/23R, 2-356 TT (2 >>)

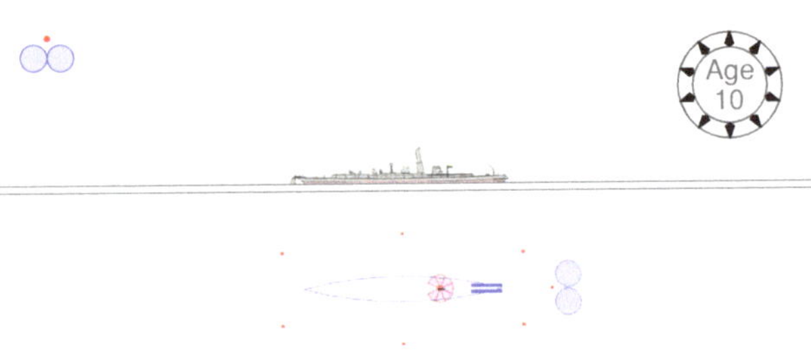

Kalithea (1880)
1880 01 01 - 1881 01 01 - 1881 12 12 | F. Ch. Médit., La Seyne

Perinika (1880)
1880 01 01 - 1881 01 01 - 1881 12 12 | F. Ch. Médit., La Seyne

Persephone (1880)
1880 01 01 - 1881 01 01 - 1881 12 12 | F. Ch. Médit., La Seyne

Terpsichore (1880)
1880 01 01 - 1881 01 01 - 1881 12 12 | F. Ch. Médit., La Seyne

Terpsithea (1880)
1880 01 01 - 1881 01 01 - 1881 12 12 | F. Ch. Médit., La Seyne

Sister ship can be found in the Danish section

4 | "Alfa" type (1879)
Spar torpedo boats, Greece

Displacement: 25 tons at draft: 1.68 m | Length: 21.95 m oa | Beam: 3.96 m | Crew: 15 men

Machinery: 1 locomotive boiler | 1 compound reciprocating E * 1 screw | 225 HP = 16 knots | Supplies: 1 ton of coal

Armament: 2 spar torpedoes

Alfa (1879)
1879 01 01 - 1880 01 01 - 1881 01 01 | F. Ch. Medit., La Seyne

Beta (1879)
1879 01 01 - 1880 01 01 - 1881 01 01 | F. Ch. Medit., La Seyne

Delta (1879)
1879 01 01 - 1880 01 01 - 1881 01 01 | F. Ch. Medit., La Seyne

Gamma (1879)
1879 01 01 - 1880 01 01 - 1881 01 01 | F. Ch. Medit., La Seyne

2 | "Ionia" type (1877)
Spar torpedo boats 2 class, Greece

Displacement: 18 tons at draft: 0.91 m | Length: 22.85 m oa | Beam: 3.2 m | Crew: 7 men

Machinery: 1 locomotive boiler | 1 compound E * 1 screw | 275 HP = 17 knots | Supplies: 1.5 tons of coal

Armament: 1 Spar torpedo

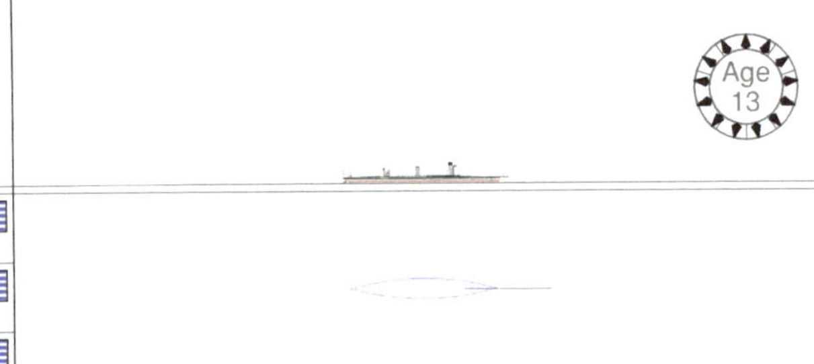

Ionia (1877)
1877 01 01 - 1878 01 01 - 1878 06 06 | Yarrow, Poplar

Sphinx (1877)
1877 01 01 - 1878 01 01 - 1878 06 06 | Yarrow, Poplar

1 | "Nordenfeldt I" type (1882)
Submarine, Greece

SUBMARINE

Displacement: 60 tons at draft: 3.4 m | 100 tons (submerged) | Length: 19.5 m oa | Beam: 2.7 m | Crew: 3 men

Machinery: 1 cylindrical boiler | 1 compound E * 1 screw | 100 HP = 9 (^) / 4 (v) knots | Supplies: 8 tons of coal

Diving depth: ?? m.

Armament: 1-25.4/40 Mg, 1-356 TT (1 >)

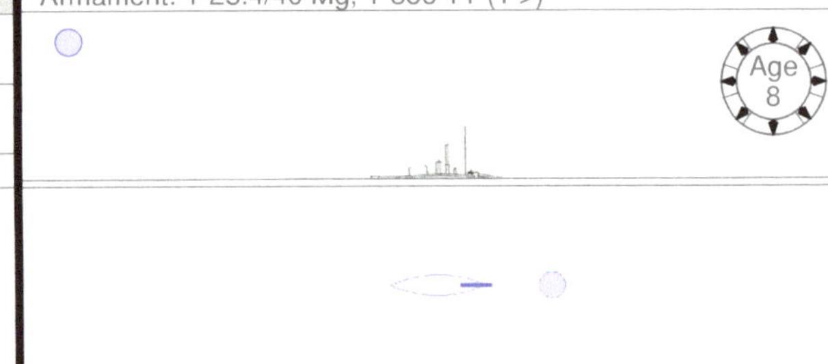

Nordenfeldt I (1882)
1882 09 09 - 1885 01 01 - 1885 09 09 | Ekensberg, Stockholm

GUNBOATS

2 | "Aidon" type (1883)
Wooden gunboats, Greece

Displacement: 86 tons at draft: 2.13 m | Length: 24.2 m oa | Beam: 4.88 m | Crew: 20 men

Machinery: 2 cylindrical boilers | 1 vertical compound E * 1 screw | 162 HP = 10 knots | Supplies: 10 tons of coal

Armament: 2-37/23R

Age 7

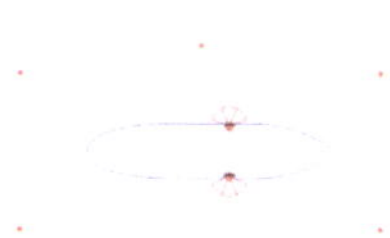

Aidon (1883)
1883 01 01 - 1884 01 01 - 1885 01 01 | Thames IW, Blackwall

Kilchi (1883)
1883 01 01 - 1884 01 01 - 1885 01 01 | Thames IW, Blackwall

Kissa (1883)
1883 01 01 - 1884 01 01 - 1885 01 01 | Thames IW, Blackwall

4 | "Achelaos" type (1883)
Composite gunboats, Greece

Displacement: 404 tons at draft: 3.5 m | Length: 40 m oa | Beam: 7.4 m | Crew: 73 men

Machinery: 2 cylindrical boilers | 1 vertical compound E * 1 screw | 400 HP = 10.5 knots | Supplies: 50 tons of coal

Armament: 2-95.3/26, 2-37/23R

Age 7

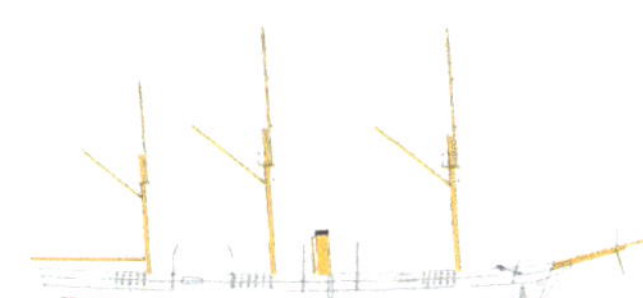

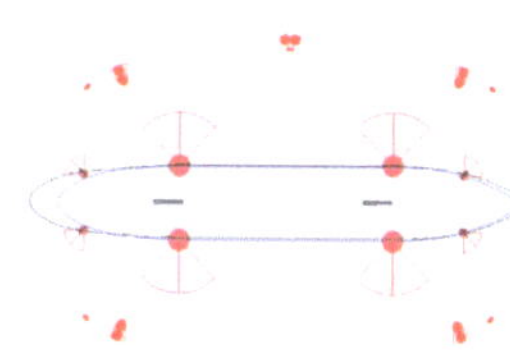

Achelaos (1883)
1883 01 01 - 1884 01 01 - 1885 01 01 | Thames IW, Blackwall

Alpheos (1883)
1883 01 01 - 1884 01 01 - 1885 01 01 | Thames IW, Blackwall

Eurotas (1883)
1883 01 01 - 1884 01 01 - 1885 01 01 | Thames IW, Blackwall

Peneos (1883)
1883 01 01 - 1884 01 01 - 1885 01 01 | Thames IW, Blackwall

2 | "Akteon" type (1880)
Gunboats, Greece

Displacement: 433 tons at draft: 2.2 m, 470 tons full | Length: 39 m pp | Beam: 7.9 m | Crew: 100 men

Machinery: 2 rectangular boilers | 1 vertical compound E * 1 screw | 380 HP = 10 knots | Supplies: 8.3 tons of coal

Armament: 1-260/27. 1-150/30, 2-37/23R, 1-356 TT (1 >)

Age 10

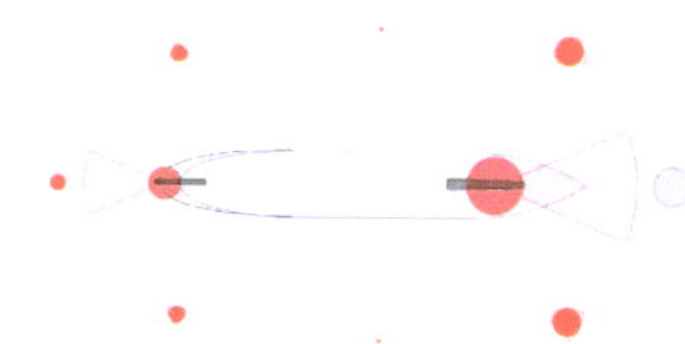

Akteon (1880)
1880 01 01 - 1881 01 01 - 1882 01 01 | Thames IW, Blackwall

Ambrakia (1880)
1880 01 01 - 1881 01 01 - 1882 01 01 | Thames IW, Blackwall

Also there were listed 5 small gunboats: Paralos, Salamnia, Afroessa, Plixavra, Syros on which there is not any exact information.

URUGUAY

General Riviera (GB), General Artigas (GB),
General Suarez (GB), Malvinas (GB)

1 | "General Riviera" type (1883)
Gunboat, Uruguay

Displacement: 300 tons at draft: 2.6 m | Length: 35.9 m oa | Beam: 6.35 m | Crew: 75 men

Machinery: 2 rectangular boilers | 1 compound * 1 screw | 360 HP = 12 knots | Supplies: app. 20 tons of coal

Armament: 1-150/25, 1-60/12

General Riviera (1883)
1883 01 01 - 1884 01 01 - 1885 01 01 | Montevideo DYd

1 | "Malvinas" type (1882)
Gunboat, Uruguay

Displacement: 400 tons at draft: 3.05 m | Length: 42.98 m oa | Beam: 6.96 m | Crew: 80 men

Machinery: 2 rectangular boilers | 1 compound * 1 screw | 280 HP = 8 knots | Supplies: app. 50 tons of coal

Armament: 2-105/22

Malvinas (1882) ex-merchantmen
bought 1887 01 01 | Schichau, Elbing

1 | "General Artigas" type (1882)
Gunboat, Uruguay

Displacement: 270 tons at draft: 2.59 m | Length: 39 m oa | Beam: 6.25 m | Crew: 80 men

Machinery: 2 cylindrical boilers | 1 compound * 1 screw | 520 HP = 12.5 knots | Supplies: app. 20 tons of coal

Armament: 2-120/30

General Artigas (1882)
1882 01 01 - 1883 01 01 - 1884 01 01 | STT, Trieste

1 | "Tactique" (1862) type
Wooden gunboat, France

Displacement: 360 tons at draft: 2.62 m | Length: 38.8 m wl | Beam: 6.71 m | Crew: 62 men

Machinery: 2 oval boilers | 1 horizontal triple expansion * 1 screws | 175 HP = 7.3 knots | Supplies: 25 / +10 tons of coal

Armament: 2-120/22

General Suarez (1862) ex-French
1862 01 01 - 1863 03 19 - 1863 10 01 | Toulon DYd

SIAM

Siam Mougkut (BZ), Ran Ruk (GB),
Maha Phchaiathep (GB), Regent (GB)

1 | "Siam Mougkut" type (1869)
Armoured battery, Siam

Displacement: 950 tons at draft: m | Length: 64 m oa | Beam: 11 m | Crew: 318 men

Machinery: 2 rectangular boilers | 1 horizontal direct acting * 1 screw | app.100 HP = 10 knots | Suppl.: app. 50 tons of coal

Armour: iron | Battery 63; Armament: 8-101.6/21

Siam Mougkut (1869)
1869 01 01 - 1870 01 01 - 1871 01 01 | ??????

1 | "Maha Phchaiathep" type (1866)
Wooden gunboat, Siam

Displacement: 580 tons at draft: 3.51 m | Length: 46.02 m oa | Beam: 7.62 m | Crew: 75 men

Machinery: 2 rectangular boilers | 1 horizontal direct acting * 1 screw | 435 HP = 10 knots | Supplies: app. 100 tons of coal

Armament: 1-140/21.6, 4-76.2/26

Maha Phchaiathep (1866)
1866 01 01 - 1867 01 01 - 1868 01 01 | ??????

1 | "Ran Ruk" type (1878)
Wooden gunboat, Siam

Displacement: 700 tons at draft: 3.31 m | Length: 38.4 m oa | Beam: 7.92 m | Crew: 75 men

Machinery: 2 rectangular boilers | 1 horizontal direct acting * 1 screw | 540 HP = 11 knots | Supplies: app. 140 tons of coal

Armament: 4-120/25, 1-57/43

Ran Ruk (1878)
1878 01 01 - 1879 01 01 - 1880 01 01 | Britain

1 | "Regent" type (1867)
Wooden gunboat, Siam

Displacement: 660 tons at draft: 3.2 m | Length: 50.6 m oa | Beam: 7.92 m | Crew: app. 100 men

Machinery: ??? * 1 screw | ??? HP = ?? knots | Supplies: ?? tons of coal

Armament: 8 guns

Regent (1867)
bought 1870 01 01

MEXICO

Democrata (GB), Independencia (GB), Libertad (GB), Mexico (GB)

2 | "Democrata" type (1874)
Gunboats, Mexico

Displacement: 445 tons at draft: 3.4 m | Length: 47.67 m oa | Beam: 7.59 m | Crew: 80 men

Machinery: 2 rectangular boilers | 1 vertical double expansion * 1 screw | 600 HP = 11 knots | Supplies: app. 70 tons of coal

Armament: 2-165/16 MLR, 2-95.3/19

Democrata (1874)
1874 01 01 - 1875 01 01 - 1876 01 01 | Rennie, London

Mexico (1874)
1874 01 01 - 1875 01 01 - 1876 01 01 | Rennie, London

2 | "Independencia" type (1873)
Gunboats, Mexico

Displacement: 480 tons at draft: 2.67 m | Length: 38.1 m oa | Beam: 7.32 m | Crew: 63 men

Machinery: 2 rectangular boilers | 1 horizontal compound * 1 screws | 425 HP = 10 knots | Supplies: 60 tons of coal

Armament: 1-165/16 MLR, 4-95.3/19

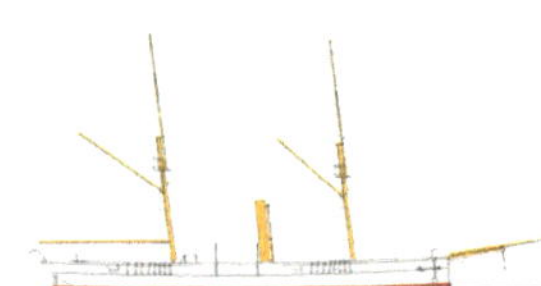

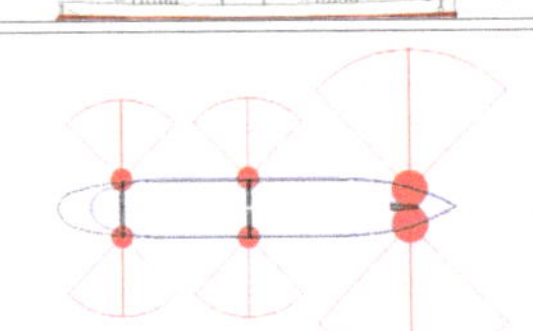

Independencia (1873)
1873 01 01 - 1874 01 01 - 1875 01 01 | Laird, Birkenhead

Libertad (1873)
1873 01 01 - 1874 01 01 - 1875 01 01 | Laird, Birkenhead

SAN DOMINGO

El Presidente (GB)

1 | "El Presidente" type (1872)
Gunboat, San-Domingo

Displacement: 389 tons at draft: 3 m | Length: 63.7 m oa | Beam: 6.5 m | Crew: 50 men

Machinery: 2 rectangular boilers | 1 horizontal triple expansion * 1 screw | 105 HP = 15 knots | Supplies: app. 30 tons of coal

Armament: 1-76.2/26, 1-37/23R

El Presidente (1872)
1872 01 01 - 1873 01 01 - 1874 01 01 | Seath & Co SB, Rutherglen

www.ingramcontent.com/pod-product-compliance
Lightning Source LLC
LaVergne TN
LVHW071514220926
848486LV00009BA/373